MW00713833

The Ecological Modernisation Reader

Environmental reform by governmental and intergovernmental agencies, private firms and industries and non-governmental organisations (NGOs) is a worldwide phenomenon. This definitive collection showcases an introduction to Ecological Modernisation Theory: state-of-the-art review essays by key international scholars and a selection of the key articles from a quarter-century of social science scholarship. It is aimed at students, researchers and policymakers interested in a deep understanding of contemporary environmental issues.

Arthur P. J. Mol is Chair and Professor in Environmental Policy in the Environmental Policy Group (ENP) of Wageningen University, The Netherlands. He has edited a number of books, and is the author of *Globalization and Environmental Reform* (2001) and *Environmental Reform in the Information Age* (2008).

David A. Sonnenfeld is Professor and Chair of Environmental Studies at the State University of New York College of Environmental Science and Forestry (SUNY-ESF), Syracuse, USA. He is co-editor of *Challenging the Chip: Labor Rights and Environmental Justice in the Global Electronics Industry* (2006) and *Ecological Modernisation Around the World* (2000).

Gert Spaargaren is part-time Professor on 'Environmental Policy for Sustainable Lifestyles and Consumption' at the Environmental Policy Group of Wageningen University, The Netherlands. His publications are in the field of (environmental) sociology, sustainable consumption and behaviour, and the globalization of environmental reform. He (co-)edited *Governing Environmental Flows* (2006).

'This is a much-needed collection which fills the gap left by theoretical and empirical work. Whatever your opinions about ecological modernisation, this promises to be a benchmark study.'

Michael Redclift, Professor of International Environmental Policy,
King's College, UK

'Ecological modernization has become the most important theory of practical attempts to deal with environmental problems. Whether it be efforts by the environmentally leading countries of northern Europe, or by the Obama Administration in the United States, or by Japan and China, the endeavors all consist of ecological modernization in one form or another. This informative reader brings together the most significant research on the subject over the three decades since its inception. The success or failure of ecological modernization remains nevertheless an open question, as indicated by global warming caused by human activities. Hence this perspective has its critics, particularly neo-Marxist and neo-Malthusian ones, and the book briefly responds to these as well. It is a must read for anyone interested in the broader aspects of environmental problem solving.'

Raymond Murphy, President, Environment and Society Research Committee of the
International Sociological Association

'This book offers the clearest and the most comprehensive introduction to ecological modernisation theory yet. It sets out the criticisms of ecological modernisation as a theory of social change and puts forward a compelling argument for a sociology of environmental reform. It shows how technological, institutional and cultural innovation concretely affect people's lives and the ways they relate to their environment. *The Ecological Modernisation Reader* will initiate a new and more sophisticated era of debate over the greening of industrial societies.'

Professor Stewart Lockie, The Australian National University

'The ideas underpinning ecological modernisation are exerting a growing influence over environmental policy-makers around the world. This valuable and timely collection brings together many of the main contributors to the development of this controversial concept, and provides a comprehensive coverage of the key theories, debates and policy developments in ecological modernisation.'

Professor Neil Carter, Department of Politics, University of York, UK

The Ecological Modernisation Reader

Environmental reform in theory and practice

Edited by Arthur P.J. Mol,
David A. Sonnenfeld and
Gert Spaargaren

Routledge
Taylor & Francis Group

LONDON AND NEW YORK

First published 2009
by Routledge
2 Park Square, Milton Park, Abingdon, Oxon OX14 4RN

Simultaneously published in the USA and Canada
by Routledge
270 Madison Ave, New York, NY 10016

Routledge is an imprint of the Taylor & Francis Group, an informa business

Typeset in Perpetua and Bell Gothic by
RefineCatch Ltd, Bungay, Suffolk
Printed and bound in Great Britain by
CPI Antony Rowe, Chippenham, Wiltshire

British Library Cataloguing in Publication Data
A catalogue record for this book is available from the British Library

Library of Congress Cataloging in Publication Data
A catalog record for this book has been requested

ISBN10: 0–415–45370–4

ISBN13: 978–0–415–45370–7

Contents

PART FOUR
Environmental reform in Asian and other emerging economies

Contributors

Mikael Skou Andersen is Professor in the Department of Policy Analysis, National Environmental Research Institute, Aarhus University, Denmark. His research interest lies in the comparative study of environmental policy, nationally and across Europe, with particular regard to policy instruments and implementation. He is co-editor of *Market-based Instruments for Environmental Management* (Edward Elgar, 2000).

Fred H. Buttel (1948–2005) was Professor of Rural Sociology and Environmental Studies at the University of Wisconsin, Madison, USA. He published numerous books and articles in the fields of environmental and rural sociology. He was co-author and editor of *Environment, Energy, and Society* (Wadsworth, 2001), *Environment and Global Modernity* (2000), and *Sociological Theory and the Environment* (Rowman and Littlefield, 2002).

Joost C.L. van Buuren is Senior Lecturer in the Department of Environmental Technology of Wageningen University and visiting researcher at Van Lang University in Ho Chi Minh City, Vietnam. His principal research interest is the implementation of sustainable technologies for urban wastewater and solid waste treatment and reuse in developing countries.

Maurie J. Cohen is Associate Professor of Environmental Policy and Sustainability in the Department of Chemistry and Environmental Science at the New Jersey Institute of Technology, USA. He is the co-editor (with Joseph Murphy) of *Exploring Sustainable Consumption: Environmental Policy and the Social Sciences* (Pergamon, 2001) and is editor of the journal *Sustainability: Science, Practice, and Policy*.

Peter Christoff is Professor, School of Anthropology, Geography and Environmental Studies, University of Melbourne. He is Vice-President of the Australian Conservation Foundation and, formerly, a director of Greenpeace Australia-Pacific. He is currently engaged in a comparative study of climate change policy across three federated states, including Australia, Canada and Germany.

John S. Dryzek is Professor and Australian Research Council Federation Fellow in the Research School of Social Sciences at the Australian National University. Recent books include *Deliberative Global Politics: Discourse and Democracy in a Divided World* (Polity, 2006).

Dana R. Fisher is an Associate Professor in the Department of Sociology at Columbia University, USA. Her research focuses on environmental policy and civic participation and activism more broadly. She has written a number of peer-reviewed articles, and is the author of *National Governance and the Global Climate Change Regime* (Rowman and Littlefield, 2004) and *Activism, Inc.* (Stanford, 2006).

Oliver Fritsch is a Ph.D. researcher at the National Environmental Research Institute, Aarhus University, Denmark. His work focuses on the institutionalisation of public participation in the European Union, and on environmental outcomes of collaborative decision making. He has journal articles forthcoming in *Environmental Management* and in *Environmental Policy and Governance*.

George A. Gonzalez is Associate Professor of Political Science at the University of Miami, USA. His area of research is US environmental politics and policy. Among his recent publications are *The Politics of Air Pollution: Urban Growth, Ecological Modernization, and Symbolic Inclusion* (SUNY Press, 2005), and *Urban Sprawl, Global Warming, and the Empire of Capital* (SUNY Press, 2009).

Andrew Gouldson is Professor of Sustainability Research and Director of the Sustainability Research Institute at the University of Leeds, UK. He is also Director of the ESRC Centre for Climate Change Economics and Policy. His research focuses on the design, delivery and impact of different forms of environmental policy and on the evolving influence of new forms of environmental governance.

Maarten A. Hajer is Director of the Netherlands Environmental Assessment Agency, and Professor of Public Policy at the University of Amsterdam. His new book is *Authoritative Governance – Policy Making in an Age of Mediatization* (Oxford UP, 2009).

Jennifer Houghton is a Lecturer in Human Geography in the School of Environmental Sciences of the University of KwaZulu-Natal, South Africa. Her main research interests are the diffusion of neoliberalism in cities of the South, urban sustainability and the capacity for sustainability implementation.

Joseph Huber is Professor and Chair of Economic and Environmental Sociology at Martin-Luther-University in Halle, Germany. He emphasises framing ecological modernisation by general social modernisation and development theory. Among his recent books is *New Technologies and Environmental Innovations* (Edward Elgar, 2004), an excerpt from which is included in this volume.

Christian Hunold is Associate Professor of Political Science at Drexel University, USA. His research interests include environmental politics, deliberative democracy, and participatory policymaking. He is profiles editor at *Environmental Politics*.

Martin Jänicke was Director of the Environmental Policy Research Center, Freie Universität Berlin from 1986–2007. Since 1999, he has served as Member of the German government Advisory Council on the Environment. He has written numerous books (in different languages) on environmental policy and on environmental innovation.

Helge Jörgens is Assistant Professor in the Department of Political Science of Konstanz University, Germany. His main research interests and publications are in the fields of comparative environmental policy, the global diffusion of policy innovations, and environmental governance. He was guest editor of the *European Environment* special issue on 'Diffusion and Convergence of Environmental Policies in Europe' (2005) and a special issue of the German *Politische Vierteljahresschrift on* 'Transfer, Diffusion and Convergence of Policies' (2007).

C.S.A. (Kris) van Koppen is Associate Professor in the Environmental Policy Group of Wageningen University, and Professor in environmental education by special appointment at the Freudenthal Institute, Utrecht University, the Netherlands. His research interests include the sociology of nature, sustainable production and consumption, and social learning.

Oluf Langhelle is Professor in Social Science at the University of Stavanger, Norway. He has edited a number of books, and is co-editor of *Arctic Oil and Gas: Sustainability at Risk?* (Routledge, 2008). His main research interests and publications are in the field of environmental politics, sustainable development and Corporate Social Responsibility (CSR).

Pieter Leroy is Professor of Political Sciences of the Environment at Radboud University Nijmegen, the Netherlands. His research covers the analysis and evaluation of environmental politics and policies. He recently co-authored *The Handbook of Environmental Policy Evaluation* (Earthscan, 2008).

Arthur P.J. Mol is Chair and Professor in Environmental Policy in the Environmental Policy Group (ENP) of Wageningen University, the Netherlands. He has edited a number of books, and is the author of *Globalization and Environmental Reform* (MIT Press, 2001) and *Environmental Reform in the Information Age* (Cambridge, 2008).

Joseph Murphy is RCUK Academic Fellow in Social Response to Environmental Change in the Sustainability Research Institute at Leeds University, UK. His recent work has focused on the politics and governance of technology including the books *Governing the Transatlantic Conflict over Agricultural Biotechnology* (Routledge, 2006, co-authored with Les Levidow) and *Governing Technology for Sustainability* (Earthscan, 2007, edited).

Nguyen Trung Viet is a senior official of the Ho Chi Minh City Department of Natural resources and Environment, and Head of the Department of Environmental Technology and Management at Van Lang University, Vietnam. He has published widely on industrial environmental management and technology.

Catherine Oelofse is a Lecturer in the School of Environmental Sciences at the University of KwaZulu-Natal, Durban, South Africa. Her main research interests are on space and sustainability, using discourses analysis to evaluate environmental policy making in South Africa and social assessment methodologies.

Gregg Oelofse is the Head of Environmental Policy and Strategy for the City of

Cape Town, South Africa. His research focuses on the relationship between people and their environment.

Phung Thuy Phuong is Senior Lecturer at the University of Science, Ho Chi Minh City, Vietnam. She was Fulbright Visiting Scholar at Portland State University, USA, in 2005–2006. Her current research interest is community-based environmental management.

Michael T. Rock is the Harvey Wexler Professor of Economics in the Department of Economics, Bryn Mawr College, USA. His current work focuses on the impact of multinational firms' standards on the environmental performance of subsidiaries and their suppliers, and on technological learning of indigenous industrial firms in Asia. He co-authored *Industrial transformation in the developing world* (Oxford, 2005) and is currently working on a book on democracy and development in Southeast Asia.

David Schlosberg is Professor of Politics and International Affairs and Director of Environmental Studies at Northern Arizona University, USA. He is the author of *Environmental Justice and the New Pluralism* (Oxford, 1999), and *Defining Environmental Justice* (Oxford, 2007), and co-editor of the forthcoming *Oxford Handbook on Climate Change and Society*.

Dianne Scott is an Associate Professor at the School of Environmental Sciences, Durban, University of KwaZulu-Natal, South Africa. She is the senior researcher for the 'Environmental Decision-Making for Sustainability' Research Programme at this centre. Her main research interests and publications are in the field of environmental policy-making, environmental management, environmental movements and networks, and urban restructuring.

David A. Sonnenfeld is Professor and Chair of Environmental Studies at the State University of New York College of Environmental Science and Forestry (SUNY-ESF), Syracuse, USA. He is co-editor of *Challenging the Chip: Labor Rights and Environmental Justice in the Global Electronics Industry* (Temple, 2006); and *Ecological Modernisation Around the World* (Frank Cass/Routledge, 2000).

Gert Spaargaren is part-time Professor on 'Environmental Policy for Sustainable Lifestyles and Consumption' at the Environmental Policy Group of Wageningen University, The Netherlands. His publications are in the field of (environmental) sociology, sustainable consumption and behaviour, and the globalization of environmental reform. He (co-)edited *Governing Environmental Flows* (MIT Press, 2006).

Jan P.M. van Tatenhove is Associate Professor at the Environmental Policy Group (ENP) of Wageningen University, the Netherlands. His main research interests are in the field of political modernization and governance (multi-level and informal governance), the innovation of policy arrangements in environmental, nature conservation and marine policy, and power and participation.

Tran Thi My Dieu is a researcher and Lecturer in the Department of Environmental Technology and Management, Van Lang University, Ho Chi Minh, Vietnam. Her main research interests and publications are in the field of solid waste and wastewater technologies and management, eco-industrial parks and eco-towns.

Preface

The first decade of the third millennium has been nothing short of tumultuous. In it, we have witnessed the rise of India and especially China as global technological and commercial powerhouses; the destruction of the World Trade Center in New York; an almost doubling of countries in the European Union; tsunamis, hurricanes and earthquakes, catastrophic in both human and environmental terms; urgent concern about global warming; and, still of great concern as this volume goes to press, the near-meltdown of the world's financial markets. Even so, environmental issues and concerns have maintained and even grown through this trying period.

The twenty-first century is not the first to face crises and turmoil. As Fernand Braudel, Clarence Glacken, Jack Goldstone, James O'Connor, Jared Diamond, and others have documented so well, social and environmental crises have gone hand-in-hand since the beginning of history. Roman, Greek, and earlier empires rose and fell with serious and widespread environmental causes and consequences. Environmental concern rose to new heights in the late eighteenth century with the advent of contemporary agricultural practices, in the early nineteenth century with intense pollution of industrializing cities in Europe, in the late nineteenth century with the widespread destruction of forests in North America, and again after two World Wars in the first half of the 20th century, with the development of the modern chemical industry, nuclear fission, and the first 'Green Revolution' of industrialized agriculture.

'Modern' environmentalism and environmental regulation were born just a few decades ago, in the 1960s. Early environmental regulations from the following decade were somewhat successful, but led to backlash, 'deregulation', and anti-environmental movements in the 1980s. In such times, Ecological Modernisation Theory (EMT) was first developed in northern Europe. EMT began to mature in the 1990s, was taken up globally in the new millennium, and is alive, well, and thriving today as never before. As documented and discussed throughout this collection, ecological modernisation perspectives have informed not only ivory-tower theories of the relationship between humans and social institutions to the natural environment, but also have been a major influence on environmental policymaking in the public, private, and civic spheres.

Along with Anthony Giddens, Ulrich Beck and others, we contend that in the late twentieth and early twenty-first centuries, even with all the other truly consequential events, dynamics, and changes happening, the environment has become irrevocably separated as a sphere of human interest, activity, engagement, and personal, cultural, and institutional development. The twenty-first century will be the first fully environmental – as well as, not coincidentally, predominantly urban – century in history. Ecological Modernisation Theory, as comprehensively discussed in this volume for the first time for readers around the world, is expected to make a lasting contribution to social theory of the environment and to environmental policymaking for decades to come.

In a collection such as this, choices must be made about what to include; inevitably, some publications we would very much have liked to include, in the end were not. This includes selections from the full range of classical contributions to Ecological Modernisation Theory, empirical studies building on and adding to the school of thought, and more recent publications which were arguably not yet 'classics' sufficient for inclusion in this volume. We have attempted to compensate for this to some extent, at least, by including a series of newly commissioned review essays by leading scholars in the field. We hope that those essays, plus the volume's introduction and conclusion, do some justice to the full range and depth of scholarship on ecological modernisation carried out by scholars around the world over the last three decades. To those whose work was not included, our apologies.

We are indebted to the authors of the review essays for their contributions to this volume. While not responsible for the design or execution of the volume as a whole, they made valuable contributions to the respective parts addressed in their essays. A special 'thank you', then, to Martin Jänicke, Joseph Huber, Dana Fisher, Oliver Fritsch, Mikael Skou Andersen, Maurie Cohen, and Michael Rock. Lisette Zewuster, graduate student in environmental policy at Wageningen University, played an essential role in securing rights for reprinting articles and book chapters. Corry Rothuizen, secretary for the Environmental Policy Group, Wageningen University, never failed in her good-natured assistance in preparing the manuscript for submission. At Routledge/Taylor & Francis, Gerhard Boomgarten, Ann Carter, Miranda Thirkettle, and James Rabson have been supportive and helpful in the development of this volume from beginning to end. Heather Dubnick, professional indexer, applied her excellent skills to make this volume even more useful. Last but not least, we would like to acknowledge the generous support of colleagues, administrators and graduate students at our respective institutions who, in helping nurture and develop stimulating intellectual environments, have made the present endeavour possible, and encouraged this critically important and dynamic area of scholarship and practice to continue to flourish.

The Editors
Wageningen, the Netherlands; Syracuse, New York
April 2009

Acknowledgements

The publishers would like to thank the following for their permission to reprint their material.

Blackwell Publishing for permission to reprint Mol, A.P.J. (2006), 'Environment and modernity in transitional China. Frontiers of ecological modernization', *Development and Change* 37, 1, 29–56.

Deutscher Fachverlag for permission to reprint Jänicke, M. (1993), 'On ecological and political modernisation', originally published as 'Über ökologische und politische Modernisierungen', *Zeitschrift für Umweltpolitik und Umweltrecht* 2, 159–175; Jänicke, M., and H. Jörgens (2004), 'New approaches to environmental governance', in M. Jänicke and K. Jacob (eds), *Environmental governance in global perspective. New approaches to ecological modernisation*, pp. 167–209, Berlin: Freie Universität Berlin Press, originally published as 'Neue Steuerungskonzepte in der Umweltpolitik', *Zeitschrift für Umweltpolitik* 27, 297–348.

Edward Elgar Publishing Ltd for permission to reprint Huber, J. (2004), 'Upstreaming environmental action', in *New Technologies and Environmental Innovation*, pp. 3–16.

Elsevier for permission to reprint Buttel, F.H. (2000), 'Ecological modernization as social theory', *Geoforum* 31, 1, 57–65; Murphy, J., and A. Gouldson (2000), 'Environmental policy and industrial innovation: integrating environment and economy through ecological modernization', *Geoforum* 31, 1, 33–44.

IWA Publishing for permission to reprint Koppen, C.S.A. (Kris) van and Arthur P.J. Mol (2002), 'Ecological modernization of industrial systems', in P. Lens *et al.* (eds), *Water recycling and resource recovery in industry: Analysis, technologies and implementation*, (ISBN 9781843390053), copyright © IWA Publishing (www.iwapublishing.com).

John Wiley & Sons Ltd for permission to reprint Langhelle, O. (2000), 'Why ecological modernisation and sustainable development should not be conflated', *Journal of Environmental Policy & Planning* 2, 4, 303–322. Reproduced with permission of John Wiley & Sons Ltd.

Lexington Books for permission to reprint Tran, M.D.T., P.T. Phung, J.C.L. van Buuren and V.T. Nguyen (2003), 'Environmental management for industrial zones in Vietnam', in A.P.J. Mol and J.C.L. van Buuren (eds), *Greening Industrialization in Asian Transitional Economies: China and Vietnam*, pp. 39–58.

M.E. Sharpe, Inc. for permission to reprint Gonzales, G.A. (2001), 'Democratic ethics and ecological modernization: The formulation of California's automobile emission standards', *Public Integrity* 3, 4 (Fall 2001): 325–44. Copyright © 2001 by American Society for Public Administration (ASPA). Reprinted by permission of M.E. Sharpe, Inc.

A.P.J. Mol for permission to reprint Huber, J. (1991), 'Ecological modernisation: beyond scarcity and bureaucracy', in A.P.J. Mol, G. Spaargaren, and B. Klapwijk (eds.), *Technologie en Milieubeheer. Tussen Sanering en Ecologische Modernisering*, pp. 167–183, Den Haag: SDU.

Oxford University Press for permission to reprint Dryzek, J.S., D. Downes, C. Hunold, and D. Schlosberg (2003), 'Ecological modernization, risk society, and the green state', in *Green States and Social Movements. Environmentalism in the United States, United Kingdom, Germany, and Norway*, pp. 164–91.

SAGE Publications for permission to reprint Hajer, M.A. (1996), 'Ecological modernization as cultural politics', in S. Lash, B. Szerszynski, and B. Wynne (eds), *Risk, Environment and Modernity. Towards a New Ecology*, © SAGE Publications, 1996; Rock, M.T. (2002), 'Integrating environmental and economic policy making in China and Taiwan', *American Behavioral Scientist* 45, 9, 1435–55, copyright © SAGE Publications, 2002. Reproduced by permission of SAGE Publications, London, Los Angeles, New Delhi and Singapore.

Taylor & Francis Books for permission to reprint Sonnenfeld, D.A. (2000), 'Contradictions of ecological modernisation: pulp and paper manufacturing in Southeast Asia', in A.P.J. Mol and D.A. Sonnenfeld (eds), *Ecological Modernisation Around the World: Critical Perspectives and Debates*, pp. 235–256, London; Portland, Ore.: Frank Cass/Routledge.

Taylor & Francis Journals for permission to reprint Spaargaren, G., and A.P.J. Mol (1992), 'Sociology, environment and modernity. Ecological modernisation as a theory of social change', *Society and Natural Resources* 5, 323–344; Christoff, P. (1996), 'Ecological modernisation, ecological modernities', *Environmental Politics* 5, 3, 476–500; Spaargaren, G. (2003), 'Sustainable consumption: A theoretical and environmental policy perspective', *Society and Natural Resources* 16, 687–701; Oelofse, C., D. Scott, G. Oelofse, and J. Houghton (2006), 'Shifts within ecological modernization in South Africa: deliberation, innovation and institutional opportunities', *Local Environment* 11, 1, 61–78. (www.informaworld.com)

The White Horse Press for permission to reprint Tatenhove, Jan van and Pieter Leroy (2003) 'Environment and Participation in a Context of Political Modernisation', *Environmental Values* 12: 155–174.

The publishers have made every effort to contact authors and copyright holders of works reprinted in this book. This has not been possible in every case however, and we would welcome correspondence from individuals or companies we have been unable to trace.

Introduction

Chapter 1

Arthur P.J. Mol, Gert Spaargaren and David A. Sonnenfeld

ECOLOGICAL MODERNISATION: THREE DECADES OF POLICY, PRACTICE AND THEORETICAL REFLECTION

The rationale for an ecological modernisation reader

WITH THE REBIRTH OF environmental concern among social scientists in the 1960s and 1970s, scholars initially were preoccupied with explaining environmental devastation. Their central concern was how human behaviour, capitalist institutions, a culture of mass consumption, failing governments and states, and industrial and technological developments, among others, contributed to the ongoing deterioration of the physical environment. In the 1970s and 1980s, there were many reasons to look for explanations of ongoing, widening and deepening environmental crises. The result was an expansive literature – both theoretical and empirical in nature – on the 'roots of the environmental crisis' (Pepper 1984). Various disciplines and schools of thought emphasised different institutional and behavioural traits as the fundamental origins and causes of this environmental crisis.

Beginning in the 1980s and maturing in the 1990s, attention in environmental sociology and politics started to change towards what sociologist Fred Buttel (2003) labelled the sociology of environmental reform. Strongly driven by empirical and ideological developments in the European environmental movement, by the practices and institutional developments in some 'environmental frontrunner states', and by developments in private companies, some European social scientists began reorienting their focus from explaining ongoing environmental devastation towards understanding processes of environmental reforms. Later and sometimes less strongly, this new environmental social science agenda was followed by American and other non-European scholars and policy analysts. By the turn of the millennium, this focus on understanding and explaining environmental reform had become mainstream in many locations around the world, not so much instead of, but rather as a complement to, studies explaining environmental deterioration.

In what we might call – after the late Fred Buttel – the social sciences of environmental reform, *ecological modernisation* has stood out as one of the strongest, most well-known, used and widely cited, and constantly debated concepts. The notion of ecological modernisation can be seen as *the social scientific interpretation of environmental reform processes at multiple scales in the contemporary world*. As a relatively young but still growing body of scholarship, ecological modernisation studies reflect on how various institutions and social actors attempt to integrate environmental concerns into their everyday functioning, development and relationships with others, including their relation with the natural world. As a result, environmental interests have become incorporated into more and more aspects of social relations and institutions, as well as into contemporary human values, cultures, and everyday practices.

From the parallel launching of the term by Martin Jänicke and Joseph Huber, respectively, around 1980 and its explicit foundation into social theory by Arthur Mol and Gert Spaargaren around 1990, ecological modernisation has been applied around the world in empirical studies, has been at the forefront in theoretical debates, and has even been used by politicians to frame environmental reform programs in, among others, Germany, the Netherlands, the UK and China. There is now wide interest and research in ecological modernisation across the globe, including in Asia (especially China, Japan, Korea, Vietnam, Thailand, Malaysia and elsewhere), North America, Australia and New Zealand, Latin America (especially Brazil, Argentina, Peru, Chile), as well as on the wider European continent (including Russia).

While a number of volumes have been published on ecological modernisation (cf. Mol and Sonnenfeld 2000; Young 2001; Barrett 2005), this is the first, comprehensive, authoritative volume bringing together both the history and current state of Ecological Modernisation theories, research, debates and policy applications in various parts of the world. This volume brings together classic ecological modernisation texts – some newly translated and published in English for the first time – and illustrative applications in different empirical and geographical contexts, together with a series of newly commissioned, topical review essays surveying critical components of a steadily developing body of literature. It is our hope that this volume will enable and facilitate further scholarship, critical debates and theoretical development in the environmental social sciences – including in sociology, political science, history and geography – with regard to processes of institutional and cultural environmental transformation.

This introduction explains the emergence and development of ecological modernisation ideas against the background of an evolving environmental discourse on environmental deterioration, sustainable development and global environmental change. In doing so, it sets the scene for the four major parts that comprise this reader, introduced at the end of this essay.

From the Club of Rome to Al Gore: waves of environmental concerns and politics

Anthony Downs (1972), in his analysis of the first major upsurge of 'modern environmentalism' that occurred in many advanced nations from 1968 to 1972, wondered if the rising tide of popular ecological consciousness at that time represented a deeply rooted structural phenomenon or instead was just a passing fashion. In retrospect, since the publication of the Club of Rome report of 1972 (Meadows et al. 1972), attention to issues of air and water pollution, land protection and (more recently) biodiversity conservation and climate change has steadily grown into an entrenched global concern for sustainable development. Over the past four decades, societies worldwide have gradually built up their institutional capacity to address environmental risks in a systematic, organised way (Jänicke 1995). Of course, this cumulative process has never been linear. We have, as Downs put it, been going 'up and down with ecology' and issue-attention cycles have evinced varying timing and dynamics in different parts of the world. Major successes have been followed by periods of stagnation, and there have also been significant setbacks and reversals of environmental reform efforts. Since the start of contemporary environmentalism in the 1960s and 1970s (with the Report to the Club of Rome, the *Blueprint for Survival* (Goldsmith and Allen 1972) and the launching of many new environmental non-governmental organisations), we can delineate three decisive moments that have accelerated innovation in and reform of environmental policies, practices and awareness. These three moments of public environmental attentiveness are key to understanding the emergence and development of ecological modernisation ideas and reflections.

First, the publication in 1987 of *Our Common Future* (also known as the Brundtland Report) by the World Commission on Environment and Development provided a landmark statement in support of sustainable development (WCSD 1987). The debate that it catalysed prompted a cadre of innovative private firms to review their environmental performance with an eye toward bridging the customary economy–ecology divide. It also contributed to the reframing of the ideology and strategies of major Northern environmental NGOs, in their efforts to become closer to the centres of economic and political decision making. And the Brundtland Report sought to transcend North–South cleavages on questions pertaining to environmental protection and economic development.[1] In all, *Our Common Future* put the relation between economic development and environmental protection strongly on public and political agendas around the world, be it framed in new terms. But these 'new terms', as codified by the Brundtland Report, reflected a number of tendencies that had already been ongoing for a number of years, since at least the early 1980s.

Second, the World Summit on Environment and Development, held in Rio de Janeiro in 1992, elevated environmental concerns into truly global phenomena. International media coverage (of the route towards and the results) of this two-week event translated biodiversity loss, climate change, desertification, and water scarcity into common and global topics of discussion. During the months and years following the Rio conclave, it became increasingly difficult for 'anyone on the planet', so Ulrich Beck (Beck and Wilms 2004: 141) argues, to maintain ignorance of the environmental side effects of modernisation and industrialisation, and of the need to do something about them. Ten years later, the follow-up summit in Johannesburg 2002 confirmed the global environmental agenda, while adding an explicit orientation on sustainable consumption. The interceding decade was formative for the negotiations, conclusions, implementation and debates of multilateral environmental agreements, but also of the North American Free Trade Agreement (NAFTA) environmental side agreements and the acceleration of European Union (EU) environmental policy-making and legislation.

Third, the urgent need for action at all levels to reduce the threat of global climate change was the key pronouncement Al Gore (2006) put forth in the impressive international campaign that led to his receiving the Nobel Peace Prize in 2007. His message both contributed to and was reinforced by the most recent wave of environmental concern, exacerbated further by concerns about 'peak oil' and rapidly rising petroleum prices worldwide. In the first decade of the 21st century, a 'new green wave' has begun to swell that is different from its two predecessors in that it includes a strong emphasis not only on sustainable production, planning and transportation, but also encompasses a growing understanding of the need for sustainable consumption practices and greener lifestyles. Al Gore's strongly media-driven campaign directly addresses citizen-consumers around the world, motivating them to build pressure throughout their respective communities and societies to green (and thus change) current behaviour and development paths.

The attempts, practices and policies of environmental reform have changed and developed over time (as is clear in the brief recap above); consequently, ecological modernisation scholarship has co-developed and matured as well during the last three decades. The following sections of this essay discuss in greater depth how ecological modernisation ideas relate to the legacy of the Brundtland Report, to the Rio process of globalising environmental concerns, and to the consumption and lifestyle focus of the most recent Gore-inspired wave of environmental concern. Together, they help explain the origin and continuing evolution of the various notions, ideas, developments and currents within the ecological modernisation literature.

Ecological modernisation and the Brundtland Report

Ecological Modernisation Theory (EMT) originated in Europe during the 1980s. In its original form, it can be regarded as the social scientific elaboration and formalisation of the underlying philosophy concerning environmental change articulated in the Brundtland Report. Or as Spaargaren and Mol put it in 1992:

> . . . the concept of sustainable development is based more on opinions than on scientifically based ideas. For this reason and because of the many possible interpretations that can be placed upon it, the concept of sustainable development is only suited to our purpose to a very limited extent. Therefore, we introduce a more analytical and sociological concept consonant with the primarily political concept of sustainable development: ecological modernisation.
>
> (Spaargaren and Mol 1992: 333)

From its inception, proponents of EMT argued for the need to transcend the ecology-economy divide internalising 'external costs' into the functions of the market and the economy in general. 'Ecologising the economy', in Huber's words, meant giving environmental issues and interests a permanent and central position in the decision-making processes of private firms and consumers with the help of life-cycle analyses (LCA), environmental reports and audits, as well as environmental management and audit and certification systems, and consumer-oriented structures such as eco-labelling, the development of consumer environmental standards, etc. (Huber 1982; Mol 1995). To make this 'structural anchoring' of environmental concerns in the market possible, it was necessary to leave behind prior tendencies within organised environmentalism that favoured vitriolic critiques of capitalism and industrialism and focused on making a fundamental break with modernity (Ullrich 1979; Schumacher 1973). To advance environmental reform efforts, this romantic yearning to revert to an agrarian past premised on 'small-is-beautiful' ideals had to be replaced by a more pragmatic posture that created space for dialogue and negotiation between professionalised environmental movements, expanding and diversifying environmental states, and increasingly engaged private sector actors. Environmental futures were not to be 'imported from the outside', but instead developed progressively from within the existing constellation of modernity in a way that reconstructed and redefined extant institutions so that environmental risks and side effects were addressed in a structural manner. During this process of deliberation, it was inevitable that 'ecology loses its innocence', because the incorporation of environmental concerns by mainstream economic actors is possible only when environmental criteria, instruments and concepts are reformulated to mesh with the logics of modern markets (Huber 1982, 1985, 1991; Spaargaren and Mol 1992; Mol and Spaargaren 1993; Hajer 1995; Mol and Sonnenfeld 2000).

An important contribution of EMT has been to reflect on the historical emergence and development of an 'ecological rationality' as a relatively independent epistemology alongside economic and political rationalities. To stress the need for an ecological sphere 'to emancipate' itself from the economic and political spheres (Giddens 1990), the language of systems theory within the social sciences (Luhmann 1995; Dryzek 1987) was used to describe and analyse processes through which ecological concerns became deeply embedded within modern societies from the 1970s onwards. Within EMT, what might be termed the 'Brundtland view' on reframing the relationship between environmental management and economic growth was analytically sharpened, since it was demonstrated in both normative and historical terms that sound ecological conditions and good economic performance should (and actually could) co-exist. Only when private firms, technologies, households and policies could simultaneously demonstrate sufficient economic performance and adequate or even superior ecological quality did they deserve to be labelled and treated as 'sustainable'.

Ecological modernisation and the post-Rio globalisation of environmental concerns

Environmental concerns have been conceptualised in global terms for some time. The Club of Rome sponsored the construction of a computerised simulation model of 'System Earth' in the late 1960s, and the 1972 Stockholm Conference was manifestly international in both its participation and agenda. The articulation of a synthesised North–South environmental perspective, one that sought to encompass the circumstances of both developed and developing countries, was a major task and achievement of the World Commission on Sustainable Development (WCSD) during the mid-1980s. It can also be argued that within the environmental sub-disciplines of the social sciences, the international dimensions of societies came to be discussed and analysed earlier and more prominently than they were in the social sciences more broadly (Caldwell 1984; Catton 1980). Environmental problems are frequently characterised by their transboundary features, and issues such as fisheries management, wildlife control and air and water pollution have been at the centre of contemporary efforts to understand the dynamics of global modernity. In sociology, Anthony Giddens (1990) and Ulrich Beck (1986, 1997, 2005) stand out as social theorists who have made frequent reference to environmental risks not just as incidental illustrations, but also to help develop their theories of reflexive modernity and the risk society, respectively (Beck et al., 1994).

EMT formulations in environmental sociology – particularly the work of Arthur Mol, Maarten Hajer, Fred Buttel and Gert Spaargaren – have relied heavily upon more general sociological theories as formulated by Anthony Giddens, Ulrich Beck, Manuel Castells, John Urry and others. Only by applying these and others' sociological analyses of globalisation has it been possible to move away from EMT's initially Eurocentric formulation, and beyond the 'methodological nationalism' (Beck and Wilms 2004) expressed in the environmental policies of the northern European countries during the 1970s and 1980s.[2] At least three factors have facilitated this global broadening of EMT. First, environmental policy making in Europe became to a considerable extent an EU affair in a relatively short period of time. In the ten years preceding the Rio Earth Summit, the EU's environmental agenda became the second major focus of European-level regulation and policy making. This process of the regionalisation of environmental policy making in Europe, referred to as 'Europeanisation', has been examined in detail by environmental sociologists and political scientists, with authors writing from an EMT perspective contributing significantly to these efforts (e.g. Liefferink 1995; Anderson 1999; Mol, Lauber and Liefferink, 2000).

Second, EMT-oriented scholars from Europe became increasingly aware of the regional outlook of their theories during confrontations with North American environmental sociologists and political scientists. Debates with Allan Schnaiberg and others working from a 'treadmill of production' (ToP) perspective (see Schnaiberg 1980; Schnaiberg et al. 2002) were not just controversies between neo-Marxist and (post-)industrial society theorists. They were exercises also in re-examining the national and regional roots of many foundational concepts and assumptions (Mol 2006a). Through such debates, EMT scholars' theoretical understanding of the role of nation-states *vis à vis* international market actors has improved markedly, with contributions by Fred Buttel, James O'Connor (1996) and several World-systems Theorists including playing active and constructive roles.

Finally, the global broadening of EMT has been the direct result of its frequent and systematic application in non-European contexts, especially in East and Southeast Asia, and to a lesser extent also Latin America (cf. Mol and Sonnenfeld 2000; Sonnenfeld and Mol 2002, 2006; Mol 2006b; Oosterveer et al. 2007). Throughout much of especially East Asia, economic growth was, and still is, booming and the environmental pressures that accompany rapid modernisation continue to accumulate. EMT scholars have been at the forefront of social scientific and policy reform efforts to understand the gains that could be realised from environmental cooperation, capacity building and networking with Asian partners, as described in Part Four of this volume. The fact that senior Chinese scholars at the national level have embraced EMT concepts and perspectives to analyse environmental dimensions of their country's process of modernisation is evidence of the reciprocal qualities of this interest (Chinese Academy of Science 2007; Zhang et al. 2007).

Since the mid-1990s, then, EMT has acquired a more international, indeed even global, outlook. The discussions surrounding a global 'local agenda' for the 21st century that were initiated at the Rio Earth Summit challenged all major theoretical constructs advanced at the time within the environmental social sciences. EMT responded actively to this challenge by adapting to the new political and institutional circumstances. In particular the dramatically changing roles of nation-states and environmental NGOs in policy making and the profound influence of globalisation as an 'attractor' in the field of environmental governance have received considerable theoretical and empirical attention from EMT scholars (Mol 2000, 2001; Spaargaren, Mol and Buttel 2000, 2006; Sonnenfeld 2002). The recent formulation of an environmental theory of networks and flows that is consistent with the core tenets of EMT (Spaargaren, Mol and Buttel 2006) is emblematic of the fact that EMT theorists continue to explore new and innovative ways to develop an approach to the study of processes of institutional and societal environmental reform that is globally relevant, without falling victim to the problems that undermined (structural functionalist) modernisation theories of the 1960s and 1970s.

Ecological modernisation and Al Gore: the consumerist turn

Al Gore's feature-length, documentary film *An Inconvenient Truth* offered an intriguing message that diffused rapidly around the globe in just a few months. The film not only presents images of 'System Earth' that are beautiful and impressive[3], but also contains a good deal of technical material generated over the years by scientific bodies such as the International Panel on Climate Change (IPCC). Although the award-winning motion picture sparked a new round in the global climate change debate and provides compelling evidence of the 'apocalyptic horizon of environmental reform', it combines a strong doomsday storyline with an equally powerful invitation for individuals to demonstrate their moral concerns and translate these commitments into concrete actions of sustainable citizenship and consumption. People who care for the environment – so it is argued – take responsibility for it, not just as citizens, but also as consumers of more environmentally friendly or 'green' products and services. For instance, environmentally conscientious con-sumers search out certification labels when buying fish, coffee or wood-based products, and look for ways to reduce the climate impact of their domestic energy and water consumption. Contemporary environmental activism by both individual consumers and NGOs puts pressure on product and service providers to deliver carbon neutral (or neutralised) food, housing, transportation (automobiles, air travel), clothing and other goods and services. It is recognised that to confront climate change, the implementation of strict government policies and the diffusion of proactive environmental practices by private firms is insufficient; it needs to be redoubled through substantial changes in consumer behaviour, including catalysing citizen-consumers' green purchasing power worldwide. Environmental reforms and improvement no longer are restricted to laws and economic tools, but engage (global) civil society and everyday life consumption practices as well.

EMT scholars began addressing consumption and lifestyle issues more closely and systematically starting in the mid-1990s (Spaargaren 1997; Cohen and Murphy 2001). However, because of the dominant focus on 'smokestack' and 'effluent pipe', production- and technology-related environmental problems and solutions in both environmental science and policy making, consumption-related topics received substantially less attention. More recently, sustainable consumption and lifestyle studies have begun to develop mature theoretical perspectives within EMT (Cohen 2001; Spaargaren 2003; Shove 2003; Carolan 2004; Mol and Spaargaren 2004; Spaargaren and Mol 2008). Also in (international) environmental policy making, the regulation of technology and production prevailed over product-policies and questions of lifestyles and consumption. Due to a growing awareness that the most persistent environmental problems – for example greenhouse gas emissions associ-ated with transportation and home heating and cooling – have a strong and obvious consumption dimension, sustainable consumption and production (SCP) policies are gaining momentum at different levels of policy making.[4] The consumerist turn in environmental practice and policy making now is unfolding rapidly, with com-panies competing for green consumers, governments developing new product policies and levelling the playing field for sustainable products and services, and third-party organisations formulating labelling and certification schemes; such trends give additional impetus to the further development and strengthening of sustainable consumption and lifestyle studies within EMT.

Outline and main arguments of the book

The bibliographies throughout this volume indicate the deepening and maturation of social scientific scholarship on environmental reform over the last three decades. At the same time, this richness presents a challenge for the selection of studies which together provide a representative introduction, foundation and overview of the EMT research tradition and 'school of thought'. Selections are structured in four main themes, together encompassing the major themes in ecological modern-isation research over the last thirty years. Each major part in this volume begins with a substantive review essay and introduction. These essays review the wider ecological modernisation studies in that area of scholarship, highlighting essential developments and debates, and introducing the readings in that part of the volume. Each part includes 'classic' and contemporary texts that are illustrative of and essential to an understanding of the development of ecological modernisation studies in that realm.

Part One of this reader establishes the theoretical foundations of ecological modernisation, including its development as a notion, idea and theory since around 1980. How does ecological modernisation relate to wider sociological and political science theories? Some of the most cited ecological modernisation texts are included here, as well as two texts made available in English for the first time. Together, the materials in Part One provide an in-depth overview of the origins and basic tenets of ecological modernisation theory.

Part Two presents ecological modernisation studies of environmental politics, governance and policy making, at various geo-political scales. Conceptualisations of the state, NGOs and other non-governmental actors, global governance regimes and processes of political modernisation are included. Selected texts range from the initial ideas of the political modernisation of environmental reform framed in 1993, to the latest ideas of environmental governance in the era of hyperglobalisation.

Part Three deals with dynamics of ecological modernisation in production and consumption processes, illustrating both earlier work on the greening of production and private firms that was at the foundation of ecological modernisation in the 1980s, as well as more recent innovative work on the greening of life styles and consumption. Special attention is given also to the – much debated – role of technology and technological development in ecological modernisation processes.

Part Four of this volume focuses on the relevance of ecological modernisation outside the core region where it was born and matured (that is: outside Europe), especially in newly industrialising economies of the world. Ecological modernisation theory did emerge in north-western Europe, and used specific developments in this region as its dominant point of orientation. Hence, from Mol's 1995 study onwards, the question of the relevance of ecological modernisation ideas for regions outside Europe has been on the research agenda. This part brings together studies of environmental reform efforts in Asia, Africa and elsewhere.

This reader's concluding chapter takes up two tasks crucial for future research, scholarship and policy making in environmental reform. First, while core debates around Ecological Modernisation are discussed throughout the reader, this epilogue addresses some of the critical debates that have come along with the development and maturation of Ecological Modernisation Theory. Secondly, the final chapter assesses the current state of ecological modernisation research, and formulates an agenda for future ecological modernisation studies. It is the hope of the editors and contributors that this reader will prove to be a milestone in the advancement of social scientific scholarship on, and policy making for, environmental reform in the years to come.

Notes

1. This cleavage between developed countries of the global North and the developing countries of the global South had become quite apparent at the first Summit

on the Environment that was held in Stockholm in 1972. The North–South divide has pertained questions regarding the main causes of environmental degradation. In some circles, the debate coalesced around the views of Paul Ehrlich who attributed environmental degradation on the population growth rate of developing countries and the contentions of Barry Commoner who identified technological developments and consumption levels in affluent countries as the main causes of environmental degradation (see Feenberg 1979 for more details).

2.   Ulrich Beck together with Anthony Giddens has criticised social sciences – sociology and political sciences in particular – for routinely taking the nation-state as its main object of theoretical analysis and point of departure for empirical research and policy making as well. They argue that with the emergence of the world network society, inter- and transnational social relations and interdependencies deserve much more attention from social scientists. The deterritorialisation and de-nationalisation of many environmental policies should follow-up on the period of the domination of nation-state politics (Beck and Wilms 2004; Spaargaren, Mol and Buttel 2006; Spaargaren and Mol 2008).

3.   As is the case with some of the follow-up movies on climate change like *The Eleventh Hour* and *Earth*.

4.   The international SCP-policies are discussed in Part Three of this book in more detail.

References

Andersen, M.S. (1999) 'Governance by green taxes: implementing clean water policies in Europe', *Environmental economics and policy studies*, **1**, 2, pp. 39–63.

Barrett, B. (ed.) (2005) *Ecological Modernization in Japan*. London: Routledge.

Beck, U. (1986) *Risikogesellschaft. Auf dem Weg in eine andere Moderne*. Frankfurt am Main: Suhrkamp.

Beck, U. (1997) *Was ist Globalisierung? Irrtümer des Globalismus – Antworten auf Globalisierung*. Frankfurt am Main: Suhrkamp.

Beck, U. (2005) *Power in the Global Age. A new global political economy*. Cambridge: Polity Press.

Beck, U., Giddens, A. and Lash, S. (1994) *Reflexive Modernisation, Politics, Tradition and Aesthetics in the Modern Social Order*. Cambridge: Polity Press.

Beck, U. and Willms, J. (2004) *Conversations with Ulrich Beck*. Cambridge: Polity Press.

Buttel, F.H. (2003) 'Environmental Sociology and the Explanation of Environmental Reform', *Organization and Environment*, **16**, 3, pp. 306–344.

Caldwell, L.K. (1984) *International Environmental Policy. Emergence and Dimensions*. Durham NC: Duke University Press.

Carolan, M. (2004) 'Ecological modernization theory: what about consumption?', *Society and Natural Resources*, **17**, 3, pp. 247–260.

Catton, W. (1980) *Overshoot: The Ecological Basis of Revolutionary Change*. Chicago, IL: University of Illinois Press.

Chinese Academy of Science (2007) *China Modernization Report Outlook 2001–2007*. Peking: Peking University Press.

Cohen, M. (2001) 'The emergent environmental policy discourse on sustainable consumption', pp. 21–37 in M. Cohen and J. Murphy (eds) *Exploring Sustainable Consumption: Environmental Policy and the Social Sciences*. New York: Elsevier.

Cohen, M. and Murphy, J. (eds) (2001) *Exploring Sustainable Consumption. Environmental Policy and the Social Sciences*. London: Pergamon Press.

Downs, A. (1972) 'Up and Down with Ecology – The "Issue-attention Cycle".' *The Public Interest*, 28, pp. 38–50.

Dryzek, J.S. (1987) *Rational Ecology. Environment and Political Economy*. Oxford/New York: Basil Blackwell.

Feenberg, A. (1979) 'Beyond the Politics of Survival', *Theory and Society*, **7**, pp. 319–360.

Giddens, A. (1990) *The Consequences of Modernity*, Cambridge: Polity Press.

Gore, A. (2006) *An Inconvenient Truth: The Crisis of Global Warming*. New York: Viking.

Goldsmith, E. and Allen, R. (1972) 'A Blueprint for Survival', *The Ecologist*, **2**, 1, pp. 1–43.

Hajer, M.A. (1995) *The Politics of Environmental Discourse: Ecological Modernisation and the Regulation of Acid Rain*. Oxford: Oxford University Press.

Huber, J. (1982) *Die verlorene Unschuld der Ökologie. Neue Technologien und superindustrielle Entwicklung*. Frankfurt am Main: Fisher Verlag.

Huber, J. (1985) *Die Regenbogengesellschaft. Ökologie und Sozialpolitik*. Frankfurt am Main: Fisher Verlag.

Huber, J. (1991) *Unternehmen Umwelt. Weichenstellungen für eine ökologische Marktwirtschaft*. Frankfurt am Main: Fisher Verlag.

Jänicke, M. (1995) *The Political System's Capacity for Environmental Policy*, FFU-report 95-4. Berlin: Forschungsstelle fur Umweltpolitik.

Liefferink, J.D. (1995) *Environmental Policy on the way to Brussels*. Wageningen: Landbouw Universiteit Wageningen (dissertation).

Luhmann, N. (1995) *Social Systems*. Stanford, CA: Stanford University Press.

Meadows, D.H. et al. (1972) *The Limits to Growth*. New York: Universe.

Mol, A.P.J. (1995) *The Refinement of Production. Ecological Modernization Theory and the Chemical Industry*. Utrecht: International Books.

Mol, A.P.J. (2000) 'The Environmental Movement in an Era of Ecological Modernisation', *Geoforum*, 31, pp. 45–57.

Mol, A.P.J. (2001) *Globalization and Environmental Reform. The ecological modernization of the global economy*. Cambridge, Massachusetts: The MIT Press.

Mol, A.P.J. (2006a) 'From Environmental Sociologies to Environmental Sociology? A Comparison of U.S. and European Environmental Sociology', *Organization & Environment*, **19**, 1, pp. 5–27.

Mol, A.P.J. (2006b) 'Environment and Modernity in Transitional China: Frontiers of Ecological Modernization', *Development and Change*, **37**, 1, pp. 29–57.

Mol, Arthur P.J., Lauber, V. and Liefferink, D. (eds) (2000) *The Voluntary Approach to Environmental Policy; Joint Environmental Policy-making in Europe*. Oxford: Oxford University Press.

Mol, A.P.J. and Sonnenfeld, D.A. (eds) (2000) *Ecological Modernization Around the World. Perspectives and Critical Debates*. Frank Cass/Routledge: London and Portland.

Mol, A.P.J. and Spaargaren, G. (1993) 'Environment, Modernity and the Risk Society. The Apocalyptic Horizon of Environmental Reform'. *International Sociology*, **8**, 4, pp. 431–459.

Mol, A.P.J. and Spaargaren, G. (2004) 'Ecological Modernization and Consumption: A Reply', *Society and Natural Resources*, **17**, pp. 261–265.

O'Connor, J. (1996) 'The Second Contradiction of Capitalism', in T. Benton (ed.), *The Greening of Marxism*. New York: Guilford.

Oosterveer, Peter, Guivant, Julia S. and Spaargaren, Gert (2007) 'Shopping for Green

Food in Globalizing Supermarkets: Sustainability at the Consumption Junction', in J. Pretty, A.S. Ball et al. (eds) *The Sage Handbook of Environment and Society*, pp. 411–429. Los Angeles: Sage.

Pepper, D. (1984) *The Roots of Modern Environmentalism*. London: Croom Helm.

Schnaiberg, A. (1980) *The Environment*. New York: Oxford University Press.

Schnaiberg, A., Weinberg, A.S. and Pellow, D.N. (2002) 'The Treadmill of Production and the Environmental State', in A.P.J. Mol and F.H. Buttel (eds) *The Environmental State Under Pressure*, pp. 15–32. London: JAI/Elsevier.

Schumacher, E.F. (1973) *Small is Beautiful*. London: Blond and Briggs.

Shove, E. (2003) *Comfort, Cleanliness and Convenience: The Social Organization of Normality*. Oxford: Berg.

Sonnenfeld, D.A. (2002) 'Social Movements and Ecological Modernization: The Transformation of Pulp and Paper Manufacturing', *Development and Change*, **33**, 1, pp. 1–27.

Sonnenfeld, D.A. and Mol, A.P.J. (2002) Special issue on 'Globalization, Governance, and the Environment', *American Behavioral Scientist*, **45**, 9, pp. 1318–1339.

Sonnenfeld, D.A. and Mol, A.P.J. (2006) Special issue on 'Environmental Reform in Asia', *Journal of Environment and Development*, **15**, 2, pp. 112–137.

Spaargaren, G. (1997) *The Ecological Modernization of Production and Consumption, Essays in Environmental Sociology*. Wageningen: WUR.

Spaargaren, G. (2003) 'Sustainable Consumption: A Theoretical and Environmental Policy Perspective', *Society and Natural Resources*, **16**, pp. 687–701.

Spaargaren, G. and Mol, A.P.J. (1992) 'Sociology, Environment and Modernity: Ecological Modernisation as a Theory of Social Change', *Society and Natural Resources*, **5**, 4, pp. 323–344.

Spaargaren, G. and Mol, A.P.J. (2008) 'Greening Global Consumption: Politics and Authority', *Global Environmental Change*, **18**, 3, pp. 350–359.

Spaargaren, G., Mol, A.P.J. and Buttel, F.H. (eds) (2000) *Environment and Global Modernity* London: Sage.

Spaargaren, G., Mol, A.P.J. and Buttel, F.H. (eds) (2006) *Governing Environmental Flows. Global Challenges to Social Theory*. Cambridge (Mass.) and London: The MIT Press.

Ullrich, O. (1979) *Weltniveau: In der Sackgasse der Industriegesellschaft*. Berlin: Rotbuch Verlag.

WCED (World Commission on Environment and Development) (1987) *Our Common Future*. Oxford: Oxford University Press.

Young, S. (ed.) (2001) *The Emergence of Ecological Modernisation*. London: Routledge.

Zhang, L., Mol, A.P.J. and Sonnenfeld, D.A. (2007) 'The Interpretation of Ecological Modernisation in China', *Environmental Politics*, **16**, 4, pp. 659–668.

Part One

Foundations of ecological modernisation theory

Chapter 2

Arthur P.J. Mol and Martin Jänicke

THE ORIGINS AND THEORETICAL FOUNDATIONS OF ECOLOGICAL MODERNISATION THEORY

Introduction

THE NOTION OF ECOLOGICAL modernisation was first launched by a member of the Berlin state parliament during debates in 1982. The speaker was the co-author of this text at that time preparing a study on 'Preventive Environmental Policy as Ecological Modernisation and Structural Policy' for the Berlin Science Center (Jänicke 1984, 1986; see also Jänicke 1978). His main objective was to argue for the need to give the modernisation processes a strong ecological twist. At the same time Joseph Huber, another Berlin environmental social scientist, published his book *Die verlorene Unschuld der Ökologie*, which actually was the first foundation of the 'greening of industry' theory (Huber 1982; Huber 1991b, which is featured in this section). As with others of the 'Berlin school' of environmental policy researchers (Simonis 1988; Zimmermann, Hartje and Ryll 1990; Prittwitz 1993; Mez and Weidner 1997) Huber later used the term 'ecological modernisation'. While Huber stimulated the (social) scientific discourse, Jänicke influenced the German policy debate. The environmental chapter of the coalition treaty of the red-green government (1998) was entitled 'ecological modernisation' (Koalitionsvertrag 1998). Both authors represented the project of a technology-based and innovation-oriented strategy focusing on the efficient use of resources and providing co-benefits both for ecology and economy. They stressed the need for fundamental ecological reform, without being trapped in some of the dead-ends of the environmental movement.

It should however be mentioned that MITI, the Japanese ministry of economic affairs, had already in 1974 developed the concept of a knowledge-intensive economy, being at the same time greener, less dependant on imported resources and much more innovative (MITI 1974; see also Kneese and Schulze 1975). This model of economic development influenced the German debate on innovation (Hauff and

Scharpf 1975), which later on had a strong impact on the early conceptualisation of 'ecological modernisation'.

At its beginning, 'ecological modernisation' was essentially a political program. It was neither a theory, nor a concept which included the social dimension of this type of modernisation. At that time, the concept was in between science and academia, and used as a practical and normative idea in pressing for far-reaching environmental reform. The coining of ecological modernisation, as well as its further development, was strongly influenced by debates with other schools of thought and belief systems on environmental degradation and reform. Also in the early 1980s, the notion expressed the need for radical change in close relation to feasible reforms.

But, of course, much has also changed during the development and maturation of ecological modernisation theory. In this maturation and further theoretical foundation, several Dutch environmental social scientists played a crucial role, followed later by UK and Scandinavian scholars. This section contains a number of key texts that can be seen as path-breaking in the coming of age of ecological modernisation theory. In this introductory essay we elaborate on the development of ecological modernisation theory in order to understand these texts in the wider development process of ecological modernisation ideas. The next section starts by focusing on the new relations between state and market that proved to be foundational in developing ecological modernisation ideas in the 1980s. The third section focuses on the conceptualisation of technology within ecological modernisation ideas, arguably one of its most criticised features. The fourth section is especially devoted to the formulation of ecological modernisation as a social theory, which took off in the early 1990s but continues until today. The fifth section deals with the various debates that have contributed to the further maturation of ecological modernisation, especially those with deep ecologists and neo-Marxists. Together then, this essay aims to provide a better understanding of the foundational contributions to and debates on ecological modernisation.

New relations between state and markets

Of key importance in the emergence and maturation of ecological modernisation ideas was the redefinition of the relation between state and market within environmental reform. While the environment is still widely conceived as a public good, in need of state action, the start of ecological modernisation was rather critical towards the role of the state in environmental reforms. In the late 1970s and the 1980s, there was widespread disappointment regarding the achievements of the environmental state in resolving the environmental crisis. Martin Jänicke's (1986 [1990]) work on state failure and the lack of preventive environmental policies of the European states, and Joseph Huber's (1982, 1985) publications on the need to provide the market and market actors with a central role in environmental reform have both been agenda-setting since the early days of ecological modernisation theory. Both scholars emphasised that in the 1970s and early 1980s the environmental state was not at all successful in bringing about the necessary environmental reforms. Ever since these early publications, ecological modernisation theory has emphasised the need to rethink and renew state–market relations in environmental reform. Ecological modernisation scholars have constantly emphasized that both the state (that is: environmental authorities and regulation modes) and the market (that is: various market actors and market mechanisms) should take up new roles in order to better contribute to environmental protection. Also today, much research within the ecological modernisation tradition is dedicated to defining these new roles of states and markets. But these new roles are defined within the framework of modern societies; that is, within the framework of modern welfare states and a market-oriented economy.

In investigating existing and new state-market relations in and for environmental reform, ecological modernisation theory – in contrast to some of the other schools of thought in the environmental social sciences – does not aim for a fundamentally different organisation of the (capitalist) economy. While the current capitalist system is seen as a major source of the environmental crisis and in this system there are clear restrictions for modern nation-states to intervene in the economy, ecological modernisationists do not put revolutionary system change as an alternative on the agenda. The agenda for state and market change within ecological modernisation theory does not move beyond a modern market economy and a modern welfare state. In that sense, ecological modernisation theory remains with the paradigm of modernity (cf. the text of Joseph Huber [1991b] in this section; see also below for a further elaboration on radicalism versus reformism in ecological modernisation).

Basically, two major innovations have developed out of this reflection on new state-market relations, which have later become much more common throughout the wider literature of environmental governance and environmental economics. First, the market and its main economic actors should not only be interpreted as forces that disturb the environment, as was the common opinion among environmental advocates in the 1970s and most of the 1980s. Major economic actors (such as producers, insurance companies, consumers, retailers, unions and credit institutions) and market institutions can also work in favour of environmental reform. This was a major deviation from the dominant line of thinking among environmentalists and environmental social science scholars in the 1970s and 1980s, but received wider support after the breakthrough of the notion of sustainable development (cf. also the text of Spaargaren and Mol [1992] on the relation between Ecological Modernisation and sustainable development). Secondly, while the environmental state remains an important institution in safeguarding environmental quality, it needs to be restructured: moving from a bureaucratic, hierarchical, reactive, command-and-control state, towards a more flexible, decentralised, and preventive institution that creates networks with other societal actors and applies a variety of approaches and instruments to guide society into directions of sustainability (cf. Weale 1993 [. . .]). Political modernisation (cf. the included text of Jänicke [1993]; Tatenhove et al. 2000; Mol 2002; Mol and Buttel, 2002) has been coined as the key, complementary concept within the ecological modernisation literature to frame such policy innovations within and beyond the field of environment. Political modernisation formed the ecological modernisation answer to state failure. The integration of environmental aspects into sectoral policies (energy, transport or agriculture) is an example for this kind of policy change.

Since the early days of reframing state market relations a rich literature has developed to further consider the role of states and markets in environmental reform, both within ecological modernisation theory, as well as in other schools-of-thought in the environmental social sciences. Environmental and ecological economists have taken the notion of economic actors and markets in environmental improvements much further, the environmental governance literature has enlarged our insights into the new roles of states, other actors and modes of governance when dealing with environmental challenges, and in various other disciplines all kind of in-between institutions (public–private partnerships, quangos (quasi-NGOs) and state-owned enterprises) have been investigated for their role in environmental problem solving.

Technological transformations

Besides states and markets, and their mutual relation, technology and technological development have been key subjects and characteristics of ecological modernisation ideas from the very beginning. Also here, we can identify a clear rupture with notions commonly held by environmental advocates in the 1960s, '70s and '80s. In those decades (and in some places, still today), environmental advocates were rather hostile towards technology and technological innovations and interpreted these rather in terms of causes of environmental crises or, at best, means for displacing environmental problems – in time, location and medium – instead of solving them. From the early 1980s onwards, ecological modernisation scholars developed a less hostile and at times – depending upon the author – a more positive and optimist assessment of the contribution of technology and technological change to environmental reform.

Arguably, Joseph Huber has been one of the first, and still most explicit, eco-modernisationists to argue for the relevance and centrality of advanced technology in any successful program of environmental reform. In his early work (cf. Huber 1982; 1985; 1991a) as well as his more recent contributions (Huber 2004), technological environmental innovation is seen as an indispensible part of any path towards sustainability. New technologies that even today meet hostile reactions from the environmental movement, such as genetically modified organisms and nanotechnologies, are – according to Huber – key elements of an ecological modernisation trajectory.

Perhaps the most common critique against ecological modernisation theory from its days of origin is related to its technological optimism and its supposed technocratic character. For example, in his initial analysis of contributions to environmental sociology Hannigan (1995: 184) claims ecological modernisation theory to be 'hobbled by an unflappable sense of technological optimism'. Many critics have preceded and followed this evaluation (see below), often without taking into account how ecological modernisation as a broad and still-evolving school-of-thought has moved to a more balanced understanding of the role of technology in environmental reform.

Maarten Hajer (1995) incorporates the debate on technocracy within the ecological modernisation project by distinguishing two variants of ecological modernisation: a techno-corporatist ecological modernisation and a reflexive ecological modernisation. While in the former, ecological reform is a purely techno-administrative affair, the latter points at practices of social learning, cultural politics and new institutional arrangements. A similar attempt is made by Christoff (1996; included as text), who identifies weak (i.e. economic–technological) and strong (institutional-democratic) ecological modernisation. Both authors – and others such as Dryzek (1997) and Neale (1997) – in fact closely resemble a distinction made in the early days of ecological modernisation theory by Joseph Huber (1985) between more technocratic and sociocratic development paths, respectively, be it that Huber himself was inconsistent in his plea for a more sociocratic version of ecological modernisation (cf. Spaargaren and Mol 1992). Arguably, Huber's Schumpeterian model of technology-induced socio-environmental change gave room for the technological optimism critique, as for instance Peter Wehling (1992) has extensively argued.

More recently, ecological modernisation theory adherents have made numerous efforts to (i) balance this Schumpeterian model and Huber's early technological optimism, and (ii) demonstrate technocracy critics' limited and/or selective reading of (later) contributions to ecological modernisation theory.

In the processes of institutional reform, technological transformations are given their place by ecological modernisation scholars, be it not as central as ecological modernisation critics want us to believe, and certainly not in the sense that technological change forms the motor of, and determines, these reforms. In addition, the conceptualisation of technology and technological change has widened considerably, from the original add-on, 'end-of-pipe' technologies that were so severely criticised in the 1970s and 1980s, via preventive technologies (e.g. pollution prevention, clean production), up to 'structural change of socio-technological systems' (cf. Mol et al. 1991; Jänicke et al. 1992; 1993 [text included in this book]; Neale 1997; Jokinen and Koskinen 1998). This also made claims on the technocratic character of ecological modernisation at least less adequate. It comes then as little surprise that in his second edition, Hannigan (2006) is much more balanced about the technological dimensions of ecological modernisation theory.

Within the technological dimensions of ecological modernisation, two innovations stand out. First, ecological modernisation scholars put on the agenda – and also increasingly identify in reality – a major development from curative, add-on/end-of-pipe technologies towards preventive, cleaner technologies. Various contributions in Simonis (1988), Zimmerman et al. (1990), Mol et al. (1991) and Prittwitz (1993) are clear examples of this. Secondly, there is a move from development and implementation of individual technologies, often restricted to what might be termed 'hardware', towards development and implementation of more complex socio-technological systems. This is for instance prevalent in the research program of industrial transformation (e.g. Olsthoorn and Wieczorek 2006), in the focus on large technological systems and in the more recent developments under the heading of transition management. Attention for wastewater treatment systems and air scrubbers is replaced by a focus on new transport systems, new (renewable) energy systems or integrated water systems, often combining technological hardware with new management concepts, new ownership relations, new prizing mechanisms, new roles of the state and the like.

In studying and conceptualising technological change, ecological modernisation scholars have relied strongly on various schools of thought including, among others, evolutionary and institutional economics, the sociology of science and technology, and technology assessment.

Ecological modernisation as social theory

While ecological modernisation concepts, and especially the agenda-setting and reformulation of state–market relations in environmental reform, originated in Germany, the development of this approach into a more formal and systematic theory was initiated from the Netherlands. In their foundational publication, Spaargaren and Mol (1992, included in this book) started to develop a coherent theory around the notion of ecological modernisation, which set off a large body of scholarly publications and debates on the key characteristics of what they have labelled ecological modernisation theory. In later publications they have further developed this theory, also in relation to and in debate with other social theories. Among others, they discussed similarities and differences between ecological modernisation theory and Ulrich Beck's risk society theory (Mol and Spaargaren 1993; Blowers 1997), compared ecological modernisation theory with deindustrialisation, neo-Marxist and post-modernity schools of thought (cf. Mol 1995; Spaargaren 1997; Mol and Spaargaren 2000; Mol and Spaargaren 2005), related ecological modernisation theory to the wider category of reflexive modernisation (Mol 1996; Hoogenboom, Mol and Spaargaren 2000), confronted it with globalization theories (Mol 2001) and consumption theories (Spaargaren 2003), and more recently renewed ecological modernisation theory by relating it to the sociology of networks and flows (Mol and Spaargaren 2005; Spaargaren et al. 2006).

But others have also entered these debates on the theoretical foundations of ecological modernisation. Wehling (1992), for instance, further developed ecological modernisation theory within the history of modernisation theories, a line of inquiry to which Seippel (2000) also contributed significantly. Weale (1992; 1993), Hajer (1995, 1996 [included in this book]) and others developed the notion of ecological modernisation especially in relation to a new discourse on the environment, in which the conventional dichotomy of environment versus economy was replaced by a new widely shared idea that the two could go together. Buttel (2000 [included in this book]) reviewed the theoretical claims and foundations of ecological modernisation theory. And Fisher (2002) has aimed to relate ecological modernisation theory to the critical theory of the Frankfurter School, most notably Habermas' work.

From these theoretical foundations several new concepts have emerged of which arguably the notion of ecological rationality stands out as of particular importance. While Dryzek (1987) coined the notion of ecological rationality already in an early phase and brought it into social theory, it was especially in the 1990s that the emergence of ecological rationality moved to the foundation of ecological modernisation theory. The growing independence of ecological rationality vis-à-vis other (e.g. economic and political) rationalities, and its increasing importance in 'governing' social practices and institutional designs, is interpreted as a core characteristic of patterns of ecological modernisation (cf. Mol and

Spaargaren 1993; Mol 1995; Spaargaren 1997). Here the essence of most writings of ecological modernisation comes together: the insight that environmental (or ecological) interests, representations and ideas move to the fore in modernisation processes, and restructure social practices and institutions in modern societies. Whether in production, consumption, dominant discourses, technological trajectories, market institutions, or civil society, environmental ideas and interests are catching up with their conventional economic, political and social equivalents.

A second main line in the more theoretical studies on ecological modernisation has been the relation (and distinction) between ecological modernisation as a more descriptive-analytical theory and ecological modernisation as a prescriptive-normative project. While initially, in the 1980s, ecological modernisation started in Germany as a normative idea, in the 1990s it developed into a descriptive-analytical theory of socio-ecological change. But application and use of this more analytical concept continued in the practices of environmental NGOs, policy makers, political parties and even market parties. As with most social science theories, there are clear interdependencies between these normative and analytical qualities of ecological modernisation (cf. Mol 1995).

But much of the clarification of the basic assumptions, concepts and notions of and in ecological modernisation theory has come from critical interactions and debates with other schools of thought in the social sciences.

Radicalism versus reformism: foundational debates

Ever since ecological modernisation was coined as a notion, concept or theory it has been in debate with other notions, concepts and theories in the environmental social sciences. This is of course common practice in the social sciences, as progress in social science goes via debate. But especially in the beginning, in the 1980s and 1990s (and even until today), these debates have been fierce, particularly with neo-Marxists.

The core of the debate with neo-Marxist and other 'radical' social scientists centres around the idea that ecological modernisation includes – in both normative and analytical connotations – a reformist trajectory for change, as it does not dissociate itself from a capitalist organisation of production and consumption. The possibilities, actuality and desirability of a green Capitalism, put first and rather provocatively on the agenda of ecological modernisation theory by Joseph Huber, have resulted in extensive debates with neo-Marxists and other 'radical' scholars. Scholars as diverse as Allan Schnaiberg (1980), David Goldblatt (1996), Peter Dickens (1998) and James O'Connor (1998) have all at some point, using different concepts, attacked the possibilities of an ecologically sound capitalism. (Dickens' more recent work [2004], is more open to the important role of environmental reforms.) James O'Connor's 'Second Contradiction of Capitalism', Schnaiberg's 'Treadmill of Production' and Goldblatt's criticisms of Giddens' limitation to the industrial dimension of modernity in understanding the environmental crisis, are used to point at the important role Capitalism plays in environmental deterioration. Neglecting Capitalism and failing to attack the fundaments of the Capitalist world order will result in superficial and cosmetic environmental reforms that are unable to resolve the ecological crisis in any fundamental way, according to these scholars. Moreover, such measures will strengthen the Capitalist mode of production as it makes Capitalism less in need of a green critique (cf. Dryzek 1997) and it promotes and facilitates the continuation of established socio-economic practices that are to the benefit of those in power (Blühdorn 2000).

Ecological modernisation theory deviates from this view, be it that its position taken towards Capitalism has changed throughout the various historical phases of ecological modernisation theory (cf. Mol and Sonnenfeld 2000). While initially the contribution of Capitalism to the 'expansion of the limits' was celebrated by some ecological modernisation scholars (e.g. Huber 1991b), more recently ecological modernisation scholars have taken more nuanced and critical positions regarding the present world-system of market economies. It is not that Capitalism is considered to be essential for environmentally sound production and consumption (as neo-liberal scholars would have eco-modernisationists believe), nor that Capitalism is believed to play no role in environmental deterioration, but rather that (i) Capitalism is changing constantly and one of the main current triggers in this change is related to environmental concerns, (ii) environmentally sound production and consumption is possible under different 'relations of production' and each mode of production requires its own environmental reform program, and (iii) until further notice, all major and fundamental alternatives for the present economic order have proved unfeasible according to various (economic, environmental and social) criteria (cf. Mol and Spaargaren 2000; Mol 2001).

Consequently, many mainstream ecological modernisation theorists today interpret Capitalism neither as an essential precondition for, nor as the key obstruction against, stringent or radical environmental reform. They focus rather on redirecting and transforming 'free market Capitalism' in such a way that it less and less obstructs, and increasingly contributes to, the preservation of society's sustenance base in a fundamental/structural way. While it can be argued – as some commentators do – that this debate should be considered as rather abstract and outdated, especially since the 'end of history', it shows at the same time continuing relevance. This fundamental controversy is connected with discussions on presumed shortcomings of ecological modernisation theory in analysing conflicts of interest related to environmental reforms. Ecological modernisation theory and theorists are believed to have undertheorised notions of power (Leroy 1994), have little attention for social contexts and ethical issues (Blowers 1997), neglect emancipatory concerns (Blühdorn 2000), and fail to link environmental reform with social injustice (as in the environmental justice literature). According to these and other scholars, ecological modernisation theory would analyse environmental reforms primarily via Schumpeterian, evolutionary models that result almost automatically in the greening of production and consumption, without paying sufficient attention to severe struggles between interests (groups) and to normative, ethical or moral reflections and debates (cf. Sarkar 1990; Leroy and Van Tatenhove 2000; Blowers 1997: 854). These observations are accurate as far as they relate to the first generation of studies in ecological modernisation theory *in the 1980s*. There is indeed considerable merit in neo-Marxist analyses of environmental conflicts and more recent versions of ecological modernisation theory have profited from engagement and debate with these and other political economy approaches. In the context of a still-young, but rapidly expanding, and theoretically maturing environmental social science school-of-thought, such (continued) critiques today are anachronistic and less relevant in relation to current ecological modernisation scholarship.

References

Blowers, A. (1997) 'Environmental policy: ecological modernization and the risk society?', *Urban Studies* **34**, 5–6, pp. 845–871.

Blühdorn, I. (2000) 'Ecological modernisation and post-ecologist politics', in G. Spaargaren, A.P.J. Mol and F. Buttel (eds), *Environment and Global Modernity.* London: Sage, pp. 209–228.

Buttel, F.H. (2000) 'Ecological modernization as social theory', *Geoforum* **31**, 1, pp. 57–65.

Christoff, P. (1996) 'Ecological modernisation, ecological modernities', *Environmental Politics* **5**, 3, pp. 476–500.

Dickens, P. (1998) 'Beyond Sociology: Marxism and the Environment', in M. Redclift and G. Woodgate (eds), *The International Handbook of Environmental Sociology.* Cheltenham: Edward Elgar, pp. 179–192.

Dickens, P. (2004) *Society and Nature.* Cambridge: Polity Press.

Dryzek, J.S. (1987) *Rational Ecology. Environment and Political Economy.* Oxford/New York: Blackwell.

Dryzek, J.S. (1997) *The Politics of the Earth: Environmental Discourses.* Oxford: Oxford University Press.

Fisher, D. (2002) 'From the Treadmill of Production to Ecological Modernization? Applying a Habermasian Framework to Society-Environment Relationships', in A.P.J. Mol and F.H. Buttel (eds), *The Environmental State under Pressure.* Amsterdam: Elsevier, pp. 53–64.

Goldblatt, D. (1996) *Social Theory and the Environment.* Cambridge: Polity Press.

Hajer, M.A. (1995) *The Politics of Environmental Discourse. Ecological Modernisation and the Policy Process.* Oxford: Clarendon Press.

Hajer, M.A. (1996) 'Ecological modernization as cultural politics', in S. Lash, B. Szerszynski and B. Wynne (eds), *Risk, Environment and Modernity. Towards a New Ecology.* London: Sage, pp. 246–268.

Hannigan, J.A. (1995) *Environmental Sociology. A Social Constructivist Perspective.* London and New York: Routledge.

Hannigan, J. (2006) *Environmental Sociology.* London and New York: Routledge (2nd edition).

Hauff, V. and Scharpf, F.W. (1975) *Modernisierung der Volkswirtschaft. Technologiepolitik als Strukturpolitik.* Frankfurt/M.: Europäische Verlagsanstalt.

Hoogenboom, J., Mol, A.P.J. and Spaargaren, G. (2000) 'Dealing with environmental risks in reflexive modernity', in M. Cohen (ed.), *Risk in the Modern Age. Social Theory, Science and Environmental Decision-making.* Basingstoke: MacMillan, pp. 83–106.

Huber, J. (1982) *Die verlorene Unschuld der Ökologie. Neue Technologien und superindustrielle Entwicklung.* Frankfurt: Fisher.

Huber, J. (1985) *Die Regenbogengesellschaft. Ökologie und Sozialpolitik.* Frankfurt am Main: Fisher Verlag.

Huber, J. (1991a) *Unternehmen Umwelt. Weichenstellungen für eine ökologische Marktwirtschaft*. Frankfurt am Main: Fisher.

Huber, J. (1991b) 'Ecological modernisation: beyond scarcity and bureaucracy'. Translation of: J. Huber (1991) 'Ecologische modernisering: weg van schaarste, soberheid en bureaucratie?', in A.P.J. Mol, G. Spaargaren and B. Klapwijk (eds), *Technologie en Milieubeheer. Tussen Sanering en Ecologische Modernisering*. Den Haag: SDU, pp. 167–183.

Huber, J. (2004) *New Technologies and Environmental Innovation*. Cheltenham: Edward Elgar Jänicke, M. (ed.) (1978) *Umweltpolitik – Beiträge zur Politologie des Umweltschutzes*. Opladen: Leske und Budrich.

Jänicke, M. (ed.) (1978) *Umweltpolitik – Beiträge zur Politologie des Umweltschutzes*. Opladen: Leske und Budrich.

Jänicke, M. (1984) *Umweltpolitische Prävention als ökologische Modernisierung und Strukturpolitik*. Discussion paper, Wissenschaftszentrum Berlin: IIUG dp. 84–1.

Jänicke, M. (1986 [1990]) *Staatsversagen. Die Ohnmacht der Politik in die Industriegesellschaft*. München: Piper. (Translated as *State Failure. The impotence of politics in industrial society*, University Park: Pennsylvania State University Press.)

Jänicke, M. (1993) 'On ecological and political modernization'. Translation of M. Jänicke (1993) 'Über ökologische und politische Modernisierungen', *Zeitschrift für Umweltpolitik und Umweltrecht* 2, pp. 159–175.

Jänicke, M., Mönch, Binder, H.M. et al. (1992) *Umweltentlastung durch industriellen Strukturwandel? Eine explorative Studie über 32 Industrieländer (1970 bis 1990)*. Berlin: Sigma.

Jokinen, P. and Koskinen, K. (1998) 'Unity in environmental discourse? The role of decision-makers, experts and citizens in developing Finish environmental policy', *Policy and Politics* **26**, 1, pp. 55–70.

Kneese/Schulze (1975) Koalitionsvereinbarung zwischen der SPD und Bündnis 90/DIE GRÜNEN (1998) *Aufbruch und Erneuerung, Deutschlands Weg ins 21. Jahrhundert*, 20 Oktober 1998.

Leroy, P. (1994) 'Nieuwe stappen in de Milieusociologie?', *Tijdschrift voor Sociologie* **15**, 2, pp. 67–80.

Leroy, P. and Tatenhove, J. van (2000) 'New policy arrangements in environmental politics: the relevance of political and ecological modernization', in G. Spaargaren, A.P.J. Mol and F. Buttel (eds), *Environment and Global Modernity*. London: Sage, pp. 187–208.

Mez, L. and Weidner, H. (eds) (1997) *Umweltpolitik und Staatsversagen – Perspektiven und Grenzen der Umweltpolitikanalyse*. Berlin: Edition Sigma.

MITI (Ministry of International Trade and Industry) (1974) *Direction for Japan's Industrial Structure*. Tokyo: MITI.

Mol, A.P.J. (1995) *The Refinement of Production. Ecological modernization theory and the chemical industry*. Utrecht: Jan van Arkel/International Books.

Mol, A.P.J. (1996) 'Ecological modernisation and institutional reflexivity. Environmental reform in the late modern age', *Environmental Politics*, **5**, 2, pp. 302–323.

Mol, A.P.J. (2001) *Globalization and Environmental Reform. The ecological modernization of the global economy*. Cambridge (Mass.)/London: MIT Press.

Mol, A.P.J. (2002) 'Political Modernisation and Environmental Governance: Between Delinking and Linking', *Europæa. Journal of the Europeanists*, **8**, 1–2, pp. 169–186.

Mol, A.P.J. and Buttel, F.H. (eds) (2002) *The Environmental State Under Pressure*. Amsterdam: Elsevier Science.

Mol, A.P.J. and Sonnenfeld, D.A. (eds) (2000) *Ecological Modernisation Around The World: Perspectives and critical debates*. London: Frank Cass/Routledge.

Mol, A.P.J. and Spaargaren, G. (1993) 'Environment, modernity and the risk-society: the apocalyptic horizon of environmental reform', *International Sociology*, **8**, 4, pp. 431–459.

Mol, A.P.J. and Spaargaren, G (2000) 'Ecological Modernization Theory in debate: a review', *Environmental Politics* **9**, 1, pp. 17–49.

Mol, A.P.J. and Spaargaren, G. (2005) 'From additions and withdrawals to environmental flows. Reframing debates in the environmental social sciences', *Organization & Environment* **18**, 1, pp. 91–107.

Mol, A.P.J., Spaargaren, G. and Klapwijk, A. (eds) (1991) *Technologie en Milieubeheer. Tussen Sanering en Ecologische Modernisering*. The Hague: SDU.

Neale, A. (1997) 'Organizing environmental self-regulation: Liberal governmentability and the pursuit of ecological modernisation in Europe', *Environmental Politics* **6**, 4, pp. 1–24.

O'Connor, J. (1998) *Natural Causes. Essays in Ecological Marxism*. New York and London: Guilford.

Olsthoorn, X. and Wieczorek, A.J. (eds) (2006) *Understanding Industrial Transformation. Views from Different Disciplines*. Dordrecht: Springer.

Prittwitz, V. von (ed.) (1993) *Umweltpolitik als Modernisierungsprozess*. Opladen: Leske u. Budrich.

Sarkar, S. (1990) 'Accommodating industrialism: A Third World view of the West German ecological movement', *The Ecologist* **20**, 4, pp. 147–152.

Schnaiberg, A. (1980) *The Environment. From Surplus to Scarcity*. Oxford/New York: Oxford University Press.

Seippel, Ø. (2000) 'Ecological modernization as a theoretical device: Strengths and weaknesses', *Journal of Environmental Policy and Planning* **2**, 4, pp. 287–302.

Simonis, U.E. (ed.) (1988) *Präventive Umweltpolitik*. Frankfurt/New York: Campus Verlag.

Spaargaren, G. (1997) *The Ecological Modernisation of Production and Consumption. Essays in Environmental Sociology*. Wageningen: Wageningen Agricultural University (dissertation).

Spaargaren, G. (2003) 'Sustainable consumption: A theoretical and environmental policy perspective', *Society and Natural Resources* **16**, 8, pp. 687–702.

Spaargaren, G. and Mol, A.P.J. (1992) 'Sociology, environment and modernity. Ecological modernisation as a theory of social change', *Society and Natural Resources* **5**, pp. 323–344

Spaargaren, G., Mol, A.P.J. and Buttel, F.H. (eds) (2006) *Governing Environmental Flows. Global Challenges for Social Theory*. Cambridge (Mass): MIT.

Tatenhove, J. van, Arts, B. and Leroy, P. (eds) (2000) *Political Modernisation and the Environment. The Renewal of Policy Arrangements*. Dordrecht: Kluwer.

Weale, A. (1992) *The New Politics of Pollution*. Manchester: Manchester University Press.

Weale, A. (1993) 'Ecological modernisation and the integration of European environmental policy', in J.D. Liefferink, P.D. Lowe, and A.P.J. Mol (eds), *European Integration and Environmental Policy*. London: Belhaven Press, pp. 196–216.

Wehling, P. (1992) *Die Moderne als Sozialmythos. Zur Kritik sozialwissenschaftlicher modernisierungstheorien*. Frankfurt/New York: Campus.

Zimmermann, K., Hartje, V.J. and Ryll, A. (1990) *Ökologische Modernisierung der Produktion – Strukturen und Trends*. Berlin: Edition Sigma.

Chapter 3

Martin Jänicke

ON ECOLOGICAL AND POLITICAL MODERNIZATION

Translation by Bettina Bluemling from: Martin Jänicke (1993), 'Über ökologische und politische Modernisierungen', *Zeitschrift für Umweltpolitik und Umweltrecht* 1993, 2, pp. 159–175

Introduction

ENVIRONMENTAL QUESTIONS PLAY A central role in the recent debates on the functional and structural changes of the state in the industrial system. This is illustrated by Luhmann's book on 'Ökologische Kommunikation'. For Luhmann, ecological problems make fully clear that 'politics are expected to do a lot but in fact can do only little' (Luhmann 1990: 169). Also Von Beyme relates the problem of political steering primarily to the ecological challenge. Mayntz treats the problem of regulatory politics, not accidentally, primarily in this similar relationship. And for Beck, ecological challenges are a constitutive characteristic of the risk society. The concept of state failure – referring to structural weaknesses in decision-making, ineffectiveness, and inefficiency – was illustrated in political science primarily with the help of environmental questions (Jänicke 1979 and 1986). It should thus not surprise us when the concepts of political modernization are also strongly influenced and shaped by the environmental problematic (cf. Beck 1986 and 1988; Jänicke 1986 and 1990; Von Prittwitz 1990).

Even more important than this simultaneity is the fact that political modernization in the 1970s and 1980s was also decisively shaped by the environmental problematic. The emerging double structure of the state as a majority-legitimized bureaucratic intervention mechanism and as a partner in consensus-legitimized negotiation systems (Bens, Scharpf and Zintl 1992) cannot be understood without reference to this problematic.

In a remarkable analysis, Helmut Weidner (1992) has made clear how strong this relationship applies in Japan, one of the frontrunners of ecological modernization. The new strategy of a state-organized form of reflexivity through which industrial society is confronted with problems that, to a large extent, have been created by industrial society itself, emerged in Japan probably in its clearest form.

But in general it can be stated that the modernization of the state – insufficient as it may be – runs parallel with the process of ecological modernization, responding to the same kind of drivers for change. As a result, we witness in many industrial societies a tendency towards:

- spatial and functional decentralization;
- increasing reliance on dialogue structures and negotiated solutions;
- expansion of participation rights (citizen participation, referenda, etc.);
- more frequent use of informational strategies;
- opening of informal preliminary decision networks (such as energy politics) for new, controversial interests;
- widening of the state's goal setting (also interesting here is the formula of the 'environmental state'; Kloepfer 1989); and
- making the juridical system more dynamic (liability rules, reversal of the onus of proof, rights of information, procedures for mediation).

Would these changes have been conceivable without the pressures originating from environmental problems and without the general perception of the incapacity of the traditional hierarchical intervention structures of the political system to adequately cope with them?

But why modernization, when the ecological problems are attributed to industries such as those in the chemical, nuclear or mobility sector – preferences of the modern state – and this 'modernity' also formed the starting point of the environmental movement's criticisms?

The increased use of the concept of modernization in the environmental debates in the 1980s can probably best be explained by the strong influence of innovation in these debates. Thus, the modernization debates addressed renewal less in terms of ongoing modernization, but more in the sense of basic innovations and paradigm shifts, as drivers of long-term development and change (Mensch 1975; Schumpeter 1942; Kuhn 1962; Kondratieff 1926). The volume 'Modernisierung der Volkswirtschaft' by Volker Hauff and Fritz Scharpf – which was so important for the debate on structural environmental and technological change – had emphasized this already in 1975. This volume already stressed as well the fact that technical innovations must be preceded by social innovations (Hauff and Scharpf 1975).

Today we must add the terminological, conceptual and methodological paradigm changes: the innovations of established 'intellectual techniques' (von Gottl-Ottlilienfeld). Modernization can then be defined as a real, social and intellectual technically innovative answer to fundamental system problems (Jänicke 1986: 154ff). In an effort to widen the old political modernization theory, with its idea of crises-induced capacity growth (Rokkan 1969; Binder et al. 1971), one can also state that modernization is the institutionalization and differentiation of a new technological, political-social, and scientific-cultural level of problem solving, based upon a fundamental paradigm change. It is not surprising that from this perspective, conventional, ecologically destructive structures are not identified with modernity, but rather with institutional sclerosis (Olson). We indeed are confronted here with the conflict between two modernities (Beck 1991). The essence of this changed modernity perspective is that progress is not a linear development of an existing phenomenon; rather, it includes a fundamental break or transition and an innovation of the direction of change.

In this way, we can understand the conception that long-term environmental protection not only includes ecological modernization and structural change of industrial societies, but also implies a modernization of the political action system (cf. Hesse and Benz 1990). It seems important to me that this approach in the end relates to a crisis-induced capacity growth. It should be remembered that the early German contributions to the environmental political sciences were often developed along such crisis theorical lines (Ronge 1972; Glasgow 1972; Weidner 1975; cf. Jänicke 1973). Perhaps it is no coincidence that one of these early environmental political scientists – Volker Ronge – was engaged quite early with para-statal forms of problem solving (in the world of banking). In his analyses, one can also detect early fears about the risks of environmental crisis management being repressed and reduced to the state of business as usual.

Ecological modernization

The concept of ecological modernization goes back to a debate in the Berlin municipal parliament on 22nd January 1982, when the environmental representative of one of the opposition political parties suggested to the government four ecological modernizations: in industry, in the energy sector, in the mobility sector, and in the construction sector. In these sectors, it was argued, employment-stimulating innovations and ecologically relevant forms of rationalization should be enhanced, which would not so much burden the production-factor labour, but rather the use of energy and natural resources (*Abgeordnetenhaus* v. *Berlin*: Plenarprotokoll 9/14: 756 ff). As a scientist, the representative had already raised the question of the political and economic modernization capacity necessary for environmental protection in 1978 (Jänicke 1978: 32). At the bottom of this was a concept influenced by the theory of political modernization and by the innovation debate; but it was especially stimulated by a search for a consensual formula that would enable a redefinition and reorientation of established modernization ideas. Along this line, the concept of ecological modernization was also used by Huber (1983) and Simonis (1985), who crucially contributed to the fact that from 1983 onwards, numerous statements appeared under this label in the labour unions and the Social-Democratic Party. The concept is now also included in OECD publications (see Hajer 1992).

Ecological modernization is a concept related primarily to economy and technology (cf. Zimmermann et al. 1990). The at-times-discussed escape or movement away from technology is eventually responded to with a technological escape forward, into the future. A flight forward equals an acceleration and simultaneous change in the development direction of technological progress: from increasing to decreasing material intensity, energy intensity, transport intensity, waste intensity, and/or risk intensity (Jänicke 1984). A 'forward escape' is needed because the employment and welfare arrangements are directly linked to the existing production system (Huber 1982, 1983); because the changing of industrialism will be much more acceptable than its abolition; because a significant environmental improvement potential can be realized through modernization; and – not the least – because a cost-reducing rationalization of environmental resource use at times proves to be a feasible alternative for the technological destruction of employment (in this respect, increasing the taxation of environmental resource use while reducing taxation rates for the factor labour forms an important argument; Binswanger et al. 1983).

Meanwhile, the first research reports have appeared which prove that a partial deindustrialization of environmentally polluting sectors (inter-sectoral change) – at least up until now – has clearly yielded fewer environmental benefits when compared to a modernization strategy within industries (intra-sectoral change). The structure of the industrial sectors of Sweden has barely developed beneficially since 1973, if we consider its total ecological effect. However, when taking into account the stagnation in the freight transport sector, it can be concluded that the industrial use of energy and water showed an absolute reduction, which resulted from the environmentally polluting heavy industries radically modernizing themselves. In Japan this technological component had an especially strong influence and caused a decoupling of energy, natural resources, water and land use, which temporarily resulted in qualitative growth. We learn from this example that little environmental improvement is realized in the long term when ecologically profitable modernization is virtually neutralized through high industrial growth. Moreover, the example of Japan shows that environmental improvements through technological progress can only be realized when during economic growth *permanent* and structural efforts are made in this direction. If the efforts decline, then the realized decoupling of industrial growth and natural resource use will come to an end, with the environmental pressure curves going upwards again (Jänicke et al. 1992). This shows that in addition to technological progress, we also must enforce (inter-sectoral) changes in the structures of environmental degradation and in the structural process of economic growth itself. In the end, the question of economic growth, or at least the role of the state as a growth engine, can no longer be ignored.

It is inevitable that ecologically sustainable development implies structural modernization. In terms of politics, this refers not only to the ecologically motivated kind of structural reforms as they were initiated in Japan. In the long term, this structural modernization involves a redefinition of the functions of the central state in ways hardly envisaged in the case of Japan. The impact of this structural modernization can be judged from the fact that ending the policy of stimulating economic growth necessitates a different policy of distribution.

The special case of environmental protection

It is not a coincidence that the disenchantment of the state ('Entzauberung des Staates'; Willke 1983) and the related postulates of political modernization are debated especially in the light of environmental problems. Several reasons can be mentioned as relevant.

1. In no other area of politics do we witness such a regular testing of the intervention capacity of the state as in the field of the environment. And

nowhere else can we see such a clear discrepancy as that which occurs between the public acknowledgement of the problems and (the lack of) countermeasures by the state.

2.   In no other area of politics is it so clear that regulatory inputs by the state (in the meaning as suggested by Luhmann) are only undertaken to the extent that they match with the codes and programs of the existing social subsystems. And also within the state (system) itself the motivation for ecological actions must be developed and sustained against opposing motives and policies.

3.   The partial erosion of the nation state (cf. Willke 1991: 182) against the background of simultaneous processes of internationalization and decentralization is further accentuated by the ecological connections existing between the local and the global (interconnections which reach beyond the environmental movement theme of 'think globally – act locally').

4.   Increasingly we witness a competition between the environmental protection institutions of the central state and decentralized interventions from the side of courts, communities, or (increasingly) from private collective actors. The decentralized level to some extent demonstrates that some of the interventions that the central state has refrained from enacting are very possible to make, for instance, when environmentally harmful products are pushed out of the market through media campaigns or through trade companies that aim to set up a green profile.

5.   The shift in time horizons – next to the complexity of societal subsystems and complexity of a globalizing world the main factor in the disenchantment of the state according to Willke (Willke 1991: 182) – is in no political area or domain so delicate as in environmental politics. The parliamentary institutions are in essence tailored to reactive politics, to learning from experience. But from an ecological point of view, there are more and more experiences that should be prevented and need to be anticipated (Jänicke 1986).

6.   This leads us to the decisive factor, and that is the nature of the environmental problem field itself. The longer industrial societies are engaged in environmental politics, the clearer it becomes that the current style of reactive environmental politics is inadequate. When considered from a long-term and global perspective, the more obvious it becomes that huge steering efforts must be made: nothing less than the transformation of the entire industrial mode of production is at stake, and this should be organized in a permanent way.

In other words, the problem of environmental politics is among others that even a considerable potential for action from the side of the political-administrative system cannot really reassure. The task for environmental politics is not only considerably larger than represented and dealt with by its current, segmented structure; the tasks are also increasingly radicalized due to both unrecognized past environmental legacies and future environmental problems that have not been coped with in a preventive way; and through the recognition that the industrial model of the welfare society over the long term will have catastrophic consequences, if this welfare model cannot be generalized to the global scale (Von Weizsäcker 1990). The tasks at hand are also more burdensome because it turns out that reductions in quantities of resource flows and emissions do not seem to really solve the problem of the environmental effects of industrialization accumulating at ever higher levels, especially in the (old!) industrial societies.

This comes together with the recognition that in the process of industrial growth, the available environmental protection strategies (such as end-of-pipe environmental protection) are quickly exhausted in their potentials. This is primarily related to the destructive consequences of exponential growth. A 3.5% economic growth rate results in an eight times increase of what is 'growing' in 60 years time (cf. Meadows et al. 1992). And this is ecologically problematic even in those cases where only office spaces increase.

An additional problem is caused by the negative environmental side effects of the existing (traditional) patterns of modernization and growth. Traditional industrial societies, as we find them for instance in Eastern Europe, suffer(ed) from the emissions of chimney industries, primarily basic industries. In principle, a modernized pattern of emission discharges can be compatible with blue skies over industrial cities and with clean rivers full of fish. But then there are other typical impacts occurring: high and growing quantities of waste, among them wastes difficult to dispose of and hazardous wastes from environmental facilities; high and increasing mobility in all modalities (passenger car transport, air transport and freight transport by road); increasing land and soil pollution with remote damages on ground water; high and still increasing levels of electricity use in production; and new toxic emissions that hamper a generalized problem perception as they occur in dispersed form and in smaller amounts.

From a long-term and global perspective, the safeguarding of the ecological status quo in the industrialized countries is not sufficient. Ecological modernization and the related restructuring of industrial societies thus entails the character of an objective necessity (cf. Von Weizsäcker 1990).

Can the state accomplish this? When taking a skeptical position on this question, this leads to the thesis of the overcharged state with respect to its environmental protection tasks (Jänicke 1992). This results in the follow-up question of which possibilities would enhance that action potential through political modernization.

Political modernization

Meanwhile, the direction of change leading to an increased action capacity of the political system becomes well apparent. Advanced industrial nations like Japan and Sweden at the beginning of the 1970s already partly showed the way with their political reforms. Over the course of the 1980s, the idea of chasing the utopia of the state's ever-increasing intervention capacity was abandoned and replaced by the idea of redistributing and transferring the intervention tasks in case they could not feasibly be reduced. As a way out of the deadlock, modes of functional and spatial decentralization, in combination with new cooperative integration mechanisms, were made the subject of discussion (Schuppert 1989). The importance of the state vis-à-vis the highly organized macro-actors should be tackled with the use of more sensitive and flexible strategies. Willke talks of 'societal steering' and 'decentral contextual steering', respectively (Willke 1983). Beck wants to give up the 'fiction of centralized state authority' and to make use of the gradual de-monopolization of politics in favour of decentralized 'sub-politics' in the decision-making domains of the 'risk society' (Beck 1986: 371).

Scharpf (1991) does provide a very clear-cut outline of the meaning of consensually legitimated, horizontally interlinked negotiation systems as a second steering entity next to the hierarchical, majority-legitimated interventions by the sovereign national state (Benz, Scharpf and Zintl 1992).

In the last two decades, a novel double structure of the political-administrative system indeed has developed, which can be represented as follows:

The novel double structure of the state: hierarchy and cooperation

Hierarchy	Cooperation
vertical intervention	'horizontal' cooperation
democratic legitimization	legitimization by consensus
majority based	minority based
imperative policy style	dialogical policy style
project based steering	proceduralisation
centrality	decentral organization
strong institutionalization	weak institutionalization

Both action types strongly bear upon each other and in no way represent exclusive alternatives. What matters after all is the expansion of the intervention range and volume: from the classical bureaucratic intervention to self-regulation initiated by the state, via pre-negotiated (therefore less opposed) state intervention, while retaining interventionist strategies in negotiated solutions. Furthermore, 'private', societal, para-governmental forms of intervention join in (see below). Particularly in the environmental sector, we are confronted with the complex coexistence and interlinkage of:

*   hierarchical intervention;
*   cooperative intervention; and
*   self-regulation.

Nowadays, there is a far-reaching consensus that the bureaucratic constitutional state no longer is nor can be the only legitimate steering authority. The expanded range of interventions certainly offers some new opportunities.

Basically, a *double paradigm change* has taken place in the brain trusts of industrial societies: one concerns the growth pattern and its ecological performance; the other concerns the political steering mechanism. Both reorientations presumably will open up real prospects for the future only if and when they are consistently realized and pursued.

Regarding the growth pattern, the paradigm of a resource preserving, emission-, waste-, transport-, and risk-poor mode of production has without doubt also received increasing acceptance in private sector (business) circles (Huber 1991; Schmidheiny 1992).

The other paradigm change, that regarding the steering potential of politics, can be said to result from the diagnosis of 'state failure', from the 'crisis of regulative politics' (Mayntz 1979) and from the 'disenchantment of the state' (Willke 1983), not to mention its roots in older critical formulas like that of 'symbolic politics' (Edelmann 1964) or the politics of 'non-decisions' (Bachrach and Baratz 1970).

We can conclude that a partial political reorientation is taking place (see Hesse 1987, Jänicke 1992):

- from bureaucratic, detailed rulemaking to an emphasis on steering the frame-work conditions and action contexts;
- from the state mode of dealing with problems to the societal mode of handling them, with inclusion of the state;
- from centralist to rather decentralized problem solving;
- from exclusive to ever more inclusive and participatory decision-making structures;
- from an imperative policy style to negotiated solutions;
- from a reactive to a more strongly anticipative policy pattern; and;
- from steering based on public expenses to strengthened steering based on public revenues (taxes, levies, tariffs, fees).

Even if this paradigm change concerns brain trusts and professional discussions rather than real politics, as a tendency, it cannot be overlooked. At its heart, it has to do with a policy model beyond liberal 'laissez faire' and bureaucratic state interventionism: a decentralized and consensus-oriented policy model that focuses the central state on strategic tasks and transfers detailed regulations more strongly to decentralized actors. Applied to the field of environmental policy, the state is attributed foremost the functions of safeguarding ecological minima as well as 'strategic' design functions. Not least, its task would be to define long-term environmental problems. The responsibility of local actors would be to go beyond the basic requirements and minima as defined by the nation state by using their specific innovation potentials.

The capacity for modernization

What does this mean for the modernization capacity of political systems in general, and for long-term environmental protection in particular?

One way to consider the emerging political paradigm change is to perceive it as the sum of temporary solutions and compromises: since the state apparatus turns out to be rather powerless and uninformed against mighty industrial interests, the imperative intervention model needs to be superseded by the organization of con-sensual and thus more realizable goals. As the mechanism of majority rule is too cumbersome and tedious, and above all determined by too many disinterested people, the recourse to parties involved and people directly concerned suggests itself. Since administrations' interventions are too ineffective and inefficient, pres-sure develops in the direction of other instruments. Since other actors partly intervene more effectively, their legitimization rises. Only gradually there emerges in this constellation the new quality of a political double structure. From a crisis theory perspective, this development of temporary solutions is the normal situation. The temporary solutions imply modernization if they are innovative and designed to work for a longer period. (Conflict of opinion then will be about whether the crisis is simply relayed and pushed back to latency or whether its root causes are solved. Hitherto approaches presumably correspond to the first variant.)

If the above definition implies that political modernization – as the institutionalization of a significantly higher problem solving power – is mainly a crisis-induced reaction, then it is basically a stepwise increase in capacity. In this sense, the 'participation capacity' (in the connotation of Binder et al. 1971), for example, also developed historically via contentious extensions of the suffrage to the current forms of civic participation, just like the 'distribution capacity' established itself through manifold distribution crises, and the legitimization capacity was developed via legitimization crises. In light of the prevailing reactive policy pattern, a new phase or quality pertaining to all the rules (only) develops under high problem pressure. At least sufficient starting capacity is required (see v. Prittwitz 1990). If this is not available, the results will be crises without consequences, which often are not even perceived as crises, as in the case of many environmental problems in Eastern Europe. The newly institutionalized capacity, however, typically develops its own dynamics, which can evolve as far as the capacities go and search for new problems (Jänicke 1979), leading an ecological helper syndrome to arise (v. Prittwitz 1990). The difficulty here is the potential institutionalization of an interest in the continued existence of the problems.

The first stages of the institutionalization of environmental policy consist of *add-on* organizations of state environmental protection policy (with a preference for add-on and curative disposal technology). In the next stage, policy making is about taking integrated, polluter-related measures (and technologies) directed at target groups (energy, transport, chemistry, etc.). In this second phase, it becomes clear that the classical bureaucratic intervention mode is hardly able to mobilize willingness to change from the side of polluters and to trigger their innovation potentials. Problem-related, ad-hoc constellations of involved actors – with or without the participation of the state – here often prove to be faster and more efficient. Such negotiation systems and 'dialogue groups' potentially have the further advantage of displaying a *lower 'institutional' stake or interest in the persistence of problems* than the established environmental protection institutions (and classical eco-industry). At any rate, they constitute capacity expansions that at the same time represent a significant potential for creativity in society.

The first attempt to circumscribe industrial societies' capacity for environmental policy modernization in light of their existing environmental policy performance already included such kinds of para-statist or government external activity levels. Next to the initial spark or trigger arising from a high problem pressure, four features were accentuated (Jänicke 1990):

- high economic performance, which not only influences the level of environmental pressure, but the resources available for its reduction as well;
- high innovative capabilities in terms of opinion and decision-making structures, which are open to new interests and innovators;

- high strategic abilities in terms of long-term policy making and a clear potential for 'inter policy cooperation' (Knoepfel et al. 1991) with regard to long-term goals; and
- high consensual competence, which, through dialogue structures, creates ample acceptance for a comprehensive change.

For some of these attributes, we can resort to empirical experiences: the openness of the political system's input side for innovators and new interests, as well as the degree to which new political objectives are realized in an integrative way play a role in general political science research (see Kitschelt 1983; Feick and Jann 1988). Weidner similarly emphasizes three basic elements of effective environmental politics: information, participation, and a level playing field (Weidner 1992).

From an international comparison of environmental and energy politics (Kitschelt 1983; Jänicke 1990; Weidner 1992), we already know that openness in terms of mechanisms for political opinion and decision-making promotes ecological innovations. This is valid in particular for:

- the openness of the *information system* (science and the media);
- the openness of the *party system*;
- *plebiscites* (in four OECD-countries, they enforced a new energy policy);
- *regions and communities*, which can go beyond national standards (see below);
- the openness of the *policy arenas* and the pluralism of policy networks in the fringe of decision-making centres;
- the openness of *jurisdiction* for new (protection-)interests; and not the least
- an *economic system* that is open to innovations as well as flexible (in contrast to an economic structure which is characterized by a strong – vertical – monopolization, sclerotisation, and concentration of economic power).

The theme of consensual competence played a big role in the corporatism debate, perhaps not so much in terms of policy styles, but more in terms of the institutional underpinning of mechanisms for concerted action. Where and when a cooperative policy style brings together opponents or counterparts both from within as well as from outside the state, we also witness a growth in the competence for an integrated implementation of political goals. Comparative studies suggest this conclusion for labor market and industrial politics (corporatism model) (see Schmidt 1986; Lehner 1988). Apparently, concerted actions and local networking – keyword 'synergy effects' – also encourage successful regional politics. (Agglomerations, as examples of institutional nobodies, typically rely on the creation of negotiation systems.) A cooperative policy style obviously also promotes environmental politics. It favours cross-section solutions. It moreover has an innovative component because in dialogical forms of policy making, innovators are rather well integrated or at least less excluded. The consideration of a broader spectrum of values is realized even if corporative structures themselves sometimes tend to develop a form of exclusiveness.

The mentioned features of the political modernization capacity were developed with reference to the example of state environmental policy. But it gradually becomes clear that political modernization nowadays foremost means that capacities are built up beyond the state, with para-statal mechanisms being institutionalized that function as additional engines of industrial transformation.

This particularly concerns spatial and societal decentralization, with all kinds of *self-regulation*, of interventions outside the reach of the state coming into existence. In a first type of case, the state is not brought into play at all. This is the case when environmental protectors and their organizations do not first go to the municipality, but directly to the polluter. One might think of the successful intervention by Greenpeace in favour of chlorine-free paper or FCKW-free refrigerators. This is also the case if the media directly attack those who cause environmental problems, or in the case that retailers exclude environmentally harmful products. These environmental improvement measures, where the state does not play a role at all, are often nowadays more effective than governmental politics. If electronic media warn against a carcinogenic product, it often disappears from the market on the following day. If wholesale companies – under the sign of consumers' growing environmental awareness – start to compete for the environmental friendliness of their products, this situation affects the producers of ecologically harmful products in a direct, effective and sustainable way. One can compare these private interventions with the long-time decision-making procedures of the political-administrative system, with all its legal and political restrictions. Patently obvious, these are actors who can far outclass traditional politics in terms of the scope as well as the speed of their impacts. *In fact, decentralised forms of policy intervention have started to compete with governmental forms of intervention* (see above). For this reason, it seems hardly useful for industrial polluters to (just) rely on a policy based on the successful lobbying of the state.

A second kind of societal self-regulation likewise takes place between private entities, with only an indirect or secondary form of participation from the state. This pertains to all kinds of private suits (especially appealing to liability laws) and to all measures in the course of which the state transfers control- and decision-making mechanisms to third parties, or itself becomes a partner in the negotiations.

Have we arrived yet at the point of 'the fading away of policy centred steering mania' (v. Beyme 1990: 462 and 473)? This claim is, in my opinion, a contestable, extreme position. From an environmental policy point of view, we would be worse off (for example, as concerns emissions of lead, dust, carbon monoxide, sulphur dioxide or nitric oxides, as well as classical water pollution), if we could not have relied as well on policy-centred forms of steering and the respective legitimacy powers mobilized by the political-administrative system. An intervention strategy of the scope and range of climate change politics likewise will have to resort to these mechanics as well.

The hierarchical state as an intervention authority legitimized by the whole of society needs to be considered indispensable (see Offe 1987), especially if one seeks to develop a modernization of the industrial system that can effectively balance the immense ecological impacts of global growth. Under the sign of pending, unpredictable forms of societal interventions in environmental protection, *negotiated forms of state intervention* can be expected to gain importance.

References

Bachrach, P. and Baratz, M.S. (1970) *Power and Poverty, Theory and Practice*. New York: Oxford University Press.

Beck, U. (1986) *Risikogesellschaft – Auf dem Weg in eine andere Moderne*. Frankfurt am Main: Suhrkamp.

Beck, U. (1991) 'Der Konflikt der zwei Modernen', in Wolfgang Zapf (ed.) *Die Modernisierung moderner Gesellschaften*, Verhandlungen des 25. Deutschen Soziologentages in Frankfurt am Main 1990, Frankfurt am Main: Campus Verlag, pp. 40–53.

Beck, U. (1991) *Gegengifte – Die organisierte Verantwortungslosigkeit*, Frankfurt am Main: Suhrkamp.

Benz, A., Scharpf, F.W. and Zintl, R. (1992) *Horizontale Politikverflechtung. Zur Theorie von Verhandlungssystemen*. Frankfurt am Main: Campus.

Beyme, K. v. (1990) 'Die vergleichende Politikwissenschaft und der Paradigmenwechsel in der politischen Theorie', in: *Politische Vierteljahresschrift*, **31**, 3, 457–474.

Binder, L. et al. (1971) *Crises and Sequences in Political Development*. Princeton, N.J.: Princeton University Press.

Blanke, B. (Hrsg.) (1991) 'Staat und Stadt. Systematische, vergleichende und problemorientierte Analysen "dezentraler" Politik', in: *Politische Vierteljahresschrift*, Sonderheft **22**, Opladen.

Edelman, M. (1964) *The Symbolic Uses of Politics*. Urbana, Illinois: University of Illinois Press.

Feick, J. and Jann, W. (1988) 'Nations matter – Vom Eklektizismus zur Integration in der vergleichenden Policy-Forschung?', in M.G. Schmidt (ed.) Staatstätigkeit, *Politische Vierteljahresschrift*, Sonderheft 19, Opladen: Westdeutscher Verlag, pp. 196–220.

Gerau, J. (1978) 'Zur politischen Ökologie der Industrialisierung des Umweltschutzes', in M. Jänicke (ed.) *Umweltpolitik*, Opladen: Leske + Budrich, pp. 114–119.

Glagow, M. (ed.) (1972) *Umweltgefährdung und Gesellschaftssystem*. München: Piper.

Görlitz, A. (ed.) (1989) *Politische Steuerung sozialer Systeme – Mediales Recht als politisches Steuerungskonzept*. Pfaffenweiler: Centaurus-Verlags-Gesellschaft.

Hauff, V. and Scharpf, F.W. (1975) *Modernisierung der Volkswirtschaft – Technologiepolitik als Strukturpolitik*. Köln: Europäische Verlagsanstalt.

Hesse, J.J. (1987) 'Aufgaben einer Staatslehre heute', in Thomas Ellwein et al (eds.) *Jahrbuch zur Staats- und Verwaltungswissenschaft*, Bd. 1, Baden-Baden: Nomos, pp. 55–87.

Hesse, J.J. and Benz, A. (1990) *Die Modernisierung der Staatsorganisation: Institutionspolitik im internationalen Vergleich: USA, Großbritannien, Frankreich, Bundesrepublik Deutschland*. Baden-Baden: Nomos.

Huber, J. (1982) *Die verlorene Unschuld der Ökologie – Neue Technologien und superindustrielle Entwicklung*. Frankfurt am Main: Fisher Verlag.

Huber, J. (1985) *Die Regenbogengesellschaft – Ökologie und Sozialpolitik*. Frankfurt am Main: Fisher Verlag.

Huber, J. (1991) *Unternehmen Umwelt – Weichenstellungen für eine ökologische Marktwirtschaft*. Frankfurt am Main: Fisher Verlag.

Jänicke, M. (1979) *Wie das Industriesystem von seinen Mißständen profitiert*. Opladen: Leske + Budrich.

Jänicke, M. (1984) *Umweltpolitischen Prävention als ökologische Modernisierung und Strukturpolitik.* Wissenschaftszentrum Berlin, IIUG dp 84-1, Berlin: WZB.

Jänicke, Martin (1986) *Staatsversagen – Die Ohnmacht der Politik in der Industriegesellschaft.* München: Piper (Überarbeitete engl. Übers. 1990).

Jänicke, M. (1990) Erfolgsbedingungen von Umweltpolitik im internationalen Vergleich, in: *Zeitschrift für Umweltpolitik und Umweltrecht*, **13**, 3, pp. 213–233.

Jänicke, M. (ed.) (1973) Die Analyse des politischen Systems aus der Krisenperspektive, in: M. Jänicke (ed.), *Politische Systemkrisen.* Köln: Kiepenheuer und Witsch, pp. 14–50.

Jänicke, M. et al. (1992) *Umweltentlastung durch industriellen Strukturwandel? – Eine explorative Studie über 32 Industrieländer (1970–1990).* Berlin: Edition Sigma.

Kitschelt, H. (1983) *Politik und Energie.* Frankfurt am Main, New York: Campus.

Kloepfer, M. (ed.) (1989) *Umweltstaat.* Berlin: Springer.

Knoepfel, P. et al. (1991) *Implementation of Environmental Policies through Interpolicy Cooperation on the Level of the Swiss Federal Government*, Beitrag zu dem internationalen Kongreß "Implementing Environmental Policies by Means of Interpolicy Cooperation", Crans-Montana, 24-09–27-09, 1991.

Kondratieff, N.D. (1926) Die langen Wellen der Konjunktur, in *Archiv für Sozialwissenschaft* Bd. 56, Tübingen (wiederabgedruckt in ders. (1972): Die langen Wellen der Konjunktur, Berlin).

Kuhn, T.S. (1962) *The Structure of Scientific Revolutions.* Chicago: University of Chicago Press.

Lehner, F. (1988) Institutionelle Determinanten der Wirtschaftspolitik in westlichen Demokratien. Ansätze und Elemente einer systemischen Theorie, in H.H. Hartwich (ed.) *Macht und Ohnmacht politischer Institutionen.* Opladen: Leske + Budrich, pp. 237–251.

Luhmann, N. (1971) *Politische Planung.* Opladen: Westdeutscher Verlag.

Luhmann, N. (1990) *Ökologische Kommunikation.* Opladen: Westdeutscher Verlag.

Mayntz, R. (1979) 'Regulative Politik in der Krise?', in J. Matthes (ed.) *Sozialer Wandel in Westeuropa.* Verhandlungen des 10. Soziologentages in Berlin, Frankfurt am Main, New York: Campus, pp. 55–81.

Mayntz, R. (ed.) (1978) *Vollzugsprobleme der Umweltpolitik.* Stuttgart.

Meadows, D.H. et al. (1992) *Die neuen Grenzen des Wachstums – Die Lage der Menschheit: Bedrohung und Zukunftschancen.* Stuttgart: Deutsche Verlags-Anstalt.

Mensch, G. (1975) *Das technologische Patt – Innovationen überwinden die Depression.* Frankfurt am Main: Umschau-Verlag.

Offe, C. (1987) 'Die Staatstheorie auf der Suche nach ihrem Gegenstand', in Thomas Ellwein et al. (eds) *Jahrbuch zur Staats- und Verwaltungswissenschaft*, Bd. I., Baden-Baden: NOMOS, pp. 309–320.

Prittwitz, V. v. (1990) *Das Katastrophenparadox – Elemente einer Theorie der Umweltpolitik.* Opladen: Leske + Budrich.

Recktenwald, H.C. (1978) 'Unwirtschaftlichkeit im Staatssektor – Elemente einer Theorie des ökonomischen Staatsversagens', in *Hamburger Jahrbuch für Wirtschafts- und Gesellschaftspolitik.* Tübingen, pp. 155–165.

Rokkan, S. (1969) 'Die vergleichende Analyse der Staaten- und Nationenbildung: Modelle und Methoden', in W. Zapf (ed.) *Theorien des sozialen Wandels.* Köln, Berlin: Kiepenheuer & Witsch, pp. 228–252.

Ronge, V. (1972) 'Die Umwelt im kapitalistischen System', in M. Glagow (ed.) *Umweltgefährdung und Gesellschaftssystem.* München: Piper, pp. 97–123.

Scharpf, F.W. (1991) 'Die Handlungsfähigkeit des Staates am Ende des zwanzigsten Jahrhunderts', in *Politische Vierteljahresschrift*, **32**, 4, pp. 621–634.

Schmalz-Bruns, R. (1992) Civil Society – ein postmodernes Kunstprodukt?, in *Politische Vierteljahresschrift*, **33**, 2, pp. 243–255.

Schmidheiny, S. (1992) *Kurswechsel – Globale unternehmerische Perspektiven für Entwicklung und Umwelt*. München: Artemis & Winkler.

Schmidt, M.G. (1986) 'Politische Bedingungen erfolgreicher Wirtschaftspolitik. Eine vergleichende Analyse westlicher Industrieländer (1960–1985)', in *Journal für Sozialforschung*, **26**, 3, pp. 251–273.

Schumpeter, J.A. (1942, 1972) *Kapitalismus, Sozialismus und Demokratie*. München: UTB.

Schuppert, G.F. (1989) 'Zur Neubelebung der Staatsdiskussion: Entzauberung des Staates oder "Bringing the State Back In",' in *Der Staat*, **28**, 1, pp. 93–104.

Simonis, U.E. (1985) *Ökologie und Ökonomie. Auswege aus einem Konflikt*. Karlsruhe: C.F. Müller-Verlag.

Weidner, H. (1975) *Die gesetzliche Regelung von Umweltfragen in hochentwikkelten kapitalistischen Industriestaaten*. Freie Universität Berlin (Schriftenreihe des FB Politische Wissenschaften, Nr. 8), Berlin: Freie Universität.

Weidner, H. (1992) *Basiselemente einer erfolgreichen Umweltpolitik – Eine Analyse der Instrumente der japanischen Umweltpolitik unter Berücksichtigung von Erfahrungen in der Bundesrepublik Deutschland*. Dissertation, Fachbereich Politische Wissenschaft der Freien Universität Berlin, Berlin: Freie Universität.

Weizsäcker, E.U. v. (1990) *Erdpolitik – Ökologische Realpolitik an der Schwelle zum Jahrhundert der Umwelt*. Darmstadt: Wissenschaftliche Buchgesellschaft.

Wicke, L. (1990) *Umweltökonomie*. München: Verlag Franz Vahlen.

Willke, H. (1983) *Entzauberung des Staates – Überlegungen zu einer sozietalen Steuerungstheorie*. Königstein/Ts: Athenäum Verlag.

Willke, Helmut (1991) *Systemtheorie*. Stuttgart, New York: Fischer Verlag.

Zilleßen, H., Dienel, P. and Strubelt, W. (eds) (1993) *Die Modernisierung der Demokratie – Internationale Ansätze*. Opladen: Leske + Budrich.

Zimmerman, K. et al. (1990) *Ökologische Modernisierung der Produktion – Strukturen und Trends*. Berlin: Sigma.

Chapter 4

Joseph Huber

ECOLOGICAL MODERNIZATION: BEYOND SCARCITY AND BUREAUCRACY

Translated from: J. Huber (1991) 'Ecologische modernisering: weg van schaarste, soberheid en bureaucratie', in: A.P.J. Mol, G. Spaargaren and A. Klapwijk (eds), *Technologie en Milieubeheer. Tussen sanering en ecologische modernisering*, Den Haag: SDU, pp. 167–183.

THIS ARTICLE PROVIDES AN extended statement regarding some of the basic points of the concept of ecological modernization. Starting from a historical analysis, the concept of ecological modernization and its relation with modern society will be elaborated. I will end with some remarks on the actors who are in positions to bring about ecological modernization: government, entrepreneurs, and new social movements.

The environmental debate: eco-frauders and eco-misers

Beginning in the 1970s, when the environmental debate revolved around the 'limits to growth', different factions could be distinguished, with the 'eco-frauders' and the 'eco-misers' at the opposite ends of the spectrum. The frauders are the people who continue to add to the bill, but refuse to pay for it. The misers are the people who do not even want to make a bill.

The frauders justify themselves by common capitalist practices based on the externalisation of, in this case, environmental costs. Economically speaking, nature is a productive factor that was not properly taken into account. Within the frame of individual and aggregate production functions, resources and environmental media were externalised factors. They did not appear in the economic bookkeeping. This type of economy and related industrial technology consumes all sorts of energy, natural resources, and sinks, but does not fully pay for them, nor does it care about reproduction or regeneration. That's fraud. But during the 1970s, many people considered this to be normal. They supported each other in considering that the 'ecological question' was largely overestimated and exaggerated by certain young, middle-class people – typically teachers or social workers rather than engineers or business managers. This has changed completely. Even the elites in industry and finance no longer need to be convinced that the ecological reconstruction of industry has top priority.

On the other side, there was, and is, the anti-growth faction, or the 'eco-misers'. The eco-misers opposed continued growth because it was economic growth and industrial development that had brought forward all the social and ecological problems of modern society. Their first reaction was: "We have to stop with it, we have to get rid of it", which resulted in ideas about zero growth, or the stationary or stable state economy, or even the shrinking of the economy. Few people in the green movement were aware that these attitudes and arguments were not new, but represented a remake of a similar criticism of capitalism, economic growth, industrial productivity, and wealth that has existed ever since Jean-Jacques Rousseau's remake of John Stuart Mill's Steady State Economy (i.e., zero growth economy) in the 1830s. At the time, as much as today, this was part of a debate between the followers of progress and their romantic counterparts.

In Germany, for example, the writings of the Scottish and the German Adams – Adam Smith and Adam Müller – confronted each other. Both contributed to economics. Adam Smith, however, relied on a model of rational self-interest and dynamic market equilibrium, whereas Adam Müller, a conservative social romanticist, referred to a supposed natural order of things, well-proven common values and institutional settings, and problems of social change, crises, and catastrophe. I would not say Smith was right and Müller wrong, nor would I maintain the opposite. Both views have their merits and shortcomings. I want to stress this: progress as seen from a Smith- or Bentham-like perspective of modernization could not have moved forward, but would have become stuck in crisis and broken down in catastrophe, without a recurrent readjustment of direction and momentum of both person-to-person and person-to-nature relations under the influence of a Rousseau- and Müller-like perspective of modernity. Both attitudes, the rationalist-utilitarian and the romanticist-idealistic one – or as Pitirim Sorokin put it, the sensate mode as well as the ideational mode – are complementary components of modern culture, and they combine to advocate a mechanism of overall self-control and societal self-adaptation in the process of co-evolution and ongoing modernization.

Nowadays, the typical controversies accompanying modernization reappear for the fourth time – first at the onset of industrialization (sentimentalism), second as social romanticism and the first stage of socialism during the first half of the 19th century, third in the decades around 1900 (in conservation and life-reform movements of various kinds together with follow-up stages of diverse currents of socialism), fourth in the social movements since the 1960s, dubbed 'new' by Alain Touraine. There is the pattern of a recurrent, deferred combined life-cycle between such social movements and the 'long waves' of industrial innovation, though this is not obvious to everyone.

Whatever the case, it seems as though every important argument in today's socio-ecological discourse can in some way be found right from the beginnings of industrialization – be it in Rousseau, or in the conservative English Tory criticism of progress, or in the social romanticism of the 'Young Europe' movement, soon leading to various revolutionary currents, such as national democracy or socialism. (Up to now, Marxism fiercely denies its roots in pan-European social romanticism of the early 19th century.)

The theory of counter-productivity has many foundations and predecessors, and its central message is: stop growing, start shrinking. A scientific foundation for this was given by the 'thermo-dynamic school of economics' related to N. Georgescu-Roegen, with a transfer of the entropy paradigm from physics to society, which has become the eco-misers' most beloved reference. The idea is that whatever is done uses up energy and resources of a higher order, which unavoidably leads to increased entropy. Each unit of economic value added with each step in the human chain of production equals a step of ecological degradation and destruction of nature. The only conclusion can be: humanity must do away with the existing rationalist, materialist, and outer-worldly oriented pursuits of happiness. We would have to slow down, retire into more spiritual inner worlds, shrinking all the way. In my analysis, both the eco-frauders' and the eco-misers' positions are wrong and in no way helpful.

From quantity to quality

New perspectives leading beyond the closed horizons of eco-frauders and eco-misers were developed in the years around 1980. One approach was introduced by environmental economics, which later on was followed by ecological economics. The basic thinking is as follows. If resources and environmental media are goods like any other goods, they need to be integrated into the calculus of production functions, and to this end, they need to have prices. Once there are prices reflecting the full 'ecological costs' of material production and consumption, actors will start to behave responsibly and try to make efficient use of ecological factors. The rationality of engineering, management and market transactions will take care of resources and environmental media, resulting in the minimization of anthropogenic ecological impact. This is, of course, a very classical idea familiar to every manager and engineer: to increase efficiency, by using less energy and natural resources in producing the same amount of goods and services at the same level of quality.

Increasing resource productivity and energy efficiency, however, is only half the picture. In the long run, the approach cannot actually work. Take the automobile, for example. Today's automobiles are more energy-efficient than their predecessors were 20 years ago. Cars then needed 15–20 litres of gasoline per 100 kilometres. Nowadays they need about 7–12 litres. Next year's new cars will need 5–8 litres, and there are certain prototypes running at 2–3 litres. This is considerable progress in increasing energy efficiency, but in a certain sense it represents progress in the wrong place because it does not solve the structural eco-problem of the technology involved – the problem of internal combustion engines fuelled by carbon-releasing gasoline. Given the increasing volume of traffic, adding a catalytic converter or raising the efficiency of combustion does not cease the release of damaging gases at dangerous levels.

So the real answer, and the necessary step beyond mere efficiency increases in old industries, is *innovation*, the introduction of different motors and of entirely new propulsion systems in vehicles, as well as the development of clean and renewable sources of energy. The focus of environmental economics is efficiency, which in fact means dealing with the problem in terms of mere quantities. But the focus must be on *quality*, on ecological quality, in this case the ecological quality of production processes and products, the ecological quality of resources used and of material flows involved. So in the example of the automobile, we need cars with different motors or different fuels that do not contain carbon, for example, electrical motors driven by electrochemical cells fuelled by hydrogen. That is why saving energy and materials, or money, is only half of the picture. The other half, and I consider it to be the more important half, is structural innovation: opening up new ways of doing things, particularly developing and introducing new, ecologically better-adapted technologies, which come with less or no harmful environmental impact.

People engaged in the ecology movement agree when the agenda is about 'saving energy', hereby intending to say 'living on less'. But they don't agree when talk is about 'increase in the turnover of environmentally adapted products', meaning economic growth on the basis of ecologically better-adapted energy and material flows. In their notion of 'ecology', which is a notion of scarcity, renunciation and withdrawal, material abundance does not exist, or is even seen as sinful. But ecology is *not* a call for 'voluntary simplicity'.

To give an analogy: the big blue and green garden Gaya, the planet earth, gives bread and water, and one can live on that. But it also provides milk and honey, and if humans can have milk and honey, why shouldn't they have it? The garden Gaya even gives a certain amount of caviar and champagne, and if that amount can be consumed without doing harm, why shouldn't it be done? The original question of the limits-to-growth movement was, 'How much is enough?' This sounds reasonable, but it turned out to have a miser's bias. A more elaborate question would be, 'What is ecologically adapted, even in large volumes?' Look at the sunlight. Day after day the earth receives huge amounts of energy from the sun. Most of this energy is, speaking in human terms, wasted completely unutilised in the depth of the universe. If this source of energy could be tapped more effectively, there would hardly exist a problem with 'wasting energy'. The conclusion is that in the first instance it is not the amount of energy that matters, but the material and technological structure of the source of energy under consideration and its ecological quality. Ecological modernization aims at bringing about high levels of ecological quality of production processes and products (while certainly not neglecting efficiency, be it just for economic reasons), thus enabling the use of large amounts of energy, resources and sinks in an ecologically sustainable way.

Ecological modernization: modernizing modernity in a sustainable way

Before explaining additional aspects of ecological modernization, let us first consider the general sociological meaning of modernization, which starts from the distinctions among primitive, traditional and modern societies. According to some scholars, modern societies have been emerging since the late Middle Ages, or since around 1500, while other scholars identify the beginning of modernization with the onset of the industrial revolution. Society as a system has a number of core 'functions' and 'structures'. Modernization is the process by which these societal functions and structures are continuously developed and upgraded to ever-higher degrees of complexity. Among the most marked of these core components are the developments of science and technology, industrial production and consumption at unheard-of high levels of productivity, trade and markets based on a well-developed money and credit economy, the establishment of the rule of modern law (public and civil), state- and nation-building, the development of 'rational' bureaucracy in administration and management and, not least, a value base giving priority to rational and utilitarian behaviour as much as to the evolvement of personal abilities and self-expression, typically reflected in some sort of individualism (pursuit of happiness), alternatively, however, also in some sort of collectivism.

Hitherto, modernization took place against the background of traditional society. For example, feudal classes and heredity largely have been replaced by some sort of meritocratic class structure on the basis of individual achievement. Peasants and craftsmen have been replaced by industrial workers and clerks, and horse and carriage by railway, steamboat, automobile and aeroplane. All the social changes involved were truly fundamental and resulted in severe crises and catastrophes, referred to since the 1840s as 'the social question', eventually resulting in the regulation and institution-building of modern work, welfare and mass-participation.

Today, we are entering into a new stage of modernization. In advanced societies, there are not many traditional features left to be modernized. From now on, the process is more about replacing earlier modern structures by the latest modern structures. For example, in today's 'knowledge economy', low-skilled, old-industrial workers are being replaced by automated machines and highly educated professionals. Yesterday's dirty smokestack industries are gradually being replaced by today's and tomorrow's clean high-tech industries. The focus is no longer on 'industrializing', but on *restructuring* industries, and restructuring them particularly with regard to ecological requirements.

The 'social question' certainly has not been solved once and for all. But nowadays and in the 21st century it is the 'ecological question' that has come to the fore. As a consequence, modernization goes on, but it is now about modernizing modernity, and in the first instance this now revolves around the 'ecological question'. Ecological modernization means modernizing modernity in a sustainable way. The important thing to understand, or rather, to accept, is that the future 'greening of industry' will come about in a continued process of modernization – in other words, continued development and innovation on the basis of new scientific and technical knowledge, advanced finance and marketing, the rule of law and regulation, and, not least, modern mindsets and lifestyles.

An example of ecological modernization can be found in agriculture. The existing system of industrial agriculture is ecologically unsustainable. It relies on heavy machinery, artificial fertilisers, pesticides that do more harm than good, and soil and groundwater degradation. This type of old-industrial agriculture cannot last. It must be changed, restructured, innovated, in brief: modernized. This will involve new information and communications technology (ICT)-enhanced machinery, as well as new, low-impact chemistry and bio-technology. In this way, agribusiness can be made ecologically more sustainable. Another option is organic farming, which avoids heavy machinery and relies on natural fertilisers and pesticides. In comparison, it tends to be more labour-intensive, rather than capital-intensive. The soil is regenerated. The produce is high quality. Productivity per hectare, though, is comparatively low. Modernizing industrial agriculture and organic farming are seen as alternatives, but I have no problems in recombining the two. And both give an example of how ecological modernization will develop: by means of modern society, in particular, by using the most advanced knowledge and technology. But modernization, and also ecological modernization, entails more than just using the most advanced science and technology.

Dimensions of modernization

Modern society can be analysed as a system made up of various subsystems. Similar to a number of familiar approaches in sociology, from Marx to Luhmann, I distinguish five or more functional subsystems (Huber 1991):

1. The Operative system, which is the realm of technology-enhanced human work, the world of formal and informal production and consumption.
2. The Economic system, which forms the realm of money, finance and markets.
3. The Ordinative system, which includes government, public administration and private management based on law and authority.
4. The Political dimension, which is about discourse, agenda-setting, will-building and decision-making (regarding all institutions and settings of society, not just government).
5. Additional Cultural or Formative dimensions, such as cosmology/religion, value base, knowledge base, expressive styles, education and ways of living or lifestyles.

These dimensions, or subsystems, co-evolve in a complex process of inter-relatedness and co-directionality. None of the subsystems could be said to be more important than the others in determining the evolution of society. But the virulent centre of impulse may be shifting from one subsystem to another over the course of time. The history of modern society so far indeed seems to fit the view that during certain epochs society is more or less preoccupied with one or more of these dimensions. In the beginning, during the Reformation and the Renaissance, the focus was on the formative realm of religion, values, knowledge – in other words, forming a modern mindset, in particular, a modern conception of the human being, God and the world. After that, during the 16th and 17th centuries, it was on state building and government, characterized by a preoccupation with how to rule, manage and 'police' effectively, for example in Machiavelli or Hobbes, resulting first in absolutism, and thereafter in the 18th century in political enlightenment. Later on in the 18th and throughout the 19th centuries, society became pre-occupied with the economy. Finally, in the 20th century, the focus shifted towards technology as the most important, productive factor of modern society.

One should note that the idea of a shifting focus in the trans-secular life cycle of modernization does at any time include the idea of *co*-evolution of *all* of the subsystems, the idea of structural and functional interrelatedness and co-directionality (at times, partial counter-directionality). All of the subsystems continue to modernize all the time – though this results overall in a complicated, crisis-prone process. It is in fact exceptional for social change to proceed in a smooth and linear way.

Modernizing modernity in a sustainable way includes an effective contribution from each subsystem – environmental awareness; environmental ethics; ecological knowledge; responsive behaviour; environment-oriented agenda-setting; will-building and decision-making; adequate lawmaking, regulation, finance and markets; and technological innovation. In the last instance (or we could equally say, in the first instance), it is technology that determines the ecological quality of human activities' impact on the bio- und geo-sphere. But environmentally benign technological innovation would not be possible without all of the other factors mentioned. Ecological modernization needs a corresponding impulse from all societal subsystems, and if only one is missing, the process of ecological modernization will be more or less blocked.

Ecological modernization and technology

One would not completely be wrong to assume that technology will continue to be the main preoccupation in advanced societies throughout the 21st century. The reason is the necessity of ecological re-adaptation of industrial society, which *in fact* can be achieved only through technological innovation – because it is technology-enhanced human work that enables human production and consumption, and it is nothing else but material production and consumption that is at the interface of 'the eternal natural necessity . . . of the metabolism between man and nature', as Marx put it in *Das Kapital*.

As the examples given above indicate, ecological modernization requires technologies or products that are specifically designed and developed: clean and renewable energy, clean technologies, low-impact chemicals, bionic design of products and selection of product materials according to recyclability and to environmental, health and safety standards.

A familiar distinction is the one between integrated technology and end-of-pipe technology, the latter also being referred to as downstream technology (e.g., in water purification) or as add-on measures.[1] The model certainly simplifies reality, and in certain cases the categories do not seem to fit with real technologies or products. The catalytic converter, for example, isn't just a simple end-of-pipe technology; it needs certain changes within the motor itself. Basically, however, the bipolar spectrum between integrated technologies and end-of-pipe measures gives a useful orientation (see Figure 4.1).

On the side of add-on measures, there are, among others, waste dumping in landfills and waste burning, the latter preferably within combined heat and power generation. Neighbouring add-on measures include technologies for reducing water- and air-borne emissions, be it at or after source. These are followed by different

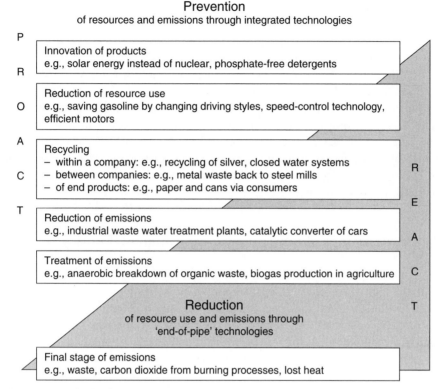

Figure 4.1 The bi-directional model of environmental technologies

types of materials recycling and reprocessing (secondary raw materials), including local micro-loops as much as cross-border macro-loops. The next category is the reduction of resource input by incremental innovations aimed at increasing efficiency or resource productivity. Finally, there is the substitution of hazardous substances in existing products and materials flows, and the conception and development of entirely new, ecologically benign technologies from scratch. On such a basis, an economy of plenty and generosity can be built, and one no longer needs to be an eco-frauder or an eco-miser. Probably, end-of-pipe and downstream technologies will remain necessary, and these technologies increasingly also include advanced 'high tech'. But the general goal clearly is to keep add-on measures to a minimum by means of integrated, innovative solutions.

Most industries in Organisation for Economic Co-operation and Development (OECD) countries started introducing end-of-pipe measures in the 1970s and 1980s because new environmental laws and regulation forced them to do so. End-of-pipe measures are readily available, and they can be added to whatever exists without prior changes. The future, however, increasingly will involve integrated, innovative solutions. These will be planned for and implemented within the frame of ecological management systems, which, in turn, are becoming integrated in the whole of a company or industry structure — from research, development and design, to regular production, finance, marketing and after-sales services. This is exactly what is happening at the moment.

Industries do not like to be forced to take end-of-pipe measures because this always represents additional costs *à fonds perdu*. However, if an entrepreneur or an innovative company finds integrated solutions to environmental problems, in developing cleaner processes and products, these represent productive and profitable investments. Take waste management, for example. There has been an explosion of costs during the last five to ten years because the space for waste dumping was made increasingly scarce, and fees ever more expensive. So in many cases it has become cheaper to recycle and reprocess used materials and products than to get rid of them. A company's productivity and competitiveness will improve even more if hazardous materials in processes and products can be avoided altogether. In this way, a business strategy of ecological modernization by integrated solutions and ecologically benign processes and product innovation completely fits into the prevalent type of economy.

The shortcomings of bureaucratic environmental policy

In the following I would like to deal with the question of the actors who will bring about the shift from old-industrial development to ecological modernization. I would like to start with the government's contribution in stimulating the process of ecological modernization, followed by the role of entrepreneurs and innovative companies, and, not least, the role of the environmental movement.

First, to be clear, the question of who will bring about ecological modernization is not a question of either government or entrepreneurs. Each has its part to play, and it should be noted that they cannot replace each other. On the one hand, the legal foundations of environmental policy and regulation by environmental authorities are in principle absolutely indispensable, even though this necessity is or ought to be limited to a certain extent. Because the snag is that beyond a critical extent, environmental regulation results in an entirely new complex of state bureaucracy. Environmental bureaucracy has had some successes in certain fields, particularly with the add-on treatments of air-borne emissions, sewage and solid waste. Environmental bureaucracy can be helpful in doing away with the first and largest environmental wastes, particularly if actors involved in industry and society are unwilling to accept their environmental responsibilities and thus prefer not to act unless everyone is subject to a general legal constraint. In any case, someone has to set obligatory environmental standards, and this is clearly the government's business.

It needs to be understood, however, that any bureaucratic approach has a number of shortcomings. The first shortcoming is that many environmental laws include a prescription for a particular state of the art (the formally acknowledged best available technology or best available knowledge). The prescribing of operationally specified and detailed procedural and technical standards is a very inflexible kind of instrument. It responds to neither business cycles nor differences between companies or sectors. Equally, bureaucratic policies cannot take special local conditions into consideration. As a result, and under the influence of lobbyism, lawmaking and regulatory authorities allow for many exceptions and exemptions, and for very long transition periods, which make it all very complicated and ineffective. Bureaucratic environmental policies tend to promise ambitious goals they realistically will not reach, or if so, only with a large time delay.

A second imperfection is that bureaucratic measures do not foster innovation. In my analysis, this is the most important criticism against environmental bureaucracy. Bureaucracy can force actors to stop doing something or require them to start doing things in a pre-specified way. But bureaucracy cannot force actors to be creative, inventive and productive. Bureaucracy is unsupportive, and quite often proves to be a hindrance, when it is about introducing new production processes, designs, products and practices. As Hayek rightly insisted, bureaucracy lacks knowledge.

At the beginning of 1990, the Federal Environmental Agency in Berlin, after extensive research, developed a list of 571 hazardous substances. Shortly afterwards, the chemical industry came up with a similar list of over 4,000 hazardous substances. This is a nice illustration of the difference in knowledge and know-how between state bureaucracies and industries – the latter undoubtedly being corporate bureaucracies themselves, but nevertheless, they are much closer to the practical problems under consideration. The places where problems are created or where problems become obvious are also the places where solutions can be found. Those who are considered to be the ones who cause problems, at the same time have the capacity to resolve these problems because they possess the necessary knowledge.

The third weakness of environmental state bureaucracy is ineffectiveness and implementation deficits. No bureaucracy can truly control what is going on in industry. That was one of the main reasons for the problems with central planning in the eastern European countries. Every day there are billions of energy and materials transformations, which no bureaucracy is capable of controlling. The control gap widens with the number of environmental laws, decrees and directives; and with the control gap widens the notorious implementation deficits on the side of the actors who are subject to regulation. Both regulators and those regulated depend on each other's willingness to cooperate, which creates a sort of statist-corporatist companionship. Unsurprisingly, environmental authorities are not keen at suing eco-sinners. Normally, it is private people from the neighbourhood or citizens' initiatives that take legal action against non-compliers.

Fourth, state bureaucracy tends to be counter-productive in certain ways. For example, once a permit for running a chemical plant has been given, this permit counts for more than the health and environmental problems that may result from running the plant. As long as the plant is run in a 'regular' way, and nothing 'irregular' occurs, the authorities are not allowed to interfere or even to shut down the plant. In this sense, there are many cases in which a permit to build and run industrial infrastructures and plants equals a license to pollute, and where state control is tantamount to state-controlled pollution. Things look different once there are environmental liability laws. Owners and managers of a plant then have an incentive to carry out reliable environmental assessments of their activities because they are interested in avoiding damages they would be made liable for.

During the 1980s, advanced industrial nations already may have reached the point where further environmental bureaucratization does more harm than good. Government alone will not be able to bring about ecological modernization.

Opportunities and risks for entrepreneurs and innovative companies

Since environmental bureaucracy has severe shortcomings and a bureaucratically imposed switch-over to ecological modernization is bound to fail, the irreplaceable contributions of entrepreneurs and innovative companies come to the fore. There are in fact several reasons for entrepreneurs to take ecological aspects into consideration, regardless of, or beyond compliance with legal standards. Three fields are particularly relevant in this respect: (i) the quality and image of products and services supplied by a company, (ii) cost performance, and (iii) questions regarding corporate organization, personnel and stakeholders. In these fields, environmentally oriented corporate behaviour opens up new opportunities, as there are certain environmental threats to businesses and entrepreneurial risks at the same time (see Figure 4.2).[2]

A company clearly can improve its public standing and its market competitiveness by ecologically modifying its production processes and products, or even introducing new ones that are ecologically more sustainable, especially in end-user

Opportunities	Dangers when not seizing opportunities	Risks when seizing opportunities
Market and competition position		
Advantage and market leadership by innovation of products and processes. Early profits. Opening up of new markets and consumer groups.	Loss of connection to markets, technically and on product quality.	Typical pioneer risks (wrong investments).
Cost structure		
Cost savings by less resource and energy use. Cost savings by less environmental burdening.	Increased costs by higher energy and resource prices. Increased costs by environmental costs, emission taxes, etc.	Deteriorated cost structure by environmental investments, especially if competitors are not following, or only 'end-of-pipe' technologies are applied.
Organization and personnel		
Secured location by timely adaptation to new environmental requirements. Flexible organizational structure. High public reputation. Highly motivated and committed employees.	Danger of losing production location by failing to adapt to environmental requirements in time. Organizational structure becomes inert. Poor public reputation. Employees lacking motivation loyalty.	Unwilling employees who cannot or refuse to adapt.

Figure 4.2 Good reasons for increased attention to resource use and environmental management within market-oriented companies

markets such as building and housing, automobiles, household and office appliances, furniture, paints, detergents, clothes, food or cosmetics. Competition over quality will increasingly include competition over eco-performance regarding production processes and products, corresponding to levels of environmental awareness among buyers and their willingness and ability to pay for eco-quality.

The main risk involved is comparable to that of any supply modification or innovation: to make the wrong investment, or to make it at the wrong time, or to communicate it in an inappropriate way. But every entrepreneur and company must deal with this type of risk all the time. Moreover, ecological attitudes continue to diffuse throughout all classes, milieus and groups in society. If companies do not respond to their customers' changing attitudes, they will lose out to competitors who moved earlier because they understood the challenge in time. It is now becoming increasingly difficult, and will be impossible in the future, to sell products tainted with environmental, health and safety risks. Of course, market competition over eco-quality is more obvious in end-of-chain industries close to end-users or consumers, and less so in heavy industries, intermediary producers and power plants upstream in the supply chain.

The second field is competition over costs. Engineers and chemists pretend they are improving technical labour productivity all the time (which tends to be true over the course of learning curves). But obviously there is a still-unrealised, considerable potential for increasing resource productivity, energy efficiency, recycling valuable materials and avoiding ever-more-expensive emissions and waste. In recent years, industry leaders increasingly have been learning the lesson and have started to develop environmental management practices rather swiftly.

Then there are questions of personnel, and stakeholders in general. For example, the workers of a company that had dumped industrial acid waste in the North Sea were so ashamed that they did not dare to say publicly that they were working for this company. The story has a clear message: environmental ethics is now part of the general value base; environmentally oriented behaviour has achieved high priority; it has become a must for everyone to take this into account. Regardless how honestly or dishonestly actors play the game, they can no longer avoid being participants, or else they will become socially excluded. A company depends on employees who identify with the company. It needs to be able to recruit good staff, as it depends on the support of a wide range of additional stakeholders, such as the neighbourhood, local polity and authorities, the mass media, financial institutions, suppliers and customers.

As always, it is a certain combination of carrot and stick, something that pulls and something that pushes, that drives entrepreneurs and companies to adopt pro-active environmental behaviour. The new opportunities opened up by ecological modernization (competitiveness on costs and eco-quality) are the intrinsic motivations that pull. Among the extrinsic factors that push are such risks as environment-health-and-safety hazards combined with the threat of high liabilities and loss of image, regular environmental costs getting out of control, and missing the right time when to join the bandwagon.

A new role for new social movements

By far the most important extrinsic factor, however, remains environmental law-making and regulation – itself embedded in a context of rising environmental awareness brought about by the environmental movement, a revitalised conservation movement, and a broad range of additional social movements, such as the student movement, the peace movement, the women's movement, the civil rights and equal-opportunities movements, the third-world-solidarity or one-world movement, the alternative movement and the lifestyle-reform movement, as well as spiritual movements of various kinds.

In the past, the prevailing constellation was one of confrontation between these social movements – strongly influencing public opinion – on the one side and industries and agribusiness on the other side, obliging government to take some action, forcing new environmental rules upon industries and agribusiness. Over time, the constellation has become less one-sidedly confrontational. Environmental awareness has spread throughout society, though it wasn't adopted in its original, deep-green biocentric, neo-romanticist mode, but was assimilated into some sort of self-enlightened utilitarianism and anthropocentrism. After all, you won't kill the cow you want to milk. Environmental action, once the domain of pioneering grassroots activists, has become institutionalised and professionalised – in non-governmental civil-society organizations; in legal structures and government institutions; in consumer organizations, mass media, institutions of research, education and training; and also in industry itself in the form of environmental management systems.

As a result, the role of social movements has undergone some changes. They are no longer merely confrontational, but include increasingly more elements of cooperation with industries and government. In the past, confrontation was unavoidable because the environmental movement had to address political and industrial elites with new challenges. By now, however, the elites have got the message. So the future focus will have to be on cooperation with research, industries, and government in favour of ecological modernization. Confrontation may remain necessary in special cases from time to time, and to that extent it is important that social movements and non-governmental organizations continue to be independent and outside of industrial and government bureaucracies. But the mission of raising environmental awareness and setting the 'ecological question' onto the general political agenda is fulfilled. Now the race is about ecological modernization: working out integrated solutions to environmental problems and developing innovative technologies to ensure that ongoing societal modernization will pursue a sustainable path into the future.

Notes

1. See also Mol (1991), Klapwijk (1991), and Cramer and Schot (1991).
2. See also Huber (1989).

References

Cramer, J. and Schot, J. (1991) 'Praktijkproblemen bij innovatie en diffusie van milieu-technologie belicht vanuit een technologie-dynamica perspectief', in A.P.J. Mol, G. Spaargaren and A. Klapwijk (eds), *Technologie en Milieubeheer. Tussen sanering en ecologische modernisering*. Den Haag: SDU, pp. 81–92.

Huber, J. (1989) 'Social Movements', *Technological Forecasting and Social Change* **35**, pp. 365–374.

Huber, J. (1991) *Unternehmen Umwelt. Weichenstellungen für eine ökologische Marktwirtschaft*. Frankfurt: S. Fischer.

Klapwijk, A. (1991) 'Waterkwaliteitsbeheer: tussen 'end-of-pipe' en procesgeïnte-greerde technologie', in A.P.J. Mol, G. Spaargaren and A. Klapwijk (eds), *Technologie en Milieubeheer. Tussen sanering en ecologische modernisering*. Den Haag: SDU, pp. 93–104.

Mol, A.P.J. (1991) 'Technologie-ontwikkeling en milieubeheer', in A.P.J. Mol, G. Spaargaren and A. Klapwijk (eds), *Technologie en Milieubeheer. Tussen sanering en ecologische modernisering*. Den Haag: SDU, pp. 55–80.

Chapter 5

Gert Spaargaren and Arthur P.J. Mol

SOCIOLOGY, ENVIRONMENT, AND MODERNITY: ECOLOGICAL MODERNIZATION AS A THEORY OF SOCIAL CHANGE

Introduction

WE ARE WITNESSING A third wave of environmental concern in the industralized countries of Western Europe, and it could be argued that this time the environment is an issue that will not wither away. It no longer seems appropriate to think of ecology as moving up and down the agendas of politicians, concerned citizens, or sociologists. It simply took most of us more than two decades to recognize that environmental problems are not just the unintended consequences of an otherwise fortuitous trajectory of modernity. These problems appear increasingly bound up with modernity in such a fundamental and "organic" way that they cannot be dealt with in isolation from it. Their "solution" is bound up with altering the major cultural, political, and economic institutions of contemporary society in certain crucial respects.

This is why environmental problems have attracted the attention of a growing number of sociologists trying to understand the fundamental character of modern society. Environmental issues are no longer absent from the debate on modernity, high, late, or even postmodernity. For example, Giddens (1990, 1991) assigned the environmental crisis a rather central position in his recent work. Can sociologists now benefit from the preparatory work conducted by a small group of self-proclaimed environmental sociologists in the United States and Europe from the early 1970s onward? Is it already possible to conceive of a distinct sociological perspective on the environmental aspects of modernity? Or is the program yet to be (re)written? These questions motivated us to write this article, which is basically an evaluation of different sociological perspectives on the relationship between environment and modernity. Special attention is given to the theory of ecological modernization or eco-restructuring, which has been developed recently in European countries such as Germany and The Netherlands.

The argument is organized as follows. The first section gives a brief and select-ive review of the literature, arguing that environmental sociology should be freed of its predominant biological-technical outlook and use sociological theory rather than general ecology as its main frame of reference. The second section tries to develop a sociological perspective on the relation between environment and society, exploring the very concept of "the environment" and discussing some sociological theories dealing with the institutional development of modern society. Environmental soci-ologists should consider the kind of analytical categories required for thinking about a sustainable buildup of modern societies. Closer examination of the political dis-course on sustainable development leads us, in the third part, to conclude that this can be interpreted in terms of a plea for an ecological modernization of the industrial sector of the rich industralized world at least. The theoretical framework for this modernization process has been developed in Germany by Huber, Simonis, and Jänicke, among others. We discuss and critically evaluate Huber's ecological modernization approach as a variant of the so-called theories of industrial society in the fourth section. Fifth, we argue that recent developments in Dutch environ-mental policy strongly correspond with the ecological modernization approach, and we show that the debate within the environmental movement mirrors this specific modernization discourse as well. However, the ecological modernization approach is lacking, in several respects, as a sociological theory of modern society and needs remodeling and adaptation for the task of analyzing the relationship between environment and modernity. One theme that needs elaboration is the significance of "intuited nature" for our understanding of the environmental crisis; this will be dealt with in a preliminary way in the concluding section.

Environmental sociology: a selective review

Figure 5.1 provides a brief introduction to the theoretical and empirical work already completed in the field of environmental sociology. As Buttel (1986, 1987)

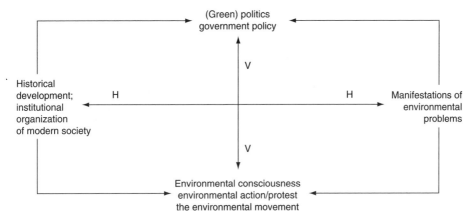

Figure 5.1 Theoretical and empirical themes in environmental sociology. Adapted from p. 7, N. J. M. Nelissen, 1979, Aanzetten tot een sociologische theorie over het milieuvraagstuk, in *Sociale Aspecten van het Milieuvraagstuk*, ed. P. Ester, pp. 5–20, Van Gorcum, Assen, The Netherlands. Used with permission

has shown in a number of illuminating reviews, most of the empirical work done in environmental sociology concerns themes situated on the vertical axis: Environmental attitudes and the environmental movement rank highest by far on the research agenda, followed by a growing number of studies in the field of environmental policies. Buttel's statement that most of the empirical work is normal science, in the sense of being scarcely theoretically informed, draws support from similar reviews by Lowe and Rüdig (1986) with respect to the European situation and by Ester and Leroy (1984) and Spaargaren (1987) with respect to the Dutch research tradition. This article leaves the empirical dimension aside and concentrates instead on the theoretical discussions within environmental sociology which, in our opinion, are of central importance to the field. Theoretical efforts in environmental sociology, almost without exception, deal with conceptualizing how the institutional development of society is related to the diverse manifestations of environmental problems (see the horizontal axis of Figure 5.1).

Environmental sociology seems to be in disarray as far as the methodology (i.e., the conceptual identity) of the field is concerned. The whole gamut of names identifying the subdiscipline already gives an indication of this: Social ecology, ecological sociology, human or new human ecology, and environmental sociology all refer to a hybrid of sociology and ecology.

Catton and Dunlap (1978, 1979, 1980) have contributed deliberately to the coquetting with ecology. In their desire to convert their general sociology colleagues, they presented their message in the form of a simple dichotomy: either you are (still) in the "human exceptionalist paradigm" (HEP), or you choose to become a disciple of the "new ecological paradigm" (NEP). The message hardly has been picked up or understood outside the small circle of environmental social scientists. Its main effect within the subdiscipline was to keep environmental sociologists busy and divided.

Buttel (1986) has taken an informative and yet ambiguous position in the debate on the relationship between sociology and ecology. Although he insistently points out the relevance of classical social theory for environmental sociology and criticizes the HEP-NEP distinction for having been drawn in too naive a way, he nonetheless identifies the "new human ecology" as the kernel of environmental sociology and seeks to clarify the troublesome relationship between sociology and biology in terms of an "inherent duality in human existence," a duality he explains in terms of the HEP-NEP distinction. Another illustration of Buttel's ambiguous position is found in the first standard introductory textbook on environmental sociology, which he wrote with Humphrey. Differences between Marx, Weber, and Durkheim are used to elucidate various possible perspectives with regard to environmental and energy issues. In the same book, however, Malthus and Darwin again figure as the founding fathers of the subdiscipline, with the classical human ecology of the Chicago school as its predecessor (Humphrey and Buttel, 1982). The wrestling between sociology, on the one hand, and biology and ecology, on the other, is not confined to the American branch of environmental sociology. One of the first Dutch environmental sociologists, Nico Nelissen, dedicated his Ph.D. thesis to the social ecology of the Chicago school, discussing, for example, the sociological themes of human agency, volunteerism versus determinism in the context of the analytical distinction between the biotic community and society (Nelissen, 1970).

Buttel (1986) seems to have gradually moved away from the ecologically inspired strand of environmental sociology in his sympathetic comments on the work of Allen Schnaiberg. We agree with Buttel that the growing influence of Schnaiberg, *vis-à-vis* those associated with Catton and Dunlap, is probably explained by two factors that distinguish his work from mainstream human ecology (Buttel, 1987). Schnaiberg (1980), notwithstanding his sympathy with the work of E. P. Odum, draws a clear analytical distinction between sociology and biology/ecology. The rules governing society, Schnaiberg argues, are basically different from those governing the ecosystems that form the sustenance base of society. We need no conceptual hybrid to consider the dangerous consequences that ecologists tell us are bound up with the enormous changes that have taken place in the sets of additions to and withdrawals from the sustenance base. Schnaiberg uses sets of additions and withdrawals to describe the interaction between society and its environment. The task of sociologists is to elucidate the developments and changes in the institutional composition of society that threaten the proper functioning of the sustenance base. A second reason that Schnaiberg's work attracts attention in the environmental field stems from his analysis of the societal dynamic behind the chronic overburdening of the sustenance base. The decisive changes that took place in the relation between modern society and its sustenance base cannot, in Schnaiberg's view, be explained by the kind of single-factor analyses (such as over-population or technology) that are characteristic of the environmental sciences field. Rather, these changes should be analyzed against the background of the overall structure of modern societies. Furthermore, Schnaiberg rejects Parsonian functionalism for describing the character of modern societies in favor of a theory that Buttel labeled neo-Weberian and neo-Marxist. Before we comment in more detail on his theory of the treadmill of production, we briefly sketch our own stance on the issues raised to this point.

We think environmental sociology would benefit from a further emancipation from the dominance of bioecological schemes and models, which form the socioecological kernel of the subdiscipline, in analyzing the relations between societies and their environments. Socioecological models should be left behind for two major reasons. First, as formal sociological theories, these models tend to lead to deterministic and functionalistic conceptualizations of human agency. Second, with regard to the analysis of historical developments, these models are usually prototypes of the kind of social-evolutionary schemes that are so convincingly criticized by Giddens as unfolding models of change (Giddens, 1984). Central to the critique of socioecological models in both instances is the fact that, as social systems, societies do not mechanically adapt to their environments. *Their members* choose to give priority to solving the environmental crisis by making it a central concern in the reflexive organization of society.[1] Environmental sociologists should orient themselves by recent debates within sociology, which center around the theme of actor and structure, to answer the question of whether and to what extent human behavior is determined by social and/or environmental structure.

Environment and modernity

In this section, we sketch the contours of a sociological approach to environmental problems. The first question, of course, deals with the very definition of "environment." What shape or form does "nature" take in relation to modernity? We argue that there are two dimensions to the man-nature relationship, which need to be distinguished analytically. The second question is how the dimension of the environmental crisis that figures most prominently within the environmental literature, nature as a sustenance base, relates to the character of modern society. With what kinds of institutional traits/properties is it intrinsically connected? We deal here with the debate about whether capitalism or industrialism is the major factor behind the environmental crisis. Looking for the institutional traits that cause environmental problems also means investigating possible solutions to those problems via institutional reform. There has always been a rather significant current of thought within the environmental debate, however, that has stressed the impossibility of reducing, let alone solving, the environmental crisis given the contemporary institutional composition of modern society. Let us turn first to this preliminary question.

Nature and (pre)modernity

The environmental movement often is said to be a demodernization movement. Central to the public image of the movement over the last 20 years has been its emphasis on premodern values, whether of a Right or Left political variety (Tellegen, 1983). Although we argue that this image of the environmental movement as one of moral protest-(ers) against modernity needs to be corrected, it nonetheless contains a point of great relevance for exploring the relationship between environment and modernity. The decisive alterations in the relations between environment and society, man and nature, and so forth obviously coincide with the emergence of modern society. Several authors, both within the field of environmental social sciences and outside it, have emphasized that modern society did not gradually come into being as the mature form of an earlier, more rudimentary society. Rejecting socioevolutionary models of historical development, they instead propose a "discontinuist interpretation of modern history" (Giddens, 1985, p. 31) to accentuate the many and crucial contrasts between modern and traditional societies. This contrast is especially relevant to the man-nature relationship, as described in the following passage:

> In class-divided societies, production does not greatly transform nature, even where, for example, major schemes of irrigation exist. The city is the main power-container and is clearly differentiated from the countryside but both partake of the "content" of the natural world, which human beings live both "in" and "with," in a connection of symbiosis. The advent of industrial capitalism alters all this. When connected to the pressures of generalized commodification, industrialism provides the means of radically altering the connections between social life and the material world (Giddens, 1985, p. 146).[2]

The eco-anarchist Murray Bookchin also portrays alterations in the relationship between society and its natural environment in an extensive and colorful way (Bookchin, 1980). Bookchin sees the advent of modern society as first and foremost the destruction of the cell-tissue society, and the replacement of complex, organic, harmonious ecosociosystems, which "yield local differences to the natural world," with simplified, inorganic systems in which the alienation of man from nature goes hand in hand with the alienation of man from man. Bookchin criticizes the "managerials" within the environmental movement for not understanding the impossibility of reconciling man and nature under conditions of generalized commodification. The only way to restore the relationship of man and nature is to "dismantle" or restructure modern society, using the Greek city-state as the example. Bookchin can be called one of the "organic intellectuals" of the environmental movement of the early 1970s. His definition of the environmental crisis is as fundamental as it is all-embracing, and the "solution" he offers points the way *out* of modern or capitalist-industrial society.

We agree with Giddens and Bookchin that a discontinuist view of history can deepen our understanding by highlighting the essential characteristics of modernity. Unlike Bookchin and many discontinuist adherents within the environmental movement, we do not think that studying premodern society implies a preference for the cell-tissue society as the basis from which to plan strategic action within the environmental movement. However, as a theoretical approach a discontinuist perspective might provide specific answers to the kinds of questions that we consider of central importance to environmental sociology.

Two dimensions of the environmental crisis

The first question for consideration concerns the man-nature relationship. William Leiss, in his book *The Domination of Nature* (inspired by the work of the Frankfurter Schüle), asserts that the mastery of nature within Western culture not only brought about an idea of separation between man and nature, but also resulted in a "bifurcation" of nature (Leiss, 1974, p. 135). Nature became split into *intuited nature*, the "experienced nature of everyday life," and the "abstract-universal, mathematized nature of the physical sciences" (Leiss, 1974, p. 136). Several authors, mostly from the cultural-philosophical sciences (Lemaire, 1970), recognized and worked on this duality of nature. Within environmental sociology, Schnaiberg contributed to our understanding of scientific nature, the dimension he considers to be more important than the cosmetic concern of intuited nature (Schnaiberg, 1980). Reserving a discussion of the intuited nature of everyday life for the last section of the paper, we begin by discussing scientific nature as the best documented and theorized dimension.

Scientific nature is nature harnessed to the ongoing rationalization and expansion of production. Despite the unveiling of nature in all her former mysteries by Baconian and Newtonian science, it could be said that there is still only a partial understanding of how the sustenance base functions. Ecology, the scientific discipline of the sustenance base, made us aware that nature can no longer be treated as a black box in relation to production. Nature as a black box would deliver inputs in the form of energy and raw materials and would absorb and process outputs in the form of waste. Clearly, nature can no longer be treated as a void in its functioning, whether as a stock of or a dump for material entities to be used endlessly and free of charge. This is the message that environmental economists such as Nicholas Georgescu-Roegen and Kenneth E. Boulding (and, in The Netherlands), Roefie Hueting, Johannes Opschoor, and Bob Goudzwaard understood 20 years ago when they tried to incorporate the environment as a production factor into their neo-classical economic models. Although much has already been done, we are only just beginning to understand the difficulties of correcting this design fault of modernity (Giddens, 1990).

There appear to be two sets of relevant questions with regard to the interrelation between societies and their sustenance base. The first focuses primarily on the sustenance base; the second directs our attention to the institutional aspects of modern society that are involved. One of the basic questions in the field of environmental sciences is whether, and to what extent, we already possess or are able to develop the scientific-technical knowledge required to bring our interaction with the sustenance base under rational control. It seems to be very difficult to grasp the consequences for the environment of human action for several reasons, including (1) the complexity of the ecosystems involved; (2) the displacement of effects in time and space; and (3) the rapidly increasing scale of the man-nature interaction, which is by now truly global. The uncertainties surrounding the predictive models of the ecotechnical scientists and the sheer lacunae of knowledge that exist, for example, in the field of the ecotoxicology, make the debate about the required adjustment of social reproduction to meet the demands of ecosystem reproduction susceptible to all kinds of mystifications. In this debate political and scientific arguments are freely intermingled. The second set of questions concerns the kind of institutional reform that is required to correct the design fault of modernity in its interaction with the sustenance base. How drastic are the changes involved? In other words, which institutions need to be reformed and how central are these changes to the overall process of reproduction of modern society, both at the level of institutional development and in terms of everyday life? These questions are of basic importance to environmental sociology and will be elaborated in the next section.

Capitalism, industrialism, and modernity

In our view, at least three schools of thought can be distinguished when the character of modernity and its relation to the environmental crisis are considered. Each emphasizes different aspects of modernity and seeks to promote different solutions to the disturbed relation between modern society and nature as its sustenance base: the neo-Marxist approach, different versions of postindustrial society theory, and the counterproductivity thesis. We begin by commenting on the neo-Marxist position that Allen Schnaiberg has taken in this debate.

On the question of which institutional traits of modern society can be held responsible for the environmental crisis, Schnaiberg is unambiguous. The treadmill of production underlies the continuing disruption of the sustenance base. This treadmill is explained in terms of the capitalistic character of the organization of production. According to Schnaiberg, a small number of powerful corporations constantly propel the process of capital accumulation. The best way to analyze how they gain and retain their control over large sectors of production and their decisive economic and political power *vis-à-vis* the labor movement and the state is by using "the broad institutional perspective of structural analysts such as Marx" (Schnaiberg, 1980, p. 209). Relying heavily on the analyses of Marxist theorists such as Baran and Sweezy, Schnaiberg seems to reduce the different aspects of the environmental crisis to the monopoly-capitalist character of modern society, leaving little room for a theoretical assessment of industrialized production in relation to environmental problems. The rather straightforward Marxist analysis used by Schnaiberg has come under attack within sociology from two different perspectives, which have in common their belief that the *industrial* rather than the *capitalist* character of modern society is the more important factor in explaining the environmental crisis. To position the ecological modernization approach (to be introduced in the fourth section) within this field of discussion, we first briefly comment on the various forms of theories of industrial and postindustrial society.

Marxist analyses have been criticized from a radical perspective by a group of authors who can be labeled "counterproductivity theorists." The ideas advanced by authors such as Barry Commoner, Ivan Illich, André Gorz, Rudolf Bahro, Otto Ullrich, and in The Netherlands, Hans Achterhuis have resonated within the environmental movement. For example, Ullrich, in his book *Weltniveau*, criticized Marx for his preoccupation with the social relations of production, leaving under-theorized the forces of production (Ullrich, 1979). We need to incorporate into our analysis the "myth of the great machine," which is embodied in the organization of the industrial system, if we are to understand why our system of production runs counter to the goals for which it was designed and to explain the increasing discrepancy of welfare as measured by a growing gross domestic product (GDP) with the well-being of man and nature. The industrial system is highly administered in an ever more centralized, hierarchical way. This centralized, hierarchical character has to be analyzed in relation to the technical systems that are omnipresent in the system of production but are no longer adapted to demands of man and nature. Finally, this model of industrial production viewed as an organizational device has become widespread, penetrating, for example, the educational and welfare sectors of modern society. Consistent with their analyses of the environmental crisis as part of an all-embracing crisis of the industrial systems, counterproductivity theorists share the belief that a solution can only be found by at least partially dismantling the existing systems of production. Bahro (1984) expresses this very clearly by his use of the term "industrial disarmament" to summarize his program of reform.

Surprisingly, none of the authors mentioned above receives any extensive treatment in Richard Badham's overview of theories of industrial and postindustrial society (Badham, 1984, 1986). However, his main focus is that branch of sociology represented by Kerr, Bell, and Aron, among others. These were sociologists who, from the 1950s onward, developed their theories of industrial society, starting from the central assumption that "the development of industry and its impact on society are the central features of modern states" (Badham, 1984, p. 2). What unites these authors, and distinguishes them from the counterproductivity theorists, is their more benign evaluation of the "all-embracing logic of industrialism." Industrial societies pass through various phases or stages in their maturation, technology being one of the prime movers and determinants of their general development. Class conflicts belong typically to the birth period of industrial society and lose their significance during later phases of its development. In dismissing Marxist theory as irrelevant for the analysis of modern society, these theories of industrial society could also be called theories of postcapitalist society. These consensus theories were long opposed to conflict theories and vice versa within sociology. We would include, within this category, some variants of industrial society theory that are usually bracketed as theories of postindustrial society. In our opinion, the later work of Daniel Bell (1976) and the writings of Alvin Toffler, Alain Touraine, and Barry Jones share the basic tenets of the first-generation theories of industrial society. The adjective "post" stands for the transition into the newest phase in the development of industrial society, characterized by a shift toward a service-sector-based economy, the displacement of blue-collar work by white-collar work, and material growth conceptions being replaced by nonmaterial values (Inglehart, 1987). Although postindustrial society is portrayed mainly through its consequences for occupational structure, the role of science and technology, and the meaning of leisure, the changes that are supposed to take place within the production structure would considerably lessen the burden on the sustenance base.

Frankel gives an illuminating assessment of the consistency and theoretical adequacy of both Left and Right political variants of postindustrial society perspectives. He considers how the proposed models of a "good" society deal with the relation between the different levels of social organization (local, national, and international), and with the role attributed to state versus market forces (Frankel, 1987). We think two main conclusions can be drawn from his analysis. First, a distinction should be made between demodernization and modernization variants of postindustrial society theory; second, both perspectives are, for different reasons, susceptible to criticism as regards the criteria formulated. Counterproductivity theorists, writing from a demodernization perspective, see local autonomy or even autarky as realizable by severing links with the world market and political relations. The Mondragon cooperatives serve as a contemporary example of this theoretical model. Apart from the questions that Frankel raises about their factual independence from the world market (Mondragon products are being sold on the Spanish market) and the difficulties in defining the proper scale of small-scale communities (Ulgor being a cooperative of 3,500 members), we think the crucial theoretical dilemma posed by the Mondragon experiment is the way local and regional levels are thought to be related to national and inter- or supranational levels. The theoretical argument against the proposed insulae within modern society is aptly summarized in Giddens' treatment of time-space distancing:

> In the modern era, the level of time-space distanciation is much higher than in any previous period, and the relations between local and distant social forms and events become correspondingly "stretched." Globalization refers essentially to that stretching process, insofar as the modes of connection between different social contexts or regions become networked across the Earth's surface as a whole. Globalization can thus be defined as the intensification of worldwide social relations which link distant localities in such a way that local happenings are

shaped by events occurring many miles away and vice versa. (Giddens, 1990, p. 64)[3]

Intermediate or convivial technology, some degree of autonomy in social relationships on the personal as well as social group level, and direct responsibility for and control over materials circulating within the ecosystems are all desirable ends in themselves. But the intensification of international social relations and the increasing level of time-space distancing within modern societies make the realization of these goals in the context of local experiments, which are thought to be exempt from power relations and market forces operating on a worldwide basis, less plausible and realistic.

Local-level processes of social change are analyzed within the hyper- or super-modernization approaches as the outcome of restructuring economic production at the international level. Frankel uses Toffler's "third wave" theorem as an example of the major weaknesses inherent in a theoretical scheme in which a multinational organization-dominated network at the global level coexists with demarketized, autonomous lifestyles at the local level. According to Frankel (1987), "it seems that Toffler espouses a naive, small liberal belief in the mutual balance and coexistence of world institutions and local democracy" (p. 39). Using arguments that are sometimes empirical and sometimes theoretical, Frankel illustrates the major problems involved in the too-simple abolition of market and state regulation at regional and national levels, and those of putting too much faith in the innovative and democratic potentials of multinational corporations. In short, his critique of industrial society theory in its superindustrial form comes down to its lack of understanding of the capitalist character of production, with state planning as a prerequisite for "taming" the treadmill.[4]

Frankel's neo-Marxist critique of industrial society theory brings us back to our starting point: the neo-Marxism of Schnaiberg. Is the opposition between Marxist analysis of capitalist society and bourgeois analysis of industrial society still the relevant dividing line within environmental sociology? Following Giddens, we think this is no longer the case. Instead, we prefer to treat industrialism and capitalism as two of the four institutional dimensions or organizational clusters of modernity that can be separated analytically (see Figure 5.2). Industrialism and capitalism are both highly relevant to understanding modernity and are defined by Giddens (1990) in the following way:

> *Capitalism* is a system of commodity production, centered upon the relation between private ownership of capital and propertyless wage labour, this relation forming the main axis of a class system. Capitalist enterprise depends upon production for competitive markets, prices being signals for investors, producers, and consumers alike. The chief characteristic of *industrialism* is the use of inanimate sources of material power in the production of goods, coupled to the central role of machinery in the production process. Industrialism presupposes the regularized social organization of production in order to coordinate human activity, machines, and the inputs and outputs of raw materials and goods (pp. 55–56).[3]

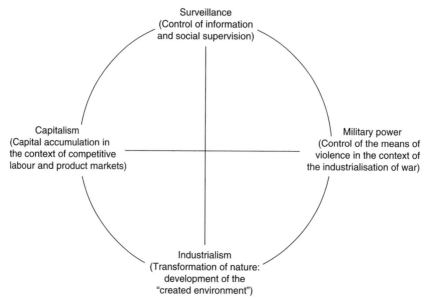

Figure 5.2 The institutional dimensions of modernity. (Reprinted from Giddens, 1990, © 1990 by Polity Press. Used with permission.)

The different theoretical perspectives considered in this section can be summarized as focusing on the relations between different institutional dimensions of modernity. Schnaiberg's analysis of the treadmill character of production indicates those institutional alignments within modernity that can be held responsible for the chronic impetus toward expansion of production and transformation of economy and technology. Theorists of industrial society point out the central role of technology and machinery and man-machine relations within modern society when they describe the different stages of industrial development. Criticizing the centralized and hierarchical character of manufacturing production, Ullrich (1979) posited a specific relation between surveillance on the one hand and industrialism and capitalism on the other.

Up to this point, our discussion can be summarized in four short statements or conclusions. First, an analytical distinction should be made between two dimensions of the environmental crisis: intuited nature as experienced in everyday life, and scientific nature functioning as a sustenance base for production in modern society. The definition of the environmental crisis prevalent within the environmental sciences is the burden on or even overexploitation of the sustenance base. Second, within the field of environmental social sciences, we are in urgent need of more refined, sophisticated theories relating the burden on the sustenance base with institutional aspects of modern society. Third, contemporary sociological theories, which deal with the institutional development of modern society and which can be said to have special relevance for environmental sociology, differ with respect to:

- their general perspective on historical development, evolutionary versus "discontinuist" models of change
- their emphasis on either the capitalist or industrial character of modern society

- their evaluation of developments within the industrial sector as theorized by theories of postindustrial society.

Fourth, in exploring the relationship between environment and modernity, Giddens's analytical distinction between four institutional dimensions of modernity can be of great assistance in assessing the value and central focus of each different theoretical perspective.

The political discourse on environment and modernity

The concept of sustainable development has a central position in the contemporary political debate on environmental issues. In political terms, sustainability deals with institutional developments in modern society with relation to its sustenance base.

There seems to be a growing consensus, at least in the industrialized countries, about sustainable development as a concept for overcoming the ecological crisis. This consensus is only possible because (1) sustainable development is a rather vague concept that allows many interpretations, and (2) the concept as introduced by the Brundtland commission (World Commission on Environment and Development, 1987) integrates ecological quality with economic growth via industrialization. Economic growth and technological development, two important institutional traits of modernity, are therefore seen as compatible with and sometimes even as a condition for sustaining the sustenance base, rather than as the main cause of environmental destruction.

As Timberlake (1989), one of the contributors to the Brundtland report, observes, the concept of sustainable development is based more on opinions than on scientifically based ideas. For this reason and because of the many possible interpretations that can be placed upon it, the concept of sustainable development is only suited to our purposes to a very limited extent. Therefore, we introduce a more analytical and sociological concept consonant with the primarily political concept of sustainable development: ecological modernization. The more analytical and sociological concept of ecological modernization highlights the relationship between the modernization process and the environment in the context of industrialized societies, whereas sustainable development also (1) pretends to be applicable to the less developed countries (Spaargaren and Mol, 1989), and (2) tries to include questions of equal development and peace. Notwithstanding these differences, both concepts originate from the same standpoint on the relationship between environment and modernity. As Simonis (1989) writes, the dominant notion in the 1990s is that the relationship between society and environment calls for "industrial restructuring for sustainable development, or 'ecological modernization'" (p. 361).

In the next section, we further elaborate on the concept of ecological modernization and situate it in the debate on environment and modernity.

Ecological modernization: a theoretical framework

The concept of ecological modernization has a short history in German and, to some extent, Dutch discussions about the institutional changes necessary in Western industrialized countries for overcoming the ecological crisis. The concept is used at two levels in these debates. First, ecological modernization is used as a theoretical concept for analyzing the necessary development of central institutions in modern societies to solve the fundamental problem of the ecological crisis (see, e.g., Huber, 1982, 1991; Spaargaren and Mol, 1991). At this level, ecological modernization can be seen as an alternative to other concepts and analyses of the relationship between institutional developments in different domains of modernity and environment. Second, on a more practical level, ecological modernization is used as a political program to direct an environmental policy. As such, it includes a certain strategy with more or less concrete measures to counter environmental problems (Jänicke, 1989; Schöne, 1987; Simonis, 1989; Zimmerman, Hartje, and Ryll, 1990). The political program of ecological modernization seems to fit rather well in recent developments in environmental politics in some Western European countries.

In this section, we have elaborated on the concept of ecological modernization as a theoretical contribution to environmental sociology and have analyzed its main theoretical problems. In the next section, we analyze developments in Dutch environmental policy, as well as in the ideology and strategy of the Dutch environmental movement, in the context of ecological modernization.

Ecological modernization as a theory of industrial society

> Die hässliche Industrieraupe werde sich im Zuge ihrer Metamorphosen noch als ökologischer Schmetterling entpuppen (Huber, 1985, p. 20).[5]
> [The dirty and ugly industrial caterpillar transforms into an ecological butterfly.]

This quote from *Die Regenbogengesellschaft* characterizes the central idea of ecological modernization. Ecological modernization stands for a major transformation, an ecological switch of the industrialization process into a direction that takes into account maintaining the sustenance base. Like the concept of sustainable development, ecological modernization indicates the possibility of overcoming the environmental crisis without leaving the path of modernization. Ecological modernization can be interpreted as the ecological restructuring of processes of production and consumption. We shall use the elaborations of Huber, one of the leading exponents of ecological modernization theory, to obtain a better understanding of it. Huber (1985) uses the concept of ecological modernization as follows:

> Das wirtschaftlich fast alles entscheidende Kernstück des ökosozialen Umbaus besteht in einer ökologischen Modernisierung der Produktions- und Konsumkreisläufe durch neue und intelligentere Technologien. (p. 174).[5]
> [The central economic theme of the ecosocial switchover will be the

ecological modernization of production and consumption cycles by new and more intelligent technologies.]

Following Huber, we see two central projects as forming the heart of the ecological switchover: the restructuring of production processes and consumption toward ecological goals. The first project is the development, inauguration, and diffusion of new technologies that are more intelligent than the older ones and that benefit the environment. From traditional end-of-pipe technologies, there is a shift toward technologies that establish clean production processes. Microelectronics, gene technology, and new materials are seen as promising technologies for disconnecting economic development from relevant resource inputs, resource use, and emissions (Simonis, 1989), and for monitoring processes of production and consumption for their effects on the environment (Huber, 1985). This must lead to the ecologization of the economy, that is, to physical change in production and consumption processes and to the possibility of monitoring these processes. Second, the concept of ecological modernization includes economizing ecology by placing an economic value on the third force of production: nature. Nature and environmental resources should regain their place in economic processes and decision making (Immler, 1989). As Simonis (1989) questions: "Apart from labour and capital, nature is the truly quiescent and exploited third production factor. How can nature's position in the 'economic game' be strengthened?" (p. 358).

Huber's (1985) elaborations on ecological modernization make it clear that this theoretical concept belongs to the industrial society theory. In the first place, Huber analyzes ecological modernization as a historical phase of industrial society. Second, he emphasizes the logic of industrialism as central to the development of modern society. We briefly elaborate on these two issues.

Huber (1982, 1985) analyzes ecological modernization as a phase in the historical development of modern societies. He sees three phases in the development of the industrial society: (1) the industrial breakthrough; (2) the construction of industrial society, which can be subdivided into three parts corresponding with the Kondratieff cycles; and (3) the ecological switchover of the industrial system through the process of superindustrialization. This historical systematization and the position of an ecological switchover is in line with other theories of industrial society (e.g., Immler, 1989). Central to changing the phases of the industrialization process are the invention, inauguration, and diffusion of new technologies. In the first phase, the key technology was the steam engine. Chip technology is what makes the ecological switchover via superindustrialization possible in the most recent phase.

Huber (1985, 1989b) differentiates three analytical categories or spheres in analyzing modern society. Apart from the industrial system (or technosphere) and the life world (or sociosphere), which are more or less in line with other social theories (e.g., Habermas, 1981), Huber introduces a third sphere: nature, or the biosphere. The main problems in the present society are, according to Huber, related to the colonization of both the sociosphere and the biosphere by the industrial system (or technosphere). These problems, interpreted as structural design faults of the industrial system, can be overcome by an ecosocial restructuring of the technosphere, which Huber calls ecological modernization. The industrial, rather than capitalist or bureaucratic, character of modernity is the point of departure for the theory of ecological modernization.

Evaluating ecological modernization

In this section, we evaluate the theory of ecological modernization with three different criteria: its view of historical development and the role of technology in it; its definition of environment/nature; and its treatment of the role of the state. We define the theoretical position of ecological modernization in the debate on environment and modernity.

What is the place of ecological modernization among theories on the relationship between modern society and the environment? Adapting Giddens's conceptual scheme of the four dimensions of modernity, ecological modernization can be said to focus primarily on the dimension of industrialism. In analyzing the main characteristics of modern society pertaining to the industrial or technological system, ecological modernization can be said to belong to the branch of society that Badham (1984, 1986) calls the industrial society theory. It highlights the industrial rather than the capitalist character of modern society. Furthermore, the ecological modernization approach stands in direct opposition to counterproductivity theory or demodernization theses in its conviction that the only possible way out of the ecological crisis is by going further into industrialization, toward hyper- or super-industrialization. Ecological modernization must be analyzed in continuity with the present system (Huber, 1985). Finally, the ecological modernization approach diverges from neo-Marxist social theories in paying little attention to changing relations of production or to altering the capitalist mode of production altogether. The focus of ecological modernization is undoubtedly on the development of the industrial system. The ecological restructuring of modern society is limited to changing the organization of production and consumption activities and does not extend to Schnaiberg's (1980) treadmill of production. On the contrary, the capitalist character of modern society is hardly questioned, as capitalist relations of production and a capitalist mode of production are seen as not relevant to overcoming the ecological problem. According to Huber (1985):

> Kapitalistisch zu sein, ist nichts Unanständiges, sondern eine Eigenschaft, die uns qua Zeitgenossenschaft zufällt, wobei fraglich ist, wie lange der kapitalismus-Begriff überhaupt noch zeitgemäss sein wird (pp. 77–78).[5]
> [Capitalism is nothing indecent, but a given quality of our age, and it might even be asked how long the concept of capitalism will be in any sense up to date.]

Focusing on technologically induced developments within the industrial system, the theory of ecological modernization exposes an evolutionary and technologically deterministic view of social development that is characteristic of the theories of industrial society. The ecological switchover is analyzed as a logical, necessary, and inevitable next stage in the development of the industrial system – the system correcting itself for the construction fault of neglecting ecology. In the "system evolutionary" view of historical developments, technology and technological innovations are the motor for socioecological change. Technological developments seem to take place autonomously and to determine the changes that take place both within the industrial system itself and those occurring in its relationship with the social and natural environments. This becomes very clear when Huber (1985, 1989b) speaks of the "technosystem" rather than the industrial system, emphasizing the central role of technology in the overall development of society. This kind of technological determinism can be questioned both from the viewpoint of recent studies concerning the social construction of technological developments (Bijker, Hughes, and Pinch, 1987; Hughes, 1986) and from a theoretical perspective that tries to combine actor- and system-oriented approaches in explaining social change.

The ecological modernization approach conceptualizes nature or the environment as one of the two spheres that are threatened by the dynamics of the industrial system, the other being the life world. In concentrating on the use that is made of nature within production, ecological modernization focuses primarily and exclusively on the relationship between the technosystem and nature. In other words, the central concerns of ecological modernization are depletion of natural resources and pollution of the environment in relation to the sustenance base. The relationship between what Huber (1989) calls the sociosphere and the deterioration of nature is not discussed. This relationship deals with what we have called "intuited nature."

The differentiation between political and economic spheres of the industrial system seems to have very little relevance within the ecological modernization approach. Both are seen as integral parts of the industrial system and are functional for the development of the industrial system. Huber (1989a) puts it as follows:

> Zu sagen, das Wissenschaft und Technik als "Produktivkräfte" zu bestimmenden formativen Faktoren geworden sind, heisst nicht, die Bedeutung von Markt und Staat zu verkennen. Es heisst, sich von der Technik einen umfassenderen Begriff zu machen und Markt und Staat selbst als Teil eines umfassenden technischen Systems zu verstehen, und zwar als institutionelles Gehäuse der technischen Entwicklung, als das Gefüge administrativer und wirtschaftlicher Rahmenbedingungen, die auf die technische Entwicklung einwirken (pp. 10–11).
>
> [To say that science and technology have changed from "forces of production" to the determining factors forming society does not denote an undervaluation of the significance of market and state. It means making technology an inclusive concept and interpreting the technological system as comprising market and state. It also means interpreting market and state as the institutional context of technological development, as the administrative and economic frameworks that influence technological developments.]

In what is seen as the autonomous development of the industrial system, propelled by technological innovation, the state plays no central role in redirecting the processes of production and consumption. Huber (1989a) even regards state intervention to promote ecological modernization as counterproductive in the long term, because it frustrates the innovation process. Huber uses Hobbes' Leviathan as an undesirable and threatening image of growing state intervention. However, there is widespread consensus in the field of environmental policies about the necessity of national and international state intervention. State intervention in the environmental field during the 1970s was primarily organized at the national level and directed toward repairing the shortcomings of free-market competition by internalizing environmental external effects within the market. Over the last decade, we have witnessed an increasing awareness of the international nature of environmental problems, inter- and supranational policies being developed to coordinate national efforts, and, especially in the context of the European Community (EC), a harmonizing of national economic interests with environmental policies. Nowadays, it seems very hard to imagine an ecological switchover without state intervention at various levels.

Having made a theoretical assessment of the concept of ecological modernization, we can draw some conclusions. The ecological modernization approach clarifies the relation between modern society and its environment with respect to one essential, institutional clustering within modernity. Huber's theory makes it possible to define the environmental crisis in more detail and is more appropriate with regard to one of its central dimensions, the burdening of the sustenance base. In this respect, it makes a contribution to our understanding of the complex relation between environment and modernity similar to that of Schnaiberg (1980). Some of the weaknesses of the ecological modernization theory include:

- paying little attention to the role of state institutions and being overoptimistic about the dynamics of the market;
- using a definition of environment or environmental crisis that is restricted to the dimension we have called the sustenance base or the burdening of the sustenance base, which pays no attention whatsoever to the experienced nature of everyday life;
- being representative of industrial society theories in using a conceptual model to analyze the historical development of industrial society, which can be said to have a technologically deterministic character.

Ecological modernization: a political program

As already noted, ecological modernization is not merely a theoretical framework for analyzing the relationship between the institutional structure of modern industrial society and environmental problems. It is also used as a political program sketching the way out of the environmental crisis. There is, of course, a very close connection between the political program and the theoretical concept.

Three political programs

Looking at the main programs of reform suggested within the field of environmental policy, a limited number of political projects can be distinguished. Jänicke (1988), for instance, discusses four different strategies toward environmental problems in modern industrial societies. In the discussion about "preventive environmental policy," other authors make different classifications or mention the same strategies but with different names or definitions. We regard the underlying structure of all these classifications as composed of two distinctive political programs toward the environmental crisis. The first one focuses on compensation for environmental damage and on the use of additional technology to minimize the effects of growing production and consumption on the environment. The second political program, which can be said to be in line with the theory of ecological modernization, focuses on altering processes of production and consumption. Common descriptive notions used in the second program include clean technology, economic valuation of environmental resources, alteration of consumption and production styles, prevention, and monitoring of compounds through the production-consumption cycles, to name but a few. Among proponents of this political program, there is some discussion about the exact meaning of policies aimed at restructuring production processes, or "umbau," versus policies that are directed toward the selective contraction, or "abbau," of the economy, for instance on the issue of synthetic materials. This, however, is a discussion within the framework of modernity.

In the proposals and measures normally put forward outside the official policy arena, a third program can be distinguished, focused on what Ullrich (1979) calls the progressive dismantling or deindustrialization of the economy and the transformation of today's production structure into small-scale, or smaller-scale than at present, units representing a closer and more direct link between production and consumption. This strategy, well known from the reform program developed in the early 1970s, must be distinguished from the limited, selective abbau programs contained within the second perspective. This third, more embracing policy of environmental/social reform, directed at demodernization rather than modernization, seems to have lost its attraction to a considerable extent.

We argue that, in The Netherlands at least, official environmental policies suggested and already partially implemented by the environmental sectors of the administration are moving from the first to the second program. At the same time, a switch from the third perspective to the second can be noticed within major sections of the environmental movement, with regard to both ideology and strategy. A broad consensus gradually seems to have come about, at least within Dutch society, as regards the general approach best suited to overcome the environmental crisis. Ecological modernization seems to be the general concept that describes this growing consensus.

The general direction of Dutch environmental policy

Developments in Dutch environmental policy concerning relations between the state and environmental polluters (often called target groups by state officials) can be divided into three phases. The initial phase, in the 1970s, can be characterized as a top-down state environmental policy, with big conflicts about individual measures, broad strategies, and final goals between the Department of the Environment, target groups (and their affiliates within the state bureaucracy), and the environmental movement. It was a sectoral and end-of-pipe-oriented policy.

This situation changed from the beginning of the 1980s. The deregulation debate after the economic crisis, the crisis of the welfare state, the poor results of environmental policy, and the depolitization of environmental issues led to discussions and initiatives to provide target groups with greater influence on and responsibilities toward environmental policy. This transitional second stage switched to the third stage by the end of the 1980s, as symbolized by the launching of the first National Environmental Policy Plan in 1989 (Ministry of Environment, 1989). Dutch environmental policy changed during this third phase, at least on paper, from an end-of-pipe strategy toward an ecological modernization perspective. The old strategy of repairing environmental deterioration after the fact and regulating environmental problems by introducing additional technology, leaving the general structure of production and consumption processes untouched, has been abandoned. The core of the new approach, which must lay the basis for sustainable development, is:

- closing substance cycles—the chain from raw material via production process to product, waste, and recycling must contain as few leaks as possible;
- conserving energy and improving the efficiency and utilization of renewable energy sources;
- improving the quality of production processes and products.

The central issue in environmental policy is the restructuring of production-consumption cycles, to be accomplished through the use of new, sophisticated, clean technologies. The private sector and the target groups (agriculture, industry, consumers, etc.), play a central role in achieving this objective. Target groups have to take responsibility for ecologizing production and consumption by innovating production technologies and products and by changing patterns of behavior. Environmental management systems and environmental audits in industry and agriculture are central instruments for attaining the structural incorporation of environmental issues in the behavior of private-sector enterprises. These environmental management systems should take into consideration all environmental issues of plant operation and products by monitoring the relevant flows of compounds and energy to minimize their effects on the environment.

The government plans to use more financial incentives to induce producers and consumers to assume their responsibilities toward the environment and to incorporate environmental costs in economic decision making. The use of fiscal measures, environmental taxes, deposits on products and materials, and economic incentives in general have been proposed and, to a limited extent, introduced until now.

Although there is considerable debate between different state departments, and between the Ministry of the Environment and some target groups, on concrete measures and the speed of realizing environmental policy goals, there seems to be a general consensus on the main approach, direction, and goals of environmental policy. When certain economic parameters are taken into consideration, proposals for restructuring production-consumption cycles do not meet with much opposition.

Dutch environmental movement

An ideological change has taken place in the Dutch environmental movement since 1980 (Cramer, 1988). Large sections of this social movement were highly critical of capitalist economic growth, the ongoing process of industrialization, and technological development before the 1980s. The climax of protest against nuclear power at the beginning of the 1980s can be seen as a turning point, both in the struggle against the dominant institutions of modern society and with regard to what Tellegen (1983) calls the antimodernity ideology of the environmental movement. The main reasons for this switch in ideology and strategy were expanding socioeconomic stagnation and crisis, the changing political climate toward realism, growing acceptance of the environmental movement as a political factor by state institutions, and concrete successes of more modest environmental strategies and ideologies.

The Dutch environmental movement is, for the most part, no longer strategically or ideologically opposing large-scale industrial production and technological innovations, as long as these are environmentally sound. A radical farewell has been said to the small is beautiful ideology, and technological developments are seen as potentially very useful in regulating environmental problems (Mol and Spaargaren, 1991). At the same time, apart from neo-Marxist analyses, the focus is not so much against capitalism or economic growth, but rather against concrete, environmentally harmful economic developments. Capitalist relations of production, operating as a treadmill in the ongoing process of economic growth, are rarely emphasized. The environmental movement has adopted an ecological modernization approach, highlighting the necessity for adaptation of the modernization process to ecological limits. Within this paradigm, a limited and selective contraction or constriction of economic growth is seen as unavoidable by environmentalists (e.g., Cramer, 1988; Nijkamp and Reijnders, 1989; Schöne, 1987; Vereniging Milieudefensie, 1991).

A pragmatic and less oppositional environmental movement was established during the 1980s. Its aims were neither changing the capitalist relations of production or the treadmill of production, nor working toward a deindustrialization or demodernization of the economy. Today, the main strategy and goals seem to be the ecologizing of processes of production and consumption within modernity. A switch has taken place from the third perspective to ecological modernization.

Conclusion

In this paper, we have tried to give a selective review of the theoretical contributions relevant to environmental sociology, when this subdiscipline deals with the relationship between environment and institutional developments in modern society.

Three schools of thought were identified as relevant to the sociological contribution to the environmental debate. One tradition works from a Marxist perspective and deals with the treadmill of production as the main cause of disturbances in the sustenance base; another school highlights the industrial dimension of modernity in analyzing environmental deterioration; and a third emphasizes the counterproductivity of the development of modern society. We conclude that there is a lack of sophisticated theories dealing explicitly with the relationship between institutional developments of modern society—whether it be capitalism, industrialism, or another development—and the burdening of the sustenance base in all three schools of thought. In that sense, environmental sociology is still in its infancy.

One of the recent sociological theories dealing explicitly with the relationship between modernity and environment has given some attention to these issues: ecological modernization. The popularity of this theoretical concept is based on its close correspondence with the idea of sustainable development, combined with recent changes in environmental policy and dominant ideologies of the environmental movement in some Western European countries.

The theory of ecological modernization is limited insofar as it deals with only the industrial dimension of modernity, neglecting dimensions of capitalism and surveillance, and because it narrows the concept of nature to the sustenance base. On this last issue, ecological modernization is consonant with most of the other relevant contributions to the debate on environment and modernity, which also ignore, to a large extent, intuited nature. We think the distinction between the two dimensions of nature, intuited nature and scientific nature, might fruitfully be further theorized. We only indicate the general direction of analysis here. In the first place, elaborating on intuited nature would necessitate an emphasis within environmental sociology on themes that, to a certain extent, have been left in the hands of philosophers and social psychologists up to now, specifically, the ways human actors deal with nature, its integrity, its intrinsic value, and its value for human agents. In the second place, and in our opinion essential for environmental sociology in the near future, it would mean giving high priority to the analysis of the risk profile (Beck, 1986; Dietz and Frey, 1992; Giddens, 1991) of modern society and the way people handle this dimension of the environmental crisis within their everyday lives.

Notes

1. We are aware that a significant group of sociologists, as well as biologists, has tried to work out a historical, nondeterministic form of evolutionary theory, giving human agency pride of place. See, for example, Musil (1990) and Dietz and Burns (1992).

2. Reprinted from A. Giddens, 1985, *The Nation-State and Violence*, Cambridge, UK: Polity Press. Used with permission.

3. Reprinted from A. Giddens, 1990, *The Consequences of Modernity*, Cambridge, UK: Polity Press. Used with permission.

4. Of course one cannot just leave matters here. A proper treatment of the role of the state in environmental planning is hardly possible without entering into the complex debate on state regulation in a period that Lash and Urry designated as the end of organized capitalism (Lash and Urry, 1987). We provide a more extended discussion in Spaargaren and Mol (1991).

5. Reprinted from Joseph Huber, 1985, *Die Regenbogengesellschaft, Okologie un Sozialpolitik* (*The Rainbow Society, Ecology and Social Policy*), copyright @ 1985 S. Fischer Verlag GmbH, Frankfurt am Main. Used with permission.

References

Badham, R. J. 1984. The sociology of industrial and post-industrial societies. *Current Sociology* 32(1): 1–141.

Badham, R. J. 1986. *Theories of Industrial Society*. London: Croom Helm.

Bahro, R. 1984. *From Red to Green*. London: Verso.

Beck, U. 1986. *Risikogesellschaft. Auf dem Weg in eine andere Moderne (Risk Society. On the Way Toward Another Modernity)*. Frankfurt am Main: Suhrkamp.

Bell, D. 1976. *The Coming of Post-Industrial Society*. Harmondsworth, UK: Penguin.

Bijker, W. E., T. P. Hughes, and T. J. Pinch. 1987. *The Social Construction of Technological Systems. New Directions in the Sociology and History of Technology*. Cambridge, MA: MIT Press.

Bookchin, M. 1980. *Toward an Ecological Society*. Quebec: Black Rose Books.

Buttel, F. H. 1986. Sociology and the environment: The winding road toward human ecology. *International Social Science Journal* 38(3):337–356.

Buttel, F. H. 1987. New directions in environmental sociology. *Annual Review of Sociology* 13:465–488.

Catton, W. R., and R. E. Dunlap. 1978. Environmental sociology: A new paradigm. *The American Sociologist* 13:41–49.

Catton, W. R., and R. E. Dunlap. 1979. Environmental sociology. *Annual Review of Sociology* 5:243–273.

Catton, W. R., and R. E. Dunlap. 1980. A new sociological paradigm for post-exuberant sociology. *American Behavioral Scientist* 24(1):14–47.

Cramer, J. 1988. *De Groene Golf. Geschiedenis en Toekomst van de Milieubeweging (The Green Wave. The History and Future of the Dutch Environmental Movement)*. Utrecht, The Netherlands: Arkel.

Dietz, T., and T. R. Burns. 1992. Human agency in evolutionary theory. In *Agency in Social Theory*, ed. B. Witrock. London: Sage.

Dietz, T., and R. S. Frey. 1992. Risk, technology, and society. In *Handbook of Environmental Sociology*, eds. R. E. Dunlap and W. Michelson. Westport, CT: Greenwood Press.

Ester, P., and P. Leroy. 1984. Sociologie en het milieuvraagstuk: Agendapunten voor sociaal wetenschappelijk milieuonderzoek (Sociology and the environmental question: Items on agenda for social science environmental research). Annu. Conf. of the Dutch Sociological and Anthropological Society (NSAV), Amsterdam.

Frankel, B. 1987. *The Post-Industrial Utopians*. Cambridge: Polity Press.

Giddens, A. 1984. *The Constitution of Society*. Cambridge: Polity Press.

Giddens, A. 1985. *The Nation-State and Violence*. Cambridge: Polity Press.

Giddens, A. 1990. *The Consequences of Modernity*. Cambridge: Polity Press.

Giddens, A. 1991. *Modernity and Self-Identity*. Cambridge: Polity Press.

Habermas, J. 1981. *Theorie des Kommunikativen Handels (Theory of Communicative Action)*. Frankfurt am Main: Suhrkamp.

Huber, J. 1982. *Die Verlorene Unschuld der Ökologie. Neue Technologien und Superindustrielle Entwicklung (The Lost Innocence of Ecology. New Technologies and Superindustrial Development)*. Frankfurt am Main: Fisher.

Huber, J. 1985. *Die Regenbogengesellschaft. Ökologie und Sozialpolitik (The Rainbow Society. Ecology and Social Policy)*. Frankfurt am Main: Fisher.

Huber, J. 1989a. *Technikbilder. Weltanschauliche Weichenstellungen der Technik- und*

Umweltpolitik (Technology Images. Propositions on Technology and Environmental Politics). Opladen: Westdeutscher.

Huber, J. 1989b. Eine sozialwissenschaftliche Interpretation der Humanökologie (A social science interpretation of human ecology). In *Grundlager Präventiver Umweltpolitik*, ed. B. Glaeser, pp. 57–75. Opladen: Westdeutscher.

Huber, J. 1991. Ecologische modernisering: Weg van schaarste, soberheid en bureaucratie? (Ecological modernization: beyond scarcity, and bureaucracy?) In *Technologie en Milieubeheer (Technology and Environmental Policy)*, ed. A. P. J. Mol. Den Haag, The Netherlands: SDU.

Hughes, T. P. 1986. The seamless web. Science, technology, etcetera, etcetera. . . . *Social Studies of Science* 16:281–292.

Humphrey, C. R., and F. H. Buttel. 1982. *Environment, Energy, and Society.* Belmont, CA: Wadsworth.

Immler, H. 1989. *Vom Wert der Natur. Zur ökologischen Reform von Wirtschaft und Gesellschaft. Natur in de ökonomischen Theorie Teil 3 (On the Value of Nature. Toward an Ecological Reform of Economy and Society. Nature in Economic Theory).* Opladen: Westdeutscher Verlag.

Inglehart, R. 1987. *Changing Values and the Rise of Environmentalism in Western Societies.* Berlin: International Institute for Environment and Society (IIUG).

Jänicke, M. 1988. Ökologische Modernisierung. Optionen und restriktionen präventiver Umweltpolitik (Ecological modernization. Options and restrictions of preventive environmental policy). In *Präventive Umweltpolitik (Preventive Environmental Policy)*, ed. U. E. Simonis. Frankfurt am Main: Verlag.

Jänicke, M. 1989. Structural change and environmental impact. *Environmental Monitoring and Assessment* 12(2):99–114.

Lash, S., and J. Urry. 1987. *The End of Organized Capitalism.* Oxford: Polity Press.

Leiss, W. 1974. *The Domination of Nature.* Boston: Beacon Press.

Lemaire, T. 1970. *Filosofie van het Landscape (Philosophy of the Landscape).* Baarn: Ambo.

Lowe, P. D., and W. Rüdig. 1986. Political ecology and the social sciences. The state of the art. *British Journal of Political Science* 16:513–550.

Ministry of the Environment. 1989. *National environmental policy plan.* Den Haag: SDU.

Mol, A. P. J. 1990. De maatschappijtheorie van Joseph Huber, een korte introductie (The social theory of Joseph Huber, a short introduction). In *Ecologische Modernisering, Milieubeweging en Maatschappijkritiek (Ecological Modernization, the Environmental Movement, and Critique of Modern Society)*, ed. A. P. J. Mol. Wageningen: Department of Sociology, Wageningen University.

Mol, A. P. J., and G. Spaargaren. 1991. Introductie: Technologie, milieubeheer en maatschappelijke verandering (Introduction: Technology, environmental policy, and social change). In *Technologie en Milieubeheer (Technology and Environmental Policy)*, ed. A. P. J. Mol. Den Haag: SDU.

Mumford, L. 1964. *The Myth of the Machine: Volumes 1 and 2.* New York: Harcourt Brace Jovanovich.

Musil, J. 1990. *Possibilities in formulating new approaches to social ecology.* Paper presented at the XIIth World Congress of Sociology, Madrid, July.

Nelissen, N. J. M. 1970. *Sociale Ecologie (Social Ecology).* Nijmegen, The Netherlands: Catholic University Nijmegen.

Nelissen, N. J. M. 1979b. Aanzetten tot een sociologische theorie over het milieuvraagstuk (First steps toward a sociological theory about the environmental question). In *Sociale Aspecten van het Milieuvraagstuk (Social Aspects of the Environmental Question)*, ed. P. Ester, pp. 5–20. Assen: Van Gorcum.

Nijkamp, P., and L. Reijnders. 1989. Ecologische modernisering een politieke nood-zaak voor de jaren negentig (Ecological modernization: A political necessity for the nineties). *Natuur en Milieu (Nature and Environment)* 2:4–7.

Schnaiberg, A. 1980. *The Environment: From Surplus to Scarcity*. Oxford: Oxford University Press.

Schöne, S. 1987. Ontwikkeling van doelen en strategie van de energiebeweging: Ecologische modernisering (Development of goals and strategy of the Dutch energy movement: Ecological modernization). In *Symposiumverslag Eenheid en Verscheidenheid in Natuur- en Milieudoelstellingen (Proceedings of the Symposium on Unity and Differences in Goals for Nature and Environment)*. Utrecht, The Netherlands: Department of Social Biology, University of Utrecht.

Simonis, U. E. 1989. Ecological modernization of industrial society: Three strategic elements. *International Social Science Journal* 121:347–361.

Spaargaren, G. 1987. Environment and society: Environmental sociology in The Netherlands. *The Netherlands' Journal of Sociology* 23(1):54–72.

Spaargaren, G., and A. P. J. Mol. 1989. Epiloog. Internationale milieusamenwerking en de toekomst (Epilogue. International environmental cooperation and the future). In *International Milieubeleid (International Environmental Policy)*, ed. G. Spaargaren. Den Haag: SDU.

Spaargaren, G., and A. P. J. Mol. 1991. Ecologie, technologie en sociale verandering: Naar een ecologisch meer rationele vorm van produktie en consumptie (Ecology, technology, and social change: Toward an ecologically more rational organization of production and consumption). In *Technologie en Milieubeheer (Technology and Environmental Policy)*, ed. A. P. J. Mol. Den Haag: SDU.

Tellegen, E. 1983. *Milieubeweging (The Environmental Movement)*. Utrecht/Antwerpen: Het Spectrum.

Timberlake, I. 1989. The role of scientific knowledge in drawing up the Brundtland Report. In *International Resource Management*, eds. S. Andresen and W. Ostreng, pp. 117–123. London/New York: Belhaven Press.

Ullrich, O. 1979. Weltniveau: In der Sackgasse der Industriegesellschaft (The Dead End of Industrial Society). Berlin: Rotbuch Verlag.

Vereniging Milieudefensie. 1991. *Zicht op een Beter Milieu (For a Better Environment)*. Amsterdam: Author.

World Commission on Environment and Development. 1987. *Our Common Future*. Oxford: Oxford University Press.

Zimmerman, K., V. Hartje, and A. Ryll. 1990. *Ökologische Modernisierung der Produktion—Strukturen und Trends (Ecological Modernization of Production—Structures and Trends)*. Berlin: Sigma.

Chapter 6

Maarten A. Hajer

ECOLOGICAL MODERNISATION AS CULTURAL POLITICS

IN HIS CELEBRATED STUDY of the US conservation movement around the turn of the century, Samuel Hays (1979) describes how the popular moral crusade for conservation of American wilderness paved the way for a group of experts that, under the veil of working for conservation, advanced their own particular programme. These 'apostles of efficiency' did not share the somewhat sentimental attitude towards wilderness that was typical of the predominantly urban movement for conservation. Above all, they were interested in applying new techniques of efficient resource management in introducing new forestry practices or in constructing and experimenting with the latest hydroelectric dams. For the American urbanites wilderness had a deeply symbolic meaning. Trees and mighty rivers were the icons of the alleged moral superiority of nature that stood in sharp contrast to the bitter reality of a rapidly industrialising society. For experts like Gifford Pinchot and his colleagues, in contrast, wilderness was a nuisance and nature was a resource: trees were merely crops and rivers were to be tamed and tapped. For the urbanites nature had to be preserved; for the experts nature had to be developed.

The story is instructive in several respects. Firstly, it shows that 'our' ecological 'problematique' most certainly is not new. The negative effects of industrialisation for nature have been thematised time and again over the last 150 years. Yet characteristically the public outcry focuses on specific 'emblems': issues of great symbolic potential that dominate environmental discourse. Examples of emblems are deforestation in the mid-nineteenth century, wilderness conservation (USA) and countryside protection (UK) at the turn of the century, soil erosion in the 1930s, urban smog in the 1950s, proliferation of chemicals in the early 1960s, resource depletion in the early 1970s, nuclear power in the late 1970s, acid rain in the early 1980s, followed by a set of global ecological issues like ozone depletion or the 'greenhouse effect' that dominate our consciousness right now. Given this sequence of issues it is better to refrain from speaking of today's predicament in terms of 'our ecological crisis' (which suggests it is time and space specific) and to speak of the *ecological dilemma* of industrial society instead.

Secondly, if we accept the thesis that environmental discourse is organised around changing emblems, we should investigate the repercussions of these subsequent orientations of the debate. After all, emblems mobilise bias in and out of environmental politics. They can be seen as specific discursive constructions or 'story lines' that dominate the perception of the nature of the ecological dilemma at a specific moment in time. Here the framing of the problem also governs the debate on necessary changes. In the case of the US conservation issue the prevailing story line framed the environmental threat as a case of 'big companies' that tried to destroy the American wilderness and rob 'the American people' of something that was constitutive of its national identity. This then paved the way for the state-controlled technocrats who established 'national parks', and seized control over rivers and pastures in the name of the common good. Hays's reinterpretation of the history of the conservation movement illuminates the often disregarded fact that technocrats subsequently used their brief to implement a comprehensive scheme of 'scientific resource management' in which wildlife and nature were largely made subordinate to their concern about achieving optimal yields, thus directly going against the original intentions of the popular movement. The word 'conservation' remained central, yet its institutional meaning changed radically. Ergo: ecological discourse is not about the environment alone. Indeed, the key question is about which social projects are furthered under the flag of environmental protection.

Thirdly, the story of the US conservation movement illustrates the complex nature of what is so often easily labelled the 'environmental movement'. Here the term 'movement' leads astray. Hays's narrative is in fact about not so much a movement as a bizarre coalition that comprised at least two rather distinct tendencies: a popular tendency that was morally motivated, and a technocratic tendency organised around a relatively confined group of experts, administrators and politicians. The important thing is that both had their own understanding of what the problem 'really' was and what sort of interventions could or should be considered as solutions. Nevertheless, together they constituted the social force behind the changes that were made. Hence, instead of speaking of a movement, we would be better to think in terms of 'coalitions'. And, as the above indicates, these coalitions are not necessarily based on shared interests, let alone shared goals, but much more on shared concepts and terms. We therefore call them 'discourse-coalitions' (see Hajer, 1995).

Fourthly, in environmental debates we can often identify implicit ideas about the appropriate role and relationship of nature, technology and society that structure implicit future scenarios. Hays sees a dialectical relation between the public outcry over the destruction of the American wilderness and the implicit critique of industrial society. Nature symbolised the unspoiled, the uncorrupted or the harmonious which was the mirror-image of the everyday reality of Chicago, Detroit or Baltimore at that time. The popular movement wanted to save nature from the effects of industrialisation but did not address the practices of industrial society head on, focusing instead on the effects on nature. In the end it thus paved the way for a programme that focused on the application of new technologies and scientific

management techniques to 'conserve nature'. Here the concern about the immoral-
ity of society was matched by a renewed appeal to forms of techno-scientific
management that were very similar to those industrialistic practices that had
motivated the moral outcry in the first place.

This chapter investigates some similar dynamics in contemporary environmental politics. It argues that environmental politics is now dominated by a discourse
that might be labelled ecological modernisation. It presents an outline of this
policy- or regulation-orientated programme and gives a brief account of its history.
Subsequently the chapter presents three ideal-typical interpretations of what ecological modernisation is about. In the fourth section this chapter then discusses
the social dynamics of ecological modernisation. Extrapolating from the developments in some countries where ecological modernisation is now put into action,
it tries to grasp the socio-political tendencies in the environmental domain in the
years to come.

What is meant by ecological modernisation?

Ecological modernisation is a discourse that started to dominate environmental
politics from about 1984 onwards.[1] Behind the text we can distinguish a complex
social project. At its centre stands the politico-administrative response to the latest
manifestation of the ecological dilemma. Global ecological threats such as ozone
layer depletion and global warming are met by a regulatory approach that starts
from the assumption that economic growth and the resolution of ecological problems can, in principle, be reconciled. In this sense, it constitutes a break with the
past. In the 1970s environmental discourse comprised a wide spectrum of – often
antagonistic – views. On one side there was a radical environmentalist tendency
that thought that the 'ecological crisis' could be remedied only through radical
social change. Its paradigmatic example was nuclear power. On the other side of the
spectrum was a very pragmatic legal-administrative response. The 'Departments
for the Environment', erected all over the Western world in the early 1970s,
worked on the basis that pollution *as such* was not the problem; the real issue was to
guarantee a certain environmental quality. Its paradigmatic example was the end-
of-pipe solution. Where ecological damage was proven and shown to be socially
unacceptable, 'pollution ceilings' were introduced and scrubbers and filters were
installed as the appropriate solution. Moderate NGOs or liberal politicians would
subsequently quarrel about the definition of the height of ceilings and whether
'enough' was being done, but they shared with the state the conviction that
ecological needs set clear limits to economic growth.

Ecological modernisation stands for a political project that breaks with both
tendencies. On the one hand it recognises the structural character of the environmental problematic, while on the other ecological modernisation differs essentially
from a radical green perspective. Radical greens or deep ecologists will argue that
the 'ecological crisis' cannot be overcome unless society breaks away from industrial modernity. They might maintain that what is needed is a new 'place-bound'
society with a high degree of self-sufficiency. This stands in contrast to ecological
modernisation which starts from the conviction that the ecological crisis can be

overcome by technical and procedural innovation. What is more, it makes the 'ecological deficiency' of industrial society into the driving force for a new round of industrial innovation. As before, society has to modernise itself out of the crisis. Remedying environmental damage is seen as a 'positive sum game': environmental damage is not an impediment for growth; quite the contrary, it is the new impetus for growth. In ecomodernist discourse environmental pollution is framed as a matter of inefficiency, and producing 'clean technologies' (clean cars, waste incinerators, new combustion processes) and 'environmentally sound' technical systems (traffic management, road pricing, cyclical product management, etc.), it is argued, will stimulate innovation in the methods of industrial production and distribution. In this sense ecological modernisation is orientated precisely towards those forces that Schumpeter once identified as producing the 'fundamental impulse that sets and keeps the capitalist engine in motion' (Schumpeter, 1961: 83).

The paradigmatic examples of ecological modernisation are Japan's response to its notorious air pollution problem in the 1970s, the 'pollution prevention pays' schemes introduced by the American company 3M, and the U-turn made by the German government after the discovery of acid rain or *Waldsterben* in the early 1980s. Ecological modernisation started to emerge in Western countries and international organisations around 1980. Around 1984 it was generally recognised as a promising policy alternative, and with the global endorsement of the Brundtland report *Our Common Future* and the general acceptance of Agenda 21 at the United Nations Conference of Environment and Development held at Rio de Janeiro in June 1992 this approach can now be said to be dominant in political debates on ecological affairs.

Making sense of ecological modernisation

How should we interpret ecological modernisation? Is it just rhetoric, 'greenspeak' devoid of any relationship with the 'material' reality of ongoing pollution and ecological destruction? Here we have to differentiate. The empirical evidence of the developments in environmental policymaking and product-innovation in Germany and Japan, the experience of the Dutch 'environmental policy planning' approach, or the emergence of Clinton's and Gore's 'win–win' strategy in environmental and conservation issues (see for example Cockburn, 1993), shows that the least we can say is that ecological modernisation has produced a real change in *thinking* about nature and society and in the *conceptualisation* of environmental problems in the circles of government and industry. This is what I call the condition of discourse structuration. One of the core ideas of ecological modernisation, 'integrating ecological concerns into the first conceptualisation of products and policies', was an abstract notion in the early 1980s but is by now a reality in many industrial practices. Especially in OECD countries, ecomodernist concepts and story lines can now be seen to act as powerful structuring principles of administration and industrial decision-making from the global down to the local levels.[2] It has produced a new ethics, since straightforward exploitation of nature (without giving thought to the ecological consequences) is, more than ever before, seen as illegitimate.

Yet one should also assess the extent to which the discourse has produced non-discursive social effects (the condition of discourse institutionalisation). Here one has to define a way to assess social change. There seems to be a consensus that in terms of classical indicators (such as energy consumption, pollution levels) one cannot come to a straightforward conclusion. There have been marked successes in some realms (say, curbing SO_2 emissions), but mostly they have been cancelled out by other developments (such as rising NO_x levels). Likewise, where energy consumption has gone down one may legitimately wonder whether these changes are the result of the new discourse or whether the 'achievements' should be attributed to some other processes (such as economic restructuring). Hence in terms of *ecological* indicators it is difficult to come to an assessment.

The question that we focus on here is the sort of *social* change that ecological modernisation has produced, a question that is neglected only too often in social-scientific research on environmental matters. Is ecological modernisation 'mercantilism with a green twist'? Has it led to a new form of 'state-managerialism'? Does ecological modernisation produce a break with previous discourses on technology and nature, or is it precisely the extension of the established technology-led social project? Or should the 'ecological question' be understood as the successor of the 'social question', and ecological modernisation as the new manifestation of progressive politics in the era of the 'risk society'?

My approach here is to first sketch three different interpretations of ecological modernisation. They are ideal-typical interpretations in the sense that one will not find them in real life in this pure form. Almost inevitably all three of them draw on certain social-scientific notions. In this respect it is important to see them merely as heuristic devices that should help us define the challenges that the social dominance of the discourse of ecological modernisation produces. Each ideal-type has its own structuring principles, its own historical narrative, its own definition of what the problem 'really' is and its own preferred socio-political arrangements.

Ecological modernisation as institutional learning

The most widespread reading of the developments in environmental discourse interprets the course of events as a process of institutional learning and societal convergence. The structuring principle of the institutional learning interpretation of ecological modernisation is that nature is 'out of control'. The historical account is framed around the sudden recognition of nature's fragility and the subsequent quasi-religious wish to 'return' to a balanced relationship with nature. Retrospectively, *Limits to Growth* here appears as the historical starting point. *Limits to Growth* first argued that we cannot endlessly exploit nature. Of course, the report was based on false premises but now we can see that the Club of Rome had a point: we should take nature seriously. Global environmental problems like global warming or the diminishing ozone layer call for decisive political interventions.

Typically, the political conflict is also seen as a learning process. 'We owe the greens something', it is argued. The dyed-in-the-wool radicals of the 1970s had a point but failed to get it through. This was partly due to the rather unqualified nature of their *Totalkritik*. The new consensus on ecological modernisation is here attributed to a process of maturation of the environmental movement: after a radical phase the issue was taken off the streets and the movement became institutionalised as so many social movements before it. With the adoption of the discourse of ecological modernisation its protagonists now speak the proper language and have been integrated in the advisory boards where they fulfil a 'tremendously important' role showing how we can design new institutional forms to come to terms with environmental problems. Likewise, the new consensus around ecological modernisation has made it possible that the arguments of individual scientists that found themselves shouting in the dark during the 1970s are now channelled into the policymaking process.

The central assumption of this paradigm is that the dominant institutions indeed *can* learn and that their learning can produce meaningful change. Following that postulate the ecological crisis comes to be seen as a primarily *conceptual* problem. Essentially, environmental degradation is seen as an 'externality' problem, and 'integration' is the conceptual solution: as economists we have too long regarded nature as a 'sink' or as a free good; as (national) politicians we have not paid enough attention to the repercussions of collective action and have failed to devise the political arrangements that could deal with 'our' global crisis. Likewise, scientists have for too long sought to understand nature in a reductionist way; what we need now is an integrated perspective. Time and again nature was defined 'outside' society, but further degradation can be prevented if we integrate nature into our conceptual apparatus. Fortunately, the sciences provide us with the tools needed: systems theory and the science of ecology show us the way. This understanding of the ecological crisis is supported by what the institutional learning perspective sees as the 'key problems' – collective action problems like the greenhouse effect, acid rain or the diminishing ozone layer. Basically, the institutional learning perspective would define ecological modernisation as the perception of nature as a new and essential subsystem and the integration of ecological rationality as a key variable in social decision-making. The hardware can be kept but the software should be changed.

The preferred socio-political arrangements in essence follow its reading of the history of ecological modernisation and its definition of the problem. Its historical narrative illustrates the strength of pluralist social arrangements. After all, ecological modernisation is the historical product of the critical interplay of opposing social forces. The fact that the World Bank has now adopted an ecomodernist stand is the best example of the radical power of rational argument: even the big institutions will change if arguments are phrased convincingly and correspond with the scientific evidence available. The institutional learning perspective would insist that we have to consider which alterations in scale and organisation we have to make to the existing institutional arrangements to improve 'communication' and make ecological concerns an 'integral part' of their thinking. On the one hand, that implies changes on the level of the firm and the nation state (that is, the stimulation of so-called 'autopoetic' or self-organising effects – for instance mineral or energy accounting in the firm, or the 'greening' of GNP and taxes on the national level). On the other hand, the need for integration finds its political translation in an increased demand for coordination which results in a preference for 'centralisation' of decision making. Global ecological problems have to be brought under political jurisdiction so what we need, above all, are new forms of global management. On the local level, ecological modernisation implies that the scenarios that have been devised to further the ecologisation of society have to be protected against the – inevitable – attacks from particular interest groups. Hence the possibilities for essentially selfish NIMBY (Not In My Back Yard) protests might have to be restricted.

The sciences should in this perspective search for the conceptual apparatus that can facilitate instrumental control over nature and minimise social disturbances. They should, first and foremost, devise a language that makes ecological decision-making possible. What is required is a specific set of social, economic and scientific concepts that make environmental issues calculable and facilitate rational social choice. Hence the natural sciences are called upon to determine 'critical loads' of how much (pollution) nature can take, and should devise 'optimal exploitation rates', as well as come up with ratings of ecological value to assist drawing up of development plans. Engineering sciences are called upon to devise the technological equipment necessary to achieve the necessary ecological quality standards respecting existing social patterns. In a similar vein, the social sciences' role in solving the puzzle of ecologisation is to come up with ideas of how behavioural patterns might be changed and to help understand how 'anti-ecological' cultural patterns might be modified.

In all, in this interpretation ecological modernisation appears as a moderate social project. It assumes that the existing political institutions can internalise ecological concerns or can at least give birth to new supranational forms of management that can deal with the relevant issues. Hence it is a sign of the strength and scope of ecological modernisation that the World Bank has become the manager of the Green Fund – it assumes that national governments can rethink their sectoral policies and that the network of corporatist interest groups can be altered in such a way that it becomes sensitive to ecological matters.

Ecological modernisation as a technocratic project

The interpretation of ecological modernisation as a technocratic project holds that the ecological crisis requires more than social learning by existing social organisations. Its structuring principle is that not nature but technology[3] is out of control. In this context it draws upon the dichotomies dominant–peripheral and material–symbolic. It holds that ecological modernisation is propelled by an elite of policy-makers, experts and scientists that imposes its definition of problems and solutions on the debate. An empirical example is the UN Brundtland Report. It is a 'nice try' but, as the Rio Conference and its aftermath show so dramatically, it falters because it is only able to generate global support by going along with the main institutional interests of national and international elites as expressed by nation states, global managerial organisations like the World Bank or the IMF, and the various industrial interests that hide behind these actors. Hence ecological modernisation is a case of 'real problems' and 'false solutions'. The material–symbolic dichotomy surfaces in the conviction that there is a deeper reality behind all the window dressing. Behind the official 'rhetoric' of ecological modernisation one can discern the silhouette of technocracy in a new disguise that stands in the way of implementing 'real solutions' for what are very 'real problems'.

Its historical narrative starts with the emergence of the 'counterculture' in the 1960s. The environmental movement is essentially seen as an offspring of that broader wave of social criticism. Environmentalism was driven by a critique of the social institutions that produced environmental degradation. Important icons are the culture of consumerism that forces people to live according to the dictum 'I shop, therefore I am', or nuclear power that would not only create a demand for more energy consumption but would also enhance the tendency towards further centralisation of power in society. Certainly, *Limits to Growth* is also seen as a milestone, but not for the environmental movement. Yet what *Limits to Growth* was for social elites and governments, so were *Small is Beautiful* and *Blueprint for Survival* for the counterculture. While in *Limits* the environmentally sound alternative was largely left implicit, the latter publications showed the way towards a truly sustainable society. In this world there would be no place for the 'big government', 'big industry' or 'big science' that, incidently, dominated the Club of Rome that published *Limits to Growth*.

The technocracy critique also has a different interpretation of the significance of the environmental movement of the 1970s. The social movements of the 1970s were not ineffective 'interest groups' that shouted loud but achieved little because they did not know the nitty-gritty of lobbying and strategic action. Quite the contrary, they were embryonic examples of new alternative democratic and ecologically benign social structures and lifestyles (Cohen, 1985). In this movement one found not only 'young well-educated middle class radicals' as some analysts would have it. It also included many scientists who were disturbed by the centralist culture that penetrated the realm of scientific inquiry but who did not necessarily share the radical political agenda of some of the activists. Likewise, it attracted many housewives and farmers who again had their own motives for participating and their own understandings of what the problem 'really' was (see the by now classic article by Offe (1985)). Many a social critic looked upon the environmental movements as one of the forerunners of a new non-technocratic society. Yet rather than learn from these movements 'the state' repressed the alternative movement, either by brute force on the squares of the cities and near nuclear power plants or by 'repressive tolerance', inviting the movements to participate in its judicial inquiries where their political message was inevitably lost in the strait-jackets of legalism (see for example Wynne, 1982). Consequently, in this perspective the emergence of ecological modernisation is not seen as product of the 'maturation' of the social movements. Ecological modernisation is much more the repressive answer to radical environmental discourse than its product. Now the ecological issue has been 'taken up by the apparatuses of power, it becomes a pretext and a means for tightening their grip on daily life and the social environment' (Gorz, 1993: 57). In this context it is not seen as a coincidence that nuclear power is conspicuously absent in the main text of the Brundtland Report. The debate has shifted from this politically explosive issue to global ecological issues that after all suggest that 'our common future' is at stake, thus obliterating old dichotomies and social alternatives.

In this perspective the ecological crisis is basically depicted as an *institutional* problem. The technocracy critique fiercely challenges the assumption that the dominant institutions can learn. How can it be that we try to resolve the ecological crisis drawing on precisely those institutional principles that brought the mess about in the first place: efficiency, technological innovation, techno-scientific management, procedural integration and coordinated management? Who believes that growth can solve the problems caused by growth? Incidently, which institutional learning processes followed the Green Revolution? Is sustainable development not the next 'top-down' model destined to bring evil while in name it intends to do good?

This interpretation would also point at the 'structural' aspects of the problem that are left unaddressed in the discourse of ecological modernisation. What ecological modernisation fails to address are those immanent features of capitalism that make waste, instability and insecurity inherent aspects of capitalist development. Surely ecological modernisation will not end the 'leapfrog' movement of capitalist innovation whereby production equipments, generations of workers or geographical areas are 'written off' periodically? In this perspective the fact that the World Bank is now in charge of the Green Fund is not seen as a sign of strength of the 'ecological turn' but precisely as evidence of the fact that ecological modernisation is really about the further advancement of technocracy. Clearly, eco-software will not save the planet if capitalist expansionism remains the name of the game.

This interpretation opens the black box of society and argues that the emergence of ecological modernisation was to be seen in the context of the increasing domination of humanity by technology, where technology refers not merely to technical 'artefacts' or machines but to social techniques as well. Consequently, the *real* problem at issue is how to stop the 'growth machine'. Only then can one set about trying to remedy the very real environmental problems.

The technocracy critique argues that the sciences have in fact to a large degree been incorporated in this technocratic project. The institutional history of the discipline of systems ecology is used as a case in point (see Kwa, 1987). As historians of science have shown, it was a paradigm on its way out that suddenly got new institutional momentum during the 1960s as NASA engineers and politicians showed an interest in the science that could be integrated in the context of a wider cybernetic perspective. Likewise, the consequence of the prevailing institutional framework is that engineers develop only those technologies that enhance control over nature and society rather than achieve ecological effects while making society more humane. The social sciences are similarly implicated and are called upon as 'social engineers' who only work to help achieve preconceived policy goals. Alternatively, new institutional arrangements in academia and 'science for policy' should be developed. 'Counterexperts' should be able to illuminate the 'technocratic bias' in the official scientific reports. Likewise, more attention, credit and space should be given to those engineers who have been working on 'soft energy paths' that would show the viability of decentralised alternatives. Finally, the social sciences should not work on puzzle-solving activities like changing individual consumer patterns but on the analysis of the immanent forces that keep the juggernaut running towards the apocalypse, so that it might be possible to steer it, or preferably to stop and dismantle it.

The preferred socio-political arrangements of this technocracy critique are those that can correct the prevailing bias towards hierarchisation and centralisation. Its initiatives to further a more democratic social choice centre on 'civil society' rather than on the state. Social movements and local initiatives need protection and attention. New political institutions that would facilitate this correction are the introduction of 'right-to-know' schemes (in Europe), the widespread use of referendums, and, above all, the decentralisation of decision making and the right to self-determination. Here the differences with the institutional learning perspective come out clearly. The fight to circumvent local NIMBY protests through centralisation and 'increased procedural efficiency', indeed the mere construction of complaints as 'NIMBY' protests, are now seen as illustrations of the tendency to take away democratic rights under the veil of environmental care. Here NIMBY protests are recognised as a building stone for an anti-technocratic coalition. After all, protests that are initially motivated by self-interest often lead to a increased awareness of the ecological problem-atique. Hence NIMBYs may become NIABYs (Not In Anybody's Back Yard) (see Schwarz and Thompson, 1990).

In all, ecological modernisation as a technocratic project is a critical interpretation that extends Habermas's argument of modernisation as the 'colonisation of the lifeworld' to include Galtung's concern over the colonisation of the future. With the demise of the radical environmental movement its hope is set on the 'triggering effect' of a few ecological disasters.

Ecological modernisation as cultural politics

The interpretation of ecological modernisation as cultural politics takes the contextualisation of the practices of ecological modernisation one step further. Here one is reminded of Mary Douglas's classic definition of pollution as 'matter out of place'. Her point was that debates on pollution are essentially to be understood as debates on the preferred social order. In the definition of certain aspects of reality as pollution, in defining 'nature', or in defining certain installations as solutions, one seeks to either maintain or change the social order. So the cultural politics perspective asks why certain aspects of reality are now singled out as 'our common problems' and wonders what sort of society is being created in the name of protecting 'nature'.

Ecological modernisation here appears as a set of claims about what the problem 'really' is. The cultural politics approach argues that some of the main political issues are hidden in these discursive constructs and it seeks to illuminate the feeble basis on which the choice for one particular scenario of development is presently made. The structuring principle in this third interpretation is that there is no coherent ecological crisis, but only story lines problematising various aspects of a changing physical and social reality. Ecological modernisation is understood as the routinisation of a new set of story lines (images, causal understandings, priorities, etc.) that provides the cognitive maps and incentives for social action. In so doing ecological modernisation 'freezes' or excludes some aspects of reality while manipulating others. Of course, reductions are inevitable for any effort to create meaningful political action in a complex society. The point is that one should be aware that this coherence is necessarily an artificial one and that the creation of discursive realities are in fact moments at which cultural politics is being made. Whether or not the actors *themselves* are aware of this is not the point. Implicitly, metaphors, categorisations, or definition of solutions always structure reality, making certain framings of reality seem plausible and closing off certain possible future scenarios while making other scenarios 'thinkable'.

To be sure, in this third interpretation there is no implicit assumption of a grand cultural design. Quite the contrary, environmental discourse is made up of 'historically constituted sets of claims' (John Forester) uttered by a variety of actors. Yet in interaction these claims 'somehow' produce new social orders. Foucault speaks in this respect of the 'polymorphous interweaving of correlations'. The analytical aim of this approach is, firstly, to reconstruct the social construction of the reductions, exclusions and choices. Secondly, it tries to come to a historical and cultural understanding of these dispositions. Hence it tries to reconstruct the social forces behind ecomodernist discourse, for instance by studying discourse-coalitions. Subsequently, this approach tries to facilitate the discussion on the various probabilities, possibilities and, above all, on the various alternative scenarios for development that could be constructed.

The historical narrative of this third perspective takes up the themes touched upon at the beginning of this chapter. It emphasises that the ecological problem is not new. It observes that the ecological dilemma of industrial society is almost constantly under discussion, be it through different emblems. What these discussions are about, it argues, are in fact the social relationships between nature, society and technology. For that reason this perspective calls attention to the 'secondary discursive reality' of environmental politics: there is a layer of mediating principles that determines our understanding of ecological problems and implicitly directs our discussion on social change. Hence it would investigate what image of nature, technology and society can be recognised in the 'story lines' that dominated environmental discourse at the time of *Limits to Growth*, or during the confrontation between the state and radical social movements in the 1970s, or in the consensual story lines that dominate ecological modernisation in the 1990s. What is the cultural meaning of the biospheric orientation that is central to present-day environmental discourse? In this respect it argues that ecological modernisation is based on objectivist, physicalist and realist assumptions, all of which are highly arbitrary. Story lines on global warming, biodiversity or the ozone layer suggest the presence of the threat of biological extinction and assert that these problems should be taken as the absolute basis for an ecological modernisation of society. But do these story lines really have the same meaning and implications for all regions? Are they as relevant for the farmers of the Himalaya as for the sunbathers on the coasts of Australia? Should we not understand the global environmental story lines as the product of 'globalised local definitions', as intellectuals from the South have suggested, since the problems have mainly been caused by the North while the solutions apparently have to come mainly from the South (Shiva, 1993)?

Rather than suggesting that there is an unequivocal (set of) ecological problem(s) the third interpretation would argue that there are only implicit future scenarios. The point here is not to doubt whether environmental change occurs. Neither would this social constructivism lead to a position in which each account is equally true or plausible. The point is primarily anti-objectivist and criticises the uncritical acceptance of certain scientific constructs as the starting point of politics. Bird (1987: 256) summarised this position most succinctly, writing that:

> scientific paradigms are socio-historical constructs – not given by the character of nature, but created out of social experience, cultural values, and political-economic structures. . . the actual objects of inquiry, the

formulation of questions and definitions, and the mythic structures of scientific theories are social constructs. Every aspect of scientific theory and practice expresses socio-political interests, cultural themes and metaphors, personal interactions, and professional negotations for the power to name the world.

Here it departs from the more traditional understandings of social action that are implicit in the previous two interpretations. To suggest that the developments in environmental discourse between the publication of *Limits* in 1972 and Rio in 1992 should be interpreted as a process of social learning does not appreciate the cultural bias of the process nor the contingency of the present definitions. But the technocracy critique has omissions too: its proponents might also reject the naive notion of social learning, but their differentiation between the material and the symbolic indicates that they too work with a naturalist understanding of what the problems 'really' are.

Whereas the previous two interpretations in fact shared a clear idea of the ecological problem, and both had their own idea about a possible remedial strategy (respectively conceptual or institutional change, and more coordination or more decentralisation), the third interpretation holds that there can be no recourse to an 'objective' truth. It suggests that the ecological crisis is first and foremost a discursive reality which is the outcome of intricate social processes. It is aware of the ambivalences of environmental discourse and would, in the first instance, not try to get 'behind' the metaphors of ecological discourse. It would try to encircle them to be able to challenge them scientifically, and to enhance consciousness of the contingency of knowledge about ecological matters. What is more, it would investigate the cultural consequences of prevailing story lines and would seek to find out which social forces propel this ecomodernist discourse coalition. Once the implicit future scenarios have been exposed, they might lead to a more reflective attitude towards certain environmental constructs and perhaps even to the formulation of alternative scenarios, the socio-political consequences of which would present a more attractive, more fair, or more responsible package. Hence the central concern of this third interpretation is with *cognitive reflectivity, argumentation* and *negotiated social choice*.

The role of academia follows from this commitment to choice and open debate. They have to help to open the black boxes of society, technology and nature. The cultural politics perspective would resist the suggestion that nature can be understood and managed by framing it in a new 'ecological' language, as for instance by giving priority to economics and systems ecology, on the basis that a pure language does not exist. Its aim would rather be to pit different languages and knowledges (for example expert knowledge versus lay knowledge) against one another to get to a higher understanding of what ecological problems could be about. Here it would assume that this interplay would lead to the recognition of the wide diversity of perspectives.

A more radical consequence of the cultural politics perspective for science would be that the ecological crisis would, potentially, be put upside down: the debate would no longer be on the protection of nature but would focus on the choice of what sort of nature and society we want. After all, once the deconstruction of, say, the biospheric discourse has exposed its naturalist and realist assumptions, the debate might take a different turn. If people have become aware of the political and economic motivations behind biospheric discourse,[4] and have come to grips with the backgrounds of their own naturalism,[5] they might become intrigued by the 'myriad ways in which we make, unmake, and remake "nature" and "human nature" ' (Bennett, 1993: 256). If technology is no longer seen as inherently problematic but also as a potential force to reconstitute the social relationships between nature, technology and society according to one's own needs and preferences, the debate might lose its simplistic modern anti-modern format and a debate on the re-creation of society might result. The consequence would, of course, be that the debate would no longer necessarily focus on environmental matters: the re-creation of society might often focus more explicitly on the conceptualisation of technologies, on the conditions of application of certain techniques, and on the preferred 'social-isation of nature' (rather than the mere protection of nature as it is – see the next section).

The preferred socio-political arrangements of this third interpretation follow from this analysis and focus on facilitating the discussion of implicit future scen-arios. It should be emphasised that in terms of concrete ideas of how this might be organised this perspective is still searching. The most conventional suggestions follow the tracks of the republican tradition, emphasising the need for explicit choice, defence and argument, for the (re)legitimation and/or rejection of certain interventions. Like the technocracy critics it would like to bring society back in. Here we should locate the idea of a 'Societal Inquiry' which would give citizens the right of initiative – for example to re-examine the policy towards acid rain that failed to bring about results.[6] The suggestion is that this initiative would be qualita-tively different from the new democratic forms that were introduced as part of the technocracy critique. A societal inquiry does not assume a clearly defined subject matter: the point of the exercise is precisely to explore and expose the contradic-tions, the reductions and exclusions and to bring into the discussion the implicit understandings of technology, nature and society as well as the implicit future scenarios. Subsequently the societal inquiry would try to create the basis for focused rhetorics, for defence and argument, for relegitimation and/or rejection, and for the reorientation of political action in the light of social debate. This would also hold true for the procedures of 'symbolic law'. The idea here is that the law should no longer be seen as a conclusive statement of dos and don'ts. It would rather have to be a set of normative arguments the meaning and consequences of which should constantly be rethought in the context of concrete cases. The role of government would thus be one of defending the operationalisation it has given to the normative commitments that were the outcomes of societal debates (although one may wonder whether these suggestions are not again based on a traditional understanding of politics).

A more radical version of the cultural politics perspective would break with the traditional understanding of politics as a centralised process. It would take the very process of the creation of discursive realities as its object. Rather than seek to develop arrangements that allow to 'get behind' the metaphors it would explore how new perspectives on society can be created. The issue would not be to 'free' the natural human identity that now suffocates under the hegemony of techno-logical applications; its aim would rather be to explore the unintended potentials of new technologies to create new identities and facilitate the awareness of affinities between various distinct identities.[7]

In all, ecological modernisation as cultural politics starts off by opening up the three black boxes of society, technology and nature and seeks to illuminate the principal openness of ecological discourse. Indeed, it would go so far as to inquire what the meaning could be of the present *ecologisation* of the risks of modernisation.

The social dynamics of ecological modernisation

In sociological theory the ecological crisis is interpreted as the confrontation of industrial society with its own latent side-effects. Zygmunt Bauman speaks about post-modernity as 'modernity coming of age'. We are, says Bauman, now able to see modernity as a 'project'. We have acquired the ability to reflect on what brought us the unprecedented wealth and we now see the (ecological) risks and dangers that we have created in the process of modernisation (see Bauman, 1991). The theory of reflexive modernisation as proposed by Ulrich Beck suggests that it is the unintentional self-dissolution or self-endangerment which he calls 'reflexivity' which has produced the ecological crisis. Reflexivity here relates to modernity as a social formation that constantly and immanently *undercuts* itself (see Beck et al., 1994; Beck [1996]). He distinguishes this 'reflexivity' from 'reflection', which relates to the knowledge we may have of the social processes taking place. Beck holds that as the modernisation of society unfolds, agents increasingly acquire the ability to reflect on the social conditions of their existence. Yet whether or not the self-endangerment of society leads to reflection remains an empirical question.

The question is of course to what extent this reflection can be shown to be present in the present discourse on ecological modernisation. Furthermore, we should assess what forms of reflexivity we can discern in the project of ecological modernisation. Which processes of deinstitutionalisation and reinstitutionalisation take place in the context of the process of ecological modernisation? Which social projects are furthered under the flag of environmental protection? Actors might now broadly share the concepts and terms of ecological modernisation but which implicit future scenarios can we discern?

It is instructive in this respect to have a look at the regulatory efforts of some of the countries that are internationally seen as the most advanced examples of ecological modernisation, such as Germany, or The Netherlands with its comprehensive policy planning approach.[8] In these Western European countries we now see a broad societal coalition working on the institutionalisation of ecomodernist ideas. By and large this institutionalisation is based on the premises of the institutional learning interpretation of ecological modernisation that was sketched above. Yet the social dynamics of ecological modernisation come out to be not as predictable as that ideal-type might suggest. We can in fact observe at least four distinct lines of development.

There can be no doubt about the fact that the main *direct* effect of ecological modernisation is the *rationalisation of ecology*, through the conceptual and institutional amendment of existing bureaucratic structures and the creation of new ones, be it by the state or by new ecocorporatist associations. There seem to be good grounds to argue that there are certain parallels between the history of the US conservation movement in the Progressive Era and the development of the discourse-coalition of ecological modernisation. What started in the late 1960s and early 1970s as a concern about the lack of care for nature seems, over the years, to have given way to a coalition of forces that produces social effects that are at odds with the original intentions of the environmental movement. This tendency towards rationalisation is well known and is in fact implicitly described in the discussion in the previous section, while its dynamics have been analysed elsewhere (Fischer, 1990; Paehlke and Torgerson, 1990; Hajer, 1995).

One should observe, however, that the attempt to rationalise ecological issues according to the prescriptions of the institutional learning paradigm does *not necessarily* produce a rationalisation of ecology. It may also produce a critical form of cultural politics. Take for instance the effects of the national political commitment to sustainable development on policymakers at the local level. They are confronted with the need to translate 'sustainable development' into new planning procedures, conservation strategies, etc. By directive they are ordered to make sustainable development their new 'cognitive map'. Yet here they are confronted with all sorts of interpretative difficulties. After all, sustainable development is merely a 'story line' that generated its global support precisely because of its ambivalence. So what happens in actual practice? Policymakers are left to themselves when it comes to the operationalisation of the notion of sustainable develpment. Here they have a great freedom. They can either make a few aesthetic alterations but basically continue with business as usual, or they can use sustainable development as a crowbar to break with previous commitments. In that case institutional learning produces cultural politics and opens the possibility for a more broadly defined reflection on the sort of problem the ecological crisis 'really' is.

Another interesting development in this respect are the quarrels that are beginning to erupt over the first ecomodernist practices, most notably the waste recycling programme. Whereas waste-reduction programmes have so far not proved to be successful, waste-recycling schemes are threatened precisely because of their success. This causes similar difficulties in Germany, The Netherlands and Austria. The unexpected quantity of materials (glass, plastics, metals) that consumers manage to bring together for recycling now exceeds the capacity of the recycling facilities. On the one hand this causes a renewed resistance of industrial firms to the waste-recycling schemes. On the other hand it is a case where the pragmatic solutions themselves produce the evidence that waste might require more fundamental changes, thus – potentially – enhancing the reflection on the meaning and consequence of the 'ecological crisis'.

A second tendency we can discern is the *technicisation of ecology*. Perhaps the most significant development is taking place behind the scenes in the leading ecomodernist countries – the striking but little discussed reorientation that has taken place in the strategic planning of big multinational firms such as Siemens, DASA and BMW. They are central in the ecomodernist discourse-coalition. They strive towards a set of clearly defined ecoindustrial innovations, they have a new idea of what the relevant actors are, and they carefully work towards a set of new institutional arrangements. They too can be seen using the threats of potential ecological disaster or climate catastrophes as a crowbar, but this time it is used to fulfil the

promise of 'intelligent' traffic systems, 'smart' highways, 'intelligent' energy savings technologies, renewable energies, and socially engineered behavioural changes.
NGOs like Greenpeace, trade unions and politicians can, albeit for varying reasons,
all be seen to help push ecological modernisation in this direction. Similar developments are well under way in the United States, where the 'Big Three' car
producers work together with firms like IBM, AT&T and the Federal government
on multi-billion dollar plans to create this new 'intelligent' transport system. What
this in fact amounts to is the amendment and extension of existing large technical
systems, a tendency which will prove to be extremely powerful.

From the perspective of the technocracy critique the danger of this technicisation of ecology is that ecological modernisation short-circuits a superficial
understanding of some emblematic ecological problems with a new technological
commitment. Essentially, microelectronic technologies are presented as the solution for the 'juggernaut effect'. The cultural politics perspective observes that the
discussion on alternative future scenarios is thus strangled in a double way: both the
debates on information technology and on the ecological crisis are shifted aside to
make way for 'efficient' and 'positive sum game' solutions.

Other examples of this tendency towards the technicisation of ecology are
not difficult to find. Compared to the 1970s the ecomodernist policy-discourse
has also caused a huge shift in the conceptualisation of environmental problems by
the NGOs. The shift in thinking about strategic solutions is of such an extent that
an NGO like Greenpeace that once started off opposing nuclear tests, and is well
known for its protection of endangered species, could recently announce its backing of the development of a 'green car' as well as its own plan to construct a tunnel
under the Öresund, creating a rail link between Sweden and Denmark. The idea
behind this latter initiative is to provide a readily available solution as an alternative
to the longstanding plans of the Danish and Swedish governments to build a bridge
for car traffic. Similarly, efforts are now being made in Japan to apply the latest
freezing techniques to conserve species that are threatened with extinction, and, as
a final example, German social scientists have found that in the debate on biotechnology ecology is now used as a justification for continued development in this
area. Unlike in the early days, when biotechnology was still proposed as the 'Eighth
day of the Creation', it is now constructed as an 'ecologically benign' technology
because it would require radically less resources (see Lau et al., 1993).

This technicisation of ecology receives its social strength of course not primarily from its beneficial effects in terms of ecological improvements. The technicisation of ecology is the translation of a social and moral issue into a market issue. It is
based on the conviction that ecology is, potentially, a new – and huge – market
which is to be created and, subsequently, to be conquered. Traditionally, sociologists would perhaps argue that this tendency is based on a great faith in the capacity
to control side-effects in advance. Yet it is questionable whether that belief really
was constitutive of this tendency in the first place.

A third line of development that ecological modernisation has produced is the
tendency towards the *ecologisation of the social*. This tendency is not so much part of
ecological modernisation as a response to it. Ecological modernisation is a strategy
that believes in further rationalisation, and in creating and maintaining the society-
wide coalitions to 'fight' environmental degradation. Despite all the critique that

one might advance of the Brundtland process, the fact remains that this was conceived as an essentially social-democratic and Western European project. Brundtland was in fact the third Western European social-democratic leader to head a UN commission on global integration (after the Brandt Commission on development and the Palme Commission on disarmament) (Ekins, 1992: chapter 2). The principle of international solidarity, and the social-democratic belief in modernist productionist solutions, therefore always figured prominently.

Inherent in the positive sum game format is the commitment to conceptualise solutions within the existing social system. That means going along with the further integration of world markets, and trying to *add* the social or, more recently, the ecological dimension. Hence *add* a 'social chapter' to the Maastricht Treaty, *add* an ecological paragraph to GATT, *add* the Green Fund to the World Bank. The contradictions that are inherent in this integrative thinking now clearly also produce their reactive counterparts. And here the alliance with the 'big institutions' and existing discourses of power backfires.

Here I would point to the important role of the ecological issue in what various others have called the new regionalism, the new localism, the new tribalism or the 'politics of place'. Those concerned about environmental degradation can rejoice in great interest from the circles of the German new right, which is pasting together a 'place'-orientated ideology. Here the economic and political uncertainties are interpreted as being the product of globalism. This is then subsequently used to recreate a new national sense of order based on a renewed appreciation of 'place' versus 'space'.

In this discourse the national environment becomes the basis for a new national identity, and care for the environment is put forward as an argument to try to regain an allegedly natural social ecology:

> Ecology opens our eyes and shows that nations are not simply human complexes, based on shared language, attitudes, culture and history. Their evolution and their inexchangable identity is also to be understood as the product of the soil from which these complexes grew, the space in which that happened and with which they are connected.[9]

Environmental problems are here seen as the product of a global political project in which a diffuse array of various 'insensitive' actors including the EC bureaucracy, multinational industries and democratic politicians participate. In this context the 'problem' of asylum seekers is drawn into the ecological debate. On the one hand they do not share concern for the regional environment (they do not 'feel' it); on the other hand it is argued that they are extra mouths to feed and people to house, which will require an increase in industrial production and hence cause more environmental destruction (see Jahn and Wehling, 1991, 1993). As Jahn and Wehling have shown for Germany, this new ecologisation of the social should not be seen as a clearly defined ideology that is confined to the new right. It is much more an 'argumentative formula' that can be found much more widely (see also Eder [1996]). Indeed, it is a formula that can be found in mainstream politics in other countries too.

The fourth line of development is the *socialisation of ecology*. Here the debate on the ecological crisis is simply recognised as being one of the few remaining places where modernity can still be reflected upon. It is in the context of environmental problems that we can discuss the new problems concerning social justice, democracy, responsibility, the preferred relation of man and nature, the role of technology in society, or indeed, what it means to be human. This gives ecological discourse a great political importance. In a way it is completely irrelevant whether emblematic problems like global warming constitute the dangers that some people argue they present. Global warming should simply be seen as one of the few possible issues in the context of which one can now legitimately raise the issue of a 'No' to further growth. Here the philosophical imperative of responsibility can be introduced in centre-stage political decision-making through a plea for a 'No regrets' scenario on global warming targets for low energy consumption.

At the same time, the socialisation of ecology perspective would hold that the 'ecological crisis' is by no means unique. Here it elaborates on the cultural politics perspective. Indeed, there is much to be said for the integration of the debates on new technologies that are generally kept separate. If one would break with the conception that nature should be understood as something 'out there', and with the idea that nature stands for the 'authentic', the pure and the good, one might create the possibility of a more vibrant sort of debate. If technology is no longer simply seen as what threatens 'nature' (and hence society) but is also seen as something which creates, at least potentially, new opportunities and new social arrangements, one might be willing to go along and take one more step and discuss the introduction of *in vitro* fertilisation, experiments on embryos or man-animal transplants as essentially ecological issues. This socialisation perspective reflects on 'nature in the age of its technical reproducibility' as the eloquent German philosopher Gernot Böhme (1992) has put it. Rather than mourn over the end of nature we might start to think about what kind of nature we really want. The struggle to protect nature 'as it is' often brings barbed wire into the prairie while the ever more frequent attempts to bring back 'nature as it was' lead to the most bizarre engineering exercises. So rather than leave modernist commitments as they are, it seeks to save Enlightenment thought from the attempts to interpret the ecological crisis as the basis for a further centralisation of power (as in the prevailing global discourses) and from the attempts to use the ecological crisis as the basis for a new intolerant regionalism.

This challenges the boundaries of what is normally understood to be the subject-matter of ecological discourse. That is precisely the point. Ecology would here become the keyword under which society discusses the issues of 'life politics' (Giddens) in a way that allows for a rethinking of existing social arrangements. The rhetorical strength of the ecological crisis would be used to reflect on the nature of modernity.

Conclusion

It will be clear that the four lines of the development sketched above are by no means equal in strength. The rationalisation and technicisation of ecology are well under way while the popular critique drifts more and more in the direction of the ecologisation of the social. The issue really seems to be how to improve the strength of the fourth perspective. Yet how would one increase the reflective awareness of the possibilities of ecological modernisation? How would we arrive at a position where reflection means more than awareness of the 'ecological crisis' alone? How would we get it to include awareness of the necessary openness of the definition of problems and solutions and, finally, how would we achieve an increased awareness of the fact that ecological modernisation should be based on a debate on the recreation of the relationships between nature, technology and society?

It should be said at the outset that there can be no naive idea about the possibilities of bringing about this discussion on 'the nature we really want'. The sociology of technology literature has given us ample evidence of how large technical systems have their own logic of development. Nevertheless, the critical task now is to devise the discursive stages where these patterns can be discussed and democratically renegotiated. In the context of the prevailing tendencies towards a rationalisation and technicisation of ecology, and in an awareness of the dangers of a ecologisation of the social, one should seriously reflect on the need to reinvent democracy. As Bauman (1991: 276) writes:

> What is left outside the confines of rational discourse is the very issue that stands a chance of making the discourse rational and perhaps even practically effective: the *political* issue of democratic control over technology and expertise, their purposes and their desirable limits – the issue of politics as self-management and collectively made choices.

Exploring the ambivalences of ecological modernisation and trying to come to institutional forms that could accommodate the increase of cognitive reflectivity, argumentation and negotiated social choice seems one of the key issues that could reconstitute the basis and meaning of environmental politics.

Notes

1. For a more elaborate analysis of ecological modernisation, see von Prittwitz, 1993; Hajer 1995.
2. This interpretation is now more widely supported. See for example Weale, 1992; Spaargaren and Mol, 1992; von Prittwitz, 1993; Harvey, 1993; Liefferink et al., 1993; Healey and Shaw, 1993; Teubner, 1994.
3. Whereby technology is conceptualised in the Schelskyan sense of the term – that is, including both technology as artefacts and 'social technologies'. See Schelsky, 1965.
4. See the various contributions in Sachs, 1992, and Sachs, 1993.
5. For this aspect compare the discussions in Harvey, 1993, and in Beck, 1995.
6. For an elaborate discussion of these ideas and others, see Hajer, 1995: chapter 6; also Zillessen et al., 1993.
7. The best examples of this can be found in Haraway, 1991; and Bennett and Chaloupka, 1993.
8. For a discussion of the development in environmental policy in these countries, see Weale, 1992.
9. Professor W. G. Haverbeck, quoted in Maegerle, 1993: 6.

References

Bauman, Zygmunt (1991) *Modernity and Ambivalence*. Cambridge: Polity Press.

Beck, Ulrich (1995) *Ecological Politics in an Age of Risk*. Cambridge: Polity Press.

Beck, Ulrich (1996) 'Risk society and the provident state', in S. Lash, B. Szerszynski and B. Wynne (eds), *Risk, Environment and Modernity. Towards a New Ecology*. London: Sage.

Beck, Ulrich, Giddens, Anthony and Lash, Scott (1994) *Reflexive Modernization: Politics, Tradition and Aesthetics in the Modern Social Order*. Cambridge: Polity Press.

Bennett, Jane (1993) 'Primate visions and alter-tales', in Jane Bennett and William Chaloupka (eds) *In the Nature of Things*, pp. 250–66.

Bennett, Jane and William Chaloupka (eds) (1993) *In the Nature of Things: Language, Politics, and the Environment*. Minneapolis: University of Minnesota Press.

Bird, Elisabeth Ann R. (1987) 'The social construction of nature: Theoretical approaches to the history of environmental problems', *Environmental Review*, 11 (4): 255–64.

Böhme, Gernot (1992) *Natürlich Natur – Die Natur im Zeitalter ihrer technischen Reproduzierbarkeit*. Frankfurt am Main: Suhrkamp.

Cockburn, Alexander (1993) ' "Win/Win" with Bruce Babbitt: The Clinton administration meets the environment', *New Left Review*, 201: 46–59.

Cohen, Jean A. (1985) 'Strategy or identity: New theoretical paradigms and contemporary social movements', *Social Research*, 52 (4): 663–716.

Eder, Klaus (1996) 'The institutionalisation of environmentalism', in S. Lash, B. Szerszynski and B. Wynne (eds), *Risk, Environment and Modernity. Towards a New Ecology*. London: Sage.

Ekins, Paul (1992) *A New World Order: Grassroots Movements for Global Change*. London: Routledge.

Fischer, Frank (1990) *Technocracy and the Politics of Expertise*. London: Sage.

Gorz, André (1993) 'Political ecology: Expertocracy versus self-limitation', *New Left Review*, 202: 55–67.

Hajer, Maarten A. (1995) *The Politics of Environmental Discourse: Ecological Modernisation and the Policy Process*. Oxford: Clarendon Press.

Haraway, Donna (1991) *Simians, Cyborgs, and Women: The Reinvention of Nature*. London: Free Association Books.

Harvey, David (1993) 'The nature of environment: Dialectics of social and environmental change', in Ralph Miliband and Leo Panitch (eds) *Socialist Register 1993*. London: Merlin Press. pp. 1–51.

Hays, Samuel P. (1979 (1959)) *Conservation and the Gospel of Efficiency: The Progressive Conservation Movement, 1890–1920*. New York: Atheneum.

Healey, P. and Shaw, T. (1993) *The Treatment of 'Environment' by Planners: Evolving Concepts and Policies in Development Plans*. Working Paper no. 31, Department of Town and Country Planning, University of Newcastle upon Tyne.

Jahn, Thomas and Wehling, Peter (1991) *Ökologie von Rechts – Nationalismus und Umweltschutz bei der Neuen Rechten und den Republikanern*. Frankfurt am Main: Campus.

Jahn, Thomas and Wehling, Peter (1993) 'Ausweg Öko-Diktatur? Umweltschütz, Demokratie und die Neue Rechte', in *Politische Ökologie*, 34: 2–6.

Kwa, C. (1987) 'Representations of nature mediating between ecology and science policy: The case of the international biological programme', *Social Studies of Science*, 17: 413–42.

Lau, Christoph et al., (1993) *Risikodiskurse-Gesellschaftliche Auseinandersetzungen um die Definition von Risiken und Gefahren*. Report to the German Federal Minister for Research and Technology.

Liefferink, J.D., Lowe, P. and Mol, A.P.J. (eds) (1993) *European Integration and Environmental Policy*. London: Belhaven.

Maegerle, Anton (1993) 'Wie das Thema Umwelt zur Modernisierung des rechtsextremen Denkens missbraucht wird', *Frankfurther Rundschau*, 21 December, p. 6.

Offe, Claus (1985) 'New social movements: Challenging the boundaries of institutional politics', *Social Research*, 52 (4): 817–68.

Paehlke, R. C. and Torgerson, D. (eds) (1990) *Managing Leviathan: Environmental Politics and the Administrative State*. Peterborough (Ontario): Broadview.

Sachs, W. (ed.) (1992) *The Development Dictionary: A Guide to Knowledge as Power*. London: Zed Books.

Sachs, W. (ed.) (1993) *Global Ecology: A New Arena of Political Conflict*. London: Zed Books.

Schelsky, H. (1965) 'Der Mensch in der wissenschaftliche Zivilisation', in *Auf der Suche nach Wirklichkeit*. Düsseldorf/Köln. pp. 439–80.

Schumpeter, J. A. (1961 (1943)) *Capitalism, Socialism and Democracy*. London: George Allen and Unwin.

Schwarz, M. and Thompson, M. (1990) *Divided We Stand: Redefining Politics, Technology and Social Choice*. London: Harvester Wheatsheaf.

Shiva, Vandana (1993) 'The greening of the global reach', in W. Sachs (ed.) *Global Ecology*, pp. 149–56.

Spaargaren, G. and Mol, A.P.J. (1992) 'Sociology, environment, and modernity: ecological modernisation as a theory of social change', in *Society and Natural Resources*, 5: 323–44.

Teubner, Gunther (ed.) (1994) *Ecological Responsibility*. London: Wiley & Sons.

von Prittwitz, V. (ed.) (1993) *Umweltpolitik als Modernisierungsprozess*. Opladen: Leske & Budrich.

Weale, Albert (1992) *The New Politics of Pollution*. Manchester: Manchester University Press.

Wynne, Brian (1982) *Rationality and Ritual: The Windscale Inquiry and Nuclear Decisions in Britain*. Chalfont St Giles: The British Society for the History of Science.

Zillessen, Horst et al. (eds) (1993) *Modernisierung der Demokratie – Internationale Ansätze*. Opladen: Westdeutscher Verlag.

Chapter 7

Peter Christoff

ECOLOGICAL MODERNISATION, ECOLOGICAL MODERNITIES

ECOLOGICAL MODERNISATION IS EMERGING as a fashionable new term to describe recent changes in environmental policy and politics.[1] Its growing popularity derives in part from the suggestive power of its combined appeal to notions of development and modernity, and to ecological critique. Yet competing definitions blur its usefulness as a concept. Does ecological modernisation refer to environmentally sensitive technological change? Does it more broadly define a style of policy discourse which serves either to foster better environmental management or to manage dissent and legitimate ongoing environmental destruction? Does it, instead, denote a new belief system or systemic change? Indeed, can it encompass all of these understandings? In this article, I want to examine current uses of the term in relation to the tensions between modernity and ecology which it evokes, and suggest ways of diminishing its ambiguity.

It is widely acknowledged that since the late 1980s significant changes have occurred in the content and style of environmental policy in most industrialised (particularly OECD) countries. The nature and extent of these changes vary between nations,[2] reflecting their distinctive political, institutional, and cultural features; the national economic importance of the sectors and industries targeted by new regulatory regimes and the extent and intensity of the environmental impact of those industries; the strength of popular environmental concern and of its political representation; the extent to which an implementation deficit (the failure to realise environmental standards and goals) exists and is recognised as a local problem, and the reasons for this deficit; and regionally distinct perceptions of the key international and global ecological problems which mobilised public concern during the 1980s.[3]

Nevertheless, despite local variations, these environmental policy changes have several generalisable features. They have aimed to shift industry beyond reactive 'end-of pipe' approaches towards anticipatory and precautionary solutions which minimise waste and pollution through increasingly efficient resource use (including through recycling). Problem displacement across media (air, water and land) and across space and time has tended to be challenged by a more integrated regulatory approach – as much to achieve greater administrative efficiency and to limit regulatory overload as to address the new environmental problems caused by such displacement. Prescriptive regulatory approaches and 'technological forcing' – applied in the 1970s as the sole or predominant strategy for achieving ongoing improvements in environmental conditions – are more often accompanied or displaced by cooperative and voluntary arrangements between government and industry: increasingly environment protection agencies seek to use industry's existing investment patterns and its capacity and need for technological innovation to facilitate improvement in environmental outcomes. A range of market-based environmental instruments have been deployed in response to the perceived exhaustion of the initial wave of regulatory intervention [*Eckersley, 1995*]. In all, the new environmental policy discourse increasingly emphasises the mutually reinforcing environmental and economic benefits of increased resource efficiency and waste minimisation.

These developments reflect an evolving international discourse in response to commonly perceived environmental problems. However, they also reflect an increasingly sophisticated political response by governments and industry to popular mobilisation around issues such as nuclear power, acid rain, biodiversity preservation, ozone depletion and induced climate change. In other words, the new policy culture and its trends are not always simply or primarily intended to resolve environmental problems. They are also shaped by a contest over political control of the environmental agenda and, separately, over the legitimacy of state regulation (predominantly in the English-speaking OECD countries). In addition, they have been influenced by the growing pressures on nation-states generated by intensified economic globalisation and by changes in the structure and nature of production towards greater flexibility and international integration.

The strengthening of linkages between environmental and economic policies is especially observable in countries such as Germany and the Netherlands – in turn raising questions about the reasons for their exceptionally good environmental performance in contrast to countries such as the United Kingdom and the United States. For instance, in the 1980s and the early 1990s German environmental policy, under pressure from the Greens, has moved rapidly to address its failure to meet targets and standards adopted during the 1970s. The promotion of design criteria enabling comprehensive re-use of materials has been accompanied by regulations requiring that 72 per cent of glass and metals and 64 per cent of paperboard, laminates and plastics be recycled by 1995 [*Moore, 1992*]. Regulations also encourage use of 'waste energy' for heating and power generation. The 1983 Large Combustion Plant Ordinance requires the retrofitting of all major power plants to cut pollutants contributing to acid rain by 90 per cent by 1995. Laws passed in 1989 ban CFC production and use by 1995. Germany has also committed itself to a unilateral reduction in carbon dioxide emissions of between 25 and 30 per cent by the year 2005 and, since 1990, has begun to articulate and implement a package of some 60 measures to enable it to meet this target [*Hatch, 1995*].

These changes have been supported by considerable government assistance. Weale [*1992*] reports that between 1979 and 1985, the German government subsidy for environmental research and development rose from $US 144.3 million to $US 236.4 million, or from 2.1 per cent of R&D to 3.1 per cent (the UK equivalent was 0.8 per cent to 1.1 per cent). This commitment is also institutionally defined: Germany has a separate Federal Ministry for Research and Technology which spends about 200 million DM per annum on research and development of environmental technologies [*Angerer, 1992: 181*]. Since 1985, the level of German public subsidy for environmental research has exceeded that of the United States in absolute terms. Public investment also provides substantial support in the energy conservation fields, including research into energy conservation devices. As a result, within a decade German industry has become a global leader in the development and/or production of solar photo-voltaics, high-efficiency turbines, hydrogen-powered cars, energy-efficient household appliances, and recyclable materials and products.

The economic advantages to countries and companies leading the field in environmental performance improvements have been recognised as considerable. It is estimated that by the year 2000, Japan will be producing some $US12 billion worth of waste incinerators, air-pollution equipment and water treatment devices, and MITI has proposed aid projects aimed at energy development in China, Indonesia and Malaysia, as a means of further tying and strengthening trade connections with these countries [*Gross, 1992*].

The extent of formal policy integration and of the diffusion of environmental principles into the practices of state, economy and society is also of particular interest. Governments in several countries – notably Australia, Canada, the Netherlands and the United Kingdom – have developed national plans for sustainable development, meta-policies aimed at the integration of national environmental and economic activity, and at encouraging greater environmental awareness in civil society. In this, the Dutch National Environmental Policy Plan (NEPP) has been significantly more successful than similar attempts elsewhere. This success is partly due to the highly corporatist nature of Dutch politics and planning, the Dutch state's acceptance of a significant role in facilitating and directing industrial development and environmental protection, and also the timing of the Plan's release in 1989, during a high point of international and national environmental concern. The NEPP has the explicit goal of achieving environmental sustainability in the Netherlands within one generation, by 2010, by recasting policies and practices in key economic sectors, including manufacturing, agriculture and transport, to limit waste production and environmental pollution [*Carley and Christie, 1992; van der Straaten, 1995*]. Despite weaknesses both in its targets and ongoing implementation [*Wintle and Reeve, 1994*], the Plan nevertheless offers a programmatic approach to working towards measurable targets against which the public, government and industry can assess its progress and iteratively adjust the Plan.[4]

Over the past decade, Germany, the Netherlands, Japan and the Scandinavian block appear to have achieved above OECD average improvements across a range of industry-related national environmental indicators, including water quality and air pollutant emissions [*OECD, 1995*]. In these countries there is now evidence of a decoupling of GNP growth from the growth of environmentally harmful effects, indicating increased economic output with decreased energy and materials consumption *per unit* of GNP. However, certain improvements in environmental conditions in the First World have been gained through displacement of high energy consuming and/or polluting industries (for example, metal processing and primary manufacturing) to newly industrialising countries (NICs) and lesser developed countries (LDCs). Meanwhile, underlying increases in total material consumption in both industrialising countries and industrialised countries continue to enhance environmental pressures, suggesting both that the pace of reform is too slow and the root cause of the implementation deficit of the 1970s has not been overcome [WRI, 1994].

The uses of ecological modernisation

Positive aspects of these recent changes have been described by academic observers as evidence of a process of 'ecological modernisation' (EM), although their uses of the term vary considerably in scope and meaning. Specifically, leading exponents of the term in the German and English literature, such as Janicke, Hajer and Weale, use it in their policy analysis, sociological analysis or political theoretical discussion in ways which are occasionally problematic, partly because of a lack of clarity about whether the term is being used descriptively, analytically or normatively. Following discussion of these distinctive uses of EM, I want to propose a typology for ecological modernisation which emphasises the normative dimensions of the term.

Ecological modernisation as technological adjustment

EM has been used narrowly to describe technological developments with environmentally beneficial outcomes – such as chlorine-free bleaching of pulp for paper and more fuel-efficient cars. These are specifically aimed at reducing emissions at source and fostering greater resource efficiency [Simonis, 1988; Janicke, 1988; Zimmermann, Hartje and Ryll, 1990; Janicke, Monch and Binder, 1992].

Janicke, who perhaps first introduced 'ecological modernisation' into the language of policy analysis (1986/1990), for instance, refers to four broadly framed 'environmental political' strategies commonly found in industrial countries [Janicke, 1988]. Two of these strategies are remedial (compensation and environmental restoration; technical pollution control) and two preventative or anticipatory (environmentally friendly technical innovation or 'ecological modernisation'; and structural change). For Janicke, EM is fundamentally a technical cost-minimisation strategy for industry and an alternative to labour-saving investment – a form of 'ecological rationalisation' which will lead simultaneously to greater 'ecological and economic efficiency' [1988: 23]. It is primarily seen as a strategy intended to maintain or improve market competitiveness, in which the environmental benefits of such technological change are incidental rather than a core concern for innovation and implementation. In this sense, such a narrow version of EM does not necessarily reflect any significant and overwhelming changes in corporate, public or political values in relation to desired ecological outcomes. Rather, it is an outcome of capital's cost-minimising responses to new pressures – such as the adoption elsewhere of post-Fordist 'lean' production methods [Best, 1990; Amin, 1994; Wallace,

1995], resource price movements and scarcities (for example, the oil crises of the 1970s); changes in consumer taste; and profit squeezes caused by taxes and regulatory strategies of the state – at a time when automation has reduced industry's capacity to increase labour's productivity. Innovation and implementation may be confined to those areas and types of technical improvements which ensure market competitiveness.[5] Consequently, such technological change may not contribute to lasting environmental improvements when viewed in the context of national or international ecological requirements.

For Janicke, moves towards sustainability depend on broader structural change, the second of his anticipatory strategies, which would lead to profound shifts in production and consumption patterns. These are not merely industrial responses to ecological symptoms (for example, resource shortages) but incorporate precautionary analysis and associated restrictions on action and lead to constrained qualitative economic growth and a decrease in absolute resource use, pollution and environmental degradation [*1988: 15–17; 1992*]. He sees the current period of multiple crises – unemployment, and accumulation, environmental and fiscal crises – which extends into the 1990s as also providing opportunities for the 'creative destruction' of old patterns and forms. The world market involves not only competition between enterprises producing new technologies but also competition between nations with stronger and weaker 'state steering capacity', a competition favouring those capable of breaking with the tendency to protect their old 'smokestack industries' and able to generate a framework for consensual transformation. Janicke's more recent empirical work documents the sites, conditions and (limited) signs of such industrial transformation [*Janicke et al., 1992*].

Janicke and his colleagues fail to identify or address potential political economic contradictions in this narrow vision of an ecological modernisation embedded in larger processes of structural transformation. At what point are the currently developing patterns of unrestricted, globalised production and trade, and the cultural demands for increasingly specialised consumption, challenged? How will the corresponding growth in international markets for new 'lean' technologies and products be restrained to ensure regional and international ecological stability rather than ongoing expansion of total resource use and waste output? And, specifically given Janicke's view [*1992*] on state failure and the limits of state action, what institutions will participate in this enhanced process of regulation? What happens, within this larger scenario, to those countries – the technological laggards – unable to compete or perform economically and ecologically?

If these new clean technologies and products *are* truly ecologically sustainable – leading to a significant absolute decrease in resource use and to effective environmental preservation – what are their ramifications for trade, employment, accumulation and wealth distribution within and between nations, particularly if they are sought according to time frames which are dictated by urgent ecological demands (for example, the potential need to cut Greenhouse gas emissions by up to 60 per cent within the next three decades)? Certainly, given its narrowly industrial focus, such EM would not necessarily serve to diminish total resource consumption or lead to the protection of 'unvalued', non-resource related ecological concerns.

Ecological modernisation as policy discourse

Others, such as Weale [1992; 1993] and Hajer [1995], have employed 'EM' more broadly to define changes in environmental policy discourse. For Hajer, the shift toward EM can be observed in at least six 'realms', namely in environmental policy-making, where anticipatory replace reactive regulatory formulae; in a new 'pro-active' and critical role for science in environmental policy-making; at the micro-economic level, in the shift from the notion that environmental protection increases cost, to the notion that 'pollution prevention pays'; at the macro-economic level, in the reconceptualisation of nature as a public good and resource rather than a free good; in the 'legislative discourse in environmental politics', where changing perceptions of the 'value' of nature mean that the burden of proof now rests with those accused as polluters rather than the damaged party; and the reconsideration of participation in policy-making practices, with the acknowledgement of new actors, 'in particular environmental organisations and to a lesser extent local residents', and the creation of 'new participatory practices' for their inclusion in a move to end the 'sharp antagonistic debate between the state and the environment movement' [Hajer, 1995: 28–9].

Hajer predominantly regards EM as a policy discourse which assumed prominence around the time of the European Community's Third Action Plan for the Environment and, more explicitly still, the 1984 OECD Conference on Environment and Economics. Such EM 'recognises the structural character of the environmental problematique but none the less assumes that existing political, economic and social institutions can internalise care for the environment' [1995: 31]. Hajer sees this discourse as being largely economistic – framing environmental problems in monetary terms, portraying environmental protection as a 'positive sum' game, and following a utilitarian logic. At the core of ecological modernisation is the idea that pollution prevention pays: it is 'essentially an efficiency-oriented approach to the environment' [1995: 101]. In other words, economic growth and the resolution of environmental problems can, in principle, be reconciled [1995: 25–6].[6] 'Ecological modernisation uses the language of business and conceptualises environmental pollution as a matter of inefficiency while operating within the bounds of cost-effectiveness and bureaucratic efficiency' [1995: 31].

Hajer is most effective where he suggests that EM is a discursive strategy useful to governments seeking to manage ecological dissent and to relegitimise their social regulatory role. It permits a critical distancing from the interventionist remedies of the 1970s which, as Hajer believes, 'did not produce satisfactory results', and may serve to legitimate moves to roll back the state and reduce its regulatory capacities in the environmental domain. It also enables governments to promote environmental protection as being economically responsible, thereby resolving the tensions created by previous perceptions that the state was acting against the logic of capital and its own interests (of functional dependency on private economic activity). He suggests that such a strategy explicitly avoids addressing basic social contradictions that other discourses might have introduced.

> Ecological modernisation is basically a modernist and technocratic approach to the environment that suggests that there is a

techno-institutional fix for present problems. Indeed ecological modernisation is based on many of the some institutional principles that were already discussed in the early 1970s: efficiency, technological innovation, techno-scientific management, procedural integration and co-ordinated management. It is also obvious that ecological modernisation as described above does not address the systemic features of capitalism that make the system inherently wasteful and unmanageable [1995: 32].

In other words, EM is not simply a technical answer to the problem of environmental degradation. It can also be seen as a strategy of political accommodation of the radical environmentalist critique of the 1970s, meshing with the deregulatory moves which typify the 1980s, with distinctive affinities with the neo-liberal ideas that dominated governments during this time, supporting their concern for structural industrial reform [1995: 32–3].

Hajer's own views here are unclear. He seems to approve of such political closure yet leaves open the question of whether or not EM might be, in the terms of the critics of Brundtland, a rhetorical ploy to take the wind out of the sails of 'real' environmentalists, one which displaces and marginalises the radical emancipatory aspects of environmental critique [1995: 34]. He is even less clear about whether or not EM 'may not in fact have a more profound meaning as the first step on a bridge that leads to a new sort of sustainable society'. Hajer mainly sees EM as a counter to the 'anti-modern' sentiments he claims are part of the critical discourse of new social movements.

> It is a policy strategy that is based on a fundamental belief in progress and the problem-solving capacity of modern techniques and skills of social engineering. Contrary to the radical environment movement that put the issue on the agenda in the 1970s, environmental degradation is no longer an anomaly of modernity. There is a renewed belief in the possibility of mastery and control, drawing on modernist policy instruments such as expert systems and science [1995: 33].

In this sense too, as it seeks to provide a soothing rhetoric promoting apparent remedial and anticipatory change, such a policy discourse may be profoundly anti-ecological in its outcomes, its narrow economism serving to devalue and work against recognition and protection of non-materialistic views of nature's 'worth'.

Ecological modernisation as belief system

Both Hajer and Weale also use the concept in more radical ways.[7] For Weale [1992], ecological modernisation represents a new belief system that explicitly articulates and organises ideas of ecological emancipation which may remain confused and contradictory in a less self-conscious discourse. It is an ideology based around, but extending beyond, the understanding that environmental protection is a precondition of long-term economic development. Weale's claims for EM as a belief system are important given the role of belief systems in organising and legitimising public policy.

Weale also sees EM as being focused on a reconceptualised relationship between environmental regulation and economic growth. It still includes an emphasis on the achievement of highest possible environmental standards as a means for developing market advantage through the integration of anticipatory mechanisms into the production process, recognition of the actual and anticipated costs of environmental externalities in economic planning, and the economic importance of strengthening consumer preferences for clean, green products [*1992: Ch.3; 1993: 206–9*]. However, 'once the conventional wisdom of the relationship between the environment and economy is challenged, other elements of the implicit belief system [which sees them in opposition] might also begin to unravel'. Regulation may 'no longer seem merely a mechanical matter'. EM thus prefigures systemic change and may, in its more radical forms, generate a broader transformation in social relations, one which leads to the ecologisation of markets and the state.

Under such circumstances, as Weale comments:

> the internalisation of externalities becomes a matter of attitude as well as finance, and a cleavage begins to open up not between business and environmentalists, but between progressive, environmentally aware business on the one hand and short-term profit takers on the other. Moreover, the behaviour of consumers becomes important, so that the role of government policy is not simply to respond to the existing wants and preferences of their citizens, but also to provide support and encouragement for forms of environmentally aware behaviour and discouragement for behaviour that threatens or damages the environment. Once this view has taken root, the line from mechanical to moral reform has been crossed. The challenge of ecological modernisation extends therefore beyond the economic point that a sound environment is a necessary condition for long-term prosperity and it comes to embrace changes in the relationship between the state, its citizens and private corporations, as well as in the relationship between states [*Weale, 1992: 31–2*].

However, in *The New Politics of Pollution*, Weale does not develop his views on the transformations of both civil society and the state necessary to achieve ecological sustainability. What limits are posed by the state's dependent relationship to private sector economic activities and how can these be overcome given the increasing political and economic vulnerability of individual nation-states to global flows of capital? To what extent would transformations of civil society and in public spheres, rather than institutional changes to the state, drive the process of ecologisation?[8]

Ecological modernisation – some unresolved issues

It is possible to illuminate problems and issues left unaddressed or unresolved by the foregoing uses of EM by asking a series of interrelated questions. In different situations (different policy forums and different countries), quite different styles of EM may prevail – ones which can be judged normatively to tend towards either weak or strong outcomes on a range of issues, such as ecological protection and democratic participation. In this sense, these questions hint at the limitations of those forms of ecological modernisation which tend toward the first rather than the second of what might seem, initially, opposing poles.

EM – economistic or ecological?

In each of the uses of EM described above, the environment is reduced to a series of concerns about resource inputs, waste and pollutant emissions. As cultural needs and non-anthropocentric values (such as are reflected in the Western interest in the preservation of wilderness) cannot be reduced to monetary terms, they tend to be marginalised or excluded from consideration. This is clearly the case for EM narrowly defined as technical innovation. But it is equally true of those interpretations of EM which see the state shaping corporate activity and markets to (re)incorporate environmental externalities into the costs of production. As has been noted, such versions of EM may remain consistent with the traditional imperatives of capital. Leading industries may welcome uniformly applied environmental regulatory regimes, as the redefinition of the boundaries of acceptable economic behaviour may represent a rationalisation of their markets which makes the rules of production and competition more certain or amenable to their entry or dominance. But ideologically and practically, such ecological modernisation may simply put a green gloss on industrial development in much the same way that the term 'sustainable development' has been co-opted – to suggest that industrial activity and resource use should be allowed as long as environmental side-effects are minimised.

Given this dominant emphasis on increasing the environmental efficiency of industrial development and resource exploitation, such EM remains only superficially or weakly *ecological*. Consideration of the integrity of ecosystems, and the cumulative impacts of industrialisation upon these, is limited and peripheral. In this sense, the entire literature is somewhat Eurocentric, deeply marked by the experience of local debates over the politics of acid rain and other outputs, rather than conflicts over biodiversity preservation. Although current uses of EM may be well adapted to describing positive environmental outcomes in certain industrialised First World countries where a version of ecological sustainability may be created in the wasteland of a vastly depleted biological world, it may be positively dangerous if taken prescriptively by those nations where the conservation of biodiversity is a more fundamental concern or opportunity and/or which depend on primary resource exploitation to fund their traditional forms of economic growth, for example as in the case in Australia, Brazil and South Africa.

EM – national or international?

The uses of EM described earlier also remain narrowly focused on changes *within* industrialised nation-states. They are therefore unable to integrate an understanding of the transformative impact of economic globalisation on environmental relations. They offer only a diminished recognition of the increasingly internationalised flows of material resources, manufactured components and goods, information and waste; of the influence of multinational corporations on investment, national industrial development and the regulatory capacities of the nation-state; and of international deregulatory developments (such as GATT) and environmental treaties (such as the Montreal Protocol). Paradoxically, each of these facets of globalisation shapes yet distorts, provokes yet inhibits and undermines the emergence of strong forms of ecological modernisation at national and regional levels. Because of their nation-statist focus, these uses of EM – including those raising broader ideological and systemic concerns – still tend to remain focussed on localised end-of-cycle issues rather than encompassing the globally integrated nature of resource extraction and manufacturing in relation to domestic consumption, overvaluing local achievements and environmental impacts while undervaluing geographically distant factors.[9]

Consider, for instance, the internationally dispersed resources and environmental impacts associated with producing and running a nation's car fleet, or with producing and using paper. Or the extent to which heavy transformative industries such as smelting, ship-building or car manufacturing have relocated to the NICs. In other words, although pollution levels and primary consumption of energy and other primary resources may have fallen in relation to GNP in certain European economies as these have become increasingly post-industrial, their *per capita* material consumption continues to grow – with environmental impacts now displaced 'overseas'.

Given this presently predominantly nation-statist view of EM, discussion of the emergent international institutions for environmental regulation and protection, and of environmental trends, remains under-developed where it occurs in the EM literature.[10] The literature fails to recognise that, because old forms of industrial activity with their associated environmental problems are being displaced to developing nations or regions and the transition to alternative technologies is occurring too slowly to prevent major global environmental problems (such as climate change), we may instead be moving towards what Everett [*1992*] has called the 'breakdown of technological escape routes' as the ecological pressure for change increases beyond the reasonable capacities for social and industrial reform.

EM – hegemonic progress or multiple possibilities?

In different ways, the types of EM described earlier are also presented as contributing to, or constituting, a unilinear path to ecological modernity. Consequently they seem to be offering a revival of mainstream development theory and of notions of uneven development and under-development, positing EM as the next necessary or even triumphant stage of an evolutionary process of industrial transformation – a stage dependent upon the hegemony of Western science, technology and consumer culture and propagated by leading Western(ised) countries. Such views of ecological modernisation may be validly subjected to the criticisms which were levelled against development theory two decades ago.

Theorists who implicitly or explicitly rely upon a simplistic division between traditional and modern societies ignore the potential for a multiplicity of paths to ecological sustainability which may rest in the diversity of non-Western cultures. They seem to suggest that all countries may undertake the great leap forward over the phase of 'dirty' industrialisation into the fully ecologically modern condition. But if ecologically modernising countries can not quite manage the great leap, then at least such nations will eventually be able to employ restorative technologies, salves and panaceas developed elsewhere to undo the ecological devastation resulting from the stage of aggressive industrialisation to which developing countries aspire or are now subject. In other words, when developing countries reach the levels of affluence which gives them the economic capacity to afford ecological modernisation, they will be able to turn to consider and repair the path of devastation which has bought them this luxury. In fact, such views of EM continue to offer a world divided by renewed or strengthened core-periphery relationships between industrialised and industrialising countries, with world markets and the motors of progress dominated by leading industrial state(s).

The problems here are most obvious when we consider the potentially disastrous local and global ecological (and social) costs of China, India, Indonesia or Brazil pursuing such a path to ecological modernity, or the perpetual mendicant status of small nation-states such as the Solomons or Vanuatu, and also much of the African continent, which would continue to be trapped in a condition of ongoing cultural and technological dependency.

EM – technocratic or democratic?

There are also tensions between what different theorists describe as the preconditions for systemic or structural ecological modernisation. Some stress the transformative impact of environmental awareness on civil society and the public sphere, and on the institutions and practices of government and industry. They emphasise the ways in which citizenship and democratic participation in planning may serve to socialise and ecologise the market and guide and limit industrial production. Others however favour a less emancipatory technocratic, neo-corporatist version of EM – one which may prove primarily a rhetorical device seeking to manage radical dissent and secure the legitimacy of existing policy while delivering limited, economically acceptable environmental improvements.

For example, Weale [1992], interpreting developments in Germany and the Netherlands, suggests that the systemic realisation of EM requires a proactive, interventionist state supporting a well-developed culture of environmental policy innovation and offering significant public investment and subsidies as a means of achieving economic advantage and environmental outcomes. Such state activity would entail an integrated regulatory environment and strong structural and process cross-linkages between different parts of the state and the development of a synoptic and reflexive use of environmental information in policy formation and implementation. In addition, this transformation is enhanced by, or indeed *depends upon*, increased public participation in political decision making, including green political pressure through both the environment movement and parliamentary politics (including green parties); and increased public influence over industry behaviour through green consumer action and the activities of environmental pressure groups and organisations.

By contrast, Hajer [1995] and Andersen [1993] seem to believe that a more technocratic relationship between state and civil society will lead more effectively to systemic EM. Andersen [1993], who is specifically concerned to define preconditions for EM through comparative analysis of national environmental performance, describes a country's *capacity* for ecological modernisation as depending upon its 'achieved level of institutional and technological problem-solving capabilities, which are critical to achieving effective environmental protection and transformation to more sustainable structures of production'. He argues, in concert with Janicke and others, that there is also a close relationship between consensus-seeking policy styles and high levels of environmental protection in industrialised countries.

Andersen suggests that four basic variables govern the capacity for such ecological modernisation. First there is *economic performance*. This is the capacity of countries to pay for environmental protection – a factor which appears directly linked to the intensity of environmental pollution. Secondly, there is *consensus ability*, which Andersen believes is best developed in countries with neo-corporatist structures, which are seen as having consensus-seeking decision-making styles that are more amenable to dealing with new ideas and interests.[11] Thirdly, there is *innovative capability*, which he describes as the capacity of both the state and the market institutions to remain open to new interests and innovations in the judicial and political system, the media and the economic system. Fourthly, there is *strategic proficiency* – the capacity to institutionalise environmental policy across sectors. He identifies federal states, which face potential fragmentation and delay in implementation, and states evidencing strong compartmentalisation of the bureaucracy – concomitant with weak environment departments or agencies – as potentially suffering problems in this area [*Andersen, 1993: 3*]. Andersen suggests that the presence of these variables or attributes seems to contribute to, or at least correlate with, the success of 'leading' European countries – such as Germany and the Scandinavian bloc – in achieving exceptional improvements in environmental conditions. But how do they apply to the NICs and LDCs? Again, what relationship between state and civil society and what forms of democratic participation are required, especially given the international dimensions of environmental problems, to enable the radical social and economic changes which ecological sustainability may require?

Insofar as EM focuses on the state and industry in terms which are narrowly technocratic and instrumental rather than on social processes in ways which are broadly integrative, communicative and deliberative, it is less likely to lead to the sorts of embedded cultural transformation which could sustain substantial reductions in material consumption levels, significant and rapid structural transformations in industrialised countries, and major international redistributions of wealth and technological capacity. In general, the extent and nature of institutional changes required to enable the full recognition of a discursive and participatory environmental politics (and to accommodate the transboundary and intertemporal nature of environmental risks and impacts) have not yet been explored in the EM literature.[12]

From weak to strong ecological modernisations

Given the range of uses to which the term has been put, can ecological modernisation be stabilised as a concept? One can differentiate between sometimes conflicting versions of ecological modernisation. These versions do not each merely describe some aspect of a more encompassing process of ecological modernisation but offer quite different real world outcomes. Some of these uses may be labelled narrow or broad, depending on the extent to which they are technological or systemic in scope or focus. More importantly, and reflecting the above discussion, it is possible to emphasise the normative dimensions of different versions of EM. I would suggest that different interpretations of what constitutes EM lie along a continuum from weak (one is tempted to write, false) to strong, according to their likely efficacy in promoting enduring ecologically sustainable transformations and outcomes across a range of issues and institutions (Table 7.1). The political contest between the environmental movement on the one hand and governments and industry on the other is predominantly over which of these types of EM should predominate.

It is essential to note that weak and strong features of EM are not always mutually exclusive binary opposites. Some features of weak or narrow EM are necessary but not sufficient preconditions for an enduring ecologically sustainable outcome. Clearly, one does not abandon technological change, economic instruments or instrumental reason in favour of institutional and systemic change or communicative rationality. In many cases – although not all (for instance, technocratic or neo-corporatist versus deliberative and open democratic systems) – aspects of narrow or weak EM need to be subsumed into and guided by the normative dimensions of strong EM.

Ecological modernisation, ecological modernities

Finally, what of the tensions and contradictions embedded in the term at the point where ecological critique challenges the ways in which simple industrial modernisation defines its relationship to Nature?[13] Perhaps the most radical use of ecological modernisation would involve its deployment against industrial modernisation itself. To understand what this might mean, it is necessary to unpack the ecological and modernising components of ecological modernisation and look at their interaction more closely.[14]

Modernity has broken or swept aside traditional forms of order and certainty: as Marx put it, 'all that is solid melts into air'. Its dynamism may be attributed to the separation of time and space into a realm that is detached from immediate experience; the disembedding of social systems; and the reflexive ordering and reordering of social relations [*Giddens, 1990: 16–17*]. By fostering relations with absent others, locationally distant from any given situation of face-to-face

Table 7.1 Types of ecological modernisation

Weak EM	Strong EM
Economistic	Ecological
Technological (narrow)	Institutional/systemic (broad)
Instrumental	Communicative
Technocratic/neo-corporatist/closed	Deliberative democratic/open
National	International
Unitary (hegemonic)	Diversifying

interaction, the process of modernisation increasingly overlays place (the immediate experience of location) with space (the abstract experience of location, into which the immediate experience of location is then fitted). It also replaces local time, based on an immediate experience of the rhythms of Nature and the requirements of one's immediate community, with abstract time – now most powerfully represented by the international acceptance of a standard differentiation of global time zones. The extreme dynamism of modernity, Giddens argues, also depends on the establishment of *disembedding* social institutions, ones which create or support the creation of abstract social relations and their associated organisations. The emergence of symbolic tokens (such as money), and expert systems, represents an essential feature of modernity and contributes centrally to this process of disembedding, which is then reflected, for instance, in increasingly global discourses as in science or law.

In addition, 'systems of technical accomplishment or professional expertise organise large areas of the material and social environments in which we live today' [*Giddens, 1990: 27*]. Crucial among these systems are those of scientific understanding and technological performance. We live in and are dependent upon – that is to say, *trust* in – them for our survival and legitimate functioning. These abstract expert systems constitute not only bodies of knowledge, but also lived forms of social relationship. We enter them whenever we turn on a light switch, fly in an aeroplane, go to the dentist or answer the telephone: their complexity and functioning are taken for granted in a socially learned, relatively unquestioned and automatic way.[15] Both types of disembedding mechanism (symbolic tokens and abstract expert systems) provide guarantees of expectations across time and space and stretch social systems as a result. They also promote a new awareness of risk, which is the product of the human-created technological and social characteristics of modernity. Risk and trust intertwine. Modernity is also notable for the development of new capacities for the reflexive appropriation of knowledge, in part born of the capacity to transmit and review which comes with the development of the book and other forms of recorded information. All knowledge and beliefs become available for scrutiny. Certainty is displaced.

These features together contribute to the emergence of the institutional dimensions of modernity. Giddens identifies four such dimensions, which are interrelated and interdependent: capital accumulation; industrialism; surveillance; and military power [*1990: 59*]. Of particular interest in relation to ecological modernisation are the first three dimensions. Industrialism seeks the transformation of nature into created or recreated (managed) environments through processes of standardisation, rationalisation and reduction. The imperatives of capital accumulation are such that the hunt for markets and resources encourages the commodification of all aspects of individual cultures and nature which remain vulnerable. The capacity for surveillance – in the broadest sense, in terms of the apparatuses of consolidated administration, monitoring and registering of social and environmental facts – has a bearing on the development of modern forms of reflexive environmental management.

The last point to note here is the globalising tendency of modernity. Its global reach is partly a result of the imperial and colonising tendencies of capital accumulation. However, as modern technologies of transport, information transfer and communication continue actively to redefine social relations, linking and integrating distant parts of the globe both as markets for commodities and abstract social networks, the notions of centre and periphery begin to blur. The flow of individuals, commodities, cultures and pollution across territorial borders is also leading to a practical redefinition of one of the other major institutions of modernity, the nation-state. Modernity brings with it the globalisation of risk by altering the scope, the type and the range of human-created environmental risks which individuals now face, and also the globalisation of the perception of these risks.

How then can we characterise the relationship between 'modernisation' and 'the ecological'? Modernity is fraught with tensions and generates its own new contradictions: nowhere is this more evident than in relation to the environment. The emergent ecological critique of untrammelled industrialism – sharpened politically in recent years by perceptions of ecological crisis and of the need for precautionary consideration of the potential consequences of development – has a paradoxical relationship to the constitutive features of modernity described above. Itself a product of simple modernity, ecological critique both depends upon and resists the modern reorganisation of time and space. It makes radically problematic and contradictory the industrialising imperative which lies at the heart of modernisation by redefining the cultural and ecological limits to the instrumental domination of nature.

Ecologically re-embedding space and time

The birth of nature has been accompanied and shaped by the simultaneous creation of technological forces which lead to what McKibben [1990] has called 'the end of Nature' through human interference with previously autonomous natural systems worldwide (through induced climate change, the global transport of pollutants, and so on). Driven by the imperatives of capital accumulation, industrialism – shaped by the alliance of science and technology – continues to transform nature in ways unimaginable to earlier generations. It does so deliberately, for instance, by introducing alien plant and animal species to new continents, by flooding valleys and levelling mountains, and by creating new relations of physical and economic dependency between the country and the city and between the First and the Third Worlds. Colonial conquests have often also led to the unintended extermination of indigenous plants and animals through destruction of their native habitat or by introduced predators. Demand for export earnings and the development of industrialised monocultural agriculture and forestry have produced a wave of extinction that continues to roll across North and South America, Australia, Asia and Africa. Yet the ecological transformation threatened by the combined impacts of induced global warming and bio-technology is more comprehensive still.

The creation of a secular, scientific understanding of nature – indeed, the development of ecology as a scientific discipline – and the triumph of technological domination over natural cycles and ecological processes, depend upon and arise from the separation of time and space discussed earlier. The 'discovery' of 'remote' regions and exotic species enabled the scientific conceptualisation of natural systems, at the same time as it involved the commodification of those environments, and the imperial domination or appropriation, of non-Western knowledge of natural systems and species.

Yet the 'so-called economy of Nature, the interrelationship of all organisms' for which Haeckel [1870] coined the term 'ecology' in 1869, depends on cycles and time scales which are generally alien to those of the political and economic institutions of industrial society. An ecological critique that recognises and respects the importance of the cycles upon which the biological world depends, and which seeks to re-embed our relationship to nature in a local place and to redefine the relationship in ecological temporal terms, often stands in opposition to the transcendent, abstracting features of modernity (and its industrial manifestations) while still to some extent depending upon its conceptual frameworks. In other words, such ecological critique tries to undo the stretching of time and space, as it seeks to limit certain aspects of industrial modernisation in order to preserve the ecological integrity of natural systems, or to preserve cultural understandings and institutions, which are locally embedded and resistant to the resource-utilitarianism of all forms of industrial modernity.[16]

Let me give several examples of such critique, each relating to primary resource use. Harvesting temperate forests on an *ecologically* sustainable cycle which also respects the needs of dependent species may involve 300- or 400-year rotations and it is probably ecologically out of the question for complex, fragile rainforest systems. As such time spans may be commercially unviable, protection of non-resource species involves fundamentally rethinking how or whether one can use these forests. The international ban on whaling, based on moral considerations, defies the industrial/instrumental belief in the potential for whales to be harvested sustainably. For similar reasons, environmentalists now campaign to preserve wilderness areas and for animal rights in general. Consider also the conflicts between the environment movement and industry over the *representations* of place versus space, struggles with profound material consequences for particular rivers, forests and wetlands. One may recognise a fierce contest over images and counter-images of sites in the diametrically opposed terms used by developers and environmentalists to represent contested terrain through the Australian media as either significant places or exploitable spaces with few specifically valuable attributes – 'the last free river' versus 'a leech-ridden ditch' for the Franklin River in Tasmania; or 'magnificent Northern wilderness' and 'sacred ground' versus 'clapped-out buffalo country' for Coronation Hill, a proposed mine site in the Northern Territory. As Harvey [1993: 23] notes, in such cases 'the cultural politics of places, the political economy of their development, and the accumulation of a sense of social power in place frequently fuse in indistinguishable ways'.

Each of these examples stresses ways in which a non-economistic ecological critique is in tension with or begins to break away from industrial modernity even as it still uses the media, scientific information and political institutions which are products of late modernity as its tools. In other words, ecological critique is not (as Hajer would suggest) naively anti-modern, seeking to dismantle all abstract relations established through modernisation. Rather, as the product of modernity and something which continues to depend upon modernity's processes for its development,[17] it aims to discipline and restrain – to put bindings, brakes and shackles [Offe, 1992] – on the over-determining effects of globalised productive systems. Beck

[*1992: 23*] writes of how the life of a blade of grass in the Bavarian forest ultimately comes to depend on the making and keeping of international agreements: ecological critique requires abstract systems and their institutions, and ecological considerations to coexist through the prioritisation of the latter.

The strongest or most radically *ecological* notion of ecological modernisation will often stand in opposition to industrial modernity's predominantly instrumental relationship to nature as exploitable resource. Recognition that over-production – the use of material resources beyond regional and global ecological capacities – must cease because of the threat of imminent ecological collapse, does not allow for the self-serving gradualism of the weak forms of ecological modernisation discussed earlier.

Reflexivity and risk, anxiety and mistrust

Giddens has noted that the forms of reflexivity involved in the continual generating of systematic self-knowledge do not stabilise the relation between expert knowledge and knowledge applied in lay actions. This is as true for scientific and technical systems as for sociology (to which Giddens was referring), for these systems also remain always at least one step away from the understanding which would control their impacts and are always on their way to creating new problems. Increased ecological awareness encourages recognition of the limits to our scientific comprehension of the physical world and therefore of the limits to our capacity to know and technically manipulate them. Our crude understanding of the interplay of biological systems and global climate is a good case in point.

However industrial modernisation has largely vanquished the traditional cultural forces which might control the abstract scientific appropriation of the environment or, more importantly, the impulse to transform nature (whether through biotechnology or *in vitro* fertilisation). At the same time, it has produced a new category of socio-technological failures – such as Chernobyl and the ozone hole – which is unprecedented in its spatial and temporal reach, respecting no territorial borders and potentially affecting future generations. The global extension of the catastrophic capacity of industrial modernisation is accompanied by the means to broadcast information about such disasters to populations which previously trusted expert systems and now become aware of these new risks, with their implications at the personal (cancer and death) and global (destruction of life on earth) levels of existence.

As Giddens, Beck and others point out, the resultant disenchantment with science and technological change, and the popular appreciation of the new risks they produce, has led to a transformation of public perceptions of progress. Optimistic notions of progress, based on uncritical belief in the benefits of the scientific and industrial appropriation of Nature, have now collapsed into anxiety and mistrust. Giddens argues that this new phase, which other theorists call postmodernity, is but an extension of modernity in process. 'We have not moved beyond modernity but are living precisely through a phase of its radicalisation', a period in which Progress is 'emptied out by continuous change' [*Giddens, 1990: 51*]. However there is good reason to suggest that this underplays the discontinuities associated with cultural disenchantment with progress and particularly its handmaidens, science and technology. This disenchantment constitutes a radical departure from simple modernity and signals the establishment of a new, more anxious phase of reflexive modernisation [*Beck, Giddens and Lash, 1994*].

Those interpretations of ecological modernisation which are still embedded in notions of industrial progress, albeit more cautious but still bearing an evolutionary sense of technological adaptation through reflexivity, do not address the extent of this corrosion of trust in simple industrial modernity.[18] They accept that modernisation has become *more* reflexive, but only in the narrow and instrumental sense of improving environmental efficiency, rather than in the broad and *reflective* manner of ecological critique which fundamentally questions the trajectories of industrial modernity. By contrast, strong ecological modernisation therefore also points to the potential for developing a range of alternative ecological modernities, distinguished by their diversity of local cultural and environmental conditions although still linked through their common recognition of human and environmental rights and a critical or reflexive relationship to certain common technologies, institutional forms and communicative practices which support the realisation of ecological rationality and values ahead of narrower instrumental forms.

In conclusion, the concept of ecological modernisation has been deployed in a range of ways – as a description of narrow, technological reforms, as a term for policy analysis, in reference to a new ideological constellation and in reference to deeply embedded and ecologically self-conscious forms of cultural transformation – and bearing quite different values. As a result, there is a danger that the term may serve to legitimise the continuing instrumental domination and destruction of the environment, and the promotion of less democratic forms of government, foregrounding modernity's industrial and technocratic discourses over its more recent, resistant and critical ecological components. Consequently there is need to identify the normative dimensions of these uses as either weak or strong, depending on whether or not such ecological modernisation is part of the problem or part of the solution for the ecological crisis.

Notes

1. For instance, see Simonis [*1988*], Janicke *et al.* [*1992*], and Zimmermann *et al.* [*1990*], Weale [*1992*], Hajer [*1995*] and Andersen [*1993*].
2. For example, see Vogel [*1986, 1990*], Vogel and Kun [*1987*], Knoepful and Weidner [*1990*], Vig and Kraft [*1990*], Yaeger [*1991*], Weale [*1992*], Feigenbaum *et al.* [*1993*] and Wintle and Reeve [*1994*].
3. For instance, while transboundary problems such as acid rain and fallout from Chernobyl shaped environmental politics, policies and institutions in Western Europe, they were of little consequence in Japan, and irrelevant to 'frontier states' such as Australia, where preservationist conflicts over the impacts of primary resource extraction – agriculture, forestry and mining – on relatively pristine environments predominated.
4. The NEPP has already undergone two four-yearly reviews, as required by legislation.
5. Consider the enormous gap between the technical capacities – which have been available for decades – to produce durable, safe, energy efficient and largely recyclable cars, and the actuality to date.

6. Hajer's use of the term varies in its elasticity. As he extends his view of 'ecological modernisation' to the point that it seems all-embracing in its cultural inclusivity, it becomes hard to see what bounds EM as a discourse – a theoretical–methodological problems common to Foucauldian approaches to policy analysis. Perhaps it is therefore better to instead regard ecological modernisation as a meta-discourse or deep cultural tendency. It then becomes possible to read EM back into the nineteenth century movement for resource conservation and forward into the growing reflexivity of science and technology. That Hajer might want to add conservative and neo-liberal opposition to state regulation to his list of the signs of EM indicates some of the problems with his own ill-defined discursive approach to 'locating' EM.

7. Towards the end of *The Politics of Environmental Discourse*, Hajer briefly touches upon an ideal form of EM which he calls 'reflexive ecological modernisation. This represents a cultural tendency rather than merely a policy discourse and stands in opposition to 'techno-corporatist ecological modernisation' in its emphasis on democratic and discursive practices.

8. See Christoff [*1996*].

9. For instance, see Weale [*1992: 78–9*].

10. Both Weale and Hajer comment on the role which international forums, such as the OECD, have played in fostering EM as a policy discourse. For instance, Weale claims 'the main bodies responsible for developing the ideology of EM were international organisations, who sought to use the new policy discourse as a way to secure acceptance of common, or at least harmonised, environmental policies, the closest example being the EC' [*Weale, 1993: 209*]. He also discusses the evolution of new international environmental regimes but does not integrate this discussion into his exploration of ecological modernisation [*Weale, 1992: Ch.7*].

11. Similarly, Jahn [*1993: 30*] notes that data seem to indicate that neo-corporatism has a positive impact on *environmental performance* and on *anti-productionist politics*. He comments that 'it seems reasonable to argue that the impact of neo-corporatist arrangements on both dependent variables is dependent upon the influence of new social movements and associated green and left-libertarian parties on established politics'. However, importantly, he also notes Offe's observation that the cost of corporatist arrangements is the marginalisation of non-organised interests, which is antithetical to the democratic principles of new social movements and of Green politics.

12. Weale [*1992: 31*] and Hajer [*1995: 280 ff*] suggest but do not explore such alternatives.

13. By 'ecological critique', I mean both the emergent scientific understanding of ecological needs which has evolved out of the biological and physical sciences, and the normative and non-instrumental (re)valuation of Nature (including its spiritual and aesthetic aspects as these manifest in concern for preservation of species and ecosystems, wilderness and visual landscape values). Both are elements increasingly dominant, motivating features of the environment movement in the late 20th Century.

14. This section draws heavily upon Giddens' elegant long essay, *The Consequences of Modernity*.

15. Of course these newer forms of trust in abstract systems may be related to pre-modern forms of trust in cultural explanatory frameworks (religion, myth and

so on.) They coexist with, and interact in, the process of identity formation with more direct forms which are essential in face-to-face communities and intimate social relations (as in families).

16.   This is not to argue for a return to essentialised and romantic, exclusionary and parochial notions of 'place', such as have been central to the campaigns of certain environmental communitarians. While arguing for the need to recognise and preserve the specific place-bounded nature of ecological relations, it is also important to note the ways in which cultural notions of identity and place have been irrevocably transformed by modernity as, globally, face-to-face communities are now infused by the information attributes and other requirements of abstract exchange.

17.   Its abstract knowledge of nature remains based on research and investigation, on the international transmission of new scientific information among scientists, environmental managers and environmentalists, as well as (potentially) upon the recovery and reauthorisation of aspects of local indigenous knowledge.

18.   See Janicke [1988] and Hajer [1995: 33].

References

Amin, A. (1994), *Post-Fordism: A Reader*, Oxford: Blackwell.

Andersen, M.S. (1993), *Ecological Modernisation: Between Policy Styles and Policy Instruments – the Case of Water Pollution Control*, Paper delivered at 1993 ECPR Conference, Leiden. University of Aarhus, Denmark: Centre for Social Science Environmental Research.

Angerer, G. (1992), 'Innovative Technologies for a Sustainable Development', in Dietz, Simonis and van der Straaten [1992: 181–90].

Beck, U. (1992), *Risk Society: Towards a New Modernity*, London/New York: Sage Publications.

Beck, U., Giddens, A. and S. Lash (1994), *Reflexive Modernisation: Politics, Tradition and Aesthetics in the Modern Social Order*, Oxford: Polity Press.

Best, M.H. (1990), *The New Competition: Institutions of Industrial Restructuring*, Oxford: Polity Press.

Carley, M. and I. Christie (1992), *Managing Sustainable Development*, London: Earthscan.

Christoff, P. (1996), 'Ecological citizenship and Ecologically Guided Democracy' in B. Doherty and M. de Geus (eds.) *Democracy and Green Political Thought*, London: Routledge, pp.151–69.

Dietz, F.J., Simonis, U.E. and J. van der Straaten (eds.), (1992), *Sustainability and Environmental Policy*, Berlin: Bohn Verlag.

Eckersley, R. (ed.) (1995), *Markets, the State and the Environment: Towards Integration*, Basingstoke: Macmillan.

Everett, M. (1992), 'Environmental Movements and Sustainable Economic Systems', in Dietz, Simonis and van der Straaten [1992: 114–28].

Feigenbaum, H., Samuels, R. and R. Kent Weaver (1993), 'Innovation, Coordination an Implementation in Energy Policy', in R. Kent Weaver and B.A. Rockman (eds.) *Do Institutions Matter? Government Capabilities in the United States and Abroad*, Washington, DC: Brookings Institute, pp. 42–109.

Giddens, A. (1990), *The Consequences of Modernity*, Oxford: Polity Press.

Gross, N. (1992), The Green Giant? It May Be Japan, *Business Week*, 24 Feb.

Haeckel, E. (1870), *Natürliche Schöpfungsgeschichte*, Berlin: Reimer.

Hajer, M.A. (1995), *The Politics of Environmental Discourse: Ecological Modernisation and the Policy Process*, Oxford: Oxford University Press.

Harvey, D. (1993), 'From Space to Place and Back Again: Reflections on the Condition of Postmodernity', in J. Bird, B. Curtis, T. Putnam, G. Robertson and L. Tickner (eds.) *Mapping the Futures: Local Cultures, Global Changes*, London: Routledge, pp.3–29.

Hatch, M.T. (1995), The Politics of Global Warming in Germany, *Environmental Politics*, Vol.4, No.3, pp.415–40.

Jahn, D. (1993), 'Environmentalism and the Impact of Green Parties in Advanced Capitalist Societies', Paper delivered at 1993 ECPR Conference, Leiden.

Janicke, M. (1986/1990), *State Failure: The Impotence of Politics in Industrial Society*, Oxford: Polity Press. (First published as *Staatsversagen: Die Ohnmacht der Politik in der Industriellgesellschaft*, Munchen: Piper GmbH.)

Janicke, M. (1988), 'Ökologische Modernisierung: Optionen und Restriktionen präventiver Umweltpolitik', in Simonis [1988: 13–26].

Janicke, M. (1990), 'Erfolgsbedingungen von Umweltpolitik im Internationalen Vergleich', *Zeitschrift für Umweltpolitik*, Nr.3, pp.213–311.

Janicke, M., Monch, H., Ranneberg, T. and U. Simonis (1988), *Economic Structure and Environmental Impact: Empirical Evidence on Thirty-One Countries in East and West*, Berlin: Wissenschaftszentrum Berlin für Sozialforschung gGmbH (WZB).

Janicke, M., Monch, H. and M. Binder (eds.) (1992), *Umweltentlastung durch industriell Struckturwandel?: Eine Explorative Studie uber 32 Industrieländer (1970 bis 1990)*, Edition Sigma, Rainer Bohn Verlag, Berlin.

Knoepful, P. and Weidner, H. 'Implementing Air Quality Programs in Europe', *Policy Studies Journal* 11, pp.103–15.

McKibben, B. (1990), *The End of Nature*, London: Penguin.

Moore, C.A. (1992), Down Germany's Road to Sustainability, *International Wildlife*, Sept./Oct., pp.24–8.

Offe, C. (1992), 'Bindings, Shackles and Brakes: On Self-Limitation Strategies', in A. Honneth, C. Offe and A. Wellmer (eds.) *Cultural-Political Interventions in the Unfinished Project of the Enlightenment*, Cambridge, MA: The MIT Press, pp.63–94.

Organisation for Economic Co-operation and Development (OECD) (1995), *OECD Environmental Data Compendium 1995*, Paris: OECD.

Simonis, U.E. (ed.) (1988), *Präventative Umweltpolitik*, Frankfurt/New York: Campus Verlag.

van der Straaten, J. (1992), The Dutch National Environmental Policy Plan: To Choose or to Lose, *Environmental Politics*, Vol.1, No.1, pp.45–71.

Vig, N.J. and M.E. Kraft (eds.) (1990), *Environmental Policy in the 1990s: Towards a New Agenda*, Washington, DC: Congressional Quarterly Press.

Vogel, D. (1986), *National Styles of Regulation: Environmental Policy in Great Britain and the United States*, Ithaca, NY: Cornell University Press.

Vogel, D. (1990), 'Environmental Policy in Europe and Japan', in Vig, and Kraft [*1990*].

Vogel, D. and Kun, V. (1987) 'The Comparative Study of Environmental Policy', in D. Meinolf, H.N. Weiler and A.B. Antal (eds.), *Comparative Policy Research: Learning from Experience*, New York: St Martin's Press.

Wallace, D. (1995), *Environmental Policy and Industrial Innovation*, London: Earthscan.

Weale, A. (1992), *The New Politics of Pollution*, Manchester: Manchester University Press.

Weale, A. (1993), 'Ecological modernisation and the Integration of European Environmental Policy', in J.D. Lieffernink, P.D. Lowe and A.P.J. Mol (eds.) *European
Integration and Environmental Policy*, London: Bellhaven, pp.198–216.

Wintle, M. and R. Reeve (eds.) (1994), *Rhetoric and Reality in Environmental Policy: The
Case of the Netherlands in Comparison with Britain*, London: Avebury Studies in
Green Research.

World Commission on Environment and Development (WCED) (1987), *Our Common
Future*, Oxford: Oxford University Press.

World Resources Institute (WRI) (1994), *World Resources 1994–95 – A Guide to the
Global* Environment, Oxford: Oxford University Press.

Yaeger, P. C. (1991), *The Limits of the Law*, Cambridge: Cambridge University Press.

Zimmermann, K, Hartje, V.J. and A. Ryll (1990), *Okologische Modernisierung der Produktion: Struktur und Trends*, Edition Sigma, Berlin: Rainer Bohn Verlag.

Chapter 8

F.H. Buttel

ECOLOGICAL MODERNIZATION AS SOCIAL THEORY

Introduction

THE RISE OF ECOLOGICAL modernization as a perspective in environmental social science[1] has been as meteoric as it has been unexpected. Ecological modernization was unknown to virtually all North American environmental social scientists half a dozen years ago, save for a small handful of comparative politics specialists who were familiar with Jänicke's (1990) work on "state failure", or environmental studies scholars who had read Simonis' (1989) paper in the *International Social Science Journal*. Now ecological modernization has come to be regarded as being on a virtual par with some of the most long-standing and influential ideas and perspectives in environmental sociology (e.g., Schnaiberg's (1980) notion of "treadmill of production", and Catton and Dunlap's (1980) notions of Human Exemptionalist and New Environmental Paradigms (see also Dunlap and Catton, 1994). Over the past two years it has come to be virtually obligatory for professional meetings of environmental social scientists to have one or more sessions devoted specifically to ecological modernization. Further, while there has been a surprising degree of acceptance of ecological modernization as one of the mainstream environmental–sociological perspectives, the pervasiveness of ecological modernization can be gauged by the fact that a broad range of environmental social scientists have found it necessary to address – even if only to critically respond to – the rising influence of this perspective (see, e.g., Benton, 1997; Harvey, 1996; Schnaiberg et al., 1999; Redclift and Woodgate, 1997a,b; also see Mol and Spaargaren, 2000; Mol, 1999; Cohen, 1997, for summaries of this critical literature and for responses to the major criticisms that have been raised). Ecological modernization has already become featured as an established perspective in the most recent environmental sociology undergraduate textbooks (Harper, 1996; Bell, 1998) and has become a particularly popular topic in the journal, *Environmental*

Politics. The publication of the present special issue of *Geoforum* testifies to the tremendous interest that ecological modernization has stimulated within geography.

A particularly important indicator of the extent to which ecological modernization thought has became influential in the environmental social sciences is the prominence given to Mol's (1997) paper in the recent and widely circulated *International Handbook of Environmental Sociology* (Redclift and Woodgate, 1997a,b). Mol's (1997) paper is one of a handful in the Redclift–Woodgate anthology devoted to a particular theoretical perspective. Not only has ecological modernization very rapidly gained a foothold in environmental sociology and environmental studies, but it has even made some inroads into general sociological scholarship. Perhaps the most telling indicator of the rising influence of ecological modernization is the fact that Giddens (1998), arguably the most well-known Anglophone social theorist of the late 20th century and a scholar interested in environmental issues and their sociological significance, has devoted 10 pages of his *The Third Way* to ecological modernization thought.

This paper will focus on some of the reasons for and implications of the extraordinary ascendance of ecological modernization thought. I will stress that its rapid rise to prominence is due less to ecological modernization having been a well-developed and highly-codified social theory, but rather because of how ecological modernization accorded particularly well with a number of intellectual and broader political–economic factors, many of which lay outside the realms of sociology and environmental sociology. I will suggest that while ecological modernization is indistinct as a social theory, ecological modernization's basic logic suggests two points. First, the most sophisticated and persuasive versions of ecological modernization revolve around the notion that political processes and practices are particularly critical in enabling ecological phenomena to be " 'moved into' the modernization process" (Mol, 1995, p. 28). Thus, a full-blown theory of ecological modernization must ultimately be a theory of politics and the state – that is, a theory of the changes in the state and political practices (and a theory of the antecedents of these changes) which tend to give rise to private eco-efficiencies and overall environmental reforms. Second, the logic of ecological modernization theory suggests that it has very close affinities to several related literatures – particularly embedded autonomy, civil society, and state–society synergy theories in political sociology – which have not yet been incorporated into the ecological modernization literature. I will conclude by arguing that ecological modernization can benefit by bringing these related – and, for that matter, more powerful – theories into its fold. Further, and perhaps most important, ecological modernization could well succeed or fail as social theory depending on the sturdiness of the bridges that can be built to these parallel theories.

The ecological modernization concept and perspective

Nearly as remarkable as ecological modernization's rising visibility and influence has been the diversity of the meanings and usages of this concept. Ecological modernization is now employed in at least four different ways. First, there is an identifiable school of ecological modernizationist/sociological thought.[2] From a

North American and British perspective Arthur Mol and Gert Spaargaren are now generally recognized as the key figures in the field, though in Germany, the Netherlands, and elsewhere on the Continent ecological modernization is still very closely associated with the work of scholars such as Joseph Huber and Martin Jänicke. Nonetheless, Mol and Spaargaren's sole- and jointly-authored works (Spaargaren and Mol, 1992; Mol and Spaargaren, 2000; Spaargaren et al., 1999; Mol and Spaargaren, 1993; Spaargaren, 1996; Mol, 1995, 1997), as well as those of close associates and colleagues (e.g., Cohen, 1997; Leroy and van Tatenhove, 1999), constitute what can be thought of as the core literature of the ecological modernization perspective. In this paper I will primarily build from Mol and Spaargaren's works because of all the scholars and researchers in this tradition (at least as far as the literature in English is concerned) they have done the most to articulate a distinctive theoretical argument.

A second respect in which ecological modernization is employed is as a notion for depicting prevailing discourses of environmental policy. The major figure associated with the political–discursive and social–constructionist perspective on ecological modernization is Hajer (1995). For Hajer (1995), ecological modernization is not so much a prediction of strong tendencies to industrial–ecological progress as it is a category for describing the dominant discourses of the environmental policy arenas of the advanced countries. In addition to Hajer's constructionism being in stark contrast with the objectivism of the core literature in ecological modernization, Hajer's view is that ecological–modernizationist environmental–political discourse may even serve to dilute the political impulse for environmental reforms by obscuring the degree to which economic expansion, growth of consumption, and capital-intensive technological change compromise the ability of states to ensure a quality environment. Thus, for many observers (including some in the core tradition of ecological modernization) Hajer's social–constructionist work is often thought of as lying outside of – or even being hostile to or incompatible with – the ecological modernization perspective per se.

Third, ecological modernization is often used as a synonym for strategic environmental management, industrial ecology, eco-restructuring, and so on (see Hawken, 1993; Ayes, 1998). Indeed, the core literature on ecological modernization has tended to give primary emphasis to environmental improvements in the private sector, particularly in relation to manufacturing industry and associated sectors (e.g., waste recycling). Social scientists from a variety of theoretical persuasions (e.g., Schnaiberg et al., 1998; Andersen, 1994) now use the notion of ecological modernization to pertain to private sector behaviors and conduct that simultaneously increase efficiency and minimize pollution and waste. Finally, there are some scholars who use the notion of ecological modernization to pertain to almost any environmental policy innovation or environmental improvement. Murphy (1997), for example, refers to state policies that make possible the internalization of environmental externalities as being instances of ecological modernization.

In addition, Mol (1999) has recently felt the need to distinguish between the first-generation of ecological modernization literature (which includes, in particular, the 1980s and early 1990s studies by German and Dutch scholars summarized in Mol, 1995) and the second generation literature that has appeared in the late 1990s. The first generation literature was based on the over-arching hypotheses that capitalist liberal democracy has the institutional capacity to reform its impact on the natural environment, and that one can predict that the further development ("modernization") of capitalist liberal democracy would tend to result in improvement in ecological outcomes. The second generation ecological modernization literature, by contrast, has increasingly revolved around identifying the specific sociopolitical processes through which the further modernization of capitalist liberal democracies leads to (or blocks) beneficial ecological outcomes. The most recent ecological modernization literature has been more concerned with comparative perspectives, including but not limited to the ways in which globalization processes might catalyze ecological modernization processes in countries in the South.

Nonetheless, the range of meanings associated with the notion of ecological modernization arguably is related to the fact that the rise of the ecological modernization perspective was not due only or even primarily to the clarity of its theoretical arguments. Indeed, the rise of ecological modernization as a concept has had to do more with the fact that ecological modernization was an effective response to a variety of circumstances or imperatives regarding social–ecological thought in the 1990s. First, the renewal of the environmental movement during the 1980s, on the grounds of global environmental change and growing recognition of ecological and technological risks, suggested to many in the environmental and ecological communities that very radical steps – significant decreases in fossil energy usage, reversal of tropical forest destruction and biodiversity loss, increasingly strict regulation of industry, the localization and decentralization [rather than globalization] of economic activity and social regulation, and so on – were necessary to address the processes of destruction of the biosphere. These impulses arguably helped to catalyze the rise of radical environmental movements in Northern Europe.[3] The rise of these environmental movements stimulated scholars such as Beck (1992) to see radical environmentalism as an enduring feature of advanced industrial politics. The growth of these counterhegemonic social–environmental views, many of the most influential of which were given visibility through publication in *The Ecologist* in the UK, led to a growing imperative to address whether they were scientifically sound or robust relative to the more managerial variants of environmental science (e.g., of the sort analyzed in Hajer, 1995). The rise of radical environmental movements also increasingly set the agenda for sociological theory and research as sizable groups of social scientists began to grapple with phenomena such as new social movements (NSMs), "the risk society", identity politics, subpolitics, and so on (Scott, 1991; Goldblatt, 1996; Martell, 1994). It thus became increasingly incumbent upon social scientists to respond to the rise and growing influence of radical environmental movements, especially in terms of whether radical environmentalism (and radical NSMs in general) would be an ascendant social force and would be a necessary precursor to effective environmental improvement and reform. Accordingly, the growing attention to ecological modernization in German social scientific circles in the 1980s had as much to do with issues that arose from the environmental sciences and from the political realm as it did with considerations from the realm of social theory per se.

Second, despite the very considerable enthusiasm and innovation which had occurred in social–scientific thought and practical policy work as a result of the more widespread use of the concepts of sustainability and sustainable development, it was becoming increasingly apparent that sustainability and sustainable development had real shortcomings in providing guidance and vision for future evolution of environmental policy. Both of these sustainability notions had originally been developed with regard to policy toward the *South*, and in addition the various notions of sustainability had been derived from experiences involving the *primary-renewable sectors* in *nonmetropolitan or rural places* in the South. Ecological modernization provided a template for new thinking about the problems and their solutions that are most urgent to address in the *transformative sectors of metropolitan regions of the advanced industrial* nations.

Third, it had become increasingly apparent that North American dominance of environmental–sociological theory had led to certain biases and blinders. The rise of ecological modernization can be seen as a response to a particularly crucial shortcoming of North American environmental sociology. While North American environmental sociology was quite diverse, most of its major theoretical works had converged on the notion that environmental degradation was intrinsically a product of the key social dynamics (be they the treadmill of production, the "growth machine", the persistence of the dominant social paradigm or of anthropocentric values, and so on) of 20th century capitalist-industrial civilization. In straining to account theoretically for why the US and other advanced industrial societies were inexorably tending toward environmental crisis, North American environmental sociology found itself in an increasingly awkward position: environmental sociologists had so overtheorized the intrinsic tendency to environmental disruption and degradation so that there was little room for recognizing that environmental improvements might be forthcoming. And the only way out of the "iron cage" of environmental despair was a rather idealistic – if not utopian – view of environmental movements as the only recourse for environmental salvation (Buttel, 1996, 1997). Ecological modernization not only provided a way for environmental sociologists to more directly conceptualize environmental improvement; ecological modernization also provided a fresh perspective on the role of environmental movements by avoiding their romanticization, and by appreciating the particularly fundamental roles that science, technology, capital, and state might play in the processes of environmental improvement.

In particular, by the mid-1990s it had become increasingly apparent that North American environmental–sociological scholarship needed to take better into account the considerable environmental progress that countries such as Germany, the Netherlands, and Switzerland had made – at least relative to the far more modest environmental progress which had been achieved in North America. Northern European environmental progress had not been confined to pollution abatement and control, but also extended to eco-efficiency improvements which had been made in manufacturing industry (Simonis, 1989; Hawken, 1993, Chapter 4).[4] But by the early 1990s these developments had remained largely ignored in mainstream North American environmental–sociological literature. Ecological modernization provided a way to understand these eco-industrial improvements while doing so in a way more satisfying than the ecological microeconomics of Hawken (1993) and the more mainstream environmental economists.

The growing embrace of ecological modernization thought by the global environmental–sociological community thus fulfilled a wide *variety* of needs and filled several gaps in social–environmental thought. Even so, this embrace has remained relatively superficial, being confined mainly to acceptance of the notion that substantial eco-efficiency gains can be made through further (or "super-") industrialization within capitalism. Thus, for example, Schnaiberg et al. (1998) have felt quite comfortable appropriating the notion of ecological modernization to depict successful instances of post-consumer waste recycling, while at the same time retaining the concept of treadmill of production (which Mol, 1995, sees as an example of deterministic neo-Marxist environmental sociology) as their main explanatory device. Most observers of the ecological modernization perspective – be they proponents, critics, or those interested in exploring the potentials of this perspective – have tended to evaluate it in terms of the third and fourth uses of the notion of ecological modernization noted above. Specifically, the questions most often asked are, "Is ecological modernization actually occurring?" or "Is there good reason to believe that we can expect trends toward ecological modernization in a significant number of economic sectors and world nations?"

The next section of the paper will be based on the notion that the first meaning of ecological modernization – that of a distinctive, though incipient social theory with the potential to create a coherent literature through hypothesis testing – is the more fundamental and useful one. Thus, while the environmental–economic and environmental–engineering conceptions of ecological modernization have tended to predominate in sociological usage of the notion of ecological modernization, I would suggest that the following are the more important postulates of a distinctive and coherent ecological modernization perspective. An ecological modernization perspective hypothesizes that while the most challenging environmental problems of this century and the next have (or will have) been caused by modernization and industrialization, their solutions must necessary lie in more – rather than less – modernization and "superindustrialization". Put somewhat differently, it is hypothesized that not only is capitalism sufficiently flexible institutionally to permit movement in the direction of "sustainable capitalism" (to turn O'Connor's, 1994 notion on its head), but its imperative of competition among capitals can – under certain political conditions – be harnessed to achieve pollution-prevention eco-efficiencies within the production process, and ultimately within consumption processes as well (Spaargaren, 1996). Thus, second, social theory must recognize and directly theorize the role that capitalist eco-efficiency and rationalization can play in environmental reform (as well as recognize their limits and the degree to which they can or must be induced by the state). Third, ecological modernization is in some sense a critical response to – if not a decisive critique of – radical environmentalism (or "countermodernity"). As Mol (1995, p. 48) notes, "the role of the environmental movement will shift from that of a critical commentator outside societal developments to that of a critical – and still independent – participant in developments aimed at ecological transformation". Fourth, an ecological modernization perspective views the environment as in potentiality or in practice being an increasingly autonomous (or "disembedded") arena of decision making (what Mol refers to as the "emancipation of ecology").

Fifth, and perhaps most fundamental, is that ecological modernization processes are a reflection of policy environments that are made possible through the restructuring (or "modernization") of the state. Thus, in Mol's (1995, pp. 46–47) words:

The ecological modernization theory has identified two options for strategies to overcome the deficiencies of the traditional bureaucratic state in environmental policymaking . . . First, a transformation of state environmental policy is necessary: from curative and reactive to preventive, from exclusive to participatory policy-making, from centralized to decentralized wherever possible, and from domineering, over-regulated environmental policy to a policy which creates favorable conditions and contexts for environmentally sound practices and behavior on the part of producers and consumers. The state will have to widen the competence of civil law in environmental policy, focus more on steering via economic mechanisms and change in its management strategy by introducing collective self-obligations for economic sectors via discursive interest mediation. The second, related, option includes a transfer of responsibilities, incentives, and tasks from the state to the market. This will advance and accelerate the ecological transformation process, mainly because the market is considered to be a more efficient and effective mechanism for coordinating the tackling of environmental problems than the state . . . The central idea is not a withering away of the state in environmental management, but rather a transformation in the relation between state and society and different accents on the steering role of the state. The state provides the conditions and stimulates social 'self-regulation', either via economic mechanisms and dynamics or via the public sphere of citizen groups, environmental NGOs and consumer organizations.

Ecological modernization as prospective social theory

As successful as ecological modernization has been as a school of environmental–sociological thought, it is at risk of ultimately suffering the same fate as its predecessor sister concept, sustainable development (SD). Though proponents of the SD notion benefited by having the imprimatur of SD being endorsed by an impressive range of institutions and international organizations (e.g., the United Nations and UNCED, the World Bank, the European Union), SD has slowly but surely begun to recede from the social-scientific radar screen. This has in large part been because of the fact that the SD concept could not overcome being seen as a nebulous knowledge claim which was too imprecise to generate a coherent set of hypotheses and body of research. Perhaps recognizing this, some of ecological modernization's most innovative thinkers, particularly Mol (1995) and Spaargaren (1996), have devoted considerable effort with the aim of anchoring ecological modernization within extant social theory.

Mol and Spaargaren's efforts at theoretical buttressing of ecological modernization have yielded certain successes. Mol and Spaargaren have noted that ecological modernization has parallels to a variety of classical theorists and influential theories (e.g., Schumpeter's and Kondratieff's notions of long cycles, Polanyi, 1957 notion of "disembedding", and Giddens, 1994 four dimensions of modernity). Arguably, however, they have tended to link ecological modernization most closely to the

work of Ulrich Beck, particularly his well-known writings on reflexive modernization and risk society (Beck, 1992; Beck et al., 1994).

There are some good reasons why Mol and Spaargaren would choose to link
ecological modernization with the work of Beck. The Netherlands and Germany
(the countries of greatest interest to Mol–Spaargaren and Beck, respectively) have
a number of structural similarities. While their political systems exhibit major
differences (e.g., the Dutch state is highly centralized while state governments
play a major role in Germany), both are parliamentary democracies within which
environmental ideologies are firmly established within their national political cultures. Beck is among the most influential and visible social theorists in Northern
Europe, and linking ecological modernization to Beck's thought would no doubt be
a plus in the mainstreaming of ecological modernization thought within European
sociological circles. Not only was Beck an influential general sociological theorist
in the 1980s, but by the early 1990s Beck was arguably beginning to displace
Schnaiberg, Dunlap, Catton and other North Americans as the most influential
environmental–sociological theorist in Europe. Thus, linking ecological modernization with Beck's work would create legitimacy and an entrée for this new
perspective within environmental sociology and sociology at large.

In some ways ecological modernization can be thought of as an instance of
Beck's (1992) notion of reflexive modernization – through which modernization
can be "turned back on to itself" in order to address the problems which it has itself
created. There is also a sense in which Mol and Spaargaren share Beck's skepticism
about the efficacy of radical environmentalism. There are additional similarities
in their views about how the role of states in advanced capitalism is changing (in
particular, the shift toward less bureaucratization and centralization). Perhaps most
fundamentally, Mol–Spaargaren and Beck agree that solutions to the problems
caused by modernization, industrialization, and science can only be solved through
more modernization, industrialization, and science.

These similarities between ecological modernization and Beck's theories of
reflexive modernization and risk society notwithstanding, there are several reasons
why I believe that ecological modernization cannot rest its main theoretical case
on reflexive modernization – or, in other words, on notions that derive directly or
indirectly (e.g., via Giddens (see Beck et al., 1994; Giddens, 1994)) from Ulrich
Beck. There are some very considerable inconsistencies between ecological modernization and Beck's notions of reflexive modernization and risk society – many
of which Mol (1995) and others readily acknowledge. Among the more salient
of these differences are the following. While Mol and Spaargaren place relatively
little emphasis on the role of radical environmental groups or new social movements (NSMs) in making possible ecological modernization processes, the lynchpin
of Beck's work is the increasingly important role being played by NSMs and
subpolitics in the restructuring of the state and political discourses. The arenas of
environmental mobilization and reform emphasized by Mol and Spaargaren also
bear little similarity to those such as anti-nuclear and anti-biotechnology protests
that are of particular concern to Beck. The very concept of "risk society" conjures
up an adherence to matters of identity politics and extra-scientific policymaking
that contrasts with the image of environmental improvement stressed by Mol and
Spaargaren. And while Beck points to a sharp distinction between "industrial

society" and "risk society", the thrust of core ecological modernization thought is that eco-efficiency gains can be achieved without radical structural changes in state and civil society. In addition to these areas of incompatibility between Mol–Spaargaren's ecological modernization perspective and Beck's theory of risk society, it is also worth noting that Beck's work has become somewhat passe in the late 1990s, and has generated very little interest in North America, so there is even less reason to anchor ecological modernization thought in the work of Beck (and of Giddens' forays into reflexive modernization).

If ecological modernization has conceptual appeal but requires more social-theoretical foundations, which way to turn? I would argue that guidance on this score can be derived from Mol and Spaargaren's own work – namely, from the stress they have placed on the types of state structures, policy networks, and policy cultures which are required to propel forward processes of ecological moderniza-tion. Their (or at least Mol's) thinking on this score is indicated quite clearly in the lengthy quote from Mol (1995) *The Refinement of Production* earlier in the paper. This lengthy quoted passage, I would argue, is strikingly compatible with the works of scholars such as Evans (1995, 1996, 1997) who have developed a set of interrelated notions of embedded autonomy and state-society synergy. In particular, Evans (1995) and the core thinkers of ecological modernization share very similar ideas about state effectiveness and state-civil society ties. Mol (1995) and Leroy and van Tatenhove (1999), for example, place a great deal of stress on the role that advocacy-coalition-type relations among state officials, corporate managers, and environmental NGOs play in making possible ecological modernization processes.

Evans (1995) work can perhaps best be characterized as a neo-Weberian per-spective on the state which at the same time is distanced from much of late 20th century neo-Weberian political sociology (as well as structuralist Marxism) through its critique of "state-centeredness" or state autonomy being primarily properties of the state itself. Prior to publication of Evans' *Embedded Autonomy*, there had been a strong consensus among "theorists of the state" (including both Weberian pro-ponents of state-centeredness as well as neo-Marxist structuralists) that large cen-tralized states that are relatively autonomous from groups and classes in civil society are best able to formulate and implement coherent and authoritative policies. In Evans' (1995, p. 22) book he argued instead that while the organization of the state does affect the capacity of states to "construct markets and promote growth" state effectiveness derives equally from the nature and quality of its relations with (rather than its autonomy or insulation from) groups in civil society. Evans (1995, Chapter 2) thus defines embedded autonomy as a state structure which combines "corporate coherence"[5] on one hand, and connectedness of, and social ties between, state agencies and officials and various groups in civil society on the other.

Evans in his *Embedded Autonomy* (1995) aims to develop evidence that the "developmental states" in the South which were successful in achieving rapid indus-trial development in the 1970s and 1980s tended to have embedded autonomous structures, involving both corporate coherence and connectedness to groups in civil society. In Evans' subsequent work (1996, 1997) on state–society synergy, which he conceptualizes as a particularly important form of embedded autonomy, he focuses on how the development of concrete sets of social ties between states and groups in society create "synergies"; on one hand, these ties between states and societies help make states more effective, and on the other hand these ties help various groups in civil society to better meet their goals. It is worth noting that while Evans' (1995) early work on embedded autonomy considered economic growth and industrial development to be the ultimate indicator of state effectiveness,[6] Evans has increasingly seen "sustainability" (particularly "urban sustainability" or "livability") as being as or more important as a dimension of state effectiveness (see Evans, 1997; Buttel, 1998).

It is also worth noting that many of the concerns of Evans and other theorists of embedded autonomy and state–society synergy (see especially the works by Evans' colleagues in his 1997 collection) were in some sense anticipated by Jänicke (1990) – a political scientist and one of the German founders of ecological modernization – in his work on "state failure". Not only does Jänicke (1990) stress the theme of the need for closer state–society ties in a manner similar to Evans, but Jänicke stresses the fact that environmental policy is among the arenas in which these ties are particularly crucial in order to achieve policy effectiveness (or, in other words to overcome state failure). Thus, not only is neo-Weberian embedded autonomy theory highly consistent with ecological modernization, but one of its founders – Martin Jänicke – has written in a parallel vein, albeit at a lower level of abstraction than achieved by Evans.

Concluding remarks

Ecological modernization has tended to be appropriated by environmental sociologists, geographers, and political scientists mainly because of its provocative and challenging views about the malleability of the institutions and technological capabilities of industrial capitalism, and because of its observations from environmental science and engineering – that eco-efficiencies can fairly readily be achieved within the framework of continued modernization of capitalism and the application of modern experimental science. Ecological modernization is a new, and in many ways an improved, synonym for sustainable development. At the same time ecological modernization is more useful than sustainable development as a macro or overarching framework for thinking about the environmental problems of metropolitan transformative industry in the North. As much as any of these factors, perhaps, ecological modernization has become attractive as a concept because it provides alternatives to the pessimistic connotations of frameworks such as the treadmill of production and the growth machine. Ecological modernization expresses hope, and makes it more readily possible to identify and appreciate the significance of environmental success stories.

Ecological modernization thought, however, has not developed to a point where one can say that it shares an identifiable set of postulates and exhibits agreement on research hypotheses and research agenda in the same way that one can do so for a theory such as the treadmill of production. In large part this is because ecological modernization did not develop primarily from a preexisting body of social-theoretical thought – as, for example, was the case with the treadmill of production (Schnaiberg, 1980) having been largely derived from O'Connor's (1973) influential theory of the accumulation and legitimation functions of the state and how their contradictions tend to become manifest in state fiscal crisis. Instead, ecological modernization thought has been more strongly driven by extra-theoretical challenges and concerns (e.g., about how to respond politically to radical environmentalism and how to conceptualize eco-efficiency improvements that are currently linked to new management practices and technical–spatial restructuring of production). Ecological modernization has essentially been an environmental science and environmental policy concept which has subsequently been buttressed with a number of citations to social–theoretical literatures, some of which are mutually quite contradictory (compare Beck vs. Jänicke, for example).

While Beck and related theorists of reflexive modernization (especially Giddens) have been cited most often within the core ecological modernization literature as theoretical exemplars, there are a number of reasons why Beck, and his notions of risk society, subpolitics, and so on, are unlikely to be sturdy theoretical foundations for ecological modernization. I would argue that ecological modernization is ultimately a *political–sociological* perspective, for reasons that are made clear in the lengthy quote from Mol (1995) earlier in the paper. And the political–sociological theory which it has closest potential relations – and, in some sense, which reflects its own origins in the work of Jänicke – is the neo-Weberian tradition of embedded autonomy and state–society synergy. I would argue that the way forward for ecological modernization is not to emphasize empirical debates over the potentials and limits of environmental engineering and industrial ecology, but rather to deepen the links to political–sociological literatures which will suggest new research problems and hypotheses. Embedded autonomy and state–society theorizing, while they are not without problems (Buttel, 1998), are particularly well suited to comparative analysis, which is a particularly exciting research frontier for ecological modernization research.

Current or prospective enthusiasts for ecological modernization driven inquiry should recognize, however, that this perspective has some important shortcomings that need to be squarely addressed. These include the perspective's (Northern) Eurocentricity (the fact that its theoretical roots and empirical examples are largely taken from a set of Northern European countries that are distinctive by world standards), the excessive stress on transformative industry, the preoccupation with efficiency and pollution control over broader concerns about aggregate resource consumption and its environmental impacts, the potentially uncritical stance toward the transformative potentials of modern capitalism, and the fact that very fundamental questions raised about modernizationism within the development studies literature (e.g., Hoogvelt, 1987; Pred and Watts, 1992) have not been addressed within ecological modernization theory.

It should also be noted that while we can agree with the ecological modernizationists that radical environmentalism may not be *directly* responsible for many of the environmental gains achieved in Northern Europe and elsewhere, these non-mainstream ecology groups arguably play a significant role in pushing mainstream environmental groups and their allies in the state and private industry to advance a more forceful ecological viewpoint. Thus, radical environmental groups, by providing alternative vocabularies and "frames" of environmentalism, stressing issues often ignored within mainstream environmentalism, and providing new loci of personal identity for citizens, will tend to strengthen the movement as a whole, and thus *indirectly* contribute to ecological modernization processes. It is worth noting, in fact, that in the US the environmental groups that are most concerned about toxics and chemicals – the primary preoccupation of ecological modernizationists – are not the mainstream environmental groups, but rather local (particularly "environmental justice" – oriented) groups which are most radical and often thought as being out of the movement mainstream (Gottlieb, 1993). In sum, as the social science community moves rapidly to explore the new ecological modernizationist viewpoint, it should do so with awareness of both its strengths and weaknesses.

Acknowledgements

A previous version of this paper was presented at the School of Natural Resources and Environment (SNRE), University of Michigan, October 1998. The author would like to thank the SNRE faculty and graduate students, William Freudenburg, two anonymous Geo-Forum reviewers, Jenny Robinson, and Joseph Murphy for their comments and suggestions on previous versions of this paper.

Notes

1. In this paper the expression environmental social science will be understood to pertain to the social science disciplines in which ecological modernization perspectives currently play a major role. Ecological modernization has become quite influential within environmental sociology, and to a lesser degree within geography and political science. Because such a large share of the ecological modernization literature (in English) has been authored by sociologists, the discussion in this paper will occasionally refer specifically to the (environmental) sociological literature.

2. Note that I use the expression ecological–modernizationist "thought" or "perspective", rather than theory, at this point in the paper because of the fact that, at least as far as the literature in English is concerned, ecological modernization is not yet a clearly-codified theory. The lack of codification has given rise to the fact that ecological modernization has been used in so many different ways by social scientists. As an example, Redclift and Woodgate (1997a,b) take the core notion of ecological modernization to be the claim that economic growth is compatible with environmental protection, and they equate the perspective primarily with the literature on industrial ecology and "industrial metabolism". While one might say that Redclift and Woodgate have simply misinterpreted ecological modernization, one can say that this type of confusion would be very unlikely to occur when environmental social scientists discuss Schnaiberg's (1980) notion of treadmill of production or O'Connor's (1994) notion of the second contradiction of capital.

3. Mol (1997, pp. 33, 58), for example, portrays radical environmentalism in terms of eco-centric ideologies which are deployed in pursuit of "de-industrialization" agendas, and mentions the "deep ecology" movement as being the prototypical radical environmental movement. While Mol acknowledges respect for radical environmentalism for its efforts to legitimize notions of ecological rationality, he suggests that the radical environmental position is not a realistic one to the

degree that it insists that ecological rationality must be substituted for (rather than being balanced with or weighed against) private economic rationality. It should also be noted that Joseph Huber's original contributions to ecological modernization thought were reactions to the anti-modernist views of key ("fundamentalist") figures such as Bahro (1984). Ecological modernization has thus been closely identified with the realist wing within the fundamentalist–realist divide within the German Green Party.

4. I am indebted to an anonymous reviewer who stressed that it is useful to recognize that there are two related, but quite separate streams of research and practice which are typically subsumed under the more general category of eco-efficiency. The first, which has been actively promoted by the World Business Council for Sustainable Development ⟨www.wbcsd.ch⟩ and is increasingly gaining attention among management consultants, is that of "strategic environmental management". Strategic environmental management is primarily concerned with achieving fewer externalities, greater eco- or environmental efficiencies, and "green marketing" advantages within the context of existing plant and equipment. By contrast, "industrial ecology" refers to a more ambitious agenda of fundamental redesign of industrial structures and processes (including industrial relocations and production synergies) aimed at achieving the dematerialization of production, and ultimately the dematerialization of society (see Ayres, 1998).

5. Evans means "corporate coherence" in the Weberian (legal rational authority) sense – that is, the cohesion among state officials which reflects commitment to the state and its goals, which in turn is made possible by meritocratic recruitment and a long-term career reward structure.

6. Evans (1995) initial work on embedded autonomy and developmentalist states stressed state ties with what he called "developmental elites", while in Evans' (1996, 1997) more recent work on state–society synergy and urban sustainability in the South he gives more stress to community and neighborhood (including shantytown) leaders and activists.

References

Andersen, M.S., 1994. Governance by Green Taxes. Manchester University Press, Manchester.

Ayres, R.U. (Ed.), 1998. Eco-Restructuring. United Nations University Press, New York.

Bahro, R., 1984. From Red to Green. Verso, London.

Beck, U., 1992. Risk Society. Sage, Beverly Hills, CA.

Beck, U., Giddens, A., Lash, S. (Eds.), 1994. Reflexive Modernization Policy, Cambridge.

Bell, M., 1998. An Invitation to Environmental Sociology. Pine Forge Books, Thousand Oaks, CA.

Benton, T., 1997. Reflexive modernization or green socialism? Paper Presented at the RC-24 Conference on Sociological Theory and the Environment. Woudschoten Conference Center, Zeist, Netherlands.

Buttel, F.H., 1996. Environmental and resource sociology: theoretical issues and opportunities for synthesis. Rural Sociology 61, 56–76.

Buttel, F.H., 1997. Social institutions and environmental change. In: Redclift, M., Woodgate, G. (Eds.), The International Handbook of Environmental Sociology, Edward Elgar, London, pp. 40–54.

Buttel, F.H., 1998. Some observations on states, world orders, and the politics of sustainability. Organization and Environment 11, 261–286.

Catton Jr., W.R., Dunlap, R.E., 1980. A new sociological paradigm for post-exuberant sociology. American Behavioral Scientist 24, 14–47.

Cohen, M., 1997. Sustainable development and ecological modernisation: national capacity for environmental reform. OCEES Research Paper No. 14, Oxford Centre for the Environment, Ethics, and Society, Oxford.

Dunlap, R.E., Catton Jr., W.R., 1994. Struggling with human exemptionalism: the rise, decline and revitalization of environmental sociology. The American Sociologist 25, 5–30.

Evans, P., 1995. Embedded Autonomy. Princeton University Press, Princeton.

Evans, P., 1996. Government action, social capital, and development: reviewing the evidence on synergy. World Development 24, 1119–1132.

Evans, P. (Ed.), 1997. State–Society Synergy. Institute for International Studies, University of California, Berkeley.

Giddens, A., 1994. Beyond Left and Right. Stanford University Press, Stanford, CA.

Giddens, A., 1998. The Third Way. Polity, Cambridge.

Goldblatt, D., 1996. Social Theory and the Environment. Westview Press, Boulder, CO.

Gottlieb, R., 1993. Forcing the Spring. Island Press, Washington, DC.

Hajer, M., 1995. The Politics of Environmental Discourse. Oxford University Press, New York.

Harper, C.L., 1996. Environment and Society. Prentice-Hall, Upper Saddle River, NJ.

Harvey, D., 1996. Justice, Nature and the Ecology of Difference. Blackwell, Oxford.

Hawken, P., 1993. The Ecology of Commerce. HarperCollins, New York.

Hoogvelt, A., 1987. The Third World in Global Development. Macmillan, London.

Jänicke, M., 1990. State Failure. Pennsylvania State University Press, University Park, PA.

Leroy, P., van Tatenhove, J., 1999. New policy arrangements in environmental politics: the relevance of political and ecological modernization. In: Spaargaren, G. et al. (Eds.), Environmental Sociology and Global Modernity, Sage, London.

Martell, L., 1994. Ecology and Society. University of Massachusetts Press, Amherst.

Mol, A.P.J., 1995. The Refinement of Production. Van Arkel, Utrecht.

Mol, A.P.J., 1997. Ecological modernization: industrial transformations and environmental reform. In: Redclift, M., Woodgate, G. (Eds.), The International Handbook of Environmental Sociology, Edward Elgar, London, pp. 138–149.

Mol, A.P.J., 1999. The environmental state in transition: exploring the contradictions between ToP and EMT. Paper presented at the International Sociological Association (RC 24) conference on The Environmental State Under Pressure, Chicago.

Mol, A.P.J., Spaargaren, G., 1993. Environment, modernity and the risk society: the apocalyptic horizon of environmental reform. International Sociology 8, 431–459.

Mol, A.P.J., Spaargaren, G., 2000. Ecological modernization theory in debate: a review, Environmental Politics (forthcoming).

Murphy, R., 1997. Sociology and Nature. Westview Press, Boulder, CO.

O'Connor, J., 1973. The Fiscal Crisis of the State. St. Martin's Press, New York.

O'Connor, J., 1994. Is sustainable capitalism possible? In: O'Connor, M. (Ed.), Is Capitalism Sustainable? Guilford, New York, pp. 152–175.

Polanyi, K., 1957. The Great Transformation. Beacon, Boston.

Pred, A., Watts, M. (Eds.), 1992. Reworking Modernity. Rutgers University Press, New Brunswick, NJ.

Redclift, M., Woodgate, G., 1997a. Sustainability and social construction. In: Redclift, M., Woodgate, G., (Eds.), The International Handbook of Environmental Sociology. Edward Elgar, London, pp. 55–70.

Redclift, M., Woodgate, G., 1997b. The International Handbook of Environmental Sociology. Edward Elgar, London.

Schnaiberg, A., 1980. The Environment. Oxford University Press, New York.

Schnaiberg, A., Weinberg, A., Pellow, D., 1998. Ecological modernization in the internal periphery of the USA. Paper presented at the annual meeting of the American Sociological Association, San Francisco.

Schnaiberg, A., Weinberg, A., Pellow, D., 1999. The treadmill of production and the environmental state. Paper presented at the International Sociological Association (RC 24) Conference on The Environmental State Under Pressure, Chicago.

Scott, A., 1991. Ideology and the New Social Movements. Unwin Hyman, London.

Simonis, U.E., 1989. Ecological modernization of industrial society: three strategic elements. International Social Science Journal 121, 347–361.

Spaargaren, G., 1996. The ecological modernization of production and consumption. Ph.D. Thesis, Wageningen University.

Spaargaren, G., Mol, A.P.J., 1992. Sociology, environment, and modernity: ecological modernization as a theory of social change. Society and Natural Resources 55, 323–344.

Spaargaren, G., Mol, A.P.J., Buttel, F.H. (Eds.), 1999. Environmental Sociology and Global Modernity. Sage, London (forthcoming).

Part Two

Transformations in environmental governance and participation

Chapter 9

Dana R. Fisher, Oliver Fritsch and Mikael Skou Andersen

TRANSFORMATIONS IN ENVIRONMENTAL GOVERNANCE AND PARTICIPATION

Introduction and overview

ALTHOUGH ECOLOGICAL MODERNISATION WAS initially introduced in the 1980s as a policy program, it has since been embraced by scholars who are interested in the way society interacts with environmental issues. Within today's literature on ecological modernisation, the research runs the gamut from social theoretical to policy prescriptive (for a full discussion, see Fisher and Freudenburg 2001). As the theory has developed, there has been a concurrent transformation in environmental policy-making, as well as in the ways that environmental issues are being addressed in and by the policy world. This chapter focuses on how ecological modernisation scholars have approached transformations taking place in environmental governance and participation by social actors beyond the state. In other words, how has ecological modernisation theory addressed changes in environmental governance and the roles that different social actors are playing in policy-making?

Since the United Nations Conference on Human Development in Stockholm, Sweden in 1972, scholars and policy-makers alike have noted that environmental problems have become increasingly globalised (e.g. Buttel 2000; Frank, Hironaka and Schofer 2000a, 2000b; Mol 2001; Spaargaren, Mol and Buttel 2000, 2006; see also Giddens 2003). Twenty years later, at the United Nations Conference on Environment and Development in Rio de Janeiro in 1992, or the 'Earth Summit', a number of potential international environmental agreements were discussed and preliminary negotiations took place. These negotiations aimed to address what Jänicke and Jörgens (2004, Part Two of this volume) call 'persistent environmental problems'. In other words, progress was made during the Summit toward beginning international negotiations on treaties that would address global environmental issues, such as biodiversity loss, deforestation and global climate change.[1] The Rio

Declaration and Agenda 21 were also created during this meeting. Together, these agreements coming out of the Earth Summit led to an era of new environmental governance, which has involved diverse social actors at multiple scales that address global environmental problems and the challenges of sustainable development.

As these international agreements have been negotiated and, in some cases, ratified and implemented, nation-states have responded by implementing additional domestic policies. Summarising these changes, Mol and Buttel suggest that now there is a general expectation that there will be an 'institutionalisation of environmental tasks in state policies and politics' (2002; see also Giddens 1998). As a further part of this transformation, market and civil society actors have joined states in the negotiation of policies that address emergent environmental issues. Jänicke and Jörgens (this volume) provide a detailed overview of the ways in which environmental governance changed during this period.

Ecological modernisation scholars have noted how these new forms of environmental governance involve the participation of new coalitions in multiple tiers of policy-making. Although ecological modernisation theory initially looked to 'modern institutions such as science and technology and state intervention' to lead the way to environmental governance (Mol and Spaargaren 1993: 454–455), more recent work has explored the role of civil society actors, including social movement organisations and non-governmental organisations (NGOs) (see Sonnenfeld, in this volume; 2002; Mol 2000; Sonnenfeld and Mol 2002; Van Tatenhove and Leroy, in this volume). Mol and Spaargaren (1998: 3), for example, suggest that ecological modernisation involves changes in the traditional roles of actors, in particular 'various transformations regarding the traditional central role of the nation-state in environmental reform . . . [with] more opportunities for non-state actors to take over traditional tasks of the nation-state'. More recently, Mol and Spaargaren (2006) have noted the role of 'hybrid arrangements' among different social actors including those from civil society in what they call 'environmental flows'.

In addition to the inclusion of civil society and other non-state social actors in environmental policy making, research has noted a broadening of the scales at which environmental policy making takes place. Ecological modernisation scholars have looked at diverse environmental issues and international environmental regimes. Some recent comparative research focuses on the international level, addressing issues related to climate change regime formation (Fisher 2004) and the effect of Europeanisation on countries in Central and Eastern Europe (Andersen 2002). Other research has focused on the national level, looking at countries as diverse as Japan (Barrett 2005), Finland (Jokinen 2000; Sairinen 2002), Hungary (Gille 2000), Lithuania (Rinkevicius 2000a, 2000b) and Vietnam (Frijns et al. 2000). At the same time, a limited number of studies have explored ecological modernisation processes taking place at the sub-national level. Gonzalez (2001, Part Two of this volume), for example, looks at automobile emission standards in California. Also looking at various localities in the United States, Scheinberg (2003) analyses recycling (but see Pellow, Schnaiberg and Weinberg 2000 for a contrasting view).

This essay aims to provide a brief overview of the ways ecological modernisation theory contextualises an understanding of processes of environmental governance and participation. The remainder of this essay is structured in three sections: first, we consider the ways ecological modernisation theory has laid the foundation for the diffusion of new approaches and policy instruments in environmental governance. Second, we discuss how public participation, stakeholder inclusion and more broadly neocorporatist arrangements relate to these innovations. Third, we conclude by discussing the contribution that more participatory modes of governance could provide for institutional ecological modernisation around the world.

Ecological modernization and governance transformations

As Albert Weale notes (1993:206), ecological modernisation not only refers to a body of theoretical literature, but also encapsulates a specific belief system. These beliefs have, in recent years, been adopted by policy elites in Europe, leading to certain transformations in environmental governance. Belief systems emphasising that environmental policy should be regarded, not as a burden on the economy, but rather as a vehicle for industrial and societal transformation, capable of improving overall welfare, have led to increased interest in new environmental policy instruments (NEPIs) capable of promoting a real transition.

When biologists first sounded alarms over the environmental impacts of the fallout of atmospheric nuclear testing, pesticides, algae blooms, toxic spills, and noxious fumes, they not only addressed decision-makers, but also mobilized lawyers in their demand for new, more elaborate laws to restrict the unwanted byproducts of industrial society. As a result, the formative phase of environmental policy legislation in the 1960s and 1970s fell victim to what Baumol and Oates dubbed the 'command and control' approach of regulation (1975). Lawyers' endeavours to detail legislation with mandatory restrictions found a strong resonance among engineers capable of fitting specific technologies to enable compliance with pre-scribed standards. This type of response became more common as biologists and engineers took on leadership roles in newly created agencies for management of the environment (Figure 9.1 provides a simple typology of policy instruments and their properties, contrasting the command-and-control approach with some NEPIs).

Private firms and industrial sectors targeted by regulators and others for environmental reform were dissatisfied with pre-specified requirements and inflexible target-setting (Marcus 1980), particularly because obtaining emission permits proved to be costly and polluters were prevented from adoption of technologies that had improved and advanced in the meantime. States frequently subsidised the adoption of new technologies to soften the command and control approach, as well as the minds of the polluters. At the same time, a call for more flexible and integrated solutions with less investment lost on passive end-of-pipe treatment of effluents continued to be heard from both proponents and opponents of command-and-control environmental regulation (for an illustrative case of end-of-pipe-failure see Andersen 1999). Despite overwhelming sympathy at the rhetorical level for more flexible approaches (Cairncross 1991), a dismantling of existing regulatory frameworks proved extremely contentious, and the use of NEPIs has made only limited progress in certain countries. Headway has been most notable in smaller and innovative northern European states, such as the Netherlands and the Nordic

	Regulator specifies the goal to be achieved	Regulator does NOT specify the goal to be achieved
Regulator specifies HOW the goal is to be achieved	Command-and-control (regulation)	Technology-based regulatory standards (e.g. use of best available technology)
Regulator does NOT specify how the goal is to be achieved	Most negotiated VAs; some mandated MBIs; some regulation	Most more purely MBIs (e.g. ecolabels); some VAs, informational devices

Figure 9.1 A typology of policy instruments (adapted from Russell and Powell, 1996, after Jordan, Wurzel, and Zito, 2003). VA = voluntary agreements; MBI = Market-Based Instruments, such as taxes, charges and emissions trading

countries, and to a much lesser extent in other countries, such as federations with complex statutes of multi-level government, such as Germany and the US (for the US, see for example Mazurek 1998). Moreover, NEPIs have tended to supplement, rather than replace, existing environmental regulations.

In their book on new instruments of environmental governance, Jordan, Wurzel and Zito (2003) discuss the adoption of NEPIs. The book also raises a fundamental question about the diversity apparent among member states within the European Union. Although the 'what' and 'where' questions relating to the diffusion of NEPIs have been addressed by 'advocacy institutions' (e.g. European Environment Agency 2006), there has been a debate among political scientists as to why patterns of diffusion vary so markedly among different countries and NEPIs (e.g. Golub 1998). Even though there is broad agreement over the basic logic, favouring flexible, adaptive and integrative approaches more in accordance with the ecological modernisation paradigm, opinions differ over the factors responsible for the variation in and shortcomings of implementation (cf. the two schools of thought mentioned below).

That different political systems have different capacities for ecological modernisation was first observed in an article by Martin Jänicke (1990b). Although the article presented a model to interpret the influencing factors, it did not address the transformation issue from the perspective of new policy instruments. Dryzek, Downes, Hunold and Schosberg (Part Two of this volume) note differences among the US, UK, Germany and Norway, and develop a comparative analysis that can account for many of the policy-related contrasts. Still, the authors treat the transformation issue in mainly descriptive terms. Readers in search of more profound explanations are referred to the broader political science literature.

One school of thought in political science argues that ideas are dominant and drive the search for new policy instruments (Hall 1993). Changes in ideas are related to processes of social learning; unexpected experiences can create a crisis for a paradigm of policy understanding and cause changes in the conception and use of policy instruments. Coalitions of actors have different perceptions of the meaning and value of policy instruments and compete to set the agenda; sudden events can tip the balance in favour of different coalitions (Sabatier 1993). Under this perspective, the advent of NEPIs would be interpreted as being supported by a belief coalition, mainly of economists and ecologists, representing ideas that rival the regulatory thinking of lawyers and engineers – and view the relative transformations as a result of changes in the power balance between these coalitions in various countries (see Pedersen 2005).

Another school of thinking emphasizes the profound character of political institutions and the ways they shape the outcomes of decision making (March and Olsen 1989). These theories assume that choice of policy instrument depends on the historical institutional trajectories in which the selection takes place. Vogel (1993) argues that countries regulate the environment in a manner consistent with how they regulate everything else: there is a path-dependency and institutions are 'sticky,' not leaving much room for decision makers to deviate in a particular setting (cf. also Van Waarden 1995). Decision makers cannot opt for the optimal solution, but tend to 'satisfice' (that is 'muddle through') within pre-existing constraints. Even major events or catastrophes only rarely trigger changes, because actors first seek out solutions by refining the existing framework. Over time, actors invest considerable time and resources in particular policies and tools, locking them in place (Pierson 1993). In their detailed, comparative study of voluntary agreements and other 'soft' policy instruments in Austria, Denmark and the Netherlands, Mol, Lauber and Liefferink (2000) follow a more institutional approach, relating political modernisation developments to the specific national policy culture and style of the country. Under the 'institutionalist' perspective described here, while it is difficult to provide an explanation for the emergence of NEPIs, there is ample evidence to emphasise restrictions on their diffusion.

Diffusion of these NEPIs is sometimes simplistically taken as a measure of the degree of institutional innovation with respect to ecological modernisation. In reality, transformations in policies and instruments reflect a more complex and muddled policy process of innovations at a tactical level, despite often deep-seated and conflicting objectives among actors in the policy process. The following section discusses how 'participation', linked with the broader ecological modernisation agenda, has recently been added to the family of NEPIs.

Public participation, stakeholder inclusion and neocorporatist arrangements

The inclusion of non-state actors in agenda setting, decision making and policy implementation constitutes one of the major transformations in local, national and international environmental governance. Still, the scope, importance and justification of such efforts differ considerably among countries and policy fields. In fact, it can be argued that the related scholarly debate follows ideational or institutional lines, in the broad context of parallel debates over NEPIs.

Ideas about the perspectives, motivations and justifications for the involvement of non-state actors in decision making and policy implementation have been discussed by scholars coming from a variety of schools of thought, including democratic theory (Barber 2004), green political thought (Smith 2003), critical theory (Habermas 1991) and conflict research (Crowfoot and Wondolleck 1990). However, for the purposes of this essay, there are good reasons to follow Van Tatenhove and Leroy (Part Two of this volume), in that current forms of inclusionary politics are a response to different eras of political modernisation in the western hemisphere. Distinguishing between the three periods of early modernisation, antimodernist backlash, and late modernisation, the authors suggest that participation was introduced first as an instrument to support institutional decision-making processes by way of consultation and citizen comments. However, the subsequent failure of institutions to incorporate adequately environmental concerns into mainstream politics and the dissatisfaction with closed-shop decision-making procedures together motivated a general move towards a more emancipatory approach experienced in the late 1960s. This move made NGO and citizen participation a key feature of the reactive, anti-modernist movement. As a consequence, new inclusionary procedures with an even wider scope of competencies were introduced that reshaped state–civil society interactions. Van Tatenhove and Leroy (this volume) argue that, in the 1980s, for societies characterised by risk and uncertainty, such forms of participation became less appropriate as participatory approaches ignored the market sphere and led to 'decision failures' on such policy issues as nuclear energy. The authors continue by arguing that, during the third phase of participatory politics, attempts were made to create new relationships between states, civil society actors and private sector actors. The latter has its own history of collaboration with government, and is discussed below. First, however, we review a functional explanation of the emergence of participatory approaches to environmental governance, one that was particularly influential in the European context.

While Van Tatenhove and Leroy touch upon the tricky relationship between the expansion of participatory forms of governance and regulators' dependence on both compliance and cooperation, scholars focusing on the critical implementation of environmental policy provide helpful insights by analysing state failure as a major challenge in eco-politics (Jänicke 1990a). 'State failure' refers to the inability of national regulators to address successfully environmental problems in the decision-making process, and effectively enforce the decisions already made (Mayntz 1993). This perspective is supported theoretically by systems theory, which claims the impossibility of directed and purposeful state action (cf. Luhmann 1989). Pressed by new social movements, global business corporations, technological risks and international regimes, the nation-state was seen as facing a two-fold loss of sovereignty: a reduced steering potential and less options for national solutions (Scharpf 1992). Cooperative and communicative modes of regulation were noted to be the most promising strategy for regaining national decision-making authority. However, forms and rationales of participation were needed to go beyond the emancipatory concepts of the 1970s that heralded the centrality of the nation-state. Focusing on substantive outcomes, apart from or in addition to fairness and other aspects, participation in an ecological modernisation context thus became a means to achieve environmental goals in a more targeted, swift and effective way (Bulkeley and Mol 2003). Specifically, participatory governance relies on the expectation that participation improves the environmental quality of decisions by incorporating knowledge of local actors (Steele 2001), providing an arena for previously excluded social actors, including for example, environmental organizations (Smith 2003) and allowing for deliberation and policy learning (Dryzek 1997). Moreover, it is expected that the involvement of non-state actors increases procedural justice, increases the acceptance of policy decisions and thus improves implementation and compliance (Schenk, Hunziker and Kienast 2007). By increasing the quality of decision making and improving the process of policy implementation, participatory forms of governance are expected to lead to better policy outcomes in terms of environmental protection than would more hierarchical modes of regulation.

According to Blowers (1998), the involvement of non-state actors in policy making has meant the establishment of partnerships between business and government. In this sense, demands for more inclusionary politics to some extent reflect and extend prevailing neocorporatist arrangements in a number of western industrialised countries (also see Scruggs 1999). Such modes of governance go beyond the opportunity to voice collective interests as offered by pluralism. They also incorporate the binding inclusion of interest groups at regular intervals into policy formulation, decision making and policy implementation through a mutual organisational communication and interaction between government actors on the one hand and strong centralised interest groups on the other (ibid., Walker 1991). In many European countries, these types of collaborations have been restricted to government and business actors. In the 1980s, however, such collaborations began to include a wider range of social actors, such as environmental and other societal groups, who were granted more systematic access to decision-making processes. These groups and organisations can bear considerable power in relation to the mobilisation of their members and supporters, and are also able to contribute expertise and experience in the policy process on environmental matters. Empirical research covers countries such as the United Kingdom (Blowers 2004) and Germany (Barthe 2001), as well as covenants and round tables in the Netherlands (Glasbergen 1995, 1998), but also countries such as Sweden or Denmark (Christiansen 1996). In the United States, negotiated agreements between the US Environmental Protection Agency, businesses and environmental groups have attracted the attention of political and legal scholars (Susskind et al. 1983).

Public and stakeholder participation was established in the early 1970s in the United States, and further institutionalised by the 1996 Model Plan for Public Participation (US EPA 1996). Related changes in other Western industrialised countries were driven by events taking place in the international arena, such as the UN Earth Summit in 1992. As a result, citizen participation in decision making and policy implementation was introduced on a larger scale. In the wake of the 1998 Aarhus Convention, four directives were passed in the EU mandating public participation in environmental decisions. These include, among others, the Public Participation Directive, a purely procedural law, and the Water Framework Directive that combines substantive requirements ('good water status') with procedural obligations, including consultation of the public as well as its 'active involvement' (Art. 14 WFD). However, previous research has indicated that cooperative forms of governance require a favourable political and societal context if they are to become institutionalised and demonstrate success in addressing environmental problems

(Jänicke and Jörgens, this volume; on 'windows of opportunity' see also Van Tatenhove and Leroy, this volume). In light of the frequently top-down approach to introducing stakeholder participation, the question arises whether European member states have the capacity and the willingness to comply with the EU's far-reaching demands and, in turn, whether the assumed environmental improvements can be expected.

To date, the ecological outcomes of neocorporatist or participatory modes of environmental governance have not been subjected to extensive research. Scholars struggle, on the one hand, with a lack of data on environmental outputs and outcomes and, on the other hand, face considerable methodological challenges (cf. Sonnenfeld and Mol 2006). These challenges include trying to understand the complex relationship between policy implementation and environmental change, and developing research designs that contribute to unravelling the causal relationships between a number of influencing factors and policy outcomes (Koontz 2006: 114). In a pioneering study, Scruggs (1999) has analysed the relationship between neocorporatist arrangements and environmental outcomes in 17 western industrialised countries. His findings suggest a positive correlation between neocorporatism and environmental improvements, which can chiefly be explained by the existence of encompassing peak organisations that represent a national constituency and are able to overcome collective action problems by means of internal mechanisms as well as consensus-seeking efforts with organised interests. Studying the effectiveness of international environmental agreements, Bernhagen (2008) draws similar conclusions. He argues that participatory forms of decision making establish communicative arenas in which knowledge asymmetries are reduced. At the same time, the author downplays the importance of power and human, organisational or financial resources for political influence. Coglianese (1997) has analysed all US EPA-negotiated rulemaking processes between 1983 and 1996 from an interest in policy implementation, disconfirming the claim that negotiated agreements are implemented significantly faster and with less litigation than hierarchical decisions. Langbein and Kerwin (2000) challenge these findings, however, on the basis of Coglianese's data. In a recent contribution, Poloni-Staudinger (2008) distinguishes several types of environmental effectiveness and compares them to different forms of democracy.

With regard to the ecological outcomes of citizen participation, Fritsch and Newig (2009) compare 55 case studies in North America and Western Europe, finding that participatory environmental policy decisions, or those that integrate bottom-up and top-down regulation, were implemented much faster and with less litigation than in purely hierarchical forms of governance. However, while participation seems to foster social learning and thus more creative solutions, these two components do not always contribute to more ecological outputs, let alone outcomes – which can be attributed to the role of the preferences of the actors involved (ibid.). Apparently, public deliberation requires a favourable societal context, such as empowerment or environmental awareness, in order to improve environmental politics – conditions that are not always met (cf. Tu 2007). As a consequence, several scholars recommend participation as an extension of hierarchical modes of governance, 'leaving the stick behind the door' (Jänicke and Jörgens, this volume; see also Ashford 2002). The number of scholars who take a critical perspective towards participation as a means of improving environmental governance is few but increasing (see e.g. Brosius, Tsing and Zerner 2005, with respect to natural resources management; and the recent collection edited by Sonnenfeld and Lockie 2008). Studying participation in Californian clean-air politics, Gonzales (this volume) argues that participation legitimised non-democratic forms of decision-making and led to the cooptation of environmental organisations. This cooptation, he maintains, could at best support weak concepts of ecological modernisation by 'greening' existing discourses, but at the same time prevents more radical strategies towards societal change and transformation of the economic system (also see Amy 1987). Blühdorn (2006) concedes that state-organised, top-down participation, which he describes as 'activation,' serves to consolidate moderate ecological aspirations and liberal consumer capitalism while, at the same time, marginalising more radical approaches that are likely to emerge in uncontrolled bottom-up 'activism'. However, interpreting top-down participation as a strategy of societal self-deception in a post-ecological era, he rejects the notion of an elitist counter movement implicit in Gonzales' (2001) account and embeds it in a general theory of reduced environmentalism, depolitication and symbolic politics.

Conclusion: potential opportunities for ecological modernisation

Transformations in environmental governance are required for the ideas and practices of ecological modernisation to penetrate the economy-environment interface more broadly around the world. How such transformations are facilitated and to what extent they will diffuse through the application of innovative policy approaches are research questions that have attracted the attention and efforts of scholars for a number of years. As described above, the new era of environmental governance was catalysed by processes taking place in the framework of the United Nations. In parallel with the transformations in environmental governance that have taken place since the Earth Summit in 1992, the science of environmental issues – predominantly regarding the issue of climate change – has achieved a remarkable level of consensus. It is possible that the release of the Intergovernmental Panel on Climate Change's Fourth Assessment Report (FAR) in 2007, as well as the Stern Report (2006) on the economic costs of climate change, will stimulate yet a new stage in environmental governance, one that may provide new momentum for the ideas and principles of ecological modernisation. Initial policy responses to the FAR can be seen in the European Union's climate package proposals for domestic measures as well as its position in international negotiations. In addition, the changes in energy systems that are required to reduce emissions of greenhouse gases, in terms of production and consumption, as well as in human behaviour, will involve deep paradigmatic changes. Such changes provide opportunities for even broader interest in the principles of ecological modernisation, as well as opportunities for policy transformation.

The learning process and debate on NEPIs reviewed above is one that has taken place predominantly in industrialised countries. However, with climate change requiring effective greenhouse gas reductions at a global level, questions arise about the role to be played by developing countries and emerging industrialised countries. Traditionally, it has been understood that it was necessary to 'grow rich in order to become clean' (Ekins 1997). With the apparently relentless laws of physics that govern our global climate, there is arguably, in absolute terms, less room for such approaches. Zhu (2007) addresses this paradox and hypothesizes that not only the usual market-based instruments, but also participatory approaches could help shortcut needs for traditional growth trajectories by promoting ecological modernisation. In his work, he uses the term 'circular economy,' which emphasizes energy and natural resource efficiency instead of 'ecological modernisation.' This concept has grown in prominence in China and other Asian countries and, despite a gap between the rhetoric and actual practices in production and consumption, the notion does suggest that new beliefs are in the making across the globe and that this new stage might extend beyond western industrialised countries.[2] It is likely that the involvement of NGOs in international policy-making around environmental issues has led to more participatory processes at all levels.

This essay has addressed the role of multi-stakeholder participation as a vehicle for understanding the transformation of environmental governance in today's globalising world. It is clear that participation is more ambiguous in terms of environmental outcomes than, for example, economic signals that punish polluters for emitting greenhouse gases. Although it appears unlikely that many current economic and technological systems will be able to adapt to climate change and incorporate the radical changes required within the relatively short time span available without citizen involvement, social and political research questions are increasing in prominence once again. Much of the participation literature to date has addressed more traditional channels for participatory processes, such as mediation with interest groups and citizen consultation at the project level (cf. Weidner 1998). The timeline that is becoming apparent for the climate issue suggests that more bottom-up types of participatory processes are needed, such as those that were common during the formative years outlined above. As explained by Mol and Spaargaren (2006), the present circumstances led to greater interest in the establishment of a 'sociology of environmental flows', i.e. a more profound exploration of the relevance of social theory for understanding environmental flows as a function of bounded human and cultural processes.

Although it is tempting to think of participation from the 'top-down' perspective as merely another policy instrument available in the tool-kit of the policy maker and regulator, the question remains how other perspectives fit into environmental governance. It is likely that ecological modernisation requires more bottom-up forms of participation. Without engagement of social actors across all sectors, contemporary societies may not be able to overcome the economic and technological structures that produced the environmental challenges facing them in the first place. Engaging social actors with environmental change in a network society with shifting forms of environmental authority (Spaargaren and Mol 2008) however is a topic that will keep ecological modernisation scholars occupied also for the years to come.

Notes

1. For a full discussion of the role of the Earth Summit in global environmental politics, see Speth and Haas 2006.
2. For more on this issue, see Environmental Reform in Asian and Other Emerging Economies, Part Four of this volume.

References

Amy, Douglas J. (1987) *The Politics of Environmental Mediation*. New York, NY: Columbia University Press.

Andersen, M.S. (1999) 'Governance by Green Taxes: Implementing Clean Water Policies in Europe', *Environmental Economics and Policy Studies* **1** (2), pp. 39–63.

Andersen, Mikael Skou (2002) 'Ecological Modernization of Subversion?', *American Behavioral Scientist* **45** (9), pp. 1394–1416.

Ashford, N.A. (2002) 'Government and Environmental Innovation in Europe and North America', *American Behavioral Scientist* **45** (9), 1417–1434.

Barber, Benjamin R. (2004) *Strong Democracy. Participatory Politics for a New Age*. Berkeley, CA: University of California Press.

Barrett, Brendan F.D. (2005) *Ecological Modernization and Japan*. London/New York: Routledge.

Barthe, Susan (2001) *Die verhandelte Umwelt. Zur Institutionalisierung diskursiver Verhandlungssysteme im Umweltbereich am Beispiel der Energiekonsensgespräche von 1993*. Baden-Baden: Nomos.

Baumol, W.J. and Oates, W.E (1975) *The Theory of Environmental Policy: Externalities, Public Outlays and the Quality of Life*. New York, NY: Prentice Hall.

Bernhagen, Patrick (2008) 'Business and International Environmental Agreements: Domestic Sources of Participation and Compliance by Advanced Industrialized Democracies', *Global Environmental Politics* **8** (1), pp. 78–110.

Blowers, Andrew (1998) 'Power, Participation and Partnership. The Limits of Cooperative Environmental Management', in Pieter Glasbergen (ed.), *Co-operative Environmental Governance. Public-Private Agreements as a Policy Strategy*, Dordrecht: Kluwer, pp. 229–249.

Blühdorn, Ingolfur (2006) 'The Third Transformation of Democracy: On the Efficient Management of Late-modern Complexity', in Ingolfur Blühdorn and Uwe Jun (eds), *Economic Efficiency – Democratic Empowerment. Contested Modernisation in Britain and Germany*. Lanham, MD: Rowman & Littlefield, pp. 299–331.

Brosius, Peter, Tsing, Anna and Zerner, Charles (eds), (2005) *Communities and Conservation: Histories and Politics of Community-Based Natural Resource Management*. Lanham, MD: AltaMira Press.

Bulkeley, Harriet and Mol, Arthur P.J. (2003) 'Participation and Environmental Governance: Consensus, Ambivalence and Debate', *Environmental Values* **12** (2), pp. 143–154.

Buttel, Fredrick H. (2000) 'World Society, the Nation-State, and Environmental Protection', *American Sociological Review* **65** (1), pp. 117–121.

Cairncross, Frances (1991) *Costing the Earth*. London: The Economist Books.

Christiansen, P.M. (ed.) *Governing the Environment: Politics, Policy, and Organization in the Nordic Countries*. Copenhagen: Nordic Council of Ministers.

Coglianese, Cary (1997) 'Assessing Consensus: The Promise and Performance of Negotiated Rulemaking', *Duke Law Journal* **46** (6), pp. 1255–1349.

Crowfoot, James E. and Wondolleck, Julia M. (1990) *Environmental Disputes. Community Involvement in Conflict Resolution*. Washington, DC: Island Press.

Dryzek, John S. (1997) *The Politics of the Earth. Environmental Discourses*. Oxford: Oxford University Press.

Dryzek, J.S., Downes, D., Hunold, C. and Schlosberg, D. (2003) 'Ecological Modernization, Risk Society, and the Green State', in *Green States and Social Movements. Environmentalism in the United States, United Kingdom, Germany, and Norway*. Oxford: Oxford University Press.

European Environment Agency (EEA) (2006) 'Using the market for cost-effective environmental policy: Market-based instruments in Europe, Report 1', Copenhagen.

Ekins, P. (1997) 'The Kuznet's Curve for the Environment and Economic Growth: Examining the Evidence', *Environment and Planning A* **29** (5), pp. 805–830.

Fisher, Dana (2004) *National Governance and the Global Climate Change Regime*. Lanham, Md.: Rowman & Littlefield.

Fisher, Dana R. and Freudenburg, William R. (2001) 'Ecological Modernization and its Critics: Assessing the Past and Looking Toward the Future', *Society & Natural Resources* **14**, pp. 701–709.

Frank, David John, Hironaka, Ann and Schofer, Evan (2000a) 'Environmentalism as a Global Institution: Reply to Buttel', *American Sociological Review* **65** (1), pp. 122–127.

Frank, David John, Hironaka, Ann and Schofer, Evan (2000b) 'The Nation-State and the Natural Environment over the Twentieth Century', *American Sociological Review* **65** (1), pp. 96–117.

Frijns, J., Phung Thuy Phuong and Mol, Arthur P.J. (2000) 'Ecological Modernization Theory and Industrializing Economies: The Case of Viet Nam', in A.P.J. Mol and D. A. Sonnenfeld (eds), *Ecological Modernisation Around the World: Perspectives and Critical Debates*. London: Frank Cass, pp. 257–291.

Fritsch, Oliver and Newig, Jens (2009) 'Under Which Conditions Does Public Participation Really Advance Sustainability Goals? Findings of a Meta-Analysis of Stakeholder Involvement in Environmental Decision Making', in Eric Brousseau, Tom Dedeurwaerdere and Bernd Siebenhüner (eds), *Reflexive Governance for Global Public Goods*. Cambridge, MA: MIT Press (forthcoming).

Giddens, Anthony (1998) *The Third Way*. Cambridge: Polity Press.

Giddens, Anthony (2003) *Runaway World: How Globalization is Reshaping our Lives*. New York, NY: Routledge.

Gille, Zsuzsa (2000) 'Legacy of Waste or Wasted Legacy? The End of Industrial Ecology in Post-Socialist Hungary', in A.P.J. Mol and D.A. Sonnenfeld (eds), *Ecological Modernisation Around the World: Perspectives and Critical Debates*. London: Frank Cass, pp. 203–234.

Glasbergen, Pieter (1998) 'Partnership as a Learning Process. Environmental Covenants in the Netherlands', in Pieter Glasbergen (Hg.), *Co-operative Environmental Governance. Public-Private Agreements as a Policy Strategy*, Dordrecht: Kluwer, pp. 133–156.

Glasbergen, Pieter (ed.) (1995) *Managing Environmental Disputes. Network Management as Alternative*, Dordrecht: Kluwer.

Golub, J. (ed.) (1998) *New Instruments for Environmental Policy in the EU*. London: Routledge.

Gonzales, G.A. (2001) 'Democratic Ethics and Ecological Modernization. The Formulation of California's Automobile Emission Standards', *Public Integrity* **3** (4), pp. 325–344.

Habermas, Jürgen (1991) *The Structural Transformation of the Public Sphere: An Inquiry into a Category of Bourgeois Society*. Cambridge, MA: MIT Press.

Hall, P. (1993) 'Policy Paradigms, Social Learning and the State', *Comparative Politics*, **25** (3), pp. 275–296.

Jokinen, Pekka (2000) 'Europeanisation and Ecological Modernisation: Agro-environmental Policy and Practices in Finland', in A.P.J. Mol and D.A. Sonnenfeld (eds), *Ecological Modernisation Around the World: Perspectives and Critical Debates*, London: Frank Cass, pp. 138–170.

Jordan, A., Wurzel, R. and Zito, A. (2003) *'New' Instruments of Environmental Governance?*. London: Routledge.

Jänicke, Martin (1990a) *State Failure. The Impotence of Politics in Industrial Society*. Cambridge: Polity Press.

Jänicke, Martin (1990b) 'Erfolgsbedingungen von Umweltpolitik im Internationalen Vergleich', *Zeitschrift für Umweltpolitik und Umweltrecht*, 3, pp. 213–231 (English translation in Jachtenfuchs and Strübel (eds), 1992, *Environmental Policy in Europe*). Baden-Baden: Nomos.

Jänicke, Martin and Jörgens, Helge (2004) 'New Approaches to Governance' (Neue Steuerungskonzepte in der Umweltpolitik), *Zeitschrift für Umweltpolitik & Umweltrecht* **27** (3), pp. 297–348 (translation from German).

Koontz, Tomas M. and Thomas, Craig W. (2006) 'What Do We Know and Need to Know about the Environmental Outcomes of Collaborative Management?', *Public Administration Review* **66**, pp. 111–121.

Langbein, Laura I. and Kerwin, Cornelius M. (2000) 'Regulatory Negotiation vs. Conventional Rule Making: Claims, Counterclaims and Empirical Evidence', *Journal of Public Administration Research and Theory* **10** (3), pp. 599–632.

Luhmann, Niklas (1989) *Ecological Communication*. Chicago: University of Chicago Press.

Mayntz, Renate (1993) 'Governing Failures and the Problem of Governability: Some Comments on a Theoretical Paradigm', in Jan Kooiman (ed.), *Modern Governance*, London: Sage, pp. 9–20.

March, J.G. and Olsen, J.P. (1989) *Rediscovering institutions*. New York, NY: Free Press.

Marcus, Alfred E. (1980) 'Promise and performance: Choosing and implementing an environmental policy'. Westport, CT: Greenwood Press.

Mazurek, J. (1998) *Making Microchips: Policy, Globalization, and Economic Restructuring in the Semiconductor Industry*. Cambridge, MA: MIT Press.

Mol, Arthur P.J. (2000) 'The Environmental Movement in an Era of Ecological Modernisation', *Geoforum* **31**, pp. 45–56.

Mol, Arthur P.J. (2001) *Globalization and Environmental Reform: The Ecological Modernization of the Global Economy*. Cambridge, MA: MIT Press.

Mol, Arthur P.J. and Buttel, Frederick H. (2002) *The Environmental State Under Pressure* (vol. 10 of *Research in Social Problems and Public Policy*). Greenwich, CT: JAI Press.

Mol, Arthur P.J. and Spaargaren, Gert (1993) 'Environment, Modernity and the Risk Society: The Apocalyptic Horizon of Environmental Reform', *International Sociology* **8**, pp. 431–459.

Mol, Arthur P.J. and Spaargaren, Gert (1998) *Ecological Modernization Theory in Debate:*

A review. Paper presented at the 14th World Congress of Sociology, Montreal, July 1998, p. 24.

Mol, A.P.J. and Spaargaren, Gert (2006) 'Towards a Sociology of Environmental Flows. A New Agenda for Twenty-first-Century Environmental Sociology', in G. Spaargaren, A.P.J. Mol and F.H. Buttel (eds), *Governing Environmental Flows. Global Challenges for Social Theory*. Cambridge, MA: MIT Press, pp. 39–83.

Mol, A.P.J., Lauber, V. and Liefferink, D. (eds), (2000) *The Voluntary Approach to Environmental Policy. Joint Environmental Policy-making in the EU and Selected Member States*. Oxford: Oxford University Press.

Pellow, David N., Schnaiberg, Allan and Weinberg, Adam S (2000) 'Putting the Ecological Modernization Thesis to the Test: The Promises and Performances of Urban Recycling', in A.P.J. Mol and D.A. Sonnenfeld (eds), *Ecological Modernisation Around the World: Perspectives and Critical Debates*. London: Frank Cass, pp. 109–137.

Pedersen, L.H. (2005) 'The Political Impact of Environmental Economic Ideas', *Scandinavian Political Studies* **28** (1), pp. 25–46.

Pierson, P. (1993) 'When Effect Becomes Cause', *World Politics*, **45** (4), pp. 598–628.

Poloni-Staudinger, Lori M. (2008) 'Are Consensus Democracies more Environmentally Effective?', *Environmental Politics* **17** (3), pp. 410–430.

Rinkevicius, Leonardas (2000a) 'Ecological Modernisation as Cultural Politics: Transformation of Civic Environmental Activisim in Lithuania', in A.P.J. Mol and D.A. Sonnenfeld (eds), *Ecological Modernisation Around the World: Perspectives and Critical Debates*, London: Frank Cass, pp. 171–202.

Rinkevicius, Leonardas (2000b) 'The Ideology of Ecological Modernization in "Double-Risk" Societies: A Case Study of Lithuanian Environmental Policy', in G. Spaargaren, A.P.J. Mol and F.H. Buttel (eds), *Environment and Global Modernity*. London: Sage Studies in International Sociology, pp. 163–186.

Russell, C. and Powell, R. (1996) 'Practical Considerations and Comparison of Instruments of Environmental Policy', in J. van der Bergh (ed.), *Handbook of Environment and Resource Economics*. Cheltenham: Edward Elgar, pp. 307–328.

Sabatier, P. (1993) 'Policy Change and Social Learning: an Advocacy Coalition Approach'. Boulder, CO: Westview Press.

Sairinen, Rauno (2002) 'Environmental Governmentality as a Basis for Regulatory Reform: The Adaptation of New Policy Instruments in Finland', in A.P.J. Mol and F.H. Buttel (eds), *The Environmental State Under Pressure*. Greenwich, CT: JAI Press, pp. 85–103.

Scharpf, Fritz W. (1992) 'Die Handlungsfähigkeit des Staates am Ende des 20. Jahrhunderts', in Beate Kohler-Koch (ed.), *Staat und Demokratie in Europa*. Opladen: Leske & Budrich, pp. 93–115.

Scheinberg, Anne (2003) 'The Proof of the Pudding: Urban Recycling in North American as a Process of Ecological Modernisation', *Environmental Politics* **12** (4), pp. 49–75.

Schenk, Anita, Hunziker, Marcel and Kienast, Felix (2007) 'Factors Influencing the Acceptance of Nature Conservation Measures – A Qualitative Study in Switzerland', *Journal of Environmental Management* **83**, pp. 66–79.

Scruggs, Lyle A. (1999) 'Institutions and Environmental Performance in Seventeen Western Democracies', *British Journal of Political Science* **29**, pp. 1–31.

Smith, Graham (2003) *Deliberative Democracy and the Environment*. London: Routledge.

Sonnenfeld, David A. (2000) 'Contradictions of Ecological Modernisation: Pulp and Paper Manufacturing in South-East Asia', *Environmental Politics* **9**, pp. 235–256.

Sonnenfeld, David A. (2002) 'Social Movements and Ecological Modernization: The Transformation of Pulp and Paper Manufacturing', *Development and Change* **33** (1), pp. 1–27.

Sonnenfeld, D.A. and Lockie, S. (2008) Special issue on 'Communities, Natural Resources, and Environments', *Local Environment* **13** (5), pp. 385–391.

Sonnenfeld, David A. and Mol, Arthur P.J. (2002) 'Globalization and the Transformation of Environmental Governance', *American Behavioral Scientist* **45** (9), pp. 1318–1339.

Sonnenfeld, David A. and Mol, Arthur P.J. (2006) 'Environmental Reform in Asia: Comparisons, Challenges, Next Steps', *Journal of Environment and Development* **15** (2), pp. 112–137.

Spaargaren, Gert and Mol, Arthur P.J. (2008) 'Greening Global Consumption: Redefining Politics and Authority', *Global Environmental Change* **18** (3), pp. 350–359.

Spaargaren, Gert, Mol, A.P.J. and Buttel, Frederick H. (2000) *Environment and Global Modernity*. London/Thousand Oaks, CA: Sage.

Spaargaren, Gert, Mol, Arthur P.J. and Buttel, Frederick H. (2006) *Governing Environmental Flows Global Challenges to Social Theory*. Cambridge, MA.: MIT Press.

Speth, James Gustave and Haas, Peter M. (2006) *Global Environmental Governance*. Washington, DC: Island Press.

Steele, Jenny (2001) 'Participation and Deliberation in Environmental Law: Exploring a Problem-solving Approach', *Oxford Journal of Legal Studies* **21** (3): pp. 415–42.

Stern, N.H. (2006) *The Economics of Climate Change: The Stern review*. Cambridge/New York: Cambridge University Press.

Susskind, Lawrence, Bacow, Lawrence and Wheeler, Michael (eds), (1983) *Resolving Environmental Regulatory Disputes*. Cambridge, MA: Schenkman Books.

Tatenhove, Jan P.M. van and Leroy, Pieter (2003) 'Environment and Participation in a Context of Political Modernisation', *Environmental Values* **12** (2), pp. 155–174.

Tu, Wenling (2007) 'IT Industrial Development in Taiwan and the Constraints on Environmental Mobilization', *Development and Change* **38** (3), pp. 505–527.

U.S. EPA (1996) 'The Model Plan for Public Participation', 300-K-96-003.

Van Waarden, F. (1995) 'Persistence of National Policy Styles: A Study of Their Institutional Foundations', in Brigitte Unger and Frans van Waarden (eds), *Convergence or Diversity: Internationalization and Economic Policy Response*, Aldershot: Avebury, pp. 333–371.

Vogel, D. (1993) 'Representing Diffuse Interests in Environmental Policy-making', in: R.K. Weaver and B.A. Rockman (eds), *Do Institutions Matter? Government Capabilities in the United States and Abroad*. Washington DC: The Brookings Institute, pp. 237–271.

Walker, Jack L. (1991) *Mobilizing Interest Groups in America: Patrons, Professions, and Social Movements*. Ann Arbor, MN: University of Michigan Press.

Weale, A. (1993) 'Ecological Modernisation and the Integration of European Environmental Policy', in: Duncan Liefferink, Philip Lowe and Arthur Mol (eds), *European Integration and Environmental Policy*. London: Belhaven Press, pp. 196–216.

Weidner, H. (ed.) (1988) *Alternative Dispute Regulation in Environmental Conflicts*. Berlin: Edition Sigma.

Zhu, D. (2007) *The Chinese Circular Economy Model*, Shanghai: Tongji University (Powerpoint lecture).

Chapter 10

Martin Jänicke and Helge Jörgens

NEW APPROACHES TO ENVIRONMENTAL GOVERNANCE

New challenges for environmental governance

THE CHALLENGES FACED BY environmental policy today differ sharply from those of past decades – both as regards the environmental problems calling for attention and the strategies available to tackle them. On the problems side, after notable successes in some sub-areas of environmental protection, the focus is now on issues where environmental policy has failed to bring about any significant improvement even over a substantial period of time (SRU 2002: 69–74; Jänicke and Volkery 2001). On the response side, there has been continuous growth in both the regulatory repertoire and the range of actors involved. Though they still predominate, traditional forms of hierarchical intervention are increasingly being supplemented by new forms of cooperative governance. This can tend to cause a weakening of state authority and democratic legitimacy, and reductions in established institutional problem-solving capacity (Pierre 2000: 2). At the same time, new policy instruments offer an opportunity to plug deficits in existing environmental policy and help address so-far unresolved environmental problems.

This paper explores how persistent environmental problems can be tackled more effectively against a backdrop of changes in the institutional and policy framework, and what part can be played here by new approaches to governance. The focus is therefore on fundamental issues of environmental policy making that come under the general heading of governance in both academic and policy debate.

The first part of the paper describes in detail the changed environmental and political situation. The second part, written in the light of experience with the Rio Process begun in 1992 and its ambitious model of multilevel governance, goes on to evaluate four central governance approaches in recent environmental policy: Target orientation, integration, cooperation and participation. New forms of environmental governance are shown to promise improvements in solving persistent environmental problems, but are highly demanding in their own right and require additional support, particularly in the form of backup from traditional hierarchical regulation. Without these added precautions, new patterns of governance can be expected to suffer in terms of efficiency and effectiveness or worse still, to deliver lower levels of environmental protection.

Persistent problems

Persistent environmental problems are problems where environmental policy has failed to bring about any significant improvement over a substantial length of time (SRU 2002: 69–74). They include globally undiminished greenhouse gas emissions, biodiversity loss, urban sprawl, soil and groundwater contamination, use of hazardous chemicals, and an array of environmental pressures on human health (EEA 2002, OECD 2001). The intractability of such persistent problems is due to three main factors.

Firstly, they often represent environmental and health risks whose sources lie outside the traditional domain of environmental policy and which instead are a product of the 'normal' functioning of other sectors of the economy and society. Unlike past environmental policy successes – such as improving surface water quality or phasing out the use of ozone-depleting substances – where a technological fix or the ability to exploit a win-win situation secured rapid progress, solving persistent environmental problems requires lasting change in the operating logic of the sectors that cause them.

This is compounded by the fact that these sectors are ones whose production activities necessarily place heavy demands on the environment. This is true of mining and the raw materials industries, and equally so of energy, transport, building and agriculture. Environmental policy has to contend here not only with sectoral pressures on the environment, but with sectoral policies that exist for each industry: Economic, energy, building and agricultural policies see their primary aim in securing production conditions for their client sectors and so improving conditions for growth and employment overall. They show a strong tendency only to consider environmental concerns where these do not oppose fundamental sector interests – as a result of which environmental policy is treated as a bolt-on extra, to be retrofitted like environmental technology. But even efficiency-boosting 'ecological modernisation' can fall foul of sectoral barriers if it adversely affects other industries' markets. A drive to save electricity, for example, only stands a chance of success if it has the power industry's backing (for example because the industry sees alternatives in new lines of business). Herein lies a special challenge for modern environmental policy.

Secondly, the majority of persistent environmental problems are highly complex. Most develop slowly, manifesting themselves as creeping forms of degradation and implicating a plethora of actors, many of them only indirectly. Substantial distances or delays between cause and effect, diffuse discharges and problematic additive impacts make reactive environmental policy a non-starter while simultaneously raising the bar for precautionary strategies targeting environmentally relevant sectors. Such problems can thus further aggravate existing governance challenges critically voiced under headings like ungovernability, governmental overload and governmental failure – challenges which moved Niklas Luhmann (1989: 85) to the acutely sceptical statement that environmental problems make it "quite clear how much politics would have to accomplish and how little it can".

Thirdly, the intractability of persistent environmental problems must be viewed alongside the relatively poor approval ratings for ambitious environmental policy action. The approval gap is partly due to the described need for incisive changes in other sectors of the economy and society. But it also arises because many of today's pressing environmental problems, such as urban sprawl, climate change and species loss, only make themselves felt gradually rather than immediately and so need to be highlighted by scientists and the media before the public will sit up and take notice (Jänicke and Jörgens 1998). This is compounded by the fact that environmental policy successes of recent decades, mostly involving visible problems like smog and surface water pollution, can give a false impression that the most urgent environmental problems are largely under control (BMU and UBA 2002: 34 f.). Finally, environmental policy's failure to master certain persistent problems has led to palpable resignation among those who have fought to bring attention to them for years or even decades. What was a heated public debate surrounding hazardous chemicals in Germany as recently as the 1980s has thus now petered out. This acceptance dilemma, with one group's false sense of security feeding the other's resignation, comes to a head when industry is asked to accept more rigorous regulation. Its severity is borne out by the fierce counterswell experienced by former environmental protection pioneers such as the USA and Japan in the 1980s and more recently countries like the Netherlands and Denmark.

Fourthly, persistent environmental problems are often global in nature. Because they can transcend national borders, effective problem-solving is only possible on an international scale. But heterogeneous interests and the many veto opportunities for opponents of radical environmental protection measures often make it harder to properly coordinate the policies of sovereign nation states than to solve geographically limited environmental problems at national level. Effectively dealing with persistent environmental problems is therefore closely linked to the difficulties of policy coordination in the multi-level international system.

Changes in the political and institutional framework

The changing nature of environmental problems is paralleled on the response side by gradual change in the political and institutional framework of environmental policy. In a shift away from the conventional debate surrounding policy instruments, these changes are mostly discussed under the heading of environmental governance. 'Governance' is an umbrella term for diverse forms of state and non-state political control exercised today at various policy levels against a backdrop of growing complexity in actor structures and the operating environment (Pierre and Peters 2000; Kooiman 2003; Hooghe and Marks 2003). The term 'governance' therefore takes in a broader range of actors and policy instruments than 'government', which is restricted to state action. Environmental policy in particular has been quick to adopt this broad concept of governance because in this policy area the legislative means and agents are diversified to a particularly advanced degree (Holzinger et al. 2003; Bressers and Kuks 2001). Ultimately, proposals for environmental governance are about how better to address intractable, predominately global problems given the large variety of policy levels (global to local), sectors (policy integration), vested interests (stakeholders) and competing policy instruments.

It is important to distinguish in the governance debate between analytical and normative uses of the governance concept (Mürle 1998: 6). In its analytical sense, the concept of governance is used to describe, without any implied value judgement, the changes in the practice and conditions of policy making outlined above. Used normatively, it denotes a variety of value-laden, sometimes conflicting visions. These range from notions of the 'minimal state', that is, the systematic reduction of state services and intervention (e.g. Osborne and Gaebler 1992) to the 'good governance' principles propagated by the World Bank and the International Monetary Fund (see, for example, Leftwich 1993).

The following sets out in greater detail the empirical changes in governance and the conditions surrounding them – changing actor groups, the increasing variety of policy levels and instruments, and changes in the institutional framework. Each section also briefly addresses key pertinent normative policy issues from the contemporary governance debate.

Actors and actor groups

The basic environmental policy actor groups have changed significantly over the last three decades, in Germany as in other industrialised nations and across the European Union (Jänicke and Weidner 1997: 146 f). The first phase of environmental policy in the late 1960s and 70s was dominated by the traditional dipole of the state as the originator of policy and industry on the receiving end. A second phase saw additional groups join the fray in the form of environmental organisations and the media. Another new feature of this phase was the emergence of interaction between actor groups. One reason for this was that the direct, usually command-and-control forms of regulation that had predominated until then began to be supplemented by cooperative approaches such as voluntary commitments by industry (SRU 1998: 130–151). Another was that many environmental organisations began to take their demands to polluters themselves – from the late 1980s mostly in the form of protests, but the 1990s increasingly saw instances of cooperation between environmental NGOs and big business (Weidner 1996; Jacob and Jörgens 2001). Finally, a partial shift of environmental policy responsibility from within environmental policy institutions and out into other policy areas can be observed in the state apparatus since the early 1990s (SRU 2002: 153–156; Lenschow 2002; Lafferty 2001).

The empirically observed expansion in the range of actors is often hastily construed as a sign of waning government and administrative influence or even a 'retreat of the state' (Schuppert 1995). Some normative policy approaches go a step further and call for a large-scale transfer of governance tasks to social actors along with comprehensive deregulation (see, for example, Osborne and Gaebler 1992). There is a misguided tendency in the context of such arguments to invoke the 'governance without government' formula coined with reference to international policy coordination (Rosenau and Czempiel 1992; Young 1999). In fact, this term is used by its authors in an analytical sense and merely denotes the empirical fact that the international system has no superordinate government with binding decision-making powers and governance therefore relies by necessity on horizontal policy coordination mechanisms (Rosenau 1992: 9). It does not infer any value judgement in favour of deregulation and reducing the role of the state.

Policy levels

As well as a more varied actor spectrum, the governance concept also denotes an increase in reciprocal influence between policy levels and their respective state and non-state actors. This effect is seen at global level with the various international institutions and environmental regimes, and also at EU level. These policy levels above the nation state have steadily gained in importance in line with the amount of policy they originate. Yet this by no means spells less influence for nation states, including EU member states. Instead, interdependencies that have arisen between the different policy levels make nation states, though no longer entirely free in their actions, indispensable as the source of power and legitimacy for international arrangements (Jänicke 2003a). The environmental policy agenda of EU member states, for example, is now largely determined by the need to implement prevailing European law and to anticipate and actively shape European measures and action plans (Héritier et al. 1994; Demmke and Unfried 2001). An array of international conventions and multilateral agreements must also be taken into account when shaping national environmental policy (Jacobsen and Brown-Weiss 2000; Lafferty and Meadowcroft 2000). Conversely, the enactment of European and international environmental measures requires input from participating nation states. Their positions and interests in turn are influenced to a large degree by national and international lobby groups and transnationally operating networks of environmental activists (Keck and Sikkink 1998) and environmental scientists (Haas 1992). Environmental governance is therefore increasingly caught up in a complex web of state and non-state actors operating and interacting at different policy levels.

Environmental issues also affect policy at the subnational levels of regional and local government. The role of the world-wide Local Agenda 21 process is just one example of this. Finally, citizens too are party to the multi-level environmental policy system (for a detailed discussion see SRU 2002: 86–122). Also, greater emphasis is now being laid on citizens' role as voters, consumers and members of nongovernmental organisations. Not only is environmental policy undergoing a proliferation of policy levels: the individual levels are gaining in importance.

Policy instruments

As with the widening range of actors and policy levels, a substantial amount of attention has been paid to the emergence of new instruments specifically in the context of environmental policy (see, for example, Golub 1998, Knill and Lenschow 2000, and De Brujin and Norberg-Bohm 2005). Information-oriented instruments like ecolabels have long supplemented the regulatory toolkit (Kern et al. 2002; Jordan et al. 2001; Winter and May 2002). Conversely, market-oriented governance approaches, which took a central place in the academic debate from the early

1970s, have been very slow to be adopted in practice (Holzinger 1987; Opschoor and Vos 1989; Zittel 1996). Since the late 1980s, an increasingly important part has also been played, both in policymaking practice and in political scientists' analysis, by cooperative instruments such as voluntary agreements between environmentally relevant sectors and the state (Glasbergen 1998; De Clercq 2002; Jordan et al. 2003a). Finally, the 1990s saw at least some degree of environmental policy deregulation. The term 'deregulation' here is taken on the one hand to mean the trend towards liberalising sectors close to the state and privatising state-owned industries as seen in telecommunications, energy, water supply and waste disposal (SRU 2002: 295–304 and 448–461) and on the other to a partial withdrawal from direct, usually command-and-control intervention by the state and increased reliance on market forces and self-regulation (Collier 1998). Only the latter variety can properly be termed deregulation, however, as liberalisation and privatisation of state activities and services are almost invariably accompanied by a significant degree of new regulation – for example in competition law or the creation of new regulatory agencies (Collier 1998: 4; for a fundamental discussion, Levi-Faur and Jordana 2005; Majone 1990).

At international level, too, elements of informational guidance or cooperative governance have gained importance alongside conventional treaties. One example of international environmental policy involving relatively little treaty law comprises United Nations Environment Programme (UNEP) and OECD activities to promote sustainable patterns of consumption. Both of these organisations rely first and foremost on informational guidance by gathering and disseminating best-practice knowledge. Their activities are based on the fourth chapter of Agenda 21, which outlines general aims and principles for "changing consumption patterns" (UNDESA 1992: Ch. 4). Even less formal governance mechanisms are described and analysed under the heading of 'governance by diffusion' (Jörgens 2004; Busch et al. 2005; Kern 2000; Kern et al. 2000). This research centres on the observation that, in developing measures and programmes, national governments increasingly look to policies already practised in other countries, and many policy innovations rapidly propagate through the international system without any need for binding international accords (Busch and Jörgens 2005a; Dolowitz and Marsh 1996 and 2000). Such diffusion processes are borne by a range of different actors – international organisations or networks. An environmental policy corollary of the governance-by-diffusion idea might be conscious reliance on the exemplar effect of policy trailblazers – nations that exert political and technological innovative and competitive pressure on other countries and so become an important element of international environmental governance (SRU 2002; Jacob and Volkery 2003; Jänicke et al. 2003).

As with environmental policy actor groups, so with policy instruments: the empirically observable emergence of cooperative, market-based or informational governance strategies is often taken to imply a decline in direct state intervention, mostly of the command-and-control variety. However, there is as yet no empirical evidence to support such a decline in the importance of 'traditional' state activity (as regards the European Union see Holzinger et al. 2003). The more likely consequence is a broadening of the governance repertoire, aiming to achieve results not so much through specific instruments as through a judiciously chosen policy mix (SRU 2002: 74–86; Jänicke 1996).

The institutional framework

Finally, the governance debate also pushes the role of institutions into the policy limelight. As relatively stable collections of formal rules and practices that structure relations between actors along predictable lines (March and Olsen 1998: 948; Hall 1986: 19), institutions affect the options available to political and social actors by precluding certain choices and making others – like giving NGOs access to justice in environmental matters – possible in the first place (Aspinwall and Schneider 2000: 4–5). Additionally, institutions shape the vested interests, preferences and also expectations of actors in their sphere of influence (DiMaggio and Powell 1991: 11). Normative policy approaches to governance – like the European Commission's White Paper on European Governance – often have an institutional dimension in that they develop detailed visions as to how to reorganise the institutional framework within which governance processes operate. They therefore also affect the power relations between different actor groups.

Interim conclusions

Pulling together what we have seen so far, recent changes in the political and institutional environmental policy framework do not readily fit under buzzwords like 'deregulation', 'retreat of the state', 'Europeanisation' or 'the end of the nation state'. These changes evidence not a zero-sum shift in political authority but increasing interconnections and interdependencies among growing numbers of actors, policy levels and policy instruments.

The complex structure of environmental governance today demands detailed analysis and appraisal of ongoing developments and tabled reform proposals. The next section therefore investigates and appraises the strengths and weaknesses of four recent approaches to environmental governance: Target orientation, integration, cooperation and participation.

Recent approaches to environmental governance

The following sections present, in the light of experience with the Rio Process and its strategic goals, an overview of key new environmental governance patterns that are potentially capable – or are claimed to be capable – of better meeting the new environmental policy challenges described so far.

Strategic environmental governance and the Rio framework

The Agenda 21 governance approach

Agenda 21 (UNDESA 1992) is a strategic governance approach to environmental and developmental problems adopted by the UN Summit at Rio de Janeiro in 1992. It is a strategy for sustainable development with overarching long-term goals and operational targets, including success monitoring. But it is also an environmental governance framework integrating key approaches to governance that have acquired importance on a national and European scale: Long-term planning, target and results-oriented governance, environmental integration, cooperative governance, self-regulation and participation (Jänicke 2003b). In the following, we undertake an appraisal of this most ambitious approach to environmental policy seen so far.

In forty chapters, Agenda 21 not only embodied the then current state of academic knowledge in matters of environmental policy. It also reflected general public sector reform trends across the industrialised world. Looking back, it was a prodigious conceptual achievement with sometimes unexpectedly far-reaching knock-on effects. It owes its special status not least to the fact that a global multi-level and multi-sectoral governance approach of this kind has so far been unique to environmental policy and has only existed in this form. In view of the complex implementation process taking effect at multiple policy levels, the German Bundestag Committee of Enquiry into Protection of Mankind and the Environment for the 1994–1998 election period deliberately picked out Agenda 21 for special mention as a "new model of governance" (Enquete-Kommission 1998: 55 ff.). A model role is sometimes also ascribed to the main substantive part of Agenda 21 – the concept of 'sustainable development'. In its three-pillar form, for example, this is recommended by the OECD as a substantive framework for global governance (GASS 2003; likewise Enquete-Kommission 2002). The Rio process can be considered the most comprehensive trial of new environmental governance approaches so far undertaken and as such can furnish key information about approaches and obstacles to solving persistent environmental problems in a changing political and institutional landscape.

The key features of the integrated model of governance put forward by Agenda 21 are as follows (SRU 2000; Jänicke and Jörgens 2000):

- Strategic approach: Consensual, broad-based target and strategy formulation with a long-term horizon (Agenda 21 Chapters 8, 37 and 38).
- Integration: Integration of environmental concerns, and in particular environment and development, into other policy areas and sectors (Chapter 8).
- Participation: Widespread participation by NGOs and citizens (Chapters 23–32).
- Cooperation: Cooperation between state and private-sector actors in environment-related decision-making and enforcement processes (a recurring theme in all chapters).
- Monitoring: Success monitoring with a diverse range of reporting obligations and indicators (Chapter 40).

Agenda 21 sets targets for key problem areas and individual policy levels. It assigns specific tasks to selected actor groups such as business and industry, the scientific and technological community, and local authorities. It aims overall to replace reactive, additive, case-by-case policy decision making to protect the environment with broad-based global, national and local efforts to achieve environmentally more sustainable and globally more equitable development.

The Rio Process

The Rio Process, structured by and designed to implement Agenda 21 ('Rio Plus 5' or the Johannesburg Summit), met with many obstacles over time (see section 2.1.3) but nonetheless delivered sometimes remarkable and unanticipated results. For example, more countries established environment ministries or central-government environment agencies during the 1990s than ever before, and today over 130 countries have institutions of this kind (Busch and Jörgens, 2005b). In addition, some 140 countries followed the non-binding Rio injunction to develop a national environment plan or national sustainability strategy. By 2002, the great majority of countries had also submitted structured reports (country profiles) on implementing Agenda 21. The OECD, too, heavily promoted national sustainability strategies for its member states (OECD 2001, 2002). Finally, some 6,400 'Local Agenda 21' processes have been completed or initiated in 113 countries (UNDP and OECD 2002: 64). The extent to which third parties are or have been at least formally involved in this process is shown among other things by the over 1,000 NGOs registered with the UN Commission on Sustainable Development (CSD) set up in 1993.

Key governance elements from Agenda 21 have also been implemented in the European Union:

- Agenda 21's strategic goal-oriented governance approach found clear expression in the Fifth EC Environmental Action Programme in 1993.
- The EU adopted its own sustainability strategy in 2001.
- The principle of environmental integration is not only enshrined in the EC Treaty (Article VI), it has also started being put into practice in what is known as the Cardiff process, an ambitious attempt to develop environmental integration strategies for different sectors.
- There are now numerous instances of cooperative governance being put to the test in environmental policy. Coregulation and industry voluntary commitments and agreements play an increasingly important role.
- Aside from the Fifth Environmental Action Programme, Agenda 21 participation principles have been taken up elsewhere, including in connection with the Aarhus Convention.

Overall, in the course of implementing the Agenda 21 approach to governance, the Rio process triggered important learning experiences at all policy levels and for key environmentally relevant sectors. Its influence essentially extended beyond the UN Johannesburg Summit in 2002. In March 2003, the UN Commission on Sustainable Development adopted a detailed working programme to implement the Johannesburg resolutions. A comprehensive appraisal of Agenda 21 and its implementation is planned in 2016/17 (BMU 2003). A number of EU countries are not only implementing their sustainability strategies as planned, but continue to develop them further. Barely a year after the Johannesburg conference, France presented a wide-ranging and in parts ambitious sustainability strategy. The German government published a preliminary assessment of its sustainability strategy in 2004 and used the opportunity to set new points of focus. At subnational level, the German states of North Rhine-Westphalia and Schleswig-Holstein recently pledged to develop their own sustainability strategies.

The spring 2003 European Council resolved to reinforce the environmental dimension of sustainable development and called for "new impetus" in this regard (Presidency Conclusions, 21 March 2003). It reaffirmed the targets of its own sustainability strategy and supplemented them with the main targets set out in the Johannesburg Plan of Implementation. For the ongoing evolution of the sustainability process, the Council underscored the importance of setting and extending "indicative targets". It resolved to strengthen the Cardiff process for environmental integration and to develop sector-specific objectives for decoupling environmental degradation and resource use from economic growth. The EU regards itself as having a "leading role in promoting sustainable development on a global scale" (loc. cit.).

In the light of the Rio process, then, the ambitious approach to governance enshrined in Agenda 21 essentially proved its worth as a guidance system for a long-term multi-level and multi-sector strategy. Multi-level governance extending from the global to the local (the latter as Local Agenda 21) proved especially effective. This is surprising to the extent that the only influence wielded here from global level consisted of describing problems and recommending strategies whose main instrument was international reporting. Multi-level governance here was ultimately a voluntary pact among nation states, including a right to not conform. The Rio model of global environmental governance thus took effect through extremely 'soft' policy instruments.

Obstacles

This effect was mostly restricted, however, to agenda-setting and strategy formulation phases at the various policy levels. A far more mixed picture emerges when we look at the quality of the strategies themselves, and even more so regarding their implementation. Overall, the strategy model pursued in the Rio Process came up against visible limitations on the way to implementation. It cannot yet be regarded as a far superior model of governance when it comes to persistent environmental problems.

Limitations to the Rio Process manifested themselves in many ways: An appraisal specifically targeting national sustainability strategies was slated for the UN Johannesburg Summit in 2002 but failed to materialise beyond academic strategy recommendations submitted at the time (UNDP and OECD 2002; World Bank 2003). This body of experience, which was also documented in country reports, thus went unexploited in terms of drawing formal conclusions. Success stories were not used as a basis for benchmarking. There was no enquiry into why many national sustainability strategies resembled routine publications kept at a general level. The Johannesburg Summit failed to make available any extra institutional capacity either to the CSD or to UNEP (UNU and IAS 2002). The Plan of Implementation that the international community adopted at the summit contained some important and specific targets but is generally considered vague and non-binding.

This tendency of the Rio Process to become bogged down – which we shall not further investigate here – had parallels at European level. Based on Agenda 21 and the Dutch National Environmental Policy Plan (NEPP) of 1989, the explicit target orientation of the Fifth EC Environmental Action Plan (1993) was not carried through to the Sixth Environmental Action Plan of 2001. The Commission itself concluded in 1999 that despite the "ambitious vision" set out in the Fifth Environmental Action Plan – its "main response to the 1992 Rio Earth Summit" – "practical progress towards sustainable development has been rather limited"

(European Commission 1999: 2, 6). Likewise, only the main outlines of a proposal for a sustainability strategy submitted by the Commission were finally adopted (in a fourteen-point resolution headed "A Strategy for Sustainable Development") by the Gothenburg European Council of June 2001. Key specific targets lacked consensus. The Cardiff process of systematically integrating environmental concerns into specific EU sectoral policies lent important impetus to agricultural and transport policy but in the majority of policy areas met with considerable resistance. Overall, the EU's strategic excursions in environmental matters show obvious signs of institutional overload (SRU 2002: 151–152).

General appraisal of key new approaches in environmental governance

Further appraisal outside of the Rio Process is needed for a subset of governance approaches that were incorporated into a unifying system under Agenda 21 but separately adopted and evolved thereafter. This relates to key governance approaches that gained importance during the 1990s, especially in the EU:

*   Target and results-oriented governance
*   Integration of environmental policy into sectoral policy
*   Cooperative governance in the narrow sense of the term (including coregulation)
*   Participation of civil actors

These overlap and combine to form hybrid approaches, and it is partly for this reason that they are discussed here as part of an overall model of environmental governance. Since Agenda 21, this model also includes multi-level governance, which will be explored to greater depth in relation to the EU. Governance techniques such as success monitoring through reporting are likewise universal.

A shared feature of recent governance approaches that have become significant to EU environmental policy in various forms (see SRU 2004) is that they differ throughout from classical, hierarchical *regulation*. This means governance through democratically legitimated action, usually on the part of the legislature, using taxes and other levies or generally applicable rules and standards, and typically directed at 'abstract' groups. New governance approaches, on the other hand, typically represent target-oriented, flexible administrative action – in the conceptual framework of New Public Management (Naschold and Bogumil 1998; von Bandemer et al. 1998) – directed at specific groups and taking them on board in various ways. Whereas conventional regulation is legitimated by democratic majority votes, the new cooperative policy instruments have other sources of legitimacy: negotiated consensus, involvement of 'affected parties', and documentary proof of effectiveness.

Whether and how far new target-oriented and cooperative policy instruments can help in better tackling the long-term environmental policy problems described cannot be fully evaluated in the present paper. In any case, they cannot be properly evaluated in isolation from their specific objectives; the abstract debate on policy instruments came to similar conclusions in the 1980s (see Klemmer et al. 1999: 52 f. and 110–115; Jänicke 1996). It is nonetheless meaningful to attempt an interpretation of the key new forms of governance with regard to their potential problem-solving capacity and, based on experience to date, to typical difficulties they present.

Target and results-oriented governance

Targets, for example in the form of environmental quality standards, are not in themselves new to environmental policy. The features distinguishing new target-oriented approaches – as tried in a number of pioneering countries' national environment plans, mostly since 1992 – are target setting, deadlines, and results monitoring (SRU 2000: 57–68). Thus delineated, target and results-oriented governance approaches are a necessary response to deficits of exclusively reactive and insufficiently effective environmental policy. Without targets, results monitoring (including efficiency measurement) is impossible. Persistent, complex long-term environmental problems in particular demand targets that enable coordinated and continuous action. In the Netherlands, target-oriented, long-term correction of complex environmental pressures has been placed for the Fourth National Environment Plan in a conceptual framework of strategic 'transition management' (Rotmans et al. 2001).

Environmental targets are best developed from a problem diagnosis and initially formulated as quality targets from which action targets can then be derived at increasing levels of detail. Whether binding or merely indicative, they can serve as guidance for a broad range of actors. Another key purpose of targets is to overcome inertia in administrations and organisations. Management by objectives (MBO) is therefore a central topic not only in administrative reform (Naschold and Bogumil 1998), but also in environmental planning in advanced OECD nations (Jänicke and Jörgens 2000). Environmental targets can also have an important instrumental function provided that target setting involves a learning and consensus-building process that makes action likelier and breaks down opposition among affected parties. Another advantage of environmental targets is that they render the policy arena more predictable for investors, stimulate adaptation processes and offer clear avenues of opportunity for innovators. Early signposting of problems to reveal the state's likely response before any decision making processes are embarked upon can be key to processes of innovation (Jacob 1999).

Target-oriented governance comprises a diverse and if anything still growing range of approaches, both as regards target setting processes and target hierarchies. Target setting can be broad-based and institutionally legitimated to a large degree – something which favours its long-term stability. But it can also be a transient outcome of ministerial decisions unlikely to survive the next election. Targets themselves can be binding or merely indicative. They can be enshrined in law (as in Germany's new Renewable Energy Act). They may consist of a results-oriented target-setting process being made binding for a specific problem area (as recently seen in the EU). They may also take the form of a precise technological requirement to be attained by a specified date (as with Japan requiring best available levels of energy efficiency for specific product groups – see Schröder 2003). This diversity can also be seen as part of a process of learning and experimentation in environment policy.

Effective environmental policy target formulation requires a bargaining process that draws upon environmental expertise, and this in turn demands professional management and a suitable institutional framework. Moreover, it is essential that the target setting process be problem-oriented. 'Creeping', persistent types of environmental problem in particular call for an input of knowledge to confront those involved with the long-standing trends. Without this, the necessary debate regarding innovations, win-win solutions and best practices cannot generate broad consensus.

To be effective, target-oriented approaches to environmental governance should build upon stakeholder interests. In view of the acceptance dilemma mentioned earlier, they need primarily to comprise target formulation processes in which a minimum of problem-related communication is assured and the arguments of the affected groups are heard. From the standpoint of industry, they need to set stable targets that provide a predictable framework for R&D processes and investment decisions while allowing flexibility to adapt, for example in line with investment cycles (SRU 2002). Unlike short-run, reactive and hence unpredictable environmental policy intervention, target-oriented approaches of this kind generally meet with a better reception from target groups in industry (see UNICE 2001: 5). In the debate surrounding a system of tradable emission permits for greenhouse gases, for example, industry circles have stressed the importance of long-term policy targets in allowing businesses to plan for the future. Priorities from an administrative standpoint include clear lines of responsibility, high-ranking institutional mandates and sufficient resources. For policymakers and the public, the emphasis is on evaluation of planned action and success monitoring.

As we have seen, target-oriented approaches to environmental governance are anything but undemanding. They require a realistic stance on dealing with foreseeable obstacles. Calls for target and results-oriented policy are by no means new, and have accompanied public sector reform attempts at least since the 1960s. The fact that they are repeatedly voiced testifies equally to their importance and to the difficulties of implementing them. It is no coincidence that environmental policy actors to date have found it easier to agree on policy instruments than on targets. Target-oriented, MBO-style environmental policy not only tends to impinge on vested interests, it also assigns monitoring powers that strong policy sectors and their economic constituencies soon try to escape. The resulting evasion tactics range from outright rejection of targets and dispensing with deadlines to adopting of various types of targets that are either irrelevant or nonbinding. Then there is always the option of subjecting unexceptional, routine tasks to a meaningless target with correspondingly little effect. In this latter scenario, targets paradoxically legitimise a status quo whose inadequacy was the reason for setting them in the first place.

Developing and implementing an appropriate target hierarchy, then, takes considerable effort. In addition to deriving operational-level targets, any target formulation process should therefore be accompanied by a capacity needs assessment. Especially with reference to persistent environment problems, target setting is mostly associated with a need to raise capacity (see below).

Policymaking processes run on scarce resources. They must therefore be allowed to focus on key targets, especially when addressing more intractable problems. Where substantial capacity gains are unfeasible, it is better to focus on a limited number of *strategic goals* or on problems whose potential impact or indeed intractability pose special challenges to society. Based on experience gained in the

Rio process, the pressing need in this regard is to set priorities for environmentally sustainable development.

Environmental policy integration and sectoral strategies

Since environmental resource use is the productive foundation of entire sectors of the economy, the need to integrate environmental concerns into such sectors and their corresponding policy areas should be considered an axiom of modern environmental policy. If key sectors of the economy are a major source of long-term pressures on the environment, environmental policies addressing causes rather than symptoms must aim to bring about change in those sectors. This includes change in the 'responsible' sectoral policy areas, which are substantially shaped by the 'logic' of the sectors concerned and usually serve to safeguard their interests. If environmentally risky sectors do not internalise environmental accountability, environmental policy tends to be additive, limited to relieving symptoms and to peripheral intervention; this insight is by no means new (see Jänicke 1979; Doran et al. 1974). Conversely, environmental policy integration means leveraging sectoral expertise and innovating capacity to attain more sustainable development paths.

Environmental policy integration is now pursued with a broad spectrum of measures. One ambitious approach consists of prompting environmentally relevant sectors such as transport or energy to adopt environmentally oriented sectoral strategies themselves; this approach has been taken up in the EU, most notably under the Cardiff Process initiated in 1998. The apportionment of sectoral obligations under Germany's Environmental Protection Programme in 2000 is another noteworthy example of environmental accountability being assigned to specific sectors (see SRU 2002: 219–220). Concepts relating to sectoral environmental impact assessment are also gaining importance. The European Environment Agency, for example, is developing a system of sectoral indicators along these lines (see EEA 2003). These endeavours focusing on appraisal of past developments and trends contrast with examples of pre-emptive policy action such as environment mainstreaming approaches and assessments targeting planned policies and measures – strategic environmental impact assessments and the like. In some European countries, environmental concerns are incorporated into the budgeting process. A further approach also pursued by the EU is the greening of government operations, for example in the form of environment-friendly procurement (OECD 2002).

The aim here, however, is not to further distinguish the possible forms of environmental policy integration or to explore in greater depth the experience gleaned so far with individual manifestations of this approach. Instead, an evaluation of the problems inherent to the approach as a whole is needed, since it has proved highly demanding in its own right. The notion of environmental policy integration itself is by no means new. Its repeated appearance on the agenda hints at implementation problems due to conflicting objectives. Under the flag of cross-sectoral policy, the approach was accorded official recognition in German environmental policy as early as the 1970s. It was ultimately a logical follow-on from the 'polluter pays' principle that accompanied environmental policy in the industrialised world from the beginning of that decade. Similarly, the concept of environmental policy integration already featured in the Third EC Environmental Action Programme, i.e. as early as 1982 (Knill 2003: 49).

There are reasons why little of this early insight fed through to environmental policy. The integration principle runs, prima facie, contrary to the internal logic of ultra-specialised policy sectors and their economic constituencies. The bias towards 'negative coordination' (Scharpf 1991) that impinges as little as possible on the vested interests of affected sectors can only be overcome with considerable institutional effort. This is certainly true when it comes to integrating environmental concerns into policy areas close to the interests of environmentally relevant industries. There are therefore also reasons why specific sectors such as mining, transport and farming can exert severe pressure on the environment, reasons that involve considerable lobbying power and path dependencies.

Sectors such as heavy industry and energy were thus first compelled – after some delay – to adopt end-of-pipe technologies. There followed more efficient technologies which prevented some of the environmental impacts from the outset. So far, sectoral environmental strategies have been primarily limited to effecting technological change. As soon as there is a need for structural change, a need to intervene not only in the technological structure but in the actual substance of sectors, their markets and even their role in society, there is a tendency for marked obstacles to emerge. Examples include traffic avoidance and power saving as environmental strategies.

Achieving structural change to reduce the environmental impact of such sectors thus entails more sweeping policy instruments than the technology-based environmental policies used to date. Environmental policy integration is therefore not only a potential solution: it is a process with major initial political and communication challenges that demands significant upgrading in its management and capacity. The importance of institutional groundwork to this end can be seen from how Canada found it necessary to back up the (advisory) Canadian Environmental Assessment Agency with a special policy integration post, the Commissioner of the Environment and Sustainable Development, who reports to parliament and appraises on an annual basis the development and implementation of sectoral sustainability strategies (for a detailed discussion, see OECD 2002: 49; SRU 2000: 97).

Sectoral strategies are backed by the arms of government apparatus responsible for them and by corresponding sectoral policies. But for such strategies to bite, sectoral departments must sway their organised stakeholder constituencies in favour of environmental policy integration. One option is dialogue strategy, where sectoral policymakers and environmental experts hold a methodologically carefully prepared, results-oriented dialogue on the joint economic and environmental long-run outlook. The objective is effectively managed discourse for long-term sectoral structural change. Examples in Germany include the consensus-based initiative to phase out nuclear power (Mez and Piening 2002) and efforts to confront the coal industry and its customers in the power generation sector with the long-standing climate trend. In the course of such endeavours, it is appropriate and necessary for sectors to be confronted with the long-term problems they cause (see SRU 2000, 2004). Sectoral strategies need this condition in place to stand a significant chance of success, although relevant scientific input is also necessary. A sectoral stake-holder dialogue is conducted, among other things, to identify the economic risks associated with increased pressure on the environment and what economic and environmental crises the sector ultimately faces as a result. A realistic appraisal of this kind must take in intervention which is avoidable in the normal course of affairs but which the state may be forced to adopt if the public is mobilised in an acute crisis. Environmental history is rich in crisis reactions of this kind, from Seveso to Chernobyl to instances of catastrophic flooding. The largely adverse economic and social impacts of crisis intervention may confront the affected sector with problems that a sectoral sustainability strategy based around long-term investment cycles can help avoid.

In formulating such a dialogue strategy, use can be made of the Dutch 'transition management' concept (Rotmans et al. 2001), of technology impact assessment-based approaches (Renn 1999; Skorupinski and Ott 2000) and the institution of 'consensus conferences' based around them (Joss and Durant 1995; Joss 1998), and of research into green structural policy (Binder et al. 2001). These approaches would, however, require further conceptual development to be useful for this purpose.

Sectoral strategies usually need a high-ranking institutional mandate (parliamentary or governmental) to take account of issues such as problem specification, lines of responsibility and procedural requirements, and also reporting obligations and monitoring. This high-ranking institutional mandate contrasts with earlier integration approaches in which a (generally weak) environment department was expected to effect horizontal coordination between (stronger) sectoral departments.

Primarily 'vertical' environmental policy integration presupposes powerful environment departments with a substantive share in both the top-level mandating process and the ensuing horizontal cooperation with sectoral departments. They must have the human and institutional resources this requires. This added capacity is also needed with regard to strategies for environmental policy integration because, in deference to affected industry interests, sectoral strategies tend to fall short of what is advocated by specialist environmental administrations.

On past experience, it is important that environmental integration and cross-sectoral policy does not result in environmental policy being implemented on an 'amateur' basis by those responsible for other policy areas. Instead, expertise residing in the relevant government departments and industries must be tapped into through appropriate networking structures. Innovative solutions in particular require the use of specialist expertise, both in environmental policy and in the affected sectoral policy departments and targeted parts of industry. Because of the differing extent to which obligations can be imposed on the lobbies involved, however, some sectors are more receptive than others for negotiating agreements with target groups.

The success of a sectoral environmental strategy can be judged among other things by how far existing environmental sections in sectoral ministries regard themselves as part of the overarching strategy rather than as inspectors on behalf of the environment ministry and its policy programmes. A large environment section, as exists for example in Germany's Federal Ministry of Economics, could be a key factor for environmental integration if its primary focus were to be on cross-cutting environmental and economic concerns and not on reporting back to the environment ministry. Capacity building here is no longer a question of (already present) specialist personnel, but of the higher-level role handed down by cabinet or parliament (Jänicke et al. 2002: 129).

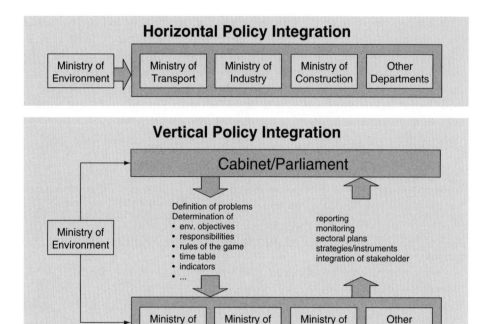

Figure 10.1 From horizontal to vertical environmental policy integration. (Source: Jänicke 2000.)

Cooperative governance

The limits of hierarchical regulation in addressing intractable problems have long been a subject of enquiry for social scientists (Schimank and Werle 2000; Prittwitz 2000; Willke 1997; Mayntz and Scharpf 1995). Depending on the theoretical standpoint, the argument centres around the prohibitively high cost of obtaining information for the desired level of fine control, the inner logic and limited reson- ance capacity of social subsystems in response to central government policy impetus (Luhmann 1989), or difficulties faced by interdependent actors in arriving at a mutually acceptable solution by negotiation.

Where command and control comes up against real or assumed limits, cooperative policy instruments – that is, ways of exerting influence that state actors agree upon with private-sector target groups as essentially equal partners – are increasingly discussed as an alternative.

Effectiveness of cooperative policy instruments in practice

Cooperative policy instruments are not effective per se. Their skilful application has been identified both in environmental policy research and elsewhere as a key success criterion (Jänicke 1996; Knoepfel 1993; Ricken 1995; Wälti 2003). Their assumed benefits include the following:

•   Direct consultation between government departments and target groups often results in better-targeted policy than regulation by legislative decree, not least because knowledge and experience gained in the process can be used as a resource.

- Government departments have a vested interest in legitimating their relatively autonomous actions with real responses to real problems; with their built-in forms of legitimisation – stakeholder participation, consensus and results orientation – cooperative policy instruments can better contribute to problem resolution.
- Policy made in consensus with industry is less likely to face opposition when measures come to be implemented.
- The long route through parliamentary decision-making processes and their constraints are bypassed and adaptive responses (such as innovations) can be stimulated at an earlier juncture.
- 'Soft', communicative and hence more readily accepted policy instruments can be applied provided the option of resorting to 'harder' – for example regulatory or fiscal – measures remains available in principle.

These potential benefits must, however, be weighed in each case against the price exacted by cooperative solutions in terms of policy rigour. The targets and rigour pursued by voluntary commitments, for example, rarely go far beyond the 'business as usual' scenario (see SRU 1998: 130–151). Empirical studies suggest that cooperative and self-regulatory policies tend if anything to deliver only mediocre results with no clear benefits compared with traditional command and control (de Bruijn and Norberg-Bohm 2005). Even the OECD has recently come round to a highly critical appraisal of voluntary commitments: "the environmental effectiveness of voluntary approaches is still questionable. (. . .) The economic efficiency (. . .) is generally low" (OECD 2003: 14). The OECD even sees a tendency for policy instruments used in combination with voluntary agreements to be weakened and therefore recommends supporting arrangements to trigger direct, credible sanctions if targets are not attained. A more recent study, moreover, questions whether voluntary agreements indeed do generally reduce the burden on the state. In one British example, it took 31 civil servants and a total of 17 man-years to negotiate 42 industry climate protection agreements (Jordan et al. 2003b). Also, excessively feeble environmental policy can trigger acceptability problems of its own – especially among affected groups with little opportunity to influence the cooperative solutions. To ensure that cooperative solutions go far enough or can be improved upon if they fall short, they must therefore be backed with ultimate state accountability and must potentially be retractable. This precondition alone shows that cooperative solutions can supplement command and control-style regulation but cannot replace it. Hierarchical regulation then assumes the role of guarantor for the soft, dialogue-based processes of cooperative governance. Soft policy instruments are only effective if state regulation is held ready as a 'stick behind the door' for the event of failure (De Clercq 2002; OECD 2003; Jordan et al. 2003a).

The ability to realise the potential of cooperative solutions also depends on their taking a form that suits the situation. For example, cooperative solutions lend themselves to greater use in some sectors than others. An evaluation of Dutch environmental planning since 1989 thus showed an important factor to be the size of industry associations and their ability to take on binding commitments. Generally speaking, negotiated solutions can usually be reached with industry associations having few major actors – as in the energy, chemicals or automotive industry – while environmentally relevant sectors with a broad and diffuse membership – for example agriculture, car drivers or consumers – can be better influenced by means of traditional regulation.

Requirements of democracy theory regarding cooperative negotiation systems are only to be mentioned here in passing. They relate to the desirable aim of linking negotiations back into the parliamentary process. Government departments should be prevented from constraining parliament with the agreements they enter into. The imperatives of transparency, pluralism and protecting bystander interests must also be observed.

In the light of the potential drawbacks of cooperative policy instruments briefly outlined above, a general reorientation towards this type of approach would not be helpful in environmental policy terms. Cooperative approaches are at best useful as a supplement to direct regulation.

Participation, self-regulation and the 'enabling state'

Under the heading of participation, the Agenda 21 governance model and also the Aarhus Convention aim to exploit the full potential of civil actors by involving them in policy formation and enforcement processes. They thus go beyond the participation envisaged by cooperative approaches, where the policymaking process takes industry on board but not environmental lobbies or interested private individuals. The aim of participatory approaches is to place policy programmes (and in particular the sustainability process) on a broad social base and to mobilise hitherto unused supporters and knowledge repositories. Involving citizens and NGOs thus offers untapped potential for environmental policy. Examples of self-regulatory influence being brought to bear on environmental pressures include activities of nature conservation organisations in designating protected areas under the Habitats Directive and buying up land, plus environmental NGOs working directly to influence planning approval decisions or the product range of retail chains (Conrad 1998: 161–182). Citizens acting in their capacity as consumers constitute a further form of participation.

Participation in environmental policy issues, like the other new policy instruments presented here, is highly demanding in its own right. It presupposes an enabling state (SRU 2002: 86–122). Citizens serving as an added resource for environmental policy need a system of incentives and an infrastructure of rights and information. This includes transparency regarding the environmental credentials of products on the market, and rights of information, participation and access to justice for citizens and environmental groups. The will to participate also presupposes a minimum of accurate and problem-focused environmental reporting in the media. There is again a wide-ranging need for capacity building.

Participation for civil organisations is a management challenge requiring personnel and skills that can by no means be taken for granted. Another management challenge is to shape participation processes so that key interests are included, sufficient expertise is present among those involved and their motivation is not squandered in fruitless debate. Motivation drain in ill-prepared discourse is among the negative experiences that resulted in waning enthusiasm for the Agenda 21 process. Whether participatory approaches strengthen or (by demotivation) weaken environmental policy generally depends on the skill applied in their design. Participatory processes are not the inherent success guarantee they are often made out to be.

Participation must neither be allowed to block the usually scarce time and personnel resources needed for effective policy and administration, nor, finally, must it overextend the capabilities of environmental NGOs by assigning universal responsibility. Formulating selective participation rules that give due account to the identified governance challenges is therefore essential if civil actors are to contribute perceptibly to improving the success of long-term environmental policy.

Multi-level governance

Our survey would be incomplete without discussing multi-level governance. With its high level of complexity, multi-level governance poses a major challenge for effective regulation. Its greatest problem is that it carries a risk of severing lines of accountability, and also offers substantial openings for evasion while leaving policy formulation structures largely opaque. On the other hand, multi-level governance also opens up new avenues of opportunity specifically when it comes to persistent environmental problems. The Rio process is a prime example of this, and one we have already explored in some depth. The 6,400 Local Agenda 21 processes resulting from a global-level strategy recommendation and for that matter the global alliances between environment-related city partnerships underscore the importance of the interplay between local and global levels in environmental policy.

From a nation-state standpoint, both concerted action on a global, European or national scale and decentralisation coupled with subsidiarity are equally effective at boosting capacity and providing flexibility to help achieve a high level of protection. The actual level attainable, however, will vary significantly according to the nature of the problem. A discerning answer to this question is still largely outstanding and represents perhaps the greatest challenge in finding a workable paradigm for environmental governance.

Decentralisation strategies work particularly well where improvements can only be decided on the basis of mostly locally available information (see Scharpf et al. 1976). Many such situations arise in the context of nature conservation, agro-environmental measures, and also transport policy (see SRU 2004). Conversely, decentralisation strategies are of limited use with broader aims (as with Natura 2000) where local decisions fall short from a national or European standpoint, in the presence of major externalities, or where too much coordination is needed between lower levels.

Past experience suggests that strategic targets coupled with reporting obligations are generally suited to the higher level, whereas lower-level implementation should leave the greatest possible scope for flexibility and above all competition. For the German context, this implies a strengthened role for federal government while allowing the sixteen German states greater flexibility in implementation, for example in a more competitive federal framework.

Success criteria for new policy instruments

The key environmental governance concepts outlined here are potentially able to deliver improved problem-solving capabilities. As has been seen, however, they are highly demanding in their own right, to the extent of being counter-productive unless certain preconditions are in place.

In the following we shall explore three important success criteria:

- Capacity building.
- Clearer definition of the role of state influence, in particular regarding guarantee mechanisms for 'soft' forms of governance.
- Improving the role of the nation state in the global and European multi-level system.

Building and 'conserving' capacity

The Rio process and ambitious governance approaches such as the development of sectoral strategies show that every strategy must begin with a capacity needs assessment from which conclusions can be drawn as to available capabilities (Bouille and McDade 2002: 192–200). More ambitious policy instruments need correspondingly greater capacity on the part of the state. This is imperative for strategic approaches at the level of importance of a sustainability strategy, but it is also a precondition for sectoral strategies. It should be recalled in this context that capacity building already featured in two chapters of Agenda 21, which also used the concept of capacity evaluation (UNDESA 1992: Ch. 34 and 37). These recommendations mostly relate to capacity building in developing countries, but the need is not unique to the developing world. Environmental policy integration and participation presuppose added capacity, as has been shown. Neglecting the need to build capacity or to improve management is a major cause of the identified difficulties (Jänicke 2003b). Capacity shortfalls do not relate to the type of measures chosen, but to what Luhmann terms the "conditions of practicability" (Luhmann 1989: 89). Capacity cannot be precisely measured, but it is noticeable by its absence: The best choice of policy instruments is to no avail if knowledge, material, personnel or political resources or institutional preconditions are wanting. In such instances, capacity building or development is inescapable unless the situation allows targets to be lowered instead (see Bouille and McDade 2002; Weidner and Jänicke 2002; OECD 1995). Environmental policy capacity and the capacity to effect sustainable development operate along three dimensions (UNDP and OECD 2002: 92):

- A human dimension, comprising the capabilities of the actors involved
- An institutional dimension, for example the ability to coordinate conflicting interests or to perform monitoring
- A systemic dimension (associated with the 'enabling environment'), for example the legal framework, widespread access to information, and networking capabilities.

Solving intractable long-term problems requires a long-term institutional focus and sufficient human and material resources. Systematically speaking, it is also a matter of social awareness. To the extent it relates to public approval for and readiness to accept demanding solutions, the capacity issue ultimately also includes the role of the media. Regarding the fickle and scarce resource known as public environmental awareness, one suitable approach might be a strategy of dialogue with major media groups, similar to dialogue conducted in Germany on depictions of violence in the media.

The ability of environmental NGOs to play a capable part in an increasing range of policy formation and decisionmaking processes also represents a capacity improvement. This is not least a matter of networking and coalition building. Strategic alliances can boost capacity. This point was well illustrated in the run-up to the adoption of Germany's Renewable Energy Act in 2000 (Bechberger 2000), which involved collaboration not only among parliamentarians with a wide range of party allegiances, but organisations like the mechanical engineering industry association (VDMA), the national metalworkers' union (IG-Metall), the national association of municipal corporations (VKU), and even support groups from the farming sector. Subject to the preconditions outlined earlier, environmental policy integration – sectoral learning about environmental concerns – can likewise help build capacity. Advocates of green solutions in environmentally risky sectors know best their industry's capacity to innovate. Their influence can be bolstered institutionally. However, this also depends on the role of the environment ministry being strengthened overall, both institutionally and in terms of human resources.

As in this case, most capacity building is needed in the state sector. But there is obvious tension here between the need for capacity building and increasingly vociferous calls for 'less state'. Modern forms of environmental governance can result in savings when it comes to detailed regulation and monitoring. The same applies for political rationalisation effects resulting, say, from the fact that negotiated solutions can bypass the lengthy institutional policy-making route altogether. However, the state as facilitator or supervisor, partner in negotiation systems and manager of target formulation processes and sectoral strategies needs extra skills as well as extra personnel. Indiscriminate layoffs and cutbacks – and pruning state activities merely because it is the done thing – can be severely counterproductive in this context.

Last but not least, the capacity to address persistent environmental problems involves the strategic capacity of actors. Strategic capacity can be understood as the capacity to enforce long-term general interests over short-term sub-interests (Jänicke et al. 2003). Its condition is precarious because unlike long-term general interests, short-term sub-interests are almost always highly organised (see Olson 1965) and are aligned to the shorter time horizon of market and electoral cycles. Above all, there is tension between calls for broad interest group participation and cooperation on the one hand and the preconditions for strategic capacity on the other. Collective strategic capacity negatively correlates with the number of organisations to be coordinated and the intensity of competition between them (Jansen 1997: 224). However broad the participation in decision making (or for that matter in enforcement), limitation of the number of decision makers is inescapable. Broad participatory consultation and restricted participation in actual decision making process are by no means mutually exclusive. The latter can also be promoted by clear assignment of decision-making powers – something that is particularly important in the intricacies of a federal system like Germany's (Benz and Lehmbruch 2001). Creating manageable decision-making structures – along with restricting the number of veto points in decision-making processes (Tsebelis 2002) – is a key capacity-related precondition for solving persistent environmental problems, but it is also one that can become compromised in a system of cooperative multi-level and multi-sector governance. The intensity of competition not only relates to cooperation with industry associations. The adversarial political style observed in the German party system especially from the 1990s constitutes another obstacle to overcome: Strategic capacity in the parliamentary system requires a minimum of consensus on basic issues (consensus that is usually further developed in smaller EU member states), otherwise long-term targets will not survive changes of government.

The tensions between ambitious environmental governance patterns and the corresponding capacity requirements raise overall questions that need resolving. In particular, the search should be stepped up for forms of governance that conserve capacity and so relieve the burden on the state and can be squared as far as possible with notions of lean government (see below).

Examples of 'capacity conserving' policy instruments

- All varieties of "negotiating in the shadow of the hierarchy" (Scharpf), which often make it possible to bypass elaborate institutional decision-making processes.
- Specifically, early signposting of problems by the state to give environmentally relevant sectors a reliable indication of impending government action while leaving them the opportunity to adapt (see Section 2.3.2).
- Operating with provisional standards that continue to apply until explicitly revoked.
- Concentrating on strategic targets.
- Reprogramming existing environmental sections in sectoral ministries (economy, transport and agriculture) from monitoring on behalf of the environment ministry to systematically championing environmental concerns.
- Using and promoting decisions at other policymaking levels.
- Exploiting opportunities for action, from environmental crises (e.g. BSE) to sudden price rises (e.g. oil prices).
- Adopting best practices from other countries (where transferable).
- Making use of the Internet, for example as an aid to participatory consultation.

Any solution addressing the causes of a problem rather than the symptoms naturally releases capacity. Above all, the governance approaches focused upon here – not only environmental policy integration, but also target orientation, cooperation and participation – can potentially help release state capacity. This was one of the reasons why they were introduced in the first place. From experience gained with them so far, however, it must be stressed that these approaches have their own capacity needs, neglect of which is a recipe for failure.

The role of the state in the changed governance model

The considerable complexity of multi-level and multi-sector environmental governance with partial or wholesale participation by private-sector actors creates substantial need for guidance. This applies both to accountability and competency structures and to the issue of government involvement and the role of the nation state.

State actors are involved at all policy levels. Policy formulation in Rio de Janeiro and Johannesburg, for example, was largely the domain of government representatives. Civil participation and cooperation with nongovernmental actors is likewise visible at all levels – but has not so far reduced the influence of the state either at national or international level (Raustiala 1997). State actors also play a part in relation to the various environmentally relevant sectors of industry and the corresponding policy sectors.

All the same, the political influence of state actors is qualified by governance approaches based on denationalisation, deregulation and new forms of regulation. Aside from efforts to simplify the law, this mostly involves coregulation in partnership between state and civil or economic actors, and in some cases self-regulation and policy ownership.

This all leads to a contradiction: The main rationale for new forms of governance was that state action fell short in terms of efficiency and effectiveness and state governance had taken on a real or assumed momentum of its own to which those on the receiving end could no longer relate. On the other hand, the new environmental governance tends to create diffuse and tangled lines of responsibility that end in less effectiveness rather than more: If everyone is responsible and accountable, then in the final analysis nobody is. At the same time, there is a so far largely unappraised need to raise state capacity. State influence must therefore manifest itself in differential institutional accountability for securing important common interests, accountability that may be delegated but as a matter of normative principle cannot be abolished. Following on from this premise, state institutions at the various policy levels must serve as guarantors if activities transferred to private-sector actors fail to deliver. Specifically with regard to the persistent environmental problems this paper is concerned with, state entities must be the first port of call and – in the case of delegated problem-solving – the final instance.

The role of the state can even be initially restricted to formally identifying the problem, which in itself can be a powerful signal to private-sector policy targets – especially environmentally relevant sectors of industry as potential sources of innovation (Jacob 1999). For this to work, the formal identification of the problem must clearly indicate that competent state authorities can be relied on to follow through with solutions of their own according to the problem if private-sector actors fail or are not expected to adapt in response. It is known from innovation studies that the mere threat of such consequences can trigger innovation processes (see SRU 2002: 74–86). Hierarchical regulation is indispensable in this role, including sanction-based regulatory law. In all events, soft, cooperative forms of governance need this guarantee function to stand a chance of success and achieve the aim of relieving the burden on the state. The more credible the guarantee function, the less likely it will be needed. For this reason it is essential to emphasise the part played by the state, particularly in the interest of enabling more flexible solutions (see SRU 2002). Jordan et al. (2003a: 222) conclude along the same lines that "environmental governance is at best supplementing, and most certainly not comprehensively supplanting, environmental government by regulatory means".

That cooperative policy instruments are by no means a general substitute for classical command and control is suggested by the fact that even in the light of new governance patterns after the environment summit at Rio, some 80 percent of all EU environmental policy measures were of the command-and-control variety (Holzinger et al. 2003: 119).

The role of the nation state

The general function and importance of the state is not the only issue in the current debate on environmental policy instruments to be in need of basic clarification. It is also necessary to elucidate the role of statehood at the specific level of the nation state – because whether including nation states in global or European multi-level governance hinders or even helps in environmental policy problem solving remains a highly contentious issue. There is also the parallel controversy regarding nation states' scope for action in the light of economic and social globalisation or Europeanisation.

The German Advisory Council on the Environment explored this aspect in detail in its Environmental Report 2002, "Towards a New Leading Role" (SRU 2002), concluding that it provided advanced industrialised nations like Germany in particular not only with major chances to score with environmental technologies in the international innovation race and so hasten the greening of international markets, but also with new opportunities and necessities for environmental policy action. In the following, we aim to recall this position and set it out in a series of propositions in an attempt to take some of the complexity out of the confusing field of environmental governance.

- Both global competition and the internationalisation of policymaking are unquestionably seen to limit the capacity and sovereignty of nation states. Taxation of mobile emission and discharge sources, global economic governance, wage levels and welfare benefits are all examples of adaptational pressure at the expense of national policies. Similar examples are known for environmental policy – for example with a view to the WTO or to EU grant aid. Nation-state environmental policy is nevertheless neither one of the 'losers' of globalisation nor has it so far been significantly obstructed in its problem-solving capacity in the framework of the EU. Indeed, the opposite experience has also been recorded. This has to do with regulatory competition becoming possible in the EU and with (technological) aspects peculiar to environmental concerns.
- Limits to national sovereignty are a logical product of integration into European or global decision-making structures. They should not, however, be equated with a loss of problem-solving capacity. On the contrary, collective action by nation states can raise the capacity to solve environmental problems. Such action is also inescapable where problems are potentially global in nature

rather being limited to a single country. Corresponding change in the operating environment of international markets is likewise only attainable through collective action.

- Environmental policy pioneering by developed industrialised nations plays a special part in this process. There have always been pioneer countries ever since the emergence of environmental policy as a separate policy portfolio in the early 1970s. If anything, the end of the East-West conflict heralded an increase in policy competition and in nation states vying for attention. Never before have even small European countries like Sweden, the Netherlands or Denmark had such influence on global policy development than has been the case in environmental policy over the last decade. Interestingly, these pioneers are closely integrated into world markets (Andersen and Liefferink 1997; Jänicke et al. 2003: Ch. 6).

- The quality of environmental regulation correlates strongly with competitiveness (World Economic Forum 2000). Even if the direction of the causal relationship is unclear, this precludes a systematic negative relationship between ambitious environmental policy and the world market integration of nation states.

- Empirical studies have not yet confirmed the existence of a race to the bottom in environmental policy (SRU 2002: 83–84). This is partly because environmental issues today are closely coupled with technical progress and have become increasingly important in quality-based competition among developed countries. As a result, national innovation systems continue to be highly important. The promotion of lead markets for innovative environmental technologies has also proved a key activity in the environmental policies of nation states (including in small countries like Denmark).

- Within the multi-level global system, nation states possess a number of important properties that have no functional equivalent at other policy levels. These include fiscal resources, a monopoly on legitimate coercion, finely differentiated sectoral expertise and highly developed network structures, including international networking by sectoral government departments. Also important is the existence of a political public and (particularly crucial for environmental concerns) pressure for legitimacy encountered neither at higher nor at sub-national levels. Cooperative governance, too, works best comparatively speaking at state level (Voelzkow 1996; Jordan 2003a: 222). Notwithstanding the broad doctrine of deregulation and denationalisation, national governments remain the public's first port of call when it comes to acute catastrophes like the 2002 floods in Germany.

The nation state level of the multi-level environmental policy system is therefore crucial, including within the EU. This necessarily extends to its European and international integration. Multi-level governance, however, itself requires a guarantor to take on ultimate responsibility if supranational or subnational solutions fail. In this respect, the nation state will have to retain and extend its role as the final instance in long-term environmental policy. For all the criticism regarding the weakening of the nation state, the preconditions for assuming this role are essentially best fulfilled at this level.

Conclusions

New approaches classified under the heading of environmental governance contrast with the traditional, hierarchical rule-based governance that still accounts for nearly 80 percent of environmental policy in the EU. The motives for seeking new approaches to governance are ambivalent. On the one hand, there is the objective of raising the effectiveness of environmental policies, which despite partial successes had failed to stabilise the state of the environment for the long term. On the other, there are the objectives, both appropriate and problematic, of relieving the state and of deregulation. There are major overlaps between these two positions: The search for more effective policy instruments is essential considering the special nature of persistent, intractable environmental problems. In principal, it is the right direction to take. However, it requires more rational management of state capacity – a topic already covered in Agenda 21 (1992) but so far blatantly neglected.

An overall conclusion emerges that it is not only the ambitious targets of a sustainability strategy or governance approaches based on environmental policy integration that stand to fail at a shortfall in state and administrative capacity. Other governance approaches – mostly of a cooperative nature – whose primary aim of reducing the burden on the state likewise imposes substantial requirements in terms of administrative capacity. This raises the question of what relief for the state new policy instruments have delivered so far and how their performance may be improved, particularly as regards persistent environmental problems.

The proliferation of international environmental policy arrangements has not so far reduced the influence of the nation state. On the contrary, nation states now play a multiple role: in solving national environmental problems and in negotiating and implementing international agreements, and in aligning national policies to increasing quantities of international law. Within the multi-level global system, nation states possess a number of important properties that have no functional equivalent at other policy levels.

References

Andersen, M. S. and Liefferink, D. (1997): European Environmental Policy: The Pioneers. Manchester: Manchester University Press.

Aspinwall, M. D. and Schneider, G. (2000): Same menu, separate tables: The institutionalist turn in political science and the study of European integration. European Journal of Political Research, Vol. 38, pp. 1–36.

Bandemer, S., Blanke, B., Nullmeier, F. and Wewer, G. (Eds.) (1998): Handbuch zur Verwaltungsreform. Opladen: Leske und Budrich.

Bechberger, M. (2000): Das Erneuerbare-Energien-Gesetz (EEG): Eine Analyse des Politikformulierungsprozesses. FFU Report 00–06. Berlin: Forschungsstelle für Umweltpolitik.

Benz, A. and Lehmbruch, G. (Eds.) (2001): Föderalismus – Analysen in entwicklungsgeschichtlicher und vergleichender Perspektive. Sonderheft 32 der Politischen Vierteljahresschrift. Opladen: Westdeutscher Verlag.

Binder, M., Jänicke, M. and Petschow, U. (Eds.) (2001): Green Industrial Restructuring: International Case Studies and Theoretical Interpretation. Berlin, Heidelberg, New York: Springer.

BMU (German Federal Ministry for the Environment, Nature Conservation and Nuclear Safety) (2003): Geschärftes Profil, verbesserte Erfolgskontrolle. UN-Kommission für nachhaltige Entwicklung beschließt Arbeitsprogramm 2004 bis 2017. UMWELT 06/2003, p. 326–327.

BMU and UBA (2002): Umweltbewusstsein in Deutschland. Ergebnisse einer repräsentativen Bevölkerungsumfrage. Berlin: BMU.

Bouille, D. and McDade, S. (2002): Capacity Development, in Johannson, T. and Goldemberg, J. (Eds.), Energy for Sustainable Development: A Policy Agenda. New York: UNDP, pp. 173–205.

Bressers, H. and Kuks, S. (2003): What does 'Governance' mean? From Conception to Elaboration, in Bressers, H. and Rosenbaum, W. (Eds.), Achieving Sustainable Development: The Challenge of Governance Across Social Scales. Westport, London: Praeger, pp. 65–88.

Busch, P. O. and Jörgens, H. (2005a): International Patterns of Environmental Policy Change and Convergence, European Environment, Vol. 15, No. 2, pp. 80–101.

Busch, P. O. and Jörgens, H. (2005b): Globale Ausbreitungsmuster umweltpolitischer Innovationen, in Tews, K. and Jänicke, M. (Eds.), Die Diffusion umweltpolitischer Innovationen im internationalen System. Wiesbaden: VS Verlag, pp. 55–193 (forthcoming).

Busch, P. O., Jörgens, H. and Tews, K. (2005): The Global Diffusion of Regulatory Instruments: The Making of a New International Environmental Regime, Annals of the American Academy of Political and Social Science, Vol. 598, pp. 146–167.

Collier, U. (1998): The Environmental Dimensions of Deregulation: An Introduction, in Collier, U. (Ed.): Deregulation in the European Union: Environmental Perspectives. London: Routledge, pp. 3–22.

Conrad, J. (Ed.) (1998): Environmental Management in European Companies. Amsterdam: Gordon & Breach.

Czada, R. and Schimank, U (2000): Institutionendynamik und politische Institutionengestaltung: Die zwei Gesichter sozialer Ordnungsbildung, in Schimank, U. and Werle, R. (Eds.): Gesellschaftliche Komplexität und kollektive Handlungsfähigkeit. Frankfurt am Main: Campus, pp. 23–43.

De Bruijn, T. and Norberg-Bohm, V. (Eds.) (2005): Industrial Transformation: Environmental Policy Innovation in the United States and Europe. Cambridge, MA: MIT Press.

De Clercq, M. (Ed.) (2002): Negotiating Environmental Agreements in Europe: Critical Factors for Success. Cheltenham, Northampton: Edward Elgar.

Demmke, C. and Unfried, M. (2001): European Environmental Policy: The Administrative Challenge for the Member States. Maastricht: European Institute of Public Administration.

DiMaggio, P. J. and Powell, W. W. (1991): Introduction, in DiMaggio, P.J. and Powell, W.W. (Eds.): The new institutionalism in organizational analysis. Chicago: University of Chicago Press.

Dolowitz, D. P. and Marsh, D. (2000): Learning from Abroad: The Role of Policy Transfer in Contemporary Policy Making, in Governance: An International Journal of Policy and Administration, Vol. 13, pp. 5–24.

Dolowitz, D. P. and Marsh, D. (1996): Who Learns What From Whom: A Review of the Policy Transfer Literature. Political Studies, Vol. 44, pp. 343–357.

Doran, C. F., Hinz, M. and Mayer-Tasch, P. C. (1974): Umweltschutz – Politik des peripheren Eingriffs. Eine Einführung in die Politische Ökologie. Darmstadt, Neuwied: Luchterhand.

EEA (European Environment Agency) (2002): Environmental Signals 2002: Benchmarking the Millenium. European Environment Agency Regular Indicator Report. Luxembourg: Office for Official Publications of the European Communities.

Enquete-Kommission (1998): Konzept Nachhaltigkeit. Vom Leitbild zur Umsetzung. Abschlußbericht der Enquete-Kommission "Schutz des Menschen und der Umwelt – Ziele und Rahmenbedingungen einer nachhaltig zukunftsverträglichen Entwicklung" des 13. Deutschen Bundestages. Zur Sache 98/4. Berlin: Deutscher Bundestag, Referat Öffentlichkeitsarbeit.

Enquete-Kommission (2002): Abschlussbericht der Enquete-Kommission "Nachhaltige Energieversorgung unter den Bedingungen der Globalisierung und der Liberalisierung". Berlin: Deutscher Bundestag, Referat Öffentlichkeitsarbeit.

European Commission (1999): Communication from the Commission: Europe's Environment: What Directions for the Future? The Global Assessment of the European Community Programme of Policy and Action in relation to the environment and sustainable development, 'Towards Sustainability'. COM(1999) 543 final. Brussels: Commission of the European Communities.

Gass, R. (2003): A Battle for World Progress. A Strategic Role for the OECD. OECD-Observer No. 236, pp. 29–31.

Glasbergen, P. (Ed.)(1998): Co-operative Environmental Governance: Public-Private Agreements as a Policy Strategy. Dordrecht, Boston, London: Kluwer Academic Publishers.

Golub, J. (Ed.) (1998): New Instruments for Environmental Policy in the EU. London and New York: Routledge.

Haas, P. M. (1992): Introduction: Epistemic Communities and International Policy Coordination, in International Organization 46 (1), pp. 1–35.

Hall, P. (1986): Governing the Economy: The Politics of State Intervention in Britain and France. Cambridge: Polity Press.

Heinelt, H. (Ed.) (2000): Prozedurale Umweltpolitik der EU. Umweltverträglichkeitsprüfungen und Öko-Audits im Ländervergleich. Opladen: Leske und Budrich.

Héritier, A., Mingers, S., Knill, C. and Becka, M. (1994): Die Veränderung von Staatlichkeit in Europa. Ein regulativer Wettbewerb. Deutschland, Großbritannien und Frankreich in der Europäischen Union. Opladen: Leske und Budrich.

Hey, C. (2003): Environmental Governance and the Commission White Paper: The Wider Background of the Debate, in Meuleman, L., Niestroy, I. and Hey, C. (Eds.), Environmental Governance in Europe. Background Study. RMNO Preliminary Studies and Background Studies No. V.02. Utrecht: Lemma Publishers, pp. 125–143.

Hey, C., Volkery, A. and Zerle, P. (2005): Neue umweltpolitische Steuerungskonzepte in der Europäischen Union, Zeitschrift für Umweltpolitik und Umweltrecht, Issue 1/2005, pp. 1–38.

Holzinger, K. (1987): Umweltpolitische Instrumente aus der Sicht der staatlichen Bürokratie. Ifo-Studien zur Umweltökonomie 6. Munich: Ifo-Institut für Wirtschaftsforschung.

Holzinger, K., Knill, C. and Schäfer, A. (2003): Steuerungswandel in der europäischen

Umweltpolitik? In Holzinger, K., Knill, C. and Lehmkuhl, D. (Eds.) (2003): Politische Steuerung im Wandel: Der Einfluss von Ideen and Problemstrukturen. – Opladen: Leske und Budrich, pp. 103–129.

Hooghe, L. and Marks, G. (2003): Unraveling The Central State, But How? Types Of Multi-level Governance. American Political Science Review 97 (2), pp. 233–243.

Jacob, K. (1999): Innovationsorientierte Chemikalienpolitik. Politische, soziale und ökonomische Faktoren des verminderten Gebrauchs gefährlicher Stoffe. Munich: Herbert Utz Verlag.

Jacob, K. and Jörgens, H. (2001): Gefährliche Liebschaften? Kommentierte Bibliografie zu Kooperationen von Umweltverbänden und Unternehmen. Berlin: Wissenschaftszentrum Berlin für Sozialforschung. Discussion Paper FS II 01–304.

Jacob, K. and Volkery, A. (2003): Instruments for Policy Integration: Intermediate Report of the RIW Project POINT. FFU Report 03–06. Berlin: Forschungsstelle für Umweltpolitik.

Jacobson, H.K. and Brown Weiss, E. (2000): Engaging Countries: Strengthening Compliance with International Environmental Accords. Cambridge: MIT Press.

Jänicke, M. (1979): Wie das Industriesystem von seinen Missständen profitiert. Kosten und Nutzen technokratischer Symptombekämpfung. Opladen: Westdeutscher Verlag.

Jänicke, M. (1996): Was ist falsch an der Umweltpolitikdebatte? Kritik des umweltpolitischen Instrumentalismus, in Altner, G., von Mettler Meibom, B., Simonis, U. E. and von Weizsäcker, E. U. (Eds.): Jahrbuch Ökologie 1997. Munich: C.H. Beck, pp. 35–46.

Jänicke, M. (2000): "Environmental plans: role and conditions for success". Presentation at the seminar of the European Economic and Social Committee, "Towards a Sixth EU Environmental Action Programme: Viewpoints from the Academic Community". Brussels: European Economic and Social Committee.

Jänicke, M. (2003a): Das Steuerungsmodell des Rio-Prozesses (Agenda 21). Jahrbuch Ökologie 2004. Munich: C. H. Beck.

Jänicke, M. (2003b): Die Rolle des Nationalstaats in der globalen Umweltpolitik. Zehn Thesen, Aus Politik und Zeitgeschichte, Issue B27/2003, pp. 6–11.

Jänicke, M. and Jörgens, H. (1998): National Environmental Policy Planning in OECD Countries: Preliminary Lessons from Cross-National Comparisons, Environmental Politics 7 (2), pp. 27–54.

Jänicke, M. and Jörgens, H. (Eds.) (2000): Umweltplanung im internationalen Vergleich: Strategien der Nachhaltigkeit. Berlin, Heidelberg und New York: Springer.

Jänicke, M. and Volkery, A. (2001): Persistente Probleme des Umweltschutzes. Natur und Kultur 2 (2), 45–59.

Jänicke, M. and Weidner, H. (1997): Germany, in Jänicke, M. and Weidner, H. (Eds.), National Environmental Policies: A Comparative Study of Capacity-Building. Berlin, Heidelberg and New York: Springer, pp.133–155.

Jänicke, M., Jörgens, H., Jörgensen, K. and Nordbeck, R. (2002): Germany, in Organisation for Economic Co-operation and Development (Ed.), Governance for Sustainable Development: Five OECD Case Studies. Paris: OECD, pp. 113–153.

Jänicke, M., Kunig, P. and Stitzel, M. (2003): Umweltpolitik. Politik, Recht und Management des Umweltschutzes in Staat und Unternehmen (2nd Ed.). Bonn: Dietz, 455 pp.

Jansen, D. (1997): Das Problem der Akteursqualität korporativer Akteure, in Benz, A.

and Seibel, W. (Eds.): Theorieentwicklung in der Politikwissenschaft – eine Zwischenbilanz. Baden-Baden: Nomos, pp. 193–235.

Jordan, A., Wurzel, R. K. W. and Zito, A. R. (Eds.) (2003a): 'New' Instruments of Environmental Governance? National Experiences and Prospects. Environmental Politics Vol. 12, No. 1. London: Frank Cass Publishers.

Jordan, A., Wurzel, R. K. W., Zito, A. R. and Brückner, L. (2001): Convergence or Divergence in European Environmental Governance: National Ecolabelling Schemes in Comparative Perspective. Paper Prepared for the International Seminar on Political Consumerism in Stockholm on May 31-June 3 2001. Manuscript.

Jordan, A., Wurzel, R. K. W., Zito, A.R. and Brückner, L. (2003b): Policy Innovation or 'Muddling Through'? 'New' Environmental Policy Instruments in the United Kingdom, in Jordan, A., Wurzel, R. K. W. and Zito, A.R. (Eds.), 'New' Instruments of Environmental Governance? National Experiences and Prospects. Environmental Politics, Vol. 12, No. 1. London: Frank Cass Publishers, 179–198.

Jörgens, H. (2004): Governance by Diffusion: Implementing Global Norms through Cross-national Imitation and Learning, in Lafferty, W. M. (Ed.), Governance for Sustainable Development: The Challenge of Adapting Form to Function. Cheltenham: Edward Elgar.

Jörrissen, J. (1997): Produktbezogener Umweltschutz und technische Normen. Zur rechtlichen und politischen Gestaltbarkeit der europäischen Normung. Cologne: Heymanns.

Joss, S. (1998): Danish Consensus Conferences as a Model of Participatory Technology Assessment: An Impact Study of Consensus Conferences on Danish Parliament and Danish Public Debate. Science and Public Policy 25, pp. 2–22.

Joss, S. and Durant, J. (1995): Public Participation in Science: The Role of Consensus Conferences in Europe. London: National Museum of Science.

Keck, M. and Sikkink, K. (1998): Activists Beyond Borders. Ithaka and London: Cornell University Press.

Kern, K. (2000): Die Diffusion von Politikinnovationen. Umweltpolitische Innovationen im Mehrebenensystem der USA. Opladen: Leske and Budrich.

Kern, K. and Kissling-Näf, I. in cooperation with S. Koenen, U. Landmann, T. Löffelsend and C. Mauch (2002): Politikkonvergenz und Politikdiffusion durch Regierungs- und Nichtregierungsorganisationen. Ein internationaler Vergleich von Umweltzeichen. WZB Discussion Paper FS II 02–302. Berlin: WZB.

Kern, K., Jörgens, H. and Jänicke, M. (2000): Die Diffusion umweltpolitischer Innovationen. Ein Beitrag zur Globalisierung von Umweltpolitik. Zeitschrift für Umweltpolitik und Umweltrecht (ZFU), 23 (4), pp. 507–546.

Klemmer, P., Lehr, U. and Löbbe, K. (Eds.) (1999): Umweltinnovationen – Anreize und Hemmnisse. Berlin: Analytica, 166 pp.

Knill, C. (2003): Europäische Umweltpolitik. Steuerungsprobleme und Regulierungsmuster im Mehrebenensystem. Opladen: Leske und Budrich.

Knill, C. and Lenschow, A. (2000): Neue Steuerungskonzepte in der europäischen Umweltpolitik: Institutionelle Arrangements für eine bessere Implementation? In von Prittwitz, V. (Ed.): Institutionelle Arrangements in der Umweltpolitik. Zukunftsfähigkeit durch innovative Verfahrenskombinationen? Opladen: Leske und Budrich, pp. 65–84.

Knoepfel, P. (1993): New Institutional Arrangements for the Next Generation of Environmental Policy Instruments: Intra- and Interpolicy Cooperation. Cahiers de l'IDHEAP no. 112. Lausanne: Institut des Hautes Etudes en Administration Publique.

Kooiman, J. (2003): Governing as Governance. London: Sage.

Lafferty, W. M. (2001): Adapting Government Practice to the Goals of Sustainable Development: The Issue of Sectoral Policy Integration. Paper prepared for presentation at the OECD seminar on "Improving Governance for Sustainable Development". Paris, 22–23 November 2001.

Lafferty, W. M. and Meadowcroft, J. (Eds.) (2000): Implementing Sustainable Development: Strategies and Initiatives in High Consumption Societies. Oxford: Oxford University Press.

Leftwich, A. (1993): Governance, Democracy and Development in the Third World, Third World Quarterly 14, 605–624.

Lenschow, A. (Ed.) (2002): Environmental Policy Integration: Greening Sectoral Policies in Europe. London: Earthscan.

Levi-Faur, D. and J. Jordana (Eds.) 2005: The Rise of Regulatory Capitalism: The Global Diffusion of a New Order, Annals of the American Academy of Political and Social Science, Vol. 588, March 2005.

Luhmann, N. (1989): Ecological Communication. Chicago: University of Chicago Press.

Majone, G. (1990): Deregulation or Re-Regulation? Regulatory Reform in Europe and the US. London: Pinter Publishers.

March, J.G. and Olsen, J.P. (1998): The Institutional Dynamics of International Political Orders. International Organization 52 (4), pp. 943–969.

Mayntz, R. and Scharpf, F. W. (Eds.) (1995): Gesellschaftliche Selbstregelung und politische Steuerung. Frankfurt am Main: Campus.

Mez, L. and Piening, A. (2002): Phasing-out Nuclear Power Generation in Germany: Policies, Actors, Issues and Non-issues. Energy and Environment, Vol. 13, No. 2, pp. 161–182.

Mürle, H. (1998): Global Governance. Literaturbericht und Forschungsfragen. INEF Report Heft 32/1998. Institut für Entwicklung und Frieden der Gerhard-Mercator-Universität-GH-Duisburg.

Naschold, F. and Bogumil, J. (1998): Modernisierung des Staates – New Public Management und Verwaltungsreform. Opladen: Leske und Budrich.

OECD (Organisation for Economic Co-operation and Development) (1995): Capacity Development in Environment – Paris: OECD.

OECD (2001): Sustainable Development: Critical Issues. Paris: OECD.

OECD (2002): Governance for Sustainable Development: Five OECD Case Studies. Paris: OECD.

OECD (2003): Voluntary Approaches for Environmental Policy. Paris: OECD.

Olson, M. (1965): The Logic of Collective Action. Cambridge, MA: Harvard University Press.

Opschoor, J. B. and Vos, H. B. (1989): The Application of Economic Instruments for Environmental Protection in OECD Countries: Final Report. Paris: OECD.

Osborne, D. and Gaebler, T. (1992): Reinventing Government. Reading, MA: Addison-Wesley.

Ott, K. (1997): Ipso Facto. Zur ethischen Begründung normativer Implikate wissenschaftlicher Praxis. Frankfurt am Main: Suhrkamp.

Pierre, J. (2000): Introduction: Understanding Governance, in Pierre, J. (Ed.): Debating Governance. Authority, Steering, and Democracy. Oxford, New York: Oxford University Press, pp. 1–10.

Pierre, J. and Peters, G. B. (2000): Governance, Politics and the State. New York: Palgrave, Macmillan.

Prittwitz, V. von (2000): Institutionelle Arrangements in der Umweltpolitik. Opladen: Leske und Budrich.

Raustiala, K. (1997): States, NGOs and International Environmental Institutions. International Studies Quarterly 41, pp. 719–740.

Renn, O. (1999): Ethische Anforderungen an den Diskurs, in Grundwald, A. and Saupe, S. (Eds.): Ethik in der Technikgestaltung. Berlin, Heidelberg, New York: Springer, pp. 63–94.

Ricken, C. (1995): Nationaler Politikstil, Netzwerkstrukturen sowie ökonomischer Entwicklungsstand als Determinanten effektiver Umweltpolitik – Ein empirischer Industrieländervergleich. Zeitschrift für Umweltpolitik und Umweltrecht Vol. 18, pp. 481–501.

Rosenau, J. N. (1992): Governance, Order and Change in World Politics, in Rosenau, J. N. and Czempiel, E. O. (Eds.), Governance without Government: Order and Change in World Politics. Cambridge: Cambridge University Press, 1–29.

Rosenau, J. N. and Czempiel, E. O. (Hrsg.) (1992): Governance without Government: Order and Change in World Politics. Cambridge: Cambridge University Press.

Rotmans, J., Kemp, R. and van Asselt, M. (2001): More Evolution than Revolution – Transition Management in Public Policy, in Foresight, 3 (1).

Scharpf, F. W. (1991): Die Handlungsfähigkeit des Staates am Ende des zwanzigsten Jahrhunderts, in Politische Vierteljahresschrift, Vol. 32, pp. 621–634.

Scharpf, F. W., Reissert, B. and Schnabel, F. (1976): Politikverflechtung. Theorie und Empirie des kooperativen Föderalismus in der Bundesrepublik. Kronberg: Scriptor Verlag.

Schimank, U. and Werle, R. (2000): Gesellschaftliche Komplexität und kollektive Handlungsfähigkeit. Frankfurt am Main: Campus.

Schröder, H. (2003): From Dusk to Dawn: Climate Change Policy in Japan. Dissertation, Fachbereich Politik- und Sozialwissenschaften, Free University of Berlin.

Schuppert, G. F. (1995). Rückzug des Staates? Zur Rolle des Staates zwischen Legitimationskrise und politischer Neubestimmung. Die Öffentliche Verwaltung, Vol. 18, pp. 761–770.

Skorupinski, B. and Ott, K. (2000): Technikfolgenabschätzung und Ethik. Zürich: Hochschulverlag.

SRU (2000): Umweltgutachten 2000. Schritte ins nächste Jahrtausend. Stuttgart: Metzler-Poeschel.

SRU (2002): Umweltgutachten 2002. Für eine neue Vorreiterrolle. Stuttgart: Metzler-Poeschel.

SRU (2004): Umweltgutachten 2004. Baden-Baden: Nomos (forthcoming).

SRU (German Advisory Council on the Environment) (1998): Umweltgutachten 1998. Umweltschutz: Erreichtes sichern – neue Wege gehen. Stuttgart: Metzler-Poeschel.

Tsebelis, G. (2002): Veto Players: How Political Institutions Work. Princeton: Princeton University Press.

UNDESA (UN Department of Economic and Social Affairs) (1992): United Nations Conference on Environment and Development, 3 to 14 June 1992: Agenda 21. New York, UNDESA.

UNDP (United Nations Development Program) and OECD (Organisation for Economic

Co-operation and Development (2002), Sustainable Development Strategies: A Resource Book. London, Sterling, VA: Earthscan.

UNICE (Union of Industrial and Employer's Confederation of Europe) (2001), European Industry's Views on EU Environmental Policy-Making for Sustainable Development. Brussels: UNICE.

UNU (United Nations University) and IAS (Institute of Advanced Studies) (2002), International Sustainable Development Governance: The Question of Reform. Key Issues and Proposals (Final Report), Tokyo: UNU.

Voelzkow, H. (1996): Private Regierungen in der Techniksteuerung. Eine sozialwissenschaftliche Analyse der technischen Normung. Frankfurt am Main: Campus.

Wälti, S. (2003): How Multi-level Structures Affect Environmental Policy in Industrialized Countries. European Journal of Political Research (forthcoming).

Weidner, H. (1996): Umweltkooperation und alternative Konfliktregelungsverfahren in Deutschland. Zur Entstehung eines neuen Politiknetzwerkes. Berlin: Wissenschaftszentrum Berlin für Sozialforschung. Discussion Paper FS II 96–302.

Weidner, H. and Jänicke, M. (Eds.) (2002): Capacity Building in National Environmental Policy: A Comparative Study of 17 Countries. Berlin, Heidelberg, New York: Springer.

Willke, H. (1997): Supervision des Staates. Frankfurt am Main: Suhrkamp.

Winter, S.C. and May, P.J. (2002): Information, Interests and Environmental Regulation. Journal of Comparative Policy Analysis: Research and Practice 4, pp. 115–142.

World Bank (2003): World Development Report 2003: Sustainable Development in a Dynamic World. Washington, D.C.: The World Bank.

World Economic Forum (2000): Global Competitiveness Report 2000. New York: Oxford University Press.

Young, O. R. (1999): Governance in World Affairs. Ithaka und London: Cornell University Press.

Zittel, T. (1996): Marktwirtschaftliche Instrumente in der Umweltpolitik. Zur Auswahl politischer Lösungsstrategien in der Bundesrepublik. Opladen: Leske und Budrich.

Chapter 11

Jan P.M. van Tatenhove and Pieter Leroy

ENVIRONMENT AND PARTICIPATION IN A CONTEXT OF POLITICAL MODERNISATION

Introduction

THE BEGINNING OF THE institutionalisation of modern environmental politics in western countries dates from the late 1960s and early 1970s. During this period environmental issues emerged on societal and political agendas. At the same time attempts were made to improve practices of political participation. Both the environmental issue and the improvement of political participation were an expression of the anti-modern critique on 'modern society'. In this article we focus on the interrelation between participation and environmental policy making within a context of political modernisation. More specifically, we look at the impact of recent changes within the domain of environmental policy upon both the debates and the practices of political participation. The central argument in this article is that participation is inextricably linked to environmental issues and can be considered both as an indicator and a motor for political change. The environmental policy domain thus functions as a laboratory for experiments with the nature of political participation.

To understand the changing discourses on political participation and the different ways of organising participation practices in environmental policy over time, we developed the concepts of *political modernisation* and *policy arrangements*. The essence of our approach is as follows: as a result of processes of political modernisation the substance and organisation of environmental policy have changed over time, resulting in the plurality and co-existence of traditional and innovative policy arrangements. The innovation of environmental politics resulting in these new policy arrangements is provoked by the emergence of new coalitions between actors, by the launching of new policy discourses, or by the capacity of actors to mobilise resources and to change and define the rules of the game. Simultaneously, these innovative environmental policy arrangements are illustrative of a general shift from primarily state-initiated regulatory strategies towards new styles and practices of governance, in which polycentric networks of actors appear to aim at the building of common visions. International campaigns such as Local Agenda 21, initiated at the Rio 92 summit, support this renewal. At the same time, though, traditional styles of governance still prevail in some domains of environmental policy making.

The innovation of environmental policy making in these new arrangements has also affected the issue of *political participation*, as issues surrounding participation were frequently the catalyst for such innovation. In general the participation of citizens, non-governmental organisations, firms and other stakeholders changed from reactive to more reflexive and pro-active ways of participation, and from legislative procedures towards extra-legal processes, often resulting in experiments with participatory (or interactive) policy making. At the same time, in other cases political participation practices have remained very traditional.

Our analysis focuses on the interrelation between participation and environmental policy making. First, we discuss the transformations within environmental policy making which have arisen as the result of the interplay between structural societal and political transformations on one hand and innovations in day-to-day policy making on the other. Second, we discuss the transformations in participation at different stages of the gradual institutionalisation of environmental policy, resulting in interactive policy making, forms of deliberative democracy and new ways of governance. In the third section, we focus on contemporary processes concerning the societalisation and marketisation of environmental politics, and their implications for political participation. Throughout, we pay attention to the co-existence of quite different discourses and practices on political participation, some of them 'late modern', but some reflecting rather earlier stages of political modernisation. In conclusion, we reflect on the relation between environmental politics and participation.

Policy arrangements, political modernisation and participation

We understand the dynamics of environmental politics and policy as related to the duality between structural processes of social and political change on one hand, and to interactions between actors in day-to-day policy processes on the other. We intend to analyse the changing practices of participation with the aid of two concepts: *political modernisation* and *policy arrangements* (Van Tatenhove, 1999; Van Tatenhove, Arts and Leroy, 2000).

Political modernisation

In order to understand change and stability in a policy domain it is necessary to combine an analysis of strategic conduct with an institutional analysis, since both strategic conduct and institutional factors may lead to the renewal of politics and policies – or hamper such a change. The concept of *political modernisation* refers to processes of transformation within the political domain of society. The actual demarcation of 'the political domain' of society depends on the degree of insulation of the subsystems state, market and civil society. In a situation where a clear distinction between state, civil society and market exists, the position of the political domain will be predominantly defined in terms of the rationales of the state. However, contemporary societies show increasing encroachment, interweaving and interference of the three subsystems and demarcation lines become rather vague. Therefore we use a broad concept of the 'political domain' (cf. Held, 1989). Essentially, the political domain of society is the setting in which different agencies and organisations (from state, market and civil society) produce and distribute resources (power and domination) and meaning (discourses) to shape public life.

To grasp the dynamics of the process of political modernisation we distinguish analytically between three 'phases', respectively labelled as 'early', 'anti-' and 'late' (cf. Alexander, 1995). Each of these three can be characterised by specific, ideal–typical relations between state, market and civil society, and by dominant discourses on governance. Therefore, each of these three relates to certain, predominant policy practices (or arrangements).

Early modernisation is characterised by great optimism about the possibility of progress by the application of rationality, on a steering and responsible state, and on the state's capacity to solve societal problems by rational policy making and comprehensive planning. Its basic beliefs are reflected in the characteristics of early environmental politics: state-initiated, taking scientifically deduced standards as goals, and presuming loyalty from both market and civil society in its actual implementation.

Anti-modernisation, in contrast, refers to scepticism about this scientistic optimism, or even to severe criticism of it that emphasises its one-sided, one-dimensional character, the limits of rationality, and the (unforeseen and neglected) external effects of a series of political decisions which affect, among others, the environment and the Third World. Or, as Alexander (1995) puts it, anti-modernisation was a reaction to the unsolved 'reality problems' of the modernisation project, such as inequity and inequality, poverty and starvation, dictatorship and post-colonialism. The anti-modernist discourse was launched by a variety of authors and by the new social movements from the late 1960s and early 1970s, focusing on issues such as inequality, emancipation, democracy and participation. The anti-modernist discourse reflects the opposition of (parts of) civil society to what was regarded as a malicious state–market coalition, and in particular to the oppressing role of the state. It is hard to label any politics as 'anti-modernist', and yet one can point at the claims for more participatory politics, resulting, for example, in policy instruments such as environmental impact assessment and involving the input of counter-expertise in some procedures. More than is apparent from concrete instruments and measures, the anti-modernist wave has given a great impetus to policy making, particularly in relation to the participation issue, as we discuss below.

Late modernisation, finally, neither presumes a synoptic rationality nor a single actor's steering capacity. Giddens (1990), Beck (1994, 1996) and other authors on late and reflexive modernity essentially argue that the side-effects of modernisation and the unforeseen consequences of modernity, such as global risks, will structure society and politics (Franklin, 1998). In this view, the side-effects of modernisation, captured by Beck as the emergence of the 'risk society', have become the pivot of governance. First of all because the state can in part be held responsible for the fact that these risks were not properly regulated, the so called 'organised irresponsibility' of the modern state. Secondly, one may argue that some modern risks cannot be dealt with by the classical, state-centred system of the industrial society. This will lead to the decreasing centrality of the state as a political actor, and to an increasing role for politicisation within other spheres of society. Therefore, late modernisation assumes an increasing interweaving of state, market and civil society, and an inevitable interference and co-operation between their respective agencies, in which the common formulation of the problem and the design of its most adequate solution strategies are part of the policy-making process. These basic features are reflected in a variety of participatory, interactive and deliberative patterns and practices of policy making that we witness throughout contemporary Europe.

The idea of conceiving the phases of political modernisation as consecutive stages in a unilinear development is seductive; and there is, as we have suggested implicitly, an overwhelming amount of empirical evidence to underpin such a stance. However we want to stress the fact that there is both a gradual transition between these stages on the one hand, and a juxtaposition of various types of political modernisation on the other. In other words, 'early' politics can be discerned in contemporary policy processes, and political institutions and practices originating from this period still remain, while paralleled by other, more recently developed styles and practices. In brief, we emphasise the juxtaposition of the three types of political modernisation distinguished, and thereby, the plurality of contemporary politics. Our empirical research in different sub-domains of environmental policy making, such as climate change, infrastructure, nature conservation and agriculture, shows indeed that–contrary to what some scholars suggest – there is no clear, univocal evolutionary path of development from one form of arrangement to another (Van Tatenhove, Arts and Leroy, 2000). At least in environmental policy making, there is no predominant movement away from traditional, (inter)statist arrangements, in which the state plays the dominant role, towards innovative policy arrangements, in which the influence of market and civil society stakeholders has increased. Instead of a unilinear evolution, we establish the juxtaposition and mixture of different types of arrangements. The latter not only differ over time, but there are also quite substantial differences between certain sub-domains of environmental policy making, one can even point some contradictory arrangements within one domain (Pestman, 2000) and, quite obviously, huge differences between different countries can be found (Arts and Van der Zouwen, 1999).

Policy arrangements

We conceive our second concept, *policy arrangement*, as the temporary stabilisation of both the substance and the organisation of a policy domain. 'Temporary' in what we regard to be an ongoing process of institutionalisation, including its construction, de- and reconstruction. The substantial and organisational characteristics of a policy domain can be analysed on the basis of four dimensions: policy coalitions, resources, 'rules of the game', and policy discourses. Policy innovations can be initiated from each of these dimensions. Policy agents may decide: (1) to allow more or new actors to participate in policy making or in coalition formation; (2) to reshape power relations, for example by adding to or withdrawing resources from a policy arrangement; (3) to reformulate the rules of the game on the basis of which policies are made; and (4) to reformulate the policy discourse concerned, for example by redefining its core concepts (Arts and Van Tatenhove, 2000). However, innovations in one dimension tend to have consequences for other dimensions, and even for the arrangement as a whole. In other words, in some cases changes have been initiated by new coalitions (e.g. the participation of citizen groups), whereas in other cases they are provoked by innovative discourses, or reinforced by rules and resources, setting off a chain reaction of changes in all aspects. Finally, this chain may lead to the change of *entire* policy arrangements. The concept of policy arrangements helps us to analyse and interpret changes and continuity in (environmental) politics, as we will illustrate in section 3 and 4.

Participation

Both the gradually changing character of policy arrangements over time and their plurality in contemporary environmental politics effect the development of political participation. We define political participation here in general as the involvement of agents, such as citizens or non-governmental organisations, in politics and the process of 'government'. The way actors participate in politics and the process of 'government' depends on the dominant model of democracy. Within the context of this article, though, it is not possible to discuss these models thoroughly. To understand the relation between environment and participation in the context of political modernisation, we discuss only the ideal types of democracy and participation that are likely to be found within each type of political modernisation. In general, early modernisation seems to be linked with classical variants of liberal (representative) democracy, whereas anti-modernisation is linked with variants of direct democracy, and late modernisation tends to be linked to forms of deliberative democracy and sub-politics.

The concept of 'early political modernisation' reflects a relative insulation of state, market and civil society, each sphere functioning according its own rationales: bureaucracy, competition and solidarity respectively. In this context democracy refers to democratic government, that is, following Warren (1999: 353), arenas of formal state-centred institutions that meet certain requirements, including a representative structure based on a broad franchise, political rights, including freedom of speech and rights to associate, protection for minorities and other related conditions. This liberal conception of democracy essentially takes the form of a cluster of rules and institutions permitting the broadest participation of the majority of citizens in the selection of representatives who alone can make political decisions (Held, 1996: 119). One can speak of 'constitutionally institutionalised' participation. It includes free and fair elections on the basis of suffrage, freedom of conscience, information and expression, associational autonomy etc.

The 1970s show both a renewal of theories of democracy and the emergence of some radical alternatives. These alternatives focused on new linkages between the state and civil society, corresponding to classical ideas of 'the political community', based on the central principle of justification of direct democracy – that 'free development of all' can only be achieved with the 'free development of each' (Held, 1996: 152) – and the New Left model of democracy referred to as 'participatory democracy' (Held, 1996: 263–73). One of the key features of 'participatory democracy' is the direct participation of citizens in the key institutions of society, including the workplace and the local community. More generally, the anti-modernisation wave promoted supplementary forms of participation to bridge the gap left by constitutionally institutionalised participation. These new forms of participation mainly focused on state–civil society interfaces. The new social movements developed a participation repertoire, ranging from public hearings and debates to demonstrations and the barricading of, for example, nuclear power plants. In addition, classical institutions, such as churches, trade unions, firms, and universities, were confronted with claims for more influence and the emancipation of their members. These forms of participation were said to be supplementary, and thereby 'corrective', to the 'constitutionally institutionalised' forms of participation.

As stated above, the emergence of late political modernisation reflects an increasing encroachment and interference of state, civil society and market, with rather vague demarcation lines between them (Van Tatenhove, Arts and Leroy, 2000: 36). In particular, the decreasing centrality of the state's political role is striking. New ways of governance have (to be) developed within and beyond the nation-state, since the state is incapable of accommodating the new and global risks of contemporary society (e.g. nuclear radiation, the greenhouse effect, the possible impact of genetic modification). On the one hand we witness globalisation – posing huge issues of participation as recently expressed by the anti-globalisation movement. On the other hand, Beck suggests that late political modernisation heralds the sub-politicisation of society, so that society is shaped from below, not only by new coalitions of actors, but also giving rise to a whole arena of hybrid sub-politics. Politics is no longer a privilege of the representative institutions of the nation-state, but also takes place in the supermarket, at schools, in the media, on the street. Sub-politicisation presupposes the intermingling of rule-directed and rule-altering politics. The former functions within the rule-system of the nation state, whereas the latter concerns altering the rules of the game.

These political developments make existing forms of participation inadequate, since they do not anticipate the decreasing role of the nation-state, nor the changing interrelations between state, civil society and market. Classical means of political participation are mainly based on a kind of passive trust which presupposes a 'thick support' of civil society for a 'strong capacity' of the state (cf. Bang and Sorensen's critique of Putnam in Akkerman, Hajer and Grin, 2000). To overcome this paradoxical situation forms of deliberative democracy (see section 4) and sub-politics have to be developed on the basis of active trust. This change in governance, from participation to deliberation, calls for new participation arrangements.

The gradual institutionalisation of political participation in environmental politics: 1970–1985

Since the emergence of environmental issues in the 1960s environment and participation have been two inextricable connected expressions of the same 'green discontent' (Lauwers, 1983). Green discontent essentially refers to the protests against environmental harm and the ways in which decisions resulting in such harm were made. In the 1960s and 1970s the green discontent was an anti-modern critique, consisting of two elements. First, a critique of the ignorance of the political and economic establishment about the environmental effects of a series of decisions, (e.g. industrial siting or infrastructure, which underpin economic development (industrial zones, airports, highways, harbours, energy plants etc.)). This critique focused upon the content (output and outcome) of political decision making. Second, there was critique about the way these decisions had been made: without sufficient participation, in some cases without even properly informing the people concerned, and in other cases overruling their protest by the so-called DAD-strategy (decide, announce, defend). This critique focused upon the process of decision making.

Both critiques, distinguished analytically here, were part of a larger anti-modernist discourse. This also encompassed other side-effects of the established politics of early modernisation (from underdevelopment via the arms race to gender discrimination), and it encompassed processes within the political system, but also at universities, in international relations etc. In brief, the anti-modernist discourse opposed the external effects of both the capitalist system and the state, the latter regarded as either politically oppressing or at least facilitating the former. Authors such as Bahro, Gorz, Illich, Schumacher, Roszak elaborated, albeit in different ways, similar critiques, and fed claims for a new understanding and practice of political participation. The claims of the newly emerging environmental movement were paralleled and even preceded by those of other so-called new social movements, including the civil rights movement (USA), students, anti-Vietnam war, and Third World movement. In fact they formed a discursive coalition protesting (a) against the establishment's ignoring of the external effects of a unilateral emphasis on economic growth and (b) against their autocratic, non-participatory way of decision making.

This double critique led to a double development in environmental politics. First, the critique on political content led, from the early 1970s onwards, to the gradual development and institutionalisation of environmental policies. This included a series of legislative initiatives, the establishment of environmental policy departments, the setting of environmental standards and their implementation in series of permits and environmental planning. It also led to the establishment of such typical features as environmental impact assessment (including cost–benefit analysis and risk assessment) and technology assessment. The latter are typical instruments of the anti-modernist stage of development, since they reflect the claim for countervailing power, e.g. for counter-expertise in environmental decision making.

Secondly, the green discontent criticised the lack of transparency and participation in environmental politics, as reflected in a series of struggles over the siting of hazardous industries, infrastructure and other facilities, nuclear power and alike. Gladwin (1980), Blowers (1984), Leroy (1979) reported on different local conflicts, as Blowers and Leroy (1994) and Gould et al. (1996) have done more recently. Apart from their specific content, the issue at stake in all these environmental conflicts was similar. Decisions that could be expected to substantially affect people, not only in their physical environment, but also with regard to their welfare and well being, were taken without participation, and in some cases even without the active knowledge of those affected.

One can discuss the actual impact of all the protesting. Authors differ in their assessment and appreciation of its influence, either on the substantial issue at stake (Huberts, 1988) or, at a wider level, to politics and participation in general (Lowe, 1983). The series of local conflicts, the increasing protest and opposition, and the gradually more powerful position of environmental groups eventually led to newly set up procedures of consultation and participation. Here, 'new' means different from and complementing classical means of political participation such as general or local elections or party membership. This included the gradual renewal of a series of public law procedures for decision making on physical planning and on environmental permits. They provided new opportunities for both citizens and environmental action groups to interfere in (specific) decision-making processes. However, they primarily, if not exclusively, dealt with state–citizen or state–civil society relations.

The nuclear case, in particular, revealed the shortcomings of these newly designed procedures: they focused on specific, mostly local processes of decision making, whereas the nuclear option was neither a specific nor a local one. They provided opportunities for citizens and citizen groups to oppose political decisions, leaving the influence of other actors, e.g. industrial monopolies (in some cases state-owned) unaffected. Therefore, these new forms of consultation did not provide a legitimate and appropriate way to deal with the nuclear issue.

Apart from those intrinsic restrictions of their scope, the functioning of those newly designed procedures largely depended upon the structural openness of the political system in different countries, which in turn influenced the strategic options of the respective environmental movements. The mutual interdependence between these factors has been partly described by scholars using the socalled 'political opportunity approach', linking structural features of a political system to strategic choices of (one of the) actors within that system (for example Kitschelt, Kriesi, Duyvendak, Van der Heijden). We restrict mention here to a brief overview of three countries based on Boehmer-Christiansen, 1991; Weale, 1992; Hajer, 1997; Van Tatenhove, Arts and Leroy, 2000.

In Germany the environmentalist movement was part of a more encompassing opposition movement (Ausserparlementarische Opposition) that largely dominated the political scene of the 1970s. Neither the United Kingdom nor the Netherlands experienced such a radical environmentalism. The German environmental movement did not succeed in getting entry to the established political arenas nor did it get real political influence on actual decision making, whether on nuclear power, airport enlargement or anything else. This lack of access and influence was caused by and in turn reinforced the political radicalism of the environmental movement, particularly during the great coalition of German Christian- and Social Democrats. In the late 1970s the political radicalisation of the environmental movement resulted in the establishment of the Green Party. Their successful electoral campaigns – and the impact of the Chernobyl accident that eventually led to the establishment of a new ministry – forced the traditional parties and the state to develop a more comprehensive environmental policy.

The British environmental movement, unlike its German counterpart, had a very small 'deep ecology' component that never reached the apogee of the German 'fundi's'. Apart from some specific and occasionally violent local and regional environmental conflicts (particularly over nuclear issues), environmental issues were only partly politicised in the UK. During the 1980s the environmental movement had a limited influence on the agenda setting of, for instance, acid rain or nuclear power. This again was partly the result of the national institutional setting, in this case the British two-party political system. On top of that, during the 1980s 'classical' political and economic issues regarding the competencies and responsibilities of state, market and civil society dominated the political agenda. They related to the future of the mining industry, the position of the trade unions, the public health service, local government and the privatisation of a series of national industries and services.

In the Netherlands the environmental movement has quite a different position. In the consensus tradition of Dutch policy making the environmental movement, right from its emergence in the early 1970s, was frequently consulted about many environmental issues. In fact there is a remarkable relationship between the movement and the Ministry of the Environment, especially from the 1980s onwards, as the environmental movement became a natural ally of the Ministry of the Environment. Not only were environmental groups subsidised, they were also invited to contribute to the policy-making process itself. Only the nuclear power debate led to some kind of radicalisation of parts of the environmental movement. Within Dutch consensus politics, an opposition strategy is unlikely to be successful. Since the 1990s onwards the agenda setting and management of environmental issues is subjected to more or less formalised negotiations between political parties, the administration, target groups and the environmental movement by a new version of Dutch consensus politics: 'the green polder model'.

In brief, the claims made by the anti-modernists in the 1970s led to some additional participatory features in the environmental policy domain. The environmental movement played a key role in both advocating and making use of them, adopting an intermediary role between politics and citizens, thereby accentuating specific interrelations between state and civil society. However, the newly designed participatory infrastructure was limited to that interrelation, and primarily conceived as a supplement (not a corrective) to representative democracy, without affecting the roles and power positions of other agencies. This seemed plausible in an era in which policy making was regarded as the responsibility of the state, with the market and particularly civil society conceived to be the mere objects of steering.

The *societalisation* and *marketisation* of environmental politics and their impact upon (the innovation of) participation

Over the last three decades several European countries have witnessed transformations of the institutions of democracy and, as a consequence, of the meaning of political participation. Among the most significant developments have been the erosion of the traditional bases of power of the democratic institutions of the nation-state and the emergence of a diversity of alternative policy arrangements. The institutional capacity of traditional democratic and governance mechanisms have been challenged by globalisation, by the transnationalisation of economic, social and cultural relationships and by the horizontalisation of politics, through which the accepted authority of the state by firms, citizens and subnational governments has been eroded.

In the literature these transformations have been captured in terms of a shift from 'government' to 'governance'. On the one hand there is a shift in the focus of democratic politics and practices, from hierarchical and well-institutionalised forms of government towards less formalised practices of governance, in which state-authority makes way for an appreciation in politics of mutual interdependence. On the other hand there is a shift in the locus of democratic politics: governance at subnational and supranational levels is gaining importance *vis à vis* the national level.

Governance refers to 'sustaining co-ordination and coherence among a wide variety of actors with different purposes and objectives, such as political actors and institutions, corporate interests, civil society and transnational organisations' (Pierre, 2000: 3–4). In this article we focus on governance as a society-centred practice, in which the focus is on co-ordination and self governance, manifested in different types of arrangements (cf. Hirst, 2000; Sbragia, 2000). Society-centred types of governance, like New Public Management and multilevel governance, take as a starting point the observation of an increasing encroachment and interference of state, civil society and market. Recently this type of governance has been linked to the notion of 'deliberative democracy' (Cohen, 1998; Warren, 1999; Hajer and Wagenaar, 2003). Essential for deliberative democracy are public reasoning and deliberation, which refer to argumentative consultation and a collective learning process in which participants (e.g. citizens, governments, NGOs) are not representatives of specific interests, but are aiming at correcting and reconsidering each other points of view. The ultimate aim is the formulation of a common understanding.

These transformations have also affected the institutionalisation of environmental politics. As in other policy domains, environmental policy is confronted with the shift from government to governance, in which the roles and positions of the state, the market, and civil society have been redefined, reflecting the gradual institutionalisation of 'interference zones' between these subsystems. As a consequence, the meaning and character of the participation of citizens, of the environmental movement, of business firms, has changed. Society is no longer seen as something separate from the state that can be governed by it. Instead, the subsystems of civil society and market and their respective agencies are now conceptualised in terms of 'networks', 'associations', 'public–private partnership' and the like, in which the state negotiates with non-state agencies, either from the market or society, in order to formulate and implement an effective and legitimate policy.

Hereafter we restrict our discussion to Dutch environmental politics and policies, and yet there are clear indications that similar evolutions are taking place elsewhere in Europe, influenced by similar mechanisms. We believe the development of Dutch environmental policy from the 1990s onwards has been particularly influenced by the institutionalisation of these 'interference zones' between state and civil society, and between state and market. We will label and analyse the former as the '*societalisation of environmental politics*', and the latter as the '*marketisation of environmental politics*'. Both '*societalisation*' and '*marketisation*' have their impact on the types of governance and, more specifically within this context, upon the form of political participation in environmental politics. In general they imply a change from 'constitutionally institutionalised' and supplementary forms of participation (typical for early- and anti-modernist politics) to more deliberative and reflexive forms of participation.

'Societalisation' of environmental politics: consequences for governance and participation

The *societalisation* of environmental politics refers to politics and types of governance as a result of the institutionalisation of the 'interference zone' between state and civil society. Where the rationales of state and civil society come together, they affect policy making, governance and participation. More specifically, in this 'interference zone' rule-directed and rule-altering arrangements intermingle, leading to a diversity of policy arrangements.

From the state's perspective, the *societalisation* of environmental politics implies innovation in rule-directed instruments, and particularly in the design and implementation of communicative instruments. They are thought to enlarge both the (organising rather than steering) capacity of the state and its legitimacy on the one hand, and to co-ordinate the role of the state in relation to the interests of societal actors on the other. This means that citizens and interest groups not only have the opportunity to influence policy reactively, but are invited to communicate proactively about policy proposals in the different stages of policy making, ranging from the formulation of views and problems to the implementation of policy.

From the perspective of civil society, the *societalisation* of environmental politics implies a change in the patterns of governance, including rule-altering arrangements. These rule-altering arrangements, reflecting the principles of self-governance, vary from policy networks management, to co-production and interactive policy making. Network management and co-production are processes aimed at collective image building, to realise a shared understanding among stakeholders. Based upon criteria such as representativeness, authority and diversity, stakeholders are selected to be partners in a negotiating policy process, the aim of which is to gain support and to realise shared definitions of the situation (Bekkers, 1996; Leroy and Van Tatenhove, 2000). The variety of forms of interactive policy making currently experimented with all over Europe, represents a 'family' of non-codified political practices in which citizens' associations and government agencies congregate to discuss politics in the early stages of policy making (Akkerman, Hajer and Grin, 2000: 3).

This *societalisation* of environmental politics, conceived as a mix of rule-directed and rule-altering arrangements, has several consequences for the meaning and practices of political participation. Firstly, the innovation of rule-directed (communicative) instruments is supplementary to official formal procedures of participation. The aim is to inform relevant actors as soon as possible in the consecutive stages of policy making. This type of participation, however, does not change the rules of policy making and hardly affects the balances of power between the state and the civil society representatives. Secondly, as a result of the development and institutionalisation of rule-altering arrangements, participation increasingly affects the rules of the game itself and the balances of power between actors. According to Gibbins and Reimer (1999: 113) interest groups and citizens develop a style of politics that embraces performativity, a set of rhetorical practices that encourages open dialogue, discussion, dissension and the sharing of information, and politicisation, the process by which previously excluded issues are brought into politics. Compared with the participation practices of the 1970s, these rule-altering arrangements show some striking differences. The experiments with participatory and interactive policy-making particularly add new policy arrangements to the existing legislative framework, co-existing side-by-side with rule-directed arrangements. A characteristic feature of these new arrangements is that citizens and interest groups are actively involved in the definition of problems and their solutions, and that they can make their competence the object of a mutual learning process. In the Netherlands there are several examples of integrative regional planning in environmental policy. In these projects actors have the possibility to change the rules of the game and are given the opportunity to mobilise resources, in order to formulate views and measures that do justice to local circumstances (Janssens and Van Tatenhove, 2000: 167–170).

'Marketisation' of environmental politics: consequences for governance and participation

The *marketisation* of environmental politics refers to the emergence of rule-directed and rule-altering arrangements in the 'interference zone' between state and market. In this zone the rationales of the state and the market intermingle, thereby also affecting policy making, governance and participation.

From the perspective of the state, there is a shift from state initiated regulation to economic instruments and types of government. Rejecting the radical anti-modern discourses and accepting the much more moderate ones on 'ecological modernisation' (Mol, 1995; Spaargaren, 1997), governments aim at pricing the environment as a common good. These efforts have resulted in a variety of economic instruments, varying from the more traditional (such as taxes, levies and subsidies), to more sophisticated and innovative ones (such as the bubble concept, tradeable emission permits etc.). The plea for economic instruments was based on normative arguments (bringing into practice 'the polluter pays' principle) as well as on functional arguments (economic instruments are expected to be more effective, since taxes and subsidies influence the weighting up of behavioural alternatives by appealing to self-interest). Other examples of rule-directed *marketisation* refer to the delegation of responsibilities and competencies towards either autonomous or privatised agencies. In environmental policy domains like energy, water management and waste management, privatisation is an important development. Within these arrangements 'the market' sets the rules and structures the relevant coalitions in terms of producer-consumer relationships. Apart from examples of mere privatisation, the *marketisation* of environmental policies, both in the Netherlands and at the European level, results in a rather neo-corporatist arrangement with the gradual institutionalisation of a so-called 'target group policy'. Target groups at first were defined as more or less homogeneous groups of polluters, such as agriculture, traffic and transport, industry and refineries, gas and electric supply. Since they are responsible for particular environmental pressures, they represented particular objects of (the state's) environmental policies. But later, the state invited them to take their share of responsibility and internalise that by being co-responsible for the setting of environmental standards and their actual implementation by the members of their particular branch of economic activities. This latter arrangement leads to a sharing of responsibilities and political authority by the state and the acknowledged industrial organisations in these policy sectors.

From the perspective of the market, several innovative policy arrangements can be distinguished, based on the intermingling of rule-directed and rule-altering politics. A first example of innovative policy arrangements are covenants and flexible policy instruments. Flexible instruments, such as Joint Implementation, the Clean Development Mechanism, and Tradeable Emissions Permits and Benchmarking, refer to a set of innovative policy mechanisms aiming at reducing greenhouse gas emissions other than through a general and common target and timetable which is imposed on all countries (Arts, 2000: 125). Covenants are voluntary agreements between the state and market parties by which producers freely adopt certain standards or targets, under the guidance of the state. Recently, a new type of covenant has emerged in environmental politics: the co-operation between civil society representatives and market agencies. Examples of these rule-altering arrangements are: the initiative of Greenpeace and the Swiss firm Wenko AG to produce the Smile (an energy extensive car); the conservation of nature in Central America by McDonalds, Conservation International and local NGOs; the initiative of Unilever and WWF to safeguard global fish stocks. Another example of rule-altering arrangements in the interference zone between 'civil society and market' is the changing nature of providers/consumers relations due to the development of green electricity and the liberalisation of the energy sector. The opening of the electricity grid to third parties resulted in a differentiation of providers (wind energy associations, NGOs or local communities). Private, public–private and self-provided types of arrangements are now joining the statist arrangements of electricity provision (see Van Vliet, Wüstenhagen and Chappels, 2000; Arts and Van Tatenhove, 2002). In Schönau, for example, citizens who felt trapped in the monopolist electricity network bought themselves out to be freed from the forced consumption of nuclear generated electricity (http://www.ews-schoenau.de). They started their own energy company (EWS), which then developed into a 'normal' utility. EWS is owner of a local grid, generates and distributes electricity with additional demand-side and peak load management tasks, and has now even entered the national electricity market with the 'Watt Ihr Volt' product (Van Vliet et al., 2000). A characteristic feature of these kinds of initiatives is that captive consumers become citizen-consumers and even participants, who have not only the possibility of choosing between different kinds of energy, but also of choosing between different suppliers or becoming a supplier themselves. As in the case of the new types of covenants between market and civil society representatives, citizens, firms and NGOs define the rules of the game and the predominant policy discourses, resulting in the renewal of politics.

Environmental politics and participation: discussion

This article has focused on the dynamics of environmental politics, and especially on the relation between the predominant style of governance and policy making on the one hand, and the organisation of political participation on the other. In order to understand this relationship we introduced the concepts of political modernisation and policy arrangements. We distinguished three ideal–typical types of political modernisation, respectively labelled as early-, anti- and late-modernisation. These three categories refer to basic discourses on politics, democracy and governance, relate to distinctive interrelations between state, market and civil society, and therefore lead to quite different policy arrangements. While there has clearly been a gradual change in western politics and society from early- to late-modernisation, the different categories also represent the contemporary variety in politics, in which early-modern features stand next to late-modern ones.

We believe that these political changes relate to a series of developments in the substantial and organisational aspects of environmental policy, which we have discussed elsewhere (Van Tatenhove, Arts and Leroy, 2000). Here, we restricted our discussion to the effects of these political changes upon the discourses and practices on political participation. This issue, it was made clear, was inextricably linked to environmental politics.

However closely linked 'environment' and participation' might be, their relation differs substantially over the various stages of the institutionalisation of environmental politics. In the formative years of environmental policies the practices of participation clearly reflected early modernisation: environmental policies were, albeit to a restricted extent, the object of classical, legal procedures by which citizens could express their opinions on applications for permits and on other specific issues. As these procedures proved not to provide sufficient influence, the emergent environmental movement severely criticised them as a part of their anti-modernist protest. This led to the complementing of these constitutional facilities by a series of measures to increase political openness, primarily with regard to state–citizen interrelations. The overall impression is that these enlarged opportunities for political participation did not alter nor affect the existing power balances substantially, particularly as far as private agencies were concerned.

From the late 1980s, we witness the emergence of new discourses and practices on participation, reflecting more fundamental changes between state, market and civil society, and particularly enabling interrelations and institutionalising interference zones between them. The gradual *societalisation* and *marketisation* of environmental policies has given rise to new mechanisms and opportunities for political participation. These latter innovations are experimental in character and, therefore, have not been institutionalised hitherto, nor can they be fully assessed as to their added value in terms of participation and power. Nevertheless, both *societalisation* and *marketisation* seem to imply more of a rule-altering potential, in that both civil society and market representatives are invited not only to bring in their points of view, but also to take their share of responsibility in the policy-making and the implementation process. In this way *societalisation* and *marketisation* represent the widening of the political sphere, far beyond that of the state's institutions, thereby confirming the *de facto* political roles of the representatives of both market and society.

However, questions can be raised about the consequences of *societalisation* for participation. Compared with the participation practices of the 1970s, the 1990s show the emergence of innovative and reflexive policy arrangements, in which citizens and interest groups are actively involved in defining problems, formulating solutions and changing the rules of the game. But, what is or will be the status of a reached consensus, resulting from deliberations and negotiations between citizens and state representatives in several domains of environmental policy? Will that consensus be taken over by the constitutional bodies, thereby replacing the policies formulated so far, or will it be treated as advice, complementing the insights that constitutional bodies have to take into account when making decisions? Some actual conflicts, e.g. on how to interpret and accommodate the outcomes of local partici-patory processes at different levels and in different domains of environmental policy in the Netherlands make clear that these experiments provoke new fundamental political questions about the nature of democracy (models) and participation.

In a similar way, the consequences of *marketisation* for participation are not unambiguous. One should distinguish here between producers and consumers and the different participation roles they might be taking. For both consumers and producers political responsibilities and market regularities intermingle. First, their access to the state is different. While consumers are hardly organised, producers such as the chemical industry are structured in branch organisations. Their inter-mediary organisations have regular consultation with governmental representatives. As to the consumers, their participation role is ambivalent, based mainly on the principle of the *homo economicus*, but also upon the principle of the *homo politicus*. As a consequence, citizens/consumers participate in diffused, if not contradictory, ways in environmental politics. On the one hand participation is oriented at the state to realise the greening of production and consumption, for example by eco-labelling. This kind of participation is based upon long-lasting and rule-directed strategies. On the other hand consumers and their organisations try in several ways to influence processes of production directly, ranging from buying eco-products to become producers themselves. Here participation is more diffused, consisting of a mix of rule-directed and rule-altering participation strategies. However, these strategies are hardly organised and still not often acknowledged by either govern-ments or producers. In short: consumers, as *homines economici*, are asked to buy eco-products based on (a state initiated) eco-labelling, but their influence on which products deserve the predicate 'eco' is very limited, if not nil. The influence of consumers as *homines politici*, in rule-altering strategies in which consumers become producers or try to influence economic decisions directly, is still very limited. Nevertheless, both the *societalisation* and *marketisation* of environmental policies call for new opportunities and for new institutions for political participation. In that sense, they might not only indicate political changes but also catalyse them.

References

Alexander, J.C. (1995). *Fin de Siècle Social Theory. Relativism, Reduction, and the Problem of Reason*. London/New York: Verso.

Akkerman, T., M. Hajer and J. Grin (2000). 'Interactive Policy Making as Deliberative Democracy? Learning from New Policy-making Practices in Amsterdam'. Paper at the Convention of the American Political Science Association, 31 August–3 September 2000, Washington DC.

Arts, B. and M. van der Zouwen (1999). 'Policy Arrangements in Nature Conservation. The Case of the Netherlands in the European Context'. *The 1999 European Environment Conference. Environmental Policy in Europe: Visions for the New Millennium. Conference Proceedings*. London: ERP Environment, pp. 1–10.

Arts, B. (2000). 'Global Environmental Policies: Between "Interstatist" and "Transnational" Arrangements', in J. van Tatenhove, B. Arts and P. Leroy (eds), *Political Modernisation and the Environment. The Renewal of Environmental Policy Arrangements*. Dordrecht: Kluwer, pp. 117–143.

Arts, B. and J. van Tatenhove (2000). 'Power and Policy. Policy Arrangements and Shifting PowerRelations, Positions and Structures'. Paper presented at panel 2, RC 36 on power at the IPSA Conference, Quebec, 1–5 August 2000.

Arts, B. and J. van Tatenhove (2002). 'Institutional Innovation and the Environment: the Interplay between "Frontstage" and "Backstage" Politics', *GAIA*, **11**(1): 64–5.

Beck, U. (1994). 'The Reinvention of Politics: Towards a Theory of Reflexive Modernization', in U. Beck, A. Giddens and S. Lash (eds), *Reflexive Modernization. Politics, Tradition and Aesthetics in the Modern Social Order*. Cambridge: Polity Press, pp. 1–55.

Beck, U. (1996). 'World Risk Society as Cosmopolitan Society? Ecological Questions in a Framework of Manufactured Uncertainties', *Theory, Culture and Society*, **13**(4): 1–32.

Bekkers, V.J.J.M. (1996). 'Coproductie in het milieubeleid', *Bestuurswetenschappen*, **3**: 177–194.

Blowers, A. (1984). *Something in the Air: Corporate Power and the Environment*. London: Harper and Row.

Blowers, A. and P. Leroy (1994). 'Power, Politics and Environmental Inequality', *Environmental Politics*, **3**(2): 197–228.

Boehmer-Christiansen, S. and J. Skea (1991). *Acid Politics. Environmental and Energy Policies in Britain and Germany*. London, New York: Belhaven Press.

Cohen, J. (1998). 'Democracy and Liberty', in Jon Elster (ed.), *Deliberative Democracy*. Cambridge: Cambridge University Press, pp. 185–231.

Franklin, J. (ed.) (1998). *Politics of Risk Society*. Cambridge: Polity Press.

Gibbins, J.R. and B. Reimer (1999). *The Politics of Postmodernity. An Introduction to Contemporary Politics and Culture*. London: Sage.

Giddens, A. (1990). *The Consequences of Modernity*. Cambridge: Polity Press.

Gladwin, T.N. (1980). Patterns of Environmental Conflict over Industrial Facilities in the United States, 1970–1980', *Natural Resources Journal*, **20**(2): 243–274.

Gould, K.A., A. Schnaiberg and A.S Weinberg (1996). *Local Environmental Struggles*. Cambridge/New York: Cambridge University Press.

Hajer, M.A. (1997). *The Politics of Environmental Discourse: Ecological Modernization and the Policy Process*. Oxford: Oxford University Press.

Hajer, M. and H. Wagenaar (eds) (2003). *Deliberative Policy Analysis: Understanding Governance in the Network Society*. Cambridge: Cambridge University Press.

Held, D. (1989). *Political Theory and the Modern State. Essays on State, Power and Democracy*. Cambridge: Polity Press.

Held, D. (1996). *Models of Democracy*. Cambridge: Polity Press.

Hirst, Paul (2000). 'Democracy and Governance', in J. Pierre (ed.) *Debating Governance: Authority, Steering and Democracy*. Oxford: Oxford University Press, pp. 13–35.

Huberts, L.W. (1988). *De politieke invloed van protest en pressie – Besluitvormingsprocessen over rijkswegen*. Leiden.

Janssens, J. and J. van Tatenhove (2000). 'Green Planning: From Sectoral to Integrative Planning Arrangements?', in J. van Tatenhove, B. Arts and P. Leroy (eds), *Political Modernisation and the Environment. The Renewal of Environmental Policy Arrangements.* Dordrecht: Kluwer, pp.145–174.

Lauwers, J. (1983). 'Het groene ongenoegen', *Tijdschrift voor Sociologie*, **4**: 431–49.

Leroy, P. (1979). *Kernenergie: milieuconflict of godsdienstoorlog?* Antwerpen/Leuven.

Leroy, P. and J. van Tatenhove (2000). 'Political Modernization Theory and Environmental Politics', in Gert Spaargaren, Arthur P.J. Mol and Frederick H. Buttel (eds), *Environment and Global Modernity.* London: Sage, pp.187–208.

Lowe, P. (1983). *Environmental Groups in Politics.* London: Allen and Unwin.

Mol, A.P.J. (1995). *The Refinement of Production. Ecological Modernization Theory and the Chemical Industry.* Utrecht: Van Arkel.

Offe, C. (1999). 'How Can we Trust our Fellow Citizens?', in M.E. Warren (ed.), *Democracy and Trust.* Cambridge: Cambridge University Press, pp. 42–87.

Pestman, P. (2000). 'Dutch Infrastructure Policies: Changing and Contradictory Policy Arrangements', in J. van Tatenhove, B. Arts and P. Leroy (eds), *Political Modernisation and the Environment. The Renewal of Environmental Policy Arrangements.* Dordrecht: Kluwer, pp. 71–96.

Pierre, J. (ed.) (2000). *Debating Governance. Authority, Steering and Democracy.* Oxford: Oxford University Press.

Sbragia, Alberta (2000). 'The European Union as Coxswain: Governance by Steering', in J. Pierre (ed.), *Debating Governance. Authority, Steering and Democracy.* Oxford: Oxford University Press, pp. 217–240.

Spaargaren, G. (1997). *The Ecological Modernization of Production and Consumption. Essays in Environmental Sociology.* Wageningen: WAU.

Van Tatenhove, J. (1999). 'Political Modernisation and the Institutionalisation of Environmental Policy', in M. Wissenburg, G. Orhan and U. Collier (eds), *European Discourses on Environmental Policy.* Aldershot: Ashgate, pp. 59–78.

Van Tatenhove, J., B. Arts and P. Leroy (eds) (2000). *Political Modernisation and the Environment. The Renewal of Environmental Policy Arrangements.* Dordrecht: Kluwer.

Van Vliet, B., R. Wüstenhagen and H. Chappells (2000). 'New Provider–Consumer Relations in the Electricity Schemes in the UK, The Netherlands, Switzerland and Germany'. Paper for the Business Strategy and the Environment Conference, September 18–19, 2000, Leeds.

Warren, M. E. (ed.) (1999). *Democracy and Trust.* Cambridge: Cambridge University Press.

Weale, A. (1992). *The New Politics of Pollution.* Manchester: Manchester University Press.

Chapter 12

George A. Gonzalez

DEMOCRATIC ETHICS AND ECOLOGICAL MODERNIZATION

THE STATE OF CALIFORNIA leads the nation in the formulation and implementation of automobile emission standards (Krier and Ursin 1977; Kamieniecki and Farrell 1991; Lowry 1992, chap. 4; Kraft 1993; Grant 1996; Perez-Pena 1999; Gonzalez 2001, chap. 6). In 1967 California became the first jurisdiction to issue such standards (Krier and Ursin 1977, chap. 11). Its automobile emission standards continue to be the toughest in the United States (Lowry 1992, chap. 4; Cone 1998; Perez-Pena 1999; Gonzalez 2001, chap. 6). These standards were most recently raised in 1998 (Cone 1998; Perez-Pena 1999).

Moreover, California's standards have been the driving force behind the nation's automobile emission standards (Kamieniecki and Farrell 1991; Pena-Perez 1999; Gonzalez 2001, chap. 6). Specifically, the strengthening of California's emission standards in 1990, in conjunction with the actions of other states, prompted the federal government to revise upward its automobile emission standards with the 1990 Clean Air Act (Gonzalez 2001, chap. 6). Similarly, events at the state level, led by California, prompted the federal government in 1999 to announce tightened emission standards (Pena-Perez 1999).

Despite California's well-developed regulatory framework and its political leadership on the issue of automobile emission standards, air pollution from automobiles continues to be a persistent and serious problem in the state. Although carbon monoxide emissions are down when compared with 1990 levels, this pollutant continues to be emitted in large and hazardous amounts into the state's air. Additionally, nitrogen oxides and particular matter in California's air remain at roughly 1990 levels. Particulate matter, moreover, is predicted to increase in the near future. The automobile (including buses and trucks) accounts for up to 75 percent of all these pollutants (California Air Resources Board 1993, 1999). Further, although the level of airborne toxins is largely unmonitored, preliminary data strongly suggest that they persist in the state's air at hazardous levels. Automobiles are a major source of airborne toxins (California Air Resources Board 1994, 1999). Although Houston, Texas, now has officially worse air quality than Los Angeles, this is the result of deteriorating air quality in Houston more than anything else (Dawson 1999; "Smog City" 1999).[1]

If California has the most stringent automobile emission standards in the country (indeed the world), why does air pollution from the automobile continue to be a persistent problem and a potentially growing problem in the future? The primary reasons are population growth, growing economic activity, and an increase in the number of automobiles (with internal combustion engines) as well as in the average number of miles driven (California Air Resources Board 1993, 1999; Lange 1999; Patton 1999). Sacramento's population, for example, is expected to grow by 50 percent in 2010 from 1987 levels, and it is expected to have an increase of 76 percent in the number of miles driven (Grant 1996, 34). Therefore the gains in emission reductions made through the application of technology to the internal combustion engine are offset in the number of automobiles on the road and an increase in the average number of miles driven by motorists (Kamieniecki and Farrell 1991; Cahn 1995, chap. 4; Warrick 1997; Cone 1999; Luger 2000; Purdum 2000). Hence, despite the increasingly onerous regulatory framework placed on the automobile and on gasoline in California, the state, especially the Los Angeles basin and the Central Valley area, will continue to have unhealthful air into the foreseeable future (California Air Resources Board 1993).

Given the relationship between growth, automobile usage, and air pollution, is the regulation of growth and automobile usage actively considered in order to reduce and remedy the considerable air pollution problem in the state? The answer is no. In his study of California's automobile and fuel emission standards, Grant (1996) analyzes the "policy community" surrounding this issue area.[2] Borrowing from Rhodes and Marsh (1992), he describes a policy community as "characterized by a limited number of participants, frequent interaction, continuity, value consensus, resource dependence, positive sum games, and regulation of members" (Grant 1996, 10). In his analysis of the California air pollution policy community, Grant (1996) concludes that issues of land management and mass transit are excluded. In contrast, technological solutions are at the center of the clean air policy community.

With the issue of growth, and the number and usage of automobiles excluded from the state government's clean air agenda (Bachrach and Baratz 1962; Lukes 1974), the goal of the state's automotive emissions regulatory regime can be characterized as the "ecological modernization" of the automobile. Moreover, with the mass production and distribution of alternative fuel automobiles becoming less likely (Borenstein 2000; Pollack 2000), the objective of this regime can in retrospect be termed the ecological modernization of the gasoline-burning internal combustion engine.

At the core of ecological modernization is the idea that environmental protection and economic growth are complementary goals. This complementary relationship can be achieved primarily through the development and application of technology. According to its proponents, the costs associated with ecological modernization are justified because an ecologically modernized society produces less pollution and hence utilizes materials more efficiently. Further, advocates hold that a cleaner environment results in greater productivity. The ecological modernizing of consumer products leads to economic growth and increases competitiveness because consumers are increasingly demanding environmentally benign products (Weale 1992; Hajer 1995; Dryzek 1996, 480; 1997, chap. 8). Weale (1992) argues that environmental politics is becoming increasingly shaped by the discourse of ecological modernization. Reflecting this trend, the central feature of former Vice President Al Gore's (1992) environmental advocacy is the ecological modernization of society.

This approach is consistent with free market environmentalism, which holds that environmental protection is consistent with the utilization of the market (Anderson and Leal 1991; Dryzek 1997, chap. 6). The advocates of ecological modernization differ from free market environmentalists in that the modernization school does not rely solely on market mechanisms to achieve a salutary environment. Instead, public regulations are often necessary to correct for market failures and advance the ecological modernization of capitalist society.

Since the objective of California's automotive emission regime is the ecological modernization of the internal combustion engine, the continued participation of environmental activists in this policy process raises ethical dilemmas. These ethical dilemmas flow from a normative framework of democracy. Specifically, the participation of environmentalists within the policymaking process that formulates California's automobile emission regulatory regime serves to undermine the potential development of a broad-based movement that could debate and challenge the narrow contours of this regime. This narrow regime is not driven by the environmentalists who are part of the policymaking process, as is widely believed (Sabatier 1987; Hajer 1995; Dryzek 1996, 1997), but primarily by the segment of the business community whose economic fortunes are tied to rising land values and investment in California.

By employing an ethical criterion to analyze the political behavior of environmental activists, I do not intend to issue an ethical or a moral judgment or condemnation. I utilize a particular framework of ethics that also serves as an analytical framework. It elucidates the role and impact of activists who choose to operate within the California polity in their pursuit of cleaner air. Moreover, the utilization of an ethical framework rooted in notions of democracy can help determine the most efficacious usage of environmentalists' limited resources.

Ecological modernization, the automobile, and symbolic inclusion

In his analysis of interest group inclusion within the policymaking process, political scientist John Dryzek explains that "oppositional groupings can only be included in the state in benign fashion when the defining interest of the group can be related quite directly to a state imperative" (1996, 479). In other words, according to Dryzek, groups that critique the status quo can only participate in the policymaking process to the extent that their goals are consistent with an objective of the state. This is reflected in the behavior of environmental groups that are active in the formulation of California's automobile emission regime.

Environmental activists involved in this formulation process are aware of the relationship between economic growth, a greater number of automobiles (with internal combustion engines), and air pollution. As a transportation analyst for the Union of Concerned Scientists (UCS) explained:

> There's definitely I think a broad agreement throughout the environmental community that the sustainable strategy for dealing with transportation, not only from an environmental perspective, but from a social perspective, is better land use management. To get people out of cars, better jobs, housing balance, renewal, urban centers, density—I think all of the buzzwords come to bear (Mark 2000).

The executive director of the California Coalition for Clean Air (CCCA) argued that in terms of improving air quality in the state, "I think there are a considerable number of environmentalists that think that limiting growth is a good idea" (Carmichael 2000). A senior staff attorney for the Natural Resources Defense Council (NRDC) noted the obvious relationship between growth, increased automobile usage, and air pollution when she stated that "more cars" in an area equals "more pollution. It's a cycle" (Hathaway 2000).

Despite the relationship observed between growth, an increasing number of automobiles (with internal combustion engines), and air pollution, these activists acknowledge that the idea of regulating growth and the number and usage of automobiles to address air pollution is not considered a policy option in the policymaking process. The official from the UCS explained that "motor vehicle [air emissions] policy is thought of relatively separately from transportation [and] land use policy" in California politics (Mark 2000). The person from CCCA stated that the government's "pollution control strategy is focused on technology" (Carmichael 2000). When asked whether regulations on growth within the policymaking process were actively considered in relationship to automobile emissions, air pollution, or other environmental concerns, the NRDC senior attorney unequivocally stated that growth "is currently, absolutely off the table" (Hathaway 2000).

With the realization that regulations on growth and on the number of automobiles have no realistic chance of being imposed to improve air quality, environmental activists limit their efforts within the policymaking process to the promotion of technology to address the automobile as a source of air pollution. This is reflected in interviews with activists from environmental groups that participate in the policy formulation process: the California Coalition for Clean Air, Natural Resources Defense Council, and the Union of Concerned Scientists (Mark 2000; Carmichael 2000; Hathaway 2000). By promoting technology, these environmental groups and activists are simply promoting the ecological modernization of the automobile, which is wholly consistent with the state imperative described above. At best, they can be viewed as the most aggressive advocates of the automobile's ecological modernization. This is congruous with Dryzek's observation that when the state has imperatives, oppositional groups within the policymaking process are limited to "influencing how imperatives are met, and how trade-offs between competing imperatives are made" (1996, 480).

Therefore, in order to be effective within the policy formulation process, environmentalists (in this specific context) must drop their critiques of growth and increased usage of the automobile. (The relative effectiveness of environmentalists in advancing the ecological modernization of the automobile is discussed below.) As explained by Dryzek, under a political process in which the state has imperatives "a high price will be paid by any [oppositional] group included [within the state] on this basis. For if state officials have no compelling reason to include the group, then presumably it must moderate its stance to fit with established state imperatives" (1996, 480). This reality was most cogently conveyed by the official within the California Air Resources Board (CARB) who is directly in charge of the state's motor vehicle emission program. He stated that environmental activists became much more effective within the policymaking process once they dropped their critiques of growth, specifically:

> In the past there was some tendency [among environmentalists] to be staking out a philosophical view and perhaps being a little more strident in looking for compromise. In the last ten years or less, I think the environmental community has become much more of a court player.

He went on to laud environmentalists because currently they are "much more focused on solutions and less on philosophical issues. I mean, for example, . . . people used to argue for . . . no growth." This official considered the current approach among environmentalists within the policymaking process to be "pragmatic." He also felt that today environmentalists "seem to look for compromises and be part of the solution" (Cackette 2000).

Rebellious politics, the environmental lobby, and democracy

Environmental activists' participation in this undemocratic policy formulation process lends legitimacy to it (Edelman 1977; Saward 1992). This process is undemocratic to the extent that certain actors or forces keep central political and economic issues off the agenda (Bachrach and Baratz 1962; Crenson 1971; Lukes 1974; Lindblom 1982; Hayward 1998). Lending legitimacy to this undemocratic process serves to undermine the potential development of a substantive and democratic politics in which such issues as growth and automobile usage can be seriously debated.

Specifically, by participating in this undemocratic process, mainstream environmental groups help prevent the development of a broader social movement in civil society that could challenge and debate the imperative of growth. Political theorist Jeffrey Isaac defines civil society as "those human networks that exist independently of . . . the political state" (1993, 356). Dryzek argues that civil society is a more democratic venue than the state because it is "relatively unconstrained." He goes on to explain that within civil society:

> discourse need not be suppressed in the interests of strategic advantage [as is the case within the state]; goals and interests need not be compromised or subordinated to the pursuit of office or access; embarrassing troublemakers need not be repressed; the indeterminacy of outcome

inherent in democracy need not be subordinated to state policy (Dryzek 1996, 486).

Thus democracy here is defined as the ability to consider and advance an indeterminate number of policy means and goals. Dryzek holds that this openness can only take place outside of the state because the state is tied to specific objectives.

To the extent that some of the possibilities considered and advanced within civil society contest and confront the state's imperatives, Isaac avers that within civil society "rebellious" politics can take place. He holds that:

> a rebellious politics is a politics of voluntary associations, independent of the state, that seeks to create spaces of opposition to remote, disempowering bureaucratic and corporate structures. Such a politics is often directed against the state, but it does not seek to control the state in the way that political parties do. Neither does it lobby the state to achieve specific advantages, as do interest groups. Rather, it is a politics of moral suasion, seeking . . . to affect the political world through the force of its example and through its very specific, proximate results (Isaac 1993, 357).

A rebellious politics that achieves critical mass is transformed into broad-based social movements. Sidney Tarrow explains that social "movements mount challenges through disruptive direct action against elites, authorities, other groups or cultural codes" (1994, 4).

Within U.S. civil society, and disconnected from the state, exists a rebellious environmental politics that challenges economic growth and the priority this growth is given above other values, such as human health, the humane treatment of animals, and environmental sustainability. This politics is led by Earth First, animal liberationists, and networks organized around toxins and environmental justice (Bullard 1990; Szasz 1994; Dowie 1995; Taylor 1995; Dryzek 1996, 480; Schlosberg 1999).

Dryzek asserts that "whether a group should choose the state, civil society, or both simultaneously depends on the particular configuration of movement interests and state imperatives" (1996, 485). He goes on to argue that the most efficacious approach for the environmental community to take is a "dualistic" approach (Cohen and Arato 1992; Wainwright 1994), in which part of the community operates within the state to advance the ecological modernization of capitalist society. The more confrontational element of this community should operate largely within civil society and confront the imperative of growth, including its attending environmental ill effects. Moreover, the activities of the more contentious part of the environmental community, by exerting outside pressure, can help the part within the state to achieve the policy goal of ecological modernization (Dryzek 1996, 483–486). The difficulty with this dualistic approach is that it fails to take into account how incorporation within the state can serve as a means to undermine rebellious politics and social movements in civil society.

The containment of rebellious politics

Historically, the state has not been passive in the face of rebellious politics and the emergence of social movements. Instead, it attempts to ensure that rebellious politics do not achieve critical mass, which could destabilize society or force the state to substantially alter its imperatives as a concession to confrontational social movements (Tarrow 1994). One means to contain rebellious politics is through coercion (Sexton 1991).

Another means is to "buy off" groups and individuals who could potentially be part of a rebellious politics. Progressives, socialists, and Marxists have historically viewed mainstream labor unions and welfare programs as overt attempts on the part of the state and corporations to blunt class conflict and politically subdue and pacify the working class to maintain internal order (Weinstein 1968; Piven and Cloward 1971; Domhoff 1998). Maintaining internal order is a key imperative of the state (Skocpol 1979).

Certain critical thinkers argue that the state manages the public's environmental concerns primarily through the dissemination of symbols (Edelman 1964; O'Connor 1994; Cahn 1995). Cahn (1995), for example, avers that post-1970 environmental regulatory policies (i.e., clean air, clean water, energy, and waste policies) are most aptly characterized as symbolic responses to the public's growing environmental concerns rather than substantive efforts to regulate corporate America. He arrives at this conclusion by analyzing the content of these policies. Furthermore, Cahn juxtaposes the content of these policies with the federal government's continued encouragement of economic growth and its continued support and subsidization of fossil fuels usage (e.g., the Persian Gulf War). These are the primary factors that cause air and water pollution, as well as waste creation.

The federal government's subsidizing and encouragement of fossil fuel usage is reflected in its 1998 appropriation of $200 billion for the maintenance and expansion of the nation's transportation infrastructure, which "amounted to the largest public-works program in the nation's history" (Andrews 1999, 303) and encouraged fossil fuel usage. More than 80 percent of these funds were dedicated to highway and bridge construction (Andrews 1999, 303–304). Thus critics like Cahn argue that U.S. environmental legislation and environmental policies designed to regulate corporate America are symbolic precisely because they do not challenge the state's imperative of economic growth and have not sought to alter the economy's reliance on highly polluting fossil fuels, especially gasoline as an automotive fuel.

In the California context, long-term regulatory planning by state agencies can also be interpreted as a symbolic response to environmental concerns (Grant 1996, 41). For example, CARB promulgated a plan in 1990 mandating that 2 percent of automobiles offered for sale in 1998 be zero emission vehicles (ZEVs), 5 percent by 2001, and 10 percent by 2003 (Kamieniecki and Farrell 1991; Grant 1996). Currently, only electrically powered vehicles have zero emissions. Similarly, California in 1989 adopted the Air Quality Management Plan (AQMP) (Kraft 1993). The state's AQMP also relied heavily on the long-term development of technology to achieve improvements in air quality. Significantly, these plans did not put forward subsidies to facilitate the development of hoped-for technologies, nor did they mandate sanctions for industrial sectors that failed to develop the necessary technologies. Commenting on the state's AQMP shortly after it was promulgated, Kamieniecki and Farrell astutely noted that "for mainly political reasons, the more difficult decisions [of the AQMP] have been postponed for a number of years, with the hope that new technologies will allow policymakers to meet federal clean-air standards with minimum disruption to . . . economic growth" (1991, 154). Notably, the 1998 and 2001 targets for the sale of ZEVs have been abandoned by CARB (Cone 1995a,b,c). Additionally, CARB has reduced the 2003 target to 2 percent (Pollack 2000).

The group mobilization incentive structure outlined by Olson (1971) offers part of the explanation as to why, even in the face of persistently poor air quality, rebellious politics in California has not been transformed into a social movement. The symbols emanated from the enactment of regulatory legislation, and unenforced regulatory frameworks, as suggested by Edelman (1964, 1988), also contribute to the public's relative political passivity on the issue of air pollution.[3] These symbols communicate to the public that something is already being done to address the issue of air pollution and that there is no need to spend time and energy in attempts to overcome the collective action barriers inherent in the mobilization of large groups.

The symbolic inclusion of environmental groups within the policymaking process that produces California's automobile emission standards becomes one of the symbols deployed against the public and works to keep it demobilized on the issue of air pollution. The participation of environmental activists in the policymaking process communicates to the broader public that this process is democratic because it putatively includes all relevant political perspectives (Edelman 1977; Wynne 1982; Saward 1992).[4] Environmental activists' participation in the California polity takes the form of both formal and informal access to the governor, the state legislature, and CARB (Mark 2000).

As already argued, this policymaking process is not democratic because the key issues of growth and automobile usage are kept off the agenda.[5] Moreover, environmental concerns are addressed within this process only to the extent that they do not conflict with the economic interests that monetarily benefit from rising land values in California and the rising sales of automobiles and gasoline in the state (Whitt 1982; Logan and Molotch 1987; Davis 1992; Grant 1996; Luger 2000). This is especially evident in CARB's backing away from policies designed to force the development, production, and distribution of alternative fuel automobiles.

The environmental lobby and California's ecological modernization of the automobile

Why would environmental activists want to lend legitimacy to an undermocratic policy formulation process, as they do when they participate in it? Or more intuitively, why would environmental activists want to participate in a process that keeps central issues from being effectively discussed? One answer is that the ecological modernization of the automobile would not be going forward were it not for the participation of these activists. All of the environmental activists who were interviewed, including the contracted lobbyist for the Sierra Club, felt that their specific role in the policymaking process was to put political pressure on government actors to prompt the strengthening of the state's automobile emission standards. Implicit in their thinking is that were it not for their participation, CARB officials would not have the same incentives to reduce emissions.

One respondent acknowledged that CARB and the automobile industry have a type of "symbiotic" relationship, with the industry being the primary source of CARB's technical knowledge on emission control technology and alternative fuel automobiles, and in return CARB only sets emission standards that the industry can comply with (White 2000; also see Cackette 2000; Scheible [executive deputy director of CARB] 2000). The same person explained that the environmental lobbying community attempts to keep these actors "honest." In other words, this lobbying community works to apply "pressure" and serve as a "check" to a potentially cozy relationship between CARB and the industry (White 2000). To the extent that California environmental lobbyists advocate policies forcing the development, mass production, and distribution of alternative fuel automobiles, my respondents from this community also felt that they were moving CARB and the automobile industry into a technological direction that these actors would not pursue on their own.

Whether the inclusion of environmental activists in the policymaking process actually results in the increasing modernization of the automobile is unknown, however. It is important to note that policies advancing the ecological modernization of the automobile preceded the inclusion of environmental activists in the process. The first California law requiring the installation of emission control technology in automobiles was enacted in 1960 (Krier and Ursin 1977, chap. 10). Environmentalists were not incorporated into the policymaking process on this issue in California until the 1970s (Roberts 1969, chap. 3; Krier and Ursin 1977, chap. 14; Fawcett 1990, 91–93; Dewey 2000, chap. 5).

Early ecological modernization efforts in California were led by economic interests that monetarily benefited from local economic and population growth and rising land values. These interests compose what Logan and Molotch refer to as a "growth coalition" made up of large land owners, land developers, local banks, real estate agents, and real estate lawyers, as well as utility firms and local media outlets (Logan and Molotch 1987). It was the *Los Angeles Times* and its owner, Norman Chandler, that led the original campaign in the 1940s to ecologically modernize business and industry in the city.[6] This campaign resulted in the creation of the state's first pollution control agency, the Los Angeles Air Pollution Control District ("Text" 1947; "Times" ' 1947; Kennedy 1954, 15; Brienes 1975, chap. 5; Krier and Ursin 1977, chap. 3; Dewey 2000, 87–88).

It was a policy-planning organization led by economic elites (Barrow 1993, chap. 1; Domhoff 1998), the Air Pollution Foundation, that in the late 1950s politically established automobiles as a major source of smog in Los Angeles and advocated the creation and application of pollution control technology to them (Air Pollution Foundation 1961; Krier and Ursin 1977, chap. 6). The foundation board of trustees was composed almost entirely of representatives from business and industry. Many of these individuals represented firms that directly benefited from economic growth in the Los Angeles basin, as well as firms involved in the production and sale of automobiles and gasoline.[7] Moreover, the foundation's list of contributors demonstrates the broad support that it enjoyed throughout the corporate community. During its seven-year existence (1954–1961) the foundation had more than two hundred donors, almost all from corporate America (Air Pollution Foundation 1961, 53–56).[8]

In order to formulate policy proposals and political strategies that relate to the environment, members of the contemporary California business community in 1973 organized and financed the California Council for Environmental and Economic Balance (CCEEB).[9] One-third of its board of directors is drawn from business and industry. Some of the firms represented on the board – the Irvine Company (real estate and land development), Pacific Telesis (regional telephone service provider), Southern California Edison (utility firm), Bank of America, and Pacific Gas and Electric Company – are economically dependent on growth in the state. Other firms, including Texaco, Chevron, and the Union Pacific Railroad, are directly affected by the state's environmental regulations.[10] The other two-thirds of the board is composed of labor union representatives and private citizens.[11] CCEEB's finances are entirely provided by its corporate members (Weisser 2000). Moreover, CCEEB disseminates its policy ideas throughout the California business community through "presentations" to organizations such as the Los Angeles Chamber of Commerce, the Santa Clara Manufacturers Group, and the Orange County Industrial League (CCEEB 2000b).

CCEEB describes its work in terms that are consistent with the concept of ecological modernization. Its president stated in our interview that a " 'healthy' environment leads to a 'good' economy." He claimed that all of CCEEB's members adhere to this belief. Moreover, he explained that, for example, a clean environment plays a specific role in maintaining and attracting high-tech industries (Weisser 2000). According to its mission statement, the council "is a coalition of California business, labor, and public leaders who work together to advance collaborative strategies for a sound economy and a healthy environment" (2000b). In its promotional literature, CCEEB describes itself as a "powerful coalition that can move the California economy forward in an environmentally responsible manner." Additionally, it claims to "understand the importance of the environment to the California business climate" (CCEEB 2000b). Finally, in terms wholly congruous with the notion of efficiency at the core of the discourse of ecological modernization (Weale 1992, 75–79; Hajer 1995, 31–36), CCEEB holds that its "work translates into . . . job creation, efficient use of tax dollars, reduced compliance costs, consolidated reporting formats, elimination of multiple agency oversight, more responsive public agencies and increased certainty for conservation and development" (CCEEB 2000a).

CCEEB's commitment to environmental protection within the context of unmitigated economic growth is reflected in its political activity. It supported the 1988 California Clean Air Act. The president of CCEEB proudly averred that the council was "instrumental" in the passage of this legislation. The act mandated a 55 percent reduction in automobile emissions, a 15 percent reduction in nitrogen oxides, and "lowering the levels of suspended particulates, carbon monoxide, and toxins as is technologically feasible," all by the year 2001 (Lamare 1994, 239). Additionally, the president of CCEEB explained that its members are "leaders on mobile source emission [reduction] efforts in California" (Weisser 2000). CCEEB also touts that it recently "co-sponsored the most comprehensive package of legislative proposals for a statewide growth strategy ever advanced; the package provided a structure for more certainty for both development and conservation" (CCEEB 2000b).

The business community, urban growth, and ecological modernization

Therefore the California public policies that ecologically modernize the state's economy and the internal combustion engine are consistent with the political activity and economic interests of that segment of the business community whose success depends on economic growth within the state. In light of the political activity of many of the members of California's growth coalition, their objective economic interests, and the acute air pollution inundating Los Angeles beginning in the 1940s (Krier and Ursin 1977, chap. 3), air pollution abatement regulations can be viewed as part of a legal infrastructure that helps attract capital to the state and ultimately facilitates and promotes local growth, much as an education and transportation infrastructure does. Although these factors may drive up the costs of production, they provide for an educated workforce, a transportation network, and a more salutary environment for workers and production (Weale 1992; Hajer 1995; Barrow 1998; Casner 1999). The most politically and environmentally significant of California's air pollution regulations relate to the automobile (Krier and Ursin 1977; Grant 1996; California Air Resources Board 1999; Dewey 2000, chap. 4; Gonzalez 2001, chap. 6).

The view that public policies that promote the ecological modernization of capitalist society arise from the economic interests and political activities of a particular segment of the corporate community is in contrast to the arguments put forward by Paul Sabatier (1987, 1999), John Dryzek (1996, 480), and Maarten Hajer (1995). Sabatier holds that the policies that moved forward the development and application of technology to address air pollution are the result of the incorporation of environmentalists and other antipollution advocates into the policymaking process, creating two adversarial political coalitions within the U.S. polity. One advocates government action to protect the environment and emphasizes the human health hazards associated with human-made pollution. The other, composed in large part of traditional economic concerns, promotes economic growth and highlights the economic costs of regulations designed to reduce human-made pollution. In the area of air pollution, Sabatier refers to the former coalition as the "Clean-Air Coalition" and the latter as the "Economic-Feasibility Coalition" (1987, 661–662). Moreover, the active engagement and interaction of these competing coalitions results in "policy learning" among actors in both political groupings. The outgrowth of this learning is a type of consensus on the utilization of technology to reduce the level of humanmade pollution, without regulating growth or automobile usage to affect pollution abatement (Sabatier 1987, 672–675). Hajer (1995, 68–72) posits that the discourse of ecological modernization arose from the configuration of political forces delineated by Sabatier, but he goes on to aver that this discourse is relatively autonomous and has an autonomous impact on policy development. Utilizing a structural Marxist approach (O'Connor 1973; Poulantzas 1973), Dryzek holds that public policies that promote the ecological modernization of society result from the state's imperatives of promoting "capital accumulation" and attaining "legitimacy" (1996, 480).

The argument is not that California public policies that ecologically modernize the automobile are the outgrowth of any particular configuration of interest group politics, nor the result of an autonomous discourse; instead, such policies are the result of the political activity of the corporate community, led by the segment whose economic interests are tied to growth in California, especially Los Angeles. Instead of resulting from interaction between the imperatives of capital accumulation and legitimation, these policies are rooted squarely in the goal of capital accumulation. As I have already argued, the state attains legitimacy primarily through the symbolism of benign inclusion, high-profile legislation, and unenforced regulatory frameworks.[12]

This view is borne out by the evidence. Environmentalism did not become important for legitimation until the late 1960s and early 1970s (Dryzek 1997); environmental activists were only incorporated into the policymaking process in the 1970s; and the discourse of ecological modernization was not widely disseminated until the 1980s and 1990s (Weale 1992). Nonetheless California's policies, which ecologically modernized business and industry as well as the automobile, were initiated in the 1940s, 1950s, and early 1960s. Moreover, they were strongly promoted by business elites. Outside of the business community, during this period the California public was relatively quiet on the issue of air quality (Roberts 1969, chap. 3; Krier and Ursin 1977, chap. 14; Dewey 2000, chap. 5).

Among the contemporary interest groups involved in the formulation and implementation of California's automobile emissions standards, the business-led and financed CCEEB most openly and aggressively embraces the discourse of ecological modernization. Its championing of the ecological modernization of California's business and industry as well as the internal combustion engine is reflected in its political activity.[13]

With the corporate community providing political support and energy to public policies that ecologically modernize the internal combustion engine, the political contribution of environmental activists incorporated into the policymaking process becomes unclear. In other words, with the business community promoting the ecological modernization of the gasoline-burning internal combustion engine, it becomes difficult to determine to what extent environmental activists are advancing the ecological modernization of the automobile through their participation in the policymaking process. Policies designed to promote ecological modernization could be the work of business political actions and not environmentalists. Furthermore, determining the actual influence of environmental lobbying efforts is complicated by the fact that, as Dowie points out (1995, 48), environmental groups, for purposes of fund-raising, are apt to take full credit for perceived legislative or regulatory victories even when they do not deserve it. Additionally, with CARB backing away from its policies designed to force the mass production and distribution of alternative fuel automobiles, the notion that incorporated environmental groups exercise significant influence over this agency appears questionable. This agency has done so in face of strong opposition from the environmental lobbying community (Carmichael 2000; Hathaway 2000).

Moreover, given the political activity of CCEEB and the economic interests of the Los Angeles and Central Valley growth coalitions, the ecological modernization of the gasoline-burning internal combustion engine will go forward with or without the incorporation of environmental activists. However, this modernization *might* not proceed at the same pace.[14]

Conclusion

In light of the factors outlined in this article, the participation of environmental activists in California's undemocratic policymaking process takes on ethical dimensions. These activists serve to enhance support for a policy formulation process that primarily abets the economic needs of the business community while also dampening the political forces that would compel the treatment of questions and issues central to a salutary and sustainable environment. Moreover, it is uncertain what environmental activists gain, in terms of environmental protection, for their participation in the policymaking process.

The environmental community as a whole, however, has affected California politics. Its gains are most readily evident in the realm of public opinion. Environmental ethicist Lester Milbrath explains that "public opinion polls show that a majority, usually a high majority, of people in most countries are aware of environmental problems and very concerned about getting them solved to ensure a decent future" (1995, 102). The environmental community deserves at least partial credit for the awareness among the citizenry of environment problems.

This success, along with the dubious nature of their participation within the policy formulation process, would suggest that the most efficacious deployment of the environmental lobbying community's resources would directly involve the public and specifically civil society. As Milbrath explains, environmental "awareness and concern does not necessarily mean that people well understand" the causes and potential solutions to society's environmental ills (1995, 102). Hence, instead of maintaining a rather hostile and contentious attitude toward confrontational environmental groups and networks that operate outside of the polity (Dowie 1995), the environmental lobbying community should leave California's polity and join with their rebellions brethren in civil society (Norton 1991). In this way the resources currently deployed in lobbying state officials could be more fruitfully directed at educating the public.

Such an effort would involve informing the residents of California that the state government utilizes a narrow, or "weak," conception of ecological modernization to address air pollution (Christoff 1996; Dryzek 1997, chap. 8). A narrow approach to ecological modernization relies heavily on technological solutions to address pollution. A more expansive, or "strong" conception of ecological modernization would involve ecologically sensitive land management. This type of land management would entail the intensive usage of land (as opposed to sprawl [Purdum 2000]), drawing residential and work areas closer together and making mass transit the primary means of transportation in urban areas. Ecologically sensitive land management would more California residents away from their dependence on the automobile (and the internal combustion engine) and toward more ecologically benign forms of transportation, such as walking and mass transit.

Additionally, such an education campaign could serve to expand environmental rebellious politics into a social movement that could potentially force the state government to abandon the narrow version of ecological modernization in favor of the more expansive version of this concept. Although it can be debated how practical, realistic, or desirable it is for the environmental lobbying community to redirect its resources away from the California polity and toward California's civil society, one thing is clear. Viewed through the lens of democracy, such a redeployment of resources is ethical.

Notes

1.   This is when the number of days that surpass federal ozone standards is compared.
2.   Reflective of the policy community approach, a substantial portion of the data for this article is composed of interviews from individuals drawn from environmental and business organizations, as well as the government officials directly involved in California's automobile emission policy community. I determined the relative importance of these organizations and individuals through a search of the Los Angeles Times computer database for the years 1990 through 1999 using the key words "vehicle emissions." The other method utilized was telephone and in-depth personal interviews. Respondents identified organizations and individuals regularly involved in the formulation of the state's automobile emission standards. Eight unstructured interviews were conducted; the averaged approximately an hour and a half in length.
3.   Here we need to analyze environmental legislation on two different planes. On the first plane, such legislation guides, influences, and/or restricts policymakers and the courts. Analysis on this plane examines how legislation affects or fails to affect the development of public policy. The second plane focuses on how the passage of environmental legislation affects public opinion. Even if legislation has no impact on policy, it can still affect public opinion. Moreover, a particular piece of environmental legislation can have substantively divergent impacts on policy and public opinion. The passage of a certain law, for example, may communicate to the public that a special interest is going to be effectively regulated, whereas the policy resulting from such legislation may actually strengthen the economic and/or political position of said special interest (Edelman 1954, 1988; McConnell 1966; Kolko 1977; Lowi 1979; Gonzalez 2001, chaps. 2–3). At this point in my argument I am focusing on the second plane of legislation.
4.   In addition to the overtly political factors outlined here, psychological, cultural, and ideological barriers exist that prevent a more robust and confrontational politics arising from civil society to challenge the state's commitment to environmentally deleterious growth (Milbrath 1989, 1995, 1996; Cahn 1995).
5.   As already explained, the definition of democracy utilized here emphasizes the ability to consider an indeterminate number of policy goals and means. Another definition of democracy could focus on the procedure through which officials are chosen. In this approach to democracy, as long as central policymakers are democratically elected, the policy outcomes of their decisions are legitimate and inherently democratic. This legitimacy would extend to policymakers appointed by the democratically elected officials. Thus, as democratically elected and legally

appointed officials, they are justified in excluding certain philosophical and policy perspectives from the policymaking process. This is because these officials can legitimately claim that they speak for the majority of citizens.

The proponents of the former version of democracy would retort that the electoral process in most cases in and of itself does not necessitate or justify elected officials in eliminating policy options from the policymaking process. Many of these theorists hold that U.S. society's reliance on the market to produce and distribute goods and services results in the effective elimination of various policy options from the policymaking process (O'Connor 1973; Lindblom 1982; Barrow 1993, chap. 2). Other thinkers hold that certain policy options are not considered in the policymaking process because particular political and economic interests are able to block their consideration (Bachrach and Baratz 1962; Barrow 1993, chap. I; Hayward 1998).

6.   Throughout the twentieth century the Chandler family has been a central political force in Los Angeles and, through its newspaper, a leading proponent of the area's economic growth (Gottlieb and Wolt 1977; Pincetl 1999, 31, 96).

7.   Among these locally oriented firms were the Southern California Gas Company, Bank of America, Broadway-Hale Stores (department stores), Western Air Lines, California Federal Savings, California Bank, Southern California Edison Company, Security-First National Bank, and Bullock's (department stores). Among the automotive-related firms represented on the foundation's board were each of the Big Three automakers, Firestone Tire and Rubber Company, U.S. Steel, and the Union Oil Company (Air Pollution Foundation 1961, 50–51).

8.   Among its financial supporters were the Automobile Manufacturers Association, the Western Oil and Gas Association, the Los Angeles Newspaper Publishers Association, the Los Angeles Clearing House Association (made up of major California banks), DuPont, Bechtel Corp. (construction), Kaiser Steel, and the Goodyear Tire and Rubber Company (Air Pollution Foundation 1961, 53–56).

9.   The policy work of the council is conducted through its "projects." According to its promotional literature, the center of CCEEB's activity is its "project committees where members design strategies which are implemented by CCEEB staff and expert consultants." This literature goes on to claim that "the result is public policy which adds value to our members rather than adding costs" (CCEEB 2000a). Project members meet once a month, and projects are composed largely of various business and industry representatives. They focus on specific policy areas of interest to the project members (Weisser 2000; Lucas 2000). The work of the individual projects is augmented with "CCEEB sponsored conferences, seminars and retreats, [in which] California legislators, regulators and administrative officials hear" what members "need" (CCEEB 2000a).

10. The CCEEB project that treats automobile emissions is known as the Transportation, Emissions, and Mobility (TEAM) project. The project is mostly composed of representatives from oil firms, utilities, and real estate concerns. Although the automobile industry is not officially part of CCEEB or the TEAM project, as the project leader explained. TEAM does have regular contact with individuals from this industry (Lucas 2000). Overall, the automobile industry, which does not have any major production facilities in the state, maintains a low political profile in California. This industry largely limits its political work on the issue of automobile emissions to its relationship with CARB. Otherwise, the automobile

industry allows the oil industry, which has several refineries in California, and local automobile dealerships to generally take the political lead on transportation issues in the state (White 2000; Hathaway 2000; Carmichael 2000).

11.   Several of the union representatives on CCEEB's board are from the building trades, which have historically been strong proponents of local growth (Logan and Molotch 1987, 81–82). Moreover, many of the private citizens on its board were formerly corporate executives. One, for example, was formerly a vice president of the Bank of America (Weisser 2000). None of CCEEB's board members are leaders from the environmental community.

12.   Substantive policies that protect human health and the environment also add to the state's legitimacy, but these policies are not as central to the state's legitimacy as are the symbols outlined here. Moreover, my contention, borrowing heavily from Edelman (1964, 1977, 1988), is that substantive policies are largely driven by specific interests seeking tangible benefits, whereas the broader public is managed primarily through symbols.

13.   CCEEB opposes any public policy that mandates the usage of any specific fuel (CCEEB 1990; Weisser 2000).

14.   It could be perceived as ethical for environmental groups to withdraw from the policymaking process because their absence may result in less stringent government environmental regulations that could in turn result in more environmental damage. Increased environmental damage could mobilize the public around antigrowth initiatives.

References

Air Pollution Foundation. 1961. *Final Report*. San Marino, Calif.: APF.

Anderson, Terry L., and Donald R. Leal. 1991. *Free-Market Environmentalism*. San Francisco: Westview.

Andrews, Richard N. L. 1999. *Managing the Environment, Managing Ourselves*. New Haven: Yale University Press.

Bachrach, Peter, and Morton Baratz. 1962. "Two Faces of Power." *American Political Science Review* 56, no. 4: 947–952.

Barrow, Clyde, 1993. *Critical Theories of the State*. Madison: University of Wisconsin Press.

——— . 1998. "State Theory and the Dependency Principle: An Institutionalist Critique of the Business Climate Concept." *Journal of Economic Issues* 32, no. 1: 107–144.

Borenstein, Seth. 2000. "Going Nowhere." *Miami Herald*, May 10, Cl.

Brienes, Marvin. 1975. "The Fight Against Smog in Los Angeles, 1943–1957." Ph.D. diss., University of California, Davis.

Bullard, Robert D. 1990. *Dumping in Dixie*. Boulder. Westview.

Cackette, Tom. 2000. Interview by author, March 14.

Cahn, Matthew A. 1995. *Environmental Deceptions*. Albany: SUNY Press.

California Air Resources Board. 1993. *California Emission Trends: 1975–2010*. Sacramento: California Environmental Protection Agency.

——— . 1994. *California's Air Toxics Program*. Sacramento: CARB.

——— . 1999. *The 1999 California Almanac of Emissions and Air Quality*. Sacramento: California Environmental Protection Agency.

California Council for Environmental and Economic Balance (CCEEB). 1990. *Alternative*

Motor Vehicle Fuels to Improve Air Quality: Options and Implications for California. San Francisco: CCEEB.

———. 2000a. *California Council for Environmental and Economic Balance*. San Francisco: CCEEB.

———. 2000b. *Mission*. San Francisco: CCEEB.

Carmichael, Tim. 2000. Interview by author, March 15.

Casner, Nicholas. 1999. "Polluter Versus Polluter." *Journal of Policy History* 11, no. 2: 179–200.

Christoff, Peter. 1996. "Ecological Modernization, Ecological Modernities." *Environmental Politics* 5, no. 3: 476–500.

Cohen, Jean L., and Andrew Arato. 1992. *Civil Society and Political Theory*. Cambridge: MIT Press.

Cone, Marla. 1995a. "State Offers to Delay Electric Car Mandate." *Los Angeles Times*, December 7, A3.

———. 1995b. "Air Panel Bending Under Pressure." *Los Angeles Times*, December 20, A3.

———. 1995c. "State Panel Puts Electric Car Mandate in Reverse." *Los Angeles Times*, December 22, A1.

———. 1998. "Air Board Seeks Tighter Auto Emission Limits." *Los Angeles Times*, July 18, A1.

———. 1999. "Vehicles Blamed for a Greater Share of Smog." *Los Angeles Times*, October 30, A1.

Crenson, Matthew A. 1971. *The Un-Politics of Air Pollution*. Baltimore: Johns Hopkins University Press.

Davis, David. 1992. *Energy Politics*. New York: St. Martin's.

Dawson, Bill. 1999. "State Points to Chevron Plant Emissions in Day That Puts Houston No. 1 in Smog." *Houston Chronicle*, November 20, A1.

Dewey, Scott. 2000. *Don't Breathe the Air*. College Station: Texas A&M University Press.

Domhoff, G. William. 1998. *Who Rules America?* 3d ed. Mountain View, Calif.: Mayfield.

Dowie, Mark. 1995. *Losing Ground*. Cambridge: MIT Press.

Dryzek, John S. 1996. "Political Inclusion and the Dynamics of Democratization." *American Political Science Review* 90, no. 3: 475–487.

———. 1997. *The Politics of the Earth*. New York: Oxford University Press.

Edelman, Murray. 1964. *The Symbolic Uses of Politics*. Urbana: University of Illinois Press.

———. 1977. *Political Language*. New York: Academic Press.

———. 1988. *Constructing the Political Spectacle*. Chicago: University of Chicago Press.

Fawcett, Jeffry. 1990. *The Political Economy of Smog in Southern California*. New York: Garland.

Gonzalez, George A. 2001. *Corporate Power and the Environment: The Political Economy of U.S. Environmental Policy*. Lanham, Md.: Rowman & Littlefield.

Gore, Albert. 1992. *Earth in the Balance*. Boston: Houghton Mifflin.

Gottlieb, Robert, and Irene Wolt. 1977. *Thinking Big*. New York: Putnam's.

Grant, Wyn. 1996. *Autos, Smog, and Pollution Control*. Brookfield, Vt.: Edward Elgar.

Hajer, Maarten A. 1995. *The Politics of Environmental Discourse*. New York: Oxford University Press.

Hathaway, Janet. 2000. Interview by author, March 16.

Hayward, Clarissa Rile. 1998. "De-Facing Power." *Polity* 31, no. 1: 1–22.

Isaac, Jeffrey. 1993. "Civil Society and the Spirit of Revolt." *Dissent* 40, summer: 356–361.

Kamieniecki, Sheldon, and Michael Farrell. 1991. "Intergovernmental Relations and Clean-Air Policy in Southern California." *Publius* 21, no. 3: 143–154.

Kennedy, Harold W. 1954. *The History, Legal and Administrative Aspects of Air Pollution Control in the County of Los Angeles*. Los Angeles: Board of Supervisors of the County of Los Angeles.

Kolko, Gabriel. 1977. *The Triumph of Conservatism*. New York: Free Press.

Kraft, Michael. 1993. "Air Pollution in the West." In Zachary Smith, ed., *Environmental Politics and Policy in the West*. Dubuque, Iowa: Kendall/Hunt.

Krier, James E., and Edmund Ursin. 1977. *Pollution and Policy*. Los Angeles: University of California Press.

Lamare, James W. 1994. *California Politics*. New York: West.

Lange, Leif Erik. 1999. "Transportation and Environmental Costs of Auto Dependency." In Tim Palmer, ed., *California's Threatened Environment*. Washington, D.C.: Island.

Lindblom, Charles. 1982. "The Market as Prison." *Journal of Politics* 44, no. 2: 324–336.

Logan, John R., and Harvey L. Molotch. 1987. *Urban Fortunes: The Political Economy of Place*. Berkeley: University of California Press.

Lowi, Theodore J. 1979. *The End of Liberalism*. New York: Norton.

Lowry, William R. 1992. *The Dimensions of Federalism: State Governments and Pollution Control Policies*. Durham: Duke University Press.

Lucas, Robert W. 2000. Telephone interview by author, March 30.

Luger, Stan. 2000. *Corporate Power, American Democracy, and the Automobile Industry*. Cambridge: Cambridge University Press.

Lukes, Steven. 1974. *Power: A Radical View*. London: Macmillan.

Mark, Jason. 2000. Interview by author, March 13.

McConnell, Grant. 1966. *Private Power and American Democracy*. New York: Knopf.

Milbrath, Lester. 1989. *Envisioning a Sustainable Society*. Albany: SUNY Press.

——— . 1995. "Psychological, Cultural, and Informational Barriers to Sustainability." *Journal of Social Issues* 51, no. 4: 101–120.

——— . 1996. *Learning to Think Environmentally*. Albany: SUNY Press.

Norton, Bryan G. 1991. *Toward Unity Among Environmentalists*. New York: Oxford University Press.

O'Connor, James. 1973. *The Fiscal Crisis of the State*. New York: St. Martin's.

O'Connor, Martin, ed. 1994. *Is Capitalism Sustainable?* New York: Guilford.

Olson, Mancur. 1971. *The Logic of Collective Action*. Cambridge: Harvard University Press.

Patton, Gary A. 1999. "Land Use and Growth Management: The Transformation of Paradise." In Tim Palmer, ed., *California's Threatened Environment*. Washington, D.C.: Island.

Perez-Pena, Richard. 1999. "Pataki to Impose Strict New Limits on Auto Emissions." *New York Times*, November 7, A1.

Pincetl, Stephanie. 1999. *Transforming California: A Political History of Land Use and Development*. Baltimore: Johns Hopkins University Press.

Piven, Frances, and Richard Cloward. 1971. *Regulating the Poor*. New York: Random House.

Pollack, Andrew. 2000. "New Plan Would Scale Back Quota for Electric Cars in California." *New York Times*, December 9. http://www.nytimes.com/2000/12/09/technology/09CAR. html.

Poulantzas, Nicos. 1973. *Political Power and Social Classes*. London: New Left Books.

Purdum, Todd. 2000. "Los Angeles Sprawl Bumps Angry Neighbor." *New York Times*, December 9, 1.

Rhodes, R. A. W., and D. Marsh. 1992. "Policy Networks in British Politics." In *Policy Networks in British Government*. Oxford: Clarendon.

Roberts, Thomas R. 1969. "Motor Vehicular Air Pollution Control in California: A Case Study in Political Unresponsiveness." Honors thesis, Harvard College.

Sabatier, Paul A. 1987. "Knowledge, Policy-Oriented Learning, and Policy Change." *Knowledge: Creation, Diffusion, Utilization* 8, no. 4: 649–692.

——— . ed. 1999. *Theories of the Policy Process*. Boulder: Westview.

Saward, Michael. 1992. *Co-optive Politics and State Legitimacy*. Dartmouth: Aldershot.

Scheible, Michael. 2000. Interview by author, March 15.

Schlosberg, David. 1999. *Environmental Justice and the New Pluralism*. New York: Oxford University Press.

Sexton, Patricia. 1991. *The War on Labor and the Left*. Boulder: Westview.

Skocpol, Theda. 1979. *States and Social Revolutions*. Cambridge: Cambridge University Press.

"Smog City." 1999. *Houston Chronicle*, November 4, YO7.

Szasz, Andrew. 1994. *Ecopopulism*. Minneapolis: University of Minnesota Press.

Tarrow, Sidney. 1994. *Power in Movement*. New York: Cambridge University Press.

Taylor, Bron R., ed. 1995. *Ecological Resistance Movements*. Albany: SUNY Press.

"Text of Report and Conclusions of Smog Expert." 1947. *Los Angeles Times*, January 19, 1.

" 'Times' Expert Offers Smog Plan." 1947. *Los Angeles Times*, January 19, 1.

Wainwright, Hilary. 1994. *Arguments for a New Left: Answering the Free-Market Right*. Cambridge, Mass.: Blackwell.

Warrick, Joby. 1997. "Greenhouse Gases Rose 3.4 Percent in 1996." *Los Angeles Times*, October 20, A8.

Weale, Albert. 1992. *The New Politics of Pollution*. New York: Manchester University Press.

Weinstein, James. 1968. *The Corporate Ideal in the Liberal State*. Boston: Beacon.

Weisser, Victor. 2000. Interview by author, March 13.

White, V. John. 2000. Interview by author, March 14.

Whitt, J. Allen. 1982. *Urban Elites and Mass Transportation*. Princeton: Princeton University Press.

Wynne, Brian. 1982. *Rationality and Ritual*. Chalfont St. Giles, U.K.: British Society for the History of Science.

Chapter 13

John S. Dryzek, David Downes, Christian Hunold and David Schlosberg, with Hans-Kristian Hernes

ECOLOGICAL MODERNIZATION, RISK SOCIETY, AND THE GREEN STATE

WE HAVE ARGUED THROUGHOUT that it matters a great deal to a social movement whether or not it can attach its defining interest to a core state imperative. If it can, then there are in principle no limits to the degree to which the movement can penetrate to the state's core once the movement has sought and achieved entry into the state. If it cannot, then the movement is likely to receive either symbolic or at best marginal rewards as a result of its engagement with the state. Whenever the movement's interest comes up against the core, the movement will lose; it is merely co-opted (see Chapter 3).

This situation long characterized environmentalism, which, locked in a zero-sum conflict with the economic imperative, with few exceptions lost whenever it approached the state's core. However, that situation may be changing. Proponents of ecological modernization now argue that economic and ecological concerns are potentially complementary; if so, environmentalism may for the first time be linked positively to the core economic imperative.[1] Ulrich Beck's risk society thesis suggests a different route for connecting movement goals and state imperatives. Beck posits a legitimation crisis in the context of risk that can be met by new forms of 'sub-politics' that effectively engage the citizenry in the selection, allocation, distribution, and amelioration of risks. If so, environmentalism may for the first time find effective linkage with the state's legitimation imperative.

These two developments could conceivably add up to a new state imperative: environmental conservation. The emergence of the economic imperative first democratized the modern state through inclusion of the bourgeoisie in the core, creating the capitalist state. The emergence of the legitimation imperative further democratized the state by including the organized working class in the core, creating the welfare state. The emergence of an environmental conservation imperative would democratize the state still further by including environmentalists in the core, creating the green state. The conservation imperative would entail rethinking both economic and legitimation imperatives—the economic with an eye towards syn-ergy with natural processes, and legitimation with an eye towards the need for public reflection on continued technological modernization. It hardly need be said that as yet there is no green state in these terms. However, we will show in this chapter that elements of its development are being played out in our four countries.

A weak form of ecological modernization is tied only to the economic impera-tive in terms of 'pollution prevention pays'. Such is the case in Norway. We argue that what Peter Christoff (1996) calls strong ecological modernization is most conducive to the convergence of environmental interests and state imperatives, which can only be secured through a connection to issues of ecological legitimacy through public participation (Christoff himself is a statist who would disagree with this analysis). With this connection, strong ecological modernization demands an informed public that reflexively and democratically addresses environmental risks. It involves reshaping public policy and political–economic structures along eco-logically sustainable lines, which in turn requires keeping movement politics alive (otherwise there will be a lapse to the weak form). Of the countries examined here, only Germany seems to be moving in the strong direction. The problem is that, historically, this sort of critical public sphere has been associated with passive exclusion, a condition that holds only for Germany of the countries considered here (and may not last there forever). Paradoxically, such states impose limits on the degree to which strong ecological modernization can penetrate state structures. However, we will show that the way history unfolds can overcome these limits—if not permanently. Also, reinforcing the sub-politics thesis, there can be more to strong ecological modernization than what is accommodated within the state.

We do not contend that environmentalists must be German to be successful. Although a green public sphere is most likely to arise and persist in passively exclusive states, variants of it exist in passively inclusive states, where its sustenance is harder work. The revitalization of the American grassroots in the 1990s and the re-emergence of radical environmental activism in the UK after the waning of active exclusion since 1989 suggests that no state abiding by liberal democratic principles is capable of permanently sidelining oppositional civil society. The grass-roots US movements focus on the legitimation imperative that we associate with strong ecological modernization. The mainstream American groups target only the economic imperative and thus only weak ecological modernization (without using this language), so far with little policy response. The Blair Labour government in the United Kingdom seems to recognize public demands and so the legitimation imperative, but gives mostly lip-service (and endless public inquiries) to these issues.

Passive states, inclusive or exclusive, willingly or not, permit enough move-ment diversity to sustain a radical programme consistent with the requirements of strong ecological modernization among a significant part of the environmental *movement*—a necessary but not sufficient condition for moving *society* in the direc-tion of strong ecological modernization.

A critical public sphere and hence strong ecological modernization are least likely to arise and hardest for environmentalists to sustain under conditions of active inclusion. However, a weak version of ecological modernization and so a pale green state *is* sustainable under active inclusion, as the case of Norway will show. This sort of case is actually consistent with the bulk of what Mol and Spaargaren

(2000: 23) call 'the first generation studies in Ecological Modernisation Theory' in the 1980s. (The literature, if not the policy practice, has now moved in a 'strong' direction.)

We pay close attention to two indicators of ecological modernization: public access to environmental information and ecological tax reform. The former is important because strong ecological modernization, linked to the legitimation imperative, requires open debate on matters that affect environmental quality, which in turn depends on the public's access to information that is analysed, interpreted, and presented in ways that non-professionals can understand (World Bank, 1998: 11).[2] Confronted with an un-cooperative state, agents in a green public sphere might organize to gather environmental information themselves, or they might emphasize green discourses that eschew the ecological sciences altogether. However, having access to information about the quality of the local water supply, soil toxicity, and the like can strengthen the hand of environmental NGOs and citizens. This access is significant because, all other things being equal, a state that tends towards openness is more likely to accept strong ecological modernization than a state that favours secrecy. Other things are not of course equal, as we will see.

Tax reform intended 'to shift from piecemeal environmental taxation to a more thorough ecological tax reform where labour taxes are replaced by environmental taxes' (European Environment Agency, 2000b: 397) is a significant aspect of the sort of restructuring advocates of weak ecological modernization have proposed. It is a central tenet of ecological modernization that pollution prevention pays, but green taxes are the other side of the coin, reaffirming the principle that the 'polluter pays'. Weak ecological modernization addresses the economic imperative, and in so doing also attaches government revenue to environmental progress. A majority of EU member states has embarked on ecological tax reforms, with the Nordic Countries and the Netherlands leading the way (European Environment Agency, 2000a). Ecotaxes alone cannot effect 'a transition from incremental to radical innovations in which ecologically problematic procedures and products are substituted by unproblematic ones' (Kemp, 1997: 9). They make little sense when the risk is acute, requiring immediate action (Jänicke, 2000: 14). And evasive actions can sometimes undercut their effectiveness (Jänicke, 2000: 15). For these reasons we again need to look at green taxes in the context of other developments, rather than use them as a simple indicator of ecological modernization (in 1998, environmentally related taxes as a percentage of total government revenue were 5.8 per cent in Germany, 8.2 per cent in Norway, 8.3 per cent in the United Kingdom, 3.1 per cent in the United States[3]).

Ecological modernization and the risk society thesis

In its basic or weak form, ecological modernization involves solving environmental problems by making capitalism less wasteful within the existing framework of production and consumption. The technical fix looms large (Street, 1992). 'Pollution prevention pays'; low pollution indicates productive efficiency in the use of resources. A clean and pleasant environment means happy healthy workers; environmental problems are solved in the present (informed by the precautionary principle—action need not await scientific proof) rather than postponed to a future where their resolution will be more expensive. The price of the broad acceptability of such a programme is that ecology is conceptualized in thin terms. Radical green critiques of technology and capitalism are muted by the claim that economic growth and environmental protection can be mutually supportive (Dryzek, 1997: 137–52).

What Hajer (1995) calls 'techno-corporatist' (weak) ecological modernization is based on the idea that 'dominant institutions indeed *can* learn and that their learning can produce meaningful change' (Hajer, 1996: 251).[4] Biologists, economists, and engineers are responsible for drawing up environmental quality standards and technologies that meet these standards. Social scientists can identify ways of modifying 'anti-ecological' cultural patterns and behaviors. Weak ecological modernization is a moderate social project (Hajer, 1996: 253; for a defence see Mol, 1996). It is attractive to state officials because it promises to meet their need to promote economic growth and improve environmental conditions with minimal disruption. It implies the progressive inclusion of mature groups that have discarded radical critique, informal organization, and protest politics in favour of pragmatic advice, professional organizational hierarchy, and interest-group activity. While radical environmentalism may have been important in getting environmental issues on the agenda, its confrontational politics and rejection of economic growth become liabilities. Here there is a comfortable fit with life-cycle theories of social movements, wherein the movement begins in inchoate radicalism and ends in the corridors of power.

We will argue that weak ecological modernization ultimately fails to deliver on its promise of securely connecting ecological aims to core state imperatives. In the words of Christoff (1996: 490) it offers little hope for 'promoting enduring ecologically sustainable transformations and outcomes across a range of issues and institutions'. The risk of co-optation is high because only modest goals can be linked to the economic imperative in the absence of a credible movement threat to destabilize existing political and economic institutions. Weak ecological modernization, linked only to the economic imperative, never challenges the legitimacy of the state's complicity in practices that generate environmental risks. A weak ecological modernity defined by managerial solutions to environmental problems (Luke, 1999) downplays the critique of industrialism by emphasizing policy learning and institutional adjustment rather than fundamental social change (Hajer and Fischer, 1999).

A particularly graphic illustration of weak ecological modernization and its limits is provided by Gonzalez (2001) in an analysis of automobile pollution control policy in California. California policy emphasizes technological changes to car engines that will reduce emissions, and is driven by powerful economic interests rather than environmentalists. However, despite the success of this policy, total emissions continue to grow as the number of cars on the road and the average per-car distance travelled per year both increase faster than technical change can reduce emissions. Land use planning that would reduce reliance on the car—the kind of structural change that strong ecological modernization implies—is simply not on the anti-pollution agenda.

In technocratic form, 'Ecological modernisation is much more the repressive answer to radical environmental discourse than its product' (Hajer, 1996: 254). The question that motivates many social movements—what kind of society do we want—is difficult to pose in pure efficiency terms. Ecological modernization thus poses a serious challenge to environmentalists wishing to keep radical critique alive. Noting the environmental movement's crucial role in bringing about industrial society's critical self-examination, Hajer wonders what will happen to the movement's 'history of reflexivity':

> The question that looms large is the extent to which the environmental movement has been able to perpetuate this reflexivity once it decided to argue its case in the appealing terms of ecological modernization. (Hajer, 1995: 102)

The answer, we suggest, depends on whether and to what effect environmental groups can show an oppositional and discursive face, and so move ecological modernization in a strong direction, problematizing norms and values driving modernization processes (Hajer, 1995: 280–1). Strong ecological modernization resists subordinating ecology to economics, is attentive to interactions among a broad array of political, economic, and social institutions, favours communicative rationality and participatory public deliberation, accepts and indeed requires movement activism, and recognizes the transnational aspects of issues. What Hajer (1996) terms 'ecological modernization as cultural politics' goes beyond encouraging public participation in order to 'explore how new perspectives on society can be created' through open public debate (Hajer, 1996: 260) concerning the actors and practices implicated in environmental degradation. Ecological modernization in its strong form requires a vibrant civil society.

Be it weak or strong, ecological modernization is attractive to many environmentalists because it provides a way for their concerns to be taken seriously in a world where economics is the first concern of governments. In other words, it facilitates a link to the economic imperative of states. Beck's theory of the risk society (1992: 1994) can be deployed to make a link to the legitimation imperative. The risk society thesis is that politics in developed societies is increasingly about the production, selection, distribution, and amelioration of risks. Such risks can relate to nuclear power, genetically modified organisms, food safety (as in the Mad Cow Disease/BSE issue in Europe), and toxic chemicals in the environment.

To Beck, the degree to which this new politics supplants the class politics of industrial society heralds a 'reflexive modernity' in which society confronts the unintended consequences generated by the combination of science, technology, and economics that has driven 'progress' since the early nineteenth century. The idea of technological progress itself is called into question, along with faith in economic growth and scientific rationality. The result is a crisis in the legitimacy of the political economy. Reflexive modernization is society's resultant self-confrontation. Although the transition from industrial to risk society is rooted in the developmental logic of industrial society, social movements play a crucial role in initiating industrial society's process of self-confrontation with its own foundations. Risk society is reflexive not only in so far as the consequences of industrial society undermine these foundations; it is also reflexive in the sense of being self-critical. As a result of environmental activism and the public awareness of risks it generates, citizens start questioning the prospects for managing environmental crises by further reliance on economic growth and technological rationality. New possibilities for social and political transformation arise from people's growing awareness that they are living in a society whose habits of production and consumption may be undermining the conditions for its future existence. Thus Beck believes that reflexive modernization is accompanied by waning influence of state structures compared to diverse 'sub-political' spaces of civil society, of the sort we discussed at length in previous chapters.

Sub-politics is consistent with the politics of ecological modernization in its strong sense.[5] The politics of ecological modernization can turn out to be contentious because 'the invasion of ecology into the economy opens it to politics' (Beck, 1999: 100). Industry and business are forced to respond to this invasion of the economic sphere with strategies that observe 'the standards of politics requiring legitimation' (Beck, 1999: 101). This politicization of the economic sphere can be interpreted as a battle over the terms of ecological modernization: environmentalists aim to ecologize economics, bringing recognition to the reality of the economic system as a subsystem of nature (Daly, 1996), while government and industry stress those aspects of ecological modernization that are most consistent with existing economic priorities. Ecologizing economics entails a more thorough local focus, in both understanding the consequences of economic action and in decision making processes. From our point of view it therefore becomes crucial to ascertain the degree to which the state (and its imperatives) are declining in importance as a site of political action in comparison with these sub-political spaces.

Both strong ecological modernization and sub-politics emphasize the role of public deliberation in democratizing and legitimating science and technology. Beck's ecological democracy would democratize the politics of expertise by rolling back the industrial coalition's colonization of politics, law, and the public sphere:

> My suggestion contains two interlocking principles: first, carrying out a division of powers and, second, the creation of a public sphere. Only a strong, competent public debate, 'armed' with scientific arguments, is capable of separating the scientific wheat from the chaff and allowing the institutions for directing technology—politics and law—to reconquer the power of their own judgement (Beck, 1999: 70).

The result would be public science wherein 'research will fundamentally take account of the public's questions and be addressed to them' (Beck 1999: 70). Public science, Beck hopes, would help lay citizens win back the competence to make their own judgement, independently of experts, by making hazards that are not really visible or otherwise directly perceivable culturally perceptible. (It is ironic that Beck's nemesis, Aaron Wildavsky, issues an identical call for 'citizen risk detectives', except that Wildavsky expects citizens to debunk alleged risks. See Wildavsky, 1995.)

Thus we can envisage two paths towards the green state, each with its own set of movement strategies. In one scenario, professional environmental NGOs pursue technocratic or weak ecological modernization within the state's decision making apparatus. In the other scenario, environmental NGOs do not limit their activities in this way, but instead follow the dual strategy of action in the state and civil society. While intrinsically attractive for all the reasons set out at the end of Chapter 6, the dual strategy presents movements with two problems. First, states are likely to prefer weak to strong ecological modernization. Second, some state configurations are not very conducive to the kind of public sphere and its relationship to the state that strong ecological modernization requires (as we will show in this chapter). We are more likely to find strong ecological modernization in states whose structures promote (if by accident rather than design) a flourishing civil society—though there are times when the odds in this respect can be beaten. Still, movement strategies will have to vary across different national contexts. The politics of strong ecological modernization is not available to all states equally, as we will now show.

Norway: weak ecological modernization and little sub-politics

Strong ecological modernization is unlikely where civil society has been depleted, either through the state's actively inclusive or actively exclusive stance towards social movements. The more problematic case is actually that of active inclusion, as liberal democratic regimes appear incapable of sustaining actively exclusive strategies for very long. States such as Norway cultivate groups that moderate their demands in exchange for state funding and guaranteed participation in policy making, to the extent they can hardly be called NGOs. Their moderation means that included groups embrace any weak ecological modernization on offer. Groups outside the state-funded sector are few and far between. Conventional and routinized forms of engagement dominate environmental politics, so ecological modernization is initiated and controlled by the state.

Contrary to conventional wisdom, we expect that Norwegian environmentalists may in the long run be little more likely to connect environmental values to state imperatives than their counterparts elsewhere. Norwegian environmental policy in the 1980s and 1990s does however fit a trajectory of weak ecological modernization. The achievements are undeniable, and Norway is rightly regarded as a leader among European countries (Sverdrup, 1997: 74). Pollution levels in Norway are relatively low, and integrated pollution control was introduced as early as 1981. Momentum intensified with the Brundtland-associated sustainable development push in the late 1980s. As Langhelle (2000) puts it, Norway then 'carried the torch' on sustainable development. The sector-encompassing ambitions of this programme and associated state of the environment reporting further promoted ecological modernization, as did the subsequent policy emphasis on sustainable consumption. In addition, Norway does very well on the informational aspects of ecological modernization. Norway's Freedom of Information Act predates the 1990 EU directive on access to environmental information by twenty years and covers all of the latter's provisions (Killingland, 1996: 309).

Perhaps the most significant accomplishment in ecological modernization came with the victory of environmentalists and research institute representatives over recalcitrant industry participants on the green tax issue in 1989–92, which led to the implementation of green taxes beginning in 1991. The argument that these taxes reconcile environmental and economic values carried the day. The Nordic countries have pioneered the use of eco-taxes, and by the end of the 1990s in Norway they accounted for 9 per cent of tax revenues. However, their popularity in Norway may be for fiscal as much as environmental reasons: income from taxes related to oil has enabled Norway to avoid deep cuts in social services of the sort common in the rest of Western Europe. Still, as we concluded in Chapter 3, there have been times when Norwegian environmentalists have succeeded in attaching their interests to the state's economic imperative. However, there are limits to this assimilation—indicating perhaps the limits of weak ecological modernization.

In the 1990s Norway consciously took the lead in an effort to curb green-house gas emissions. But this policy ran up against 'the most long-standing principle of Norwegian environmental policy: that is, the principle that Norwegian policy should not put Norwegian industry and commerce in a disadvantageous position compared with foreign competitors' (Langhelle, 2000: 208, citing Jansen and Osland, 1996). The centrality to Norway's economy of energy-intensive industry, notably aluminium (relying on cheap hydro-electricity) and oil and gas production, also imposes limits. In contrast to other Nordic countries, energy and environment remain in separate ministries, and old energy-related political networks can resist the encroachment of ecological modernization (Midttun and Kamfjord, 1999). Hydropower and then North Sea oil have been at the heart of Norway's economic policy (especially after the energy shocks of the 1970s). Green taxes too had their reach limited when their implementation continued to be opposed by companies in key industrial and energy sectors (Kasa, 2000), preventing energy taxes from contributing to ecological restructuring. The opposition of industry to a carbon tax in the 1990s revealed the limits to consensual policy action of the sort that weak ecological modernization requires (Reitan, 1998).

In light of its undeniable accomplishments, why do we remain sceptical about Norway's capacity to pursue strong ecological modernization. The answer is that when limits are encountered, such as those imposed by the international position of industry and commerce or the centrality of hydro-power and oil, there is no countervailing pressure from civil society that might inspire a search for creative solutions. Episodes in the politics of hydro-power (and pollution) suggest that sub-politics has existed in Norway. Opposition to hydro-power was a focal point for Norway's movement until the early 1980s, but since then the issue has been dormant and depoliticized. Environmental activism has declined since inclusion reached its zenith around the issue of sustainable development in the late 1980s and early 1990s, when groups such as the Nature Conservation Society experienced a short-lived increase in membership. Although public access to environmental information has been very good by international standards, it remains something of an empty gesture in the absence of an autonomous green public sphere. Further, 'in actual practice the authorities do not always respond fully and in proper form to all information requests' (Killingland, 1996: 310).

Waning public involvement in Norway's established environmental NGOs in the 1990s has coincided with only a limited increase in direct action. The resurgence of grassroots environmentalism in the UK and the US in the 1990s, let alone the anti-statist environmentalism of Germany in the 1970s and 1980s, has no parallel here. The Bellona Foundation, established in 1986, became in the 1990s Norway's highest-profile environmental group. Bellona is not tied to the state after the fashion of the mainstream groups, and its style is occasionally confrontational. Another sub-politics aspect appears in its consultancy activities. But like Greenpeace, Bellona operates more like a company than a charity, still less a movement. A portion of the group's revenue has come from business interests, and as we pointed out in Chapter 6 by the end of the 1990s it was accepting project grants from the government. This type of 'environmental enterprise' is not unique to Norway, but this organizational form may be the only viable environmentalist response to the co-optation and decline of older groups in an actively inclusive state.

Norway also offers little scope for connecting environmental concerns to the legitimation imperative via any risk society scenario. In our other three countries, chemical pollution, biotechnology, and nuclear power have raised the need for legitimation in the context of risk, opening the door to oppositional action and sub-politics. In Norway, these issues are largely absent, and legitimation plays out in more old-fashioned terms, tied to a popular but expensive welfare state. Norway's export-oriented oil industry supports that welfare state, so legitimation actually points in an anti-environmental direction.

Perhaps the muted quality of Norwegian environmental politics in the 1990s simply reflects the reality that environmental degradation is less severe in Norway than in more heavily industrialized countries. But it also reveals, we argue, the analgesic and even anti-democratic face of the actively inclusive state. Ecological modernization in Norway is weak and set to remain a moderate, top-down project. There are no significant actors raising critical questions about the basic structure of the political economy or about the basic direction of public policy, no social movements to constitute an autonomous public sphere that can act as both a source of ideas and a reminder to policy makers of the seriousness of concerns that need to be addressed.

The United States: some sub-politics, no ecological modernization

The days when the United States was an environmental pioneer (enabled, as we argued in Chapter 3, by a link between environmentalism and the legitimacy imperative in the early 1970s) are a distant memory. Of course, there has remained plenty of activity in the periphery of the state, where movement access remained. But the term ecological modernization is not part of US policy discourse, nor is there much in policy practice to suggest pursuit under any other name (beyond the California emissions case we mentioned earlier). And although there are sporadic initiatives that join business and environmentalists, such as those associated with Amory Lovins and his *Factor Four* (the idea that prosperity can be doubled while halving resource use; see also Lovins and Lovins, 1999), these do not find their way into government.

The terms of debate in Washington in the 1990s were set by a conservative

Congressional majority that saw environmental regulation as a drag on the economy, with economics and ecology remaining cast in old-fashioned zero-sum terms. 'Rather than asking more fundamental questions about how to balance and integrate economic growth and ecological sustainability, policy makers are mired in efforts to defend or attack the regulatory system that has been in place since the 1970s' (Bryner, 2000: 277). The approach to pollution regulation remains very much 'end of pipe'. And with few exceptions the politics of land management and wilderness protection still pits old adversaries against one another (loggers, miners, and ranchers on one side, environmentalists on the other). The environmental justice movement may in Lois Gibbs's words have succeeded in 'plugging the toilet' on toxic wastes, but this has not led to the creative redesign of production processes to minimize waste generation as sought even by weak ecological modernization. In short, if we seek ecological modernization in the United States we find very little, certainly when it comes to the national state and the mainstream of US business. As Andrews (1997: 41) puts it, when it comes to pollution control, 'The basic structure of U.S. environmental policy . . . has remained entrenched in the paradigm of detailed federal standards and technology-based permits, rather than evolving further towards pollution prevention and ecological modernisation principles.'

Green taxes have not been widely applied in the United States, where *any* new taxes are extremely unpopular. An energy tax proposed by the Clinton administration failed by two votes in Congress, although several hundred eco-taxes exist at the state and local levels (Beck et al., 1998: 31). The economic logic associated with environmental initiatives appears under the heading of 'free market environmentalism' in the United States (for example, Anderson and Leal, 2001), not ecological modernization. This may explain the greater popularity of tradable permit schemes, such as the SO_2 trading scheme introduced by the 1990 Clean Air Act Amendments to deal with regional acid rain pollution (Lovei and Weiss, 1998: ix). The absence of so much as a public *debate* on ecological tax reform underscores our point concerning the irrelevance of ecological modernization in the United States. Significantly, presidential candidate Al Gore renounced his earlier commitment to eco-taxes during the 2000 election campaign.

Gore once believed in something like weak ecological modernization, as indicated by his 1992 book, *Earth in the Balance*. President George W. Bush thinks of the economy and the environment in inflexible zero-sum terms, and so never got close to that. In an exchange during the second presidential debate in October 2000, Gore supported tax incentives to encourage the development of renewable energy production technologies while asserting the mutual compatibility of environmental conservation and economic growth; though he retreated from his earlier support for energy taxes. Gore called Bush's plan to promote the domestic production of fossil fuels and his preference for voluntary agreements with industry to regulate pollution outdated ways of thinking about the environment/economy relationship. Bush, in turn, questioned the scientific merit of global warming projections, foreshadowing his decision as president to take the United States out of the Kyoto Treaty. Defending this decision to a European audience in July 2001, the president said: 'My job is to represent my country. We want to reduce greenhouse gases. . . . But first things first, as far as I'm concerned. Our strategy must make sure working people in America aren't thrown out of work' (Hutcheson, 2001). It is hard to imagine a clearer refutation of the central tenet of ecological modernization. The George W. Bush administration's position on global warming received strong political support from resource-intensive industries. The response from industries that operate in multinational political and regulatory regimes and are thus wary of an international backlash against US unilateral action was somewhat less enthusiastic.

Presidential administrations and congressional majorities come and go. But if we look at the underlying capacity for ecological modernization, we find a number of deficiencies. Jänicke and Weidner (1997a) and their collaborators explain comparative environmental policy performance in terms of such a capacity. Aside from the strength and competence of actors inside and outside government committed to environmental values, this capacity refers to the framework of political institutions and the structure of the economy. On the positive side, the United States does very well when it comes to scientific and technical aspects of capacity, mainly because agencies must justify regulatory proposals to the courts and to the public (Andrews, 1997: 35). But overall, Andrews (1997: 25) interprets the history of environmental policy in the United States in terms of 'swings of the pendulum . . . between initiatives to create and initiatives to destroy such a capacity'. These swings are themselves indicative of both the zero-sum framing of environmental issues and a lack of underlying capacity—one aspect of which is the capability for consensual action. Co-operative action of the sort weak ecological modernization requires is problematic in a system where business is hostile to labour and to the idea of government (while lapping up government contracts and subsidies), resource-intensive businesses attack environmentalists (often sponsoring anti-environmental groups and campaigns), and environmentalists distrust business. There is no government agency in a position to promote policy integration and consensual action. The EPA has little budgetary discretion, and its hands are tied by congressional direction to implement single-medium statutes. This situation is indicative of a more general lack of environmental policy integration in government (Andrews, 1997: 38–9).

The way passive inclusion plays out helps to explain the lack of ecological modernization in the United States. Once included, groups engage in adversarial politics—be it in lobbying or in court actions. To an outsider, the range of acceptable positions within the pluralist system may look narrow, but within this range adversarial politics is still the norm. The traditional picture of American pluralism is in terms of public policy as the resultant of competing groups pulling in different directions. It is difficult to frame matters in the positive-sum terms required by weak ecological modernization in such a situation. Citing a long list of precursors, Jänicke (1997: 13) sees the ability to undertake consensual or co-operative action as a key aspect of state environmental policy capacity. (For quantitative cross-national evidence of a positive association between degree of corporatism—the main form of consensual policy making—and environmental policy performance, see Scruggs, 1999, 2001.) Co-operative exercises do exist in the United States, and designs such as environmental mediation, policy dialogues, and regulatory negotiation have been pioneered here (Dryzek, 1990). However, such forums are normally billed as forms of conflict resolution: typically, they come into play only after the various parties to a conflict have fought each other to a standstill in more adversarial institutions, especially the court system (Weber, 1998). More truly co-operative structures exist—but only at the local level. Examples include the Resource Advisory Councils for the management of Western grazing lands that involve ranchers, government officials, and environmentalists (Welsh, 2000); collaborative management processes in the same region (Brick, *et al.*, 2000); and the kind of 'backyard' structures celebrated by Sabel, Fung, and Karkkainen, 2000).

In previous chapters we pointed to the diversity of the US environmental movement, and the resurgence of activism in the 1980s and 1990s among those who felt themselves excluded from the kind of relationships that insider groups had established in government. Thus the United States may well have sufficient vitality in the green public sphere to help turn weak ecological modernization into the strong form, especially as the focus is so often on legitimacy, reflexivity, and democratic participation. The only minor difficulty here would come with the historically difficult relationships between the leadership of the major groups and grassroots activists. The sort of mutual hostility and contempt in such relations in the United States is largely absent from say the German movement, where mainstream and grassroots groups still share a sense of tilling different sections of the same field.[6] Still, as we pointed out in Chapter 6, mutual hostility has given way to more productive relationships between insiders and radical activists in the United States. The mainstream's inclusion in the state has brought many disappointments, but it has not robbed the movement as a whole of its ability to 'preserve its reflexivity'—even though specific organizations may have lost theirs. Such a dynamic follows the developmental logic of the passively inclusive state, which absorbs interests faster than exclusive states, but more slowly than actively inclusive states such as Norway. As older groups are absorbed, newer, more radical ones can emerge to take their place in the public sphere.

But these positive features of the contemporary American movement turn out to be beside the point when it comes to ecological modernization. The real difficulty in the United States comes not with any deficiencies in the character of its civil society that would preclude *strong* ecological modernization (the Norwegian problem), but much earlier, with the system's inability to turn zero-sum conflicts between economic and environmental interests into *weak* ecological modernization. Here, movement vitality is no help. The same might be said of the country's strong freedom of information and right-to-know toxics laws (the Freedom of Information Act was first passed in 1966 and substantially revised in 1974; Adler, 1997). They would help turn weak into strong ecological modernization, but are no help in the absence of the weak form.

Issues of risk are more salient in the United States than in Norway, and they do produce some sub-politics, though they have not occasioned quite the crises that have been observable in the United Kingdom and Germany. The simple fact that the United States is a big place means that many risks are localized, so it is no surprise that the major risk-related movements, anti-toxics and environmental justice, begin as series of local actions. These movements are risk society phenomena in Beck's terms. They call economic and technological progress into question, and refuse to accept that hazards can be managed through established policy making structures deploying scientific expertise. There are numerous critiques of the technocentric process of risk assessment, and of the lack of democratic participation in both the direction of scientific inquiry and the application of science to policy making (see e.g. Fischer, 1995; O'Brien, 2001). Moreover, the environmental justice, anti-toxics, and public-interest science movements engage a kind of sub-politics that involves co-operation with activists in other localities, direct action protests and campaigns against polluters and government, shading into more conventional action such as lobbying, consultation, and court challenges.

In the United States, science actually lacks the authority it possesses in environmental policy in Germany and the United Kingdom. Science often holds itself apart from the public and its desires, and communication is seen as only unidirectional, from the experts to the public. The scientific community has not accepted that its own legitimation and effectiveness could depend on the information and support it receives from the public (Sarewitz, 1996). Further, the science carried out by government agencies is influenced by powerful political actors. For example, the first EPA study on dioxin was completed in 1985; it was a review of the available literature on animal studies and cancer. A risk assessment was done based on this available evidence, and the EPA developed a recommended exposure level. But this level was the lowest ever recommended by any scientific or policy body in the world, and was immediately challenged by industry. So the EPA began a 're-evaluation' of dioxin exposure, which ultimately raised the acceptable risk level. Challenged by both grassroots groups and the scientific community, the EPA tried again. When the final report (or set of reports) came out in 1995, members of Congress attacked the conclusions because they were not peer reviewed. There are still no updated exposure guidelines from the EPA; in the meantime, not only the victims of dioxin exposure, but also the scientific process and the public's opinion of it have taken a beating.

A related reason for the loss of scientific authority is the adversarial nature of most rule making. Each side presents its own scientists, making it hard to maintain the appearance of scientific neutrality (Weiland, 2001: 14). Science loses its neutrality and become another commodity for sale. Adversarial policy making also forces science into a peculiar position in environmental impact assessment. According to the courts' interpretation of NEPA, impact assessments can be challenged only on the basis of the adequacy of the document, not its integration into policymaking. This forces government agencies preparing the documents to employ scientists to overstate the negative environmental impacts of the developments they are proposing.

Mistrust of science and experts has led to sub-political practices, such as 'popular epidemiology' (Fischer, 2000). In the most cited case (the basis for the popular film *A Civil Action* starring John Travolta), citizens in the vicinity of toxic waste sites in Woburn, Massachusetts, carried out their own study of the incidence of leukemia and birth defects after refusing to accept government reassurances. Despite the rejection of the results by state and government agencies, they were used in a lawsuit against one of the companies, which was settled out of court (Brown and Mikkelsen, 1990). Another example of this public science is the 'Bucket Brigades' organized by the California environmental group Communities for a Better Environment (CBE). Communities located near oil refineries have a common problem: accidental or deliberate 'toxic release events' or 'airborne toxic events'. Most states just require industry to self-report such events, or send inspectors out (usually after the smoke has cleared). Citizens may use video cameras to show releases, but those can only show smoke, not whether or not regulated pollutants are involved. So CBE developed an easy-to-use air sampling kit; citizens take the sample and send it to an independent laboratory. CBE then sends the documentation to the community organizations and local, state, and federal monitoring agencies. This system uses public involvement in the scientific process to put some serious pressure on industry that has, in the past, avoided scrutiny. CBE now makes available its 'Bucket Brigade' kits to any interested communities, so they can do the scientific monitoring of toxic events. There is a growing 'public interest' science in the US as well. 'Science shops', where communities bring scientific questions to researchers willing to focus on public concerns, have expanded (see Sclove, 1995, or the Loka Institute at www.loka.org).

The localized character of most risk issues in the United States does not preclude a more pervasive sub-politics that transcends the local level, but retains local roots. The vehicle here is the network, as we discussed in Chapter 5. It is noteworthy that environmental justice as a movement is quite successful when it sticks to local actions and network associations. It runs into major difficulties as soon as it renounces sub-politics and adopts more conventional trappings. In Chapter 3 we reported on the disappointments of environmental justice activists who had made the transition into government, and the meagre impact of the Office of Environmental Justice and National Environmental Justice Advisory Committee within the EPA. In a passively inclusive state, the lure of this sort of inclusion may therefore need resisting if sub-politics and reflexivity are to be protected.

The United States does, then, feature some sub-politics in the presence of risk issues. However, the ecological transformation of its state and political economy remains unlikely so long as the structural and political–economic obstacles to even weak ecological modernization that we have identified persist.

The United Kingdom: sub-politics by default and belated ecological modernization

We are interested in the United Kingdom primarily because it is the best approximation we can find to an actively exclusive state, though as we have pointed out, active exclusion began to wane in 1989. The extreme form of market liberalism that defined the 1979–89 era had little time for environmental values under any rubric, and stressed the need for deregulation in order to promote economic growth (even the 1995 Environment Act had a deregulatory impetus in requiring cost–benefit analysis for regulatory proposals). It is no surprise that ecological modernization failed to make any inroads. In a way, this contingent anti-environmental feature makes it hard to discern the impact on the prospects for ecological modernization of active exclusion *per se*, though there are good reasons to suppose that active exclusion is detrimental. In particular, it destroys any potential for consensual environmental policy making of the sort that is key to capacity building. The conflictual terms in which the main axes of politics are cast by active exclusion precludes any co-operative search for positive sum measures that would involve environmentalists and business.

By the 1990s, environmental groups could re-enter the corridors of power as the excesses of market liberalism eased and the British state reverted towards a more passive orientation, mixing inclusion and exclusion. But the government did not adopt ecological modernization's components as policy. As Weale (1997: 105) puts it:

> there is little policy development on what might be regarded as the central tenets of ecological modernization. There is: little use of pollution charges in accordance with the polluter pays principle; only haphazard encouragement of research and development for cleaner technologies; virtually no work on reconceptualising measures of economic well-being to net out from national income accounts defensive environmental expenditures; and a poor record on energy conservation.

Moreover, there was no policy integration across central government departments. Policy making continued to be dominated by a requirement to demonstrate scientific proof of a hazard prior to policy response—the antithesis of the precautionary principle (Hajer, 1995). Environmental regulation that did exist was very much an end-of-pipe affair. Road-building continued to be the focus of transportation policy, lip service to alternatives by the Blair Labour government after 1997 notwithstanding.

Proclaimed commitments to sustainable development in the wake of the 1992 Earth Summit generated several White Papers in the 1990s, but little of policy substance. In 1997 a Sustainable Development Unit was established in the new Environment Agency, and it began pushing a cross-sector approach, but it faced an uphill struggle against other departments (S. Young, 2000: 254–5). These developments did open doors to government for the environmental movement, but as we saw in previous chapters, inclusion often acted as a burden upon environmental groups. Jacobs (1997: 5) casts doubt on the value of sustainable development in the United Kingdom:

> on occasion the urge of environmentalists to join the political mainstream has threatened to give utopianism a good name. But for all its success in uniting former opponents on a more or less common environmental platform, sustainable development has not succeeded in making the environment into a central political issue.

As opportunities for consultation became more plentiful in the areas of sustainable development and biodiversity in particular, sections of the movement appeared to be settling into an increasingly moderate, institutionalized relationship with government. Following Jänicke, this development ought to have increased the potential for consensual action and so environmental policy capacity—itself instrumental to the pursuit of ecological modernization. Yet this did not happen. For the relationship with government often involved consultation without influence, with a large element of co-optation of environmentalists (see Chapter 3). Pollution control policy in the United Kingdom has traditionally been a matter of co-operation—a far cry indeed from the legalistic, politicized, and adversarial character of regulation in the United States. But the co-operation in question is between government regulators and corporations at the micro level of implementation; it is done in secret with no possibility for environmentalist participation or public scrutiny. In the absence of any commitment to ecological modernization on the part of regulators or polluters, there is no structural facilitation of ecological modernization resulting from such co-operation.

Aside from having to escape the legacy of active exclusion, ecological modernization (especially its stronger version) is impeded by the British tradition of secrecy in government. The UK's response to the EU's 1990 environmental right-to-know directive has been contradictory. Hallo (1996: 6) assigns the UK to a middle group of countries situated between those with good general access to information prior to the adoption of the directive and those that lacked such laws and have progressed most slowly. While today's Britons have better access to environmental information than their parents, one observer cautions that the 'concept and language of rights have always found acceptance with rhetoricians in the UK more easily than they have been accommodated within its legal systems' (Roderick, 1996: 274). The Official Secrets Act of 1911, which prevented officials from releasing information to the public unless specifically permitted to do so, remains part of the ethos of government. Prior to the Environmental Protection Act of 1990 and the Environmental Information Regulations Act of 1992, most information held by regulatory agencies about pollution and the state of the environment was kept secret (Davis, 1996). Today acceptance in principle of greater openness clashes with officials' hostility to, or at least unfamiliarity with, public participation in environmental policy making.

The United Kingdom did however see two significant developments at the end of the decade that indicated a belated arrival of some elements of ecological modernization. Until 2000 there was little in the way of green taxation, though revenue from these taxes appeared high because of the traditionally high taxes on fuel and transportation, in place for purely fiscal reasons, not environmental ones. But in 2001 a Climate Change Levy came into operation, imposing a per-kilowatt charge on energy generated by fossil fuels used by business and government bodies. The charge was opposed by the Conservative Party (which promised to abolish it in its manifesto for the 2001 general election) and the Confederation of British Industry, and its implementation was accompanied by the negotiation of a complex system of discounts and exemptions for many industries. Still, this tax broke one major barrier to ecological modernization in the United Kingdom—the resistance of Treasury to any such use of the taxation system (on the strength of which see Voisey and O'Riordan, 1997: 49; Weale, 1997: 106).

More radical still was a report issued in 1998 by the Royal Commission on Environmental Pollution entitled *Setting Environmental Standards*, which recommended abandoning the traditional British secretive and informal approach to regulation. The report recommended 'that the whole process of analysis, deliberation and synthesis should take place in a context in which the articulation of public values should be included at all stages of the policy process' (Weale, 2001: 362–3). Democratization would go further than mere consultation of stakeholders, but instead seek to involve *citizens* in the process (Weale, 2001: 368). In its discursive democratic prescriptions, the report is actually consistent with strong ecological modernization. The report met with a belated and lukewarm response from the government, and at the time of writing it remains unclear whether any of its recommendations will be adopted, especially in light of what Voisey and O'Riordan (1997: 49) identify as a major barrier to sustainable development in the United Kingdom: 'the role of citizen is seen as a rhetorical focus, but a practical nightmare.'

The early 1990s also saw a growing dialogue between NGOs and business. Borrowing from Beck, Rose (1993: 294) describes this as 'a sort of unpolitics,' though a better description is 'paragovernmental activity', in the terms we established in Chapter 6. In 1996 Greenpeace organized a conference for business groups to emphasize the opportunities for strategic alliances with environmental interests and the market opportunities for environmentally benign technologies. Of the mainstream environmental organizations, Greenpeace has often stood out as the least willing to take part in formal governmental processes. Greenpeace had challenged Thatcher's claims of international leadership on ozone and climate change issues, dubbing Britain the 'Dirty Man of Europe' (Lowe, interview, 1999). In the early 1990s Greenpeace dealt quite closely with government over climate change issues, but then strategically shifted its position, electing to remain outside of the Sustainable Development Roundtable. Both Greenpeace and the World Wide Fund for Nature have worked directly with major companies on issues such as CFC production and industrial fishing (Grove-White, 1997: 18). Friends of the Earth has moved into environmental consultancy with private companies and engaged in a 'green consumerism' campaign (Byrne, 1997: 134). New environmental groups have emerged to facilitate relationships with the business community, such as the Environmental Council and the Forum for the Future.[7] Unlike the Norwegian Environmental Home Guard campaign, however, these initiatives occurred without state involvement let alone state funding. These developments can be described as weak paragovernmental ecological modernization.

We have seen that belated moves in the direction of ecological modernization can be discerned in the United Kingdom, beginning in the late 1990s. The United Kingdom is still behind Germany and Norway in this respect, but in advance of the United States. But any movement from weak to strong ecological modernization requires an active oppositional civil society, capable of engaging sub-politics in Beck's terms. As pointed out in earlier chapters, the 1990s did see oppositional movement politics arise for the first time in the United Kingdom. While we have emphasized the importance of anti-roads protests, Greenpeace, alone of the more organized and established groups, has also played a role. The organization maintained media-based campaigns and reasserted its role as 'environmental policeman' through highly symbolic protests, the most dramatic and influential of these being the 1995 occupation of Shell's Brent Spar platform, discussed in Chapter 3. Grove-White (1997: 17) believes that the Brent Spar controversy essentially 'rewrote the rules' in British environmental politics, bypassing the government through an effective consumer campaign and illustrating 'the mounting significance of public opinion for emerging new concerns about corporate social responsibility.' An editorial in the *Independent* (21 June 1995, p. 20) similarly concluded from the campaign's success: 'It is now clear that neither governments nor big business are strong enough to withstand a new phenomenon: an alliance of direct action with public opinion.' In Beck's terms, the Brent Spar issue was resolved through sub-politics within civil society (see also Wapner, 1996). It is an example of paragovernmental action (though not of ecological modernization).

The British environmental movement has had some success in recasting the relationship between certain development initiatives and the state's economic imperative. In both the nuclear and transport inquiries of the 1970s environmental groups presented economic arguments. The cost of nuclear power development was subsequently—and quite ironically—highlighted through the Conservatives' privatization programme. The reluctance of the corporate world to acquire nuclear power resources emphasized their true cost, and it is this that has led to the demise of nuclear development. With road development too, government policy in the late 1990s began to shift as the economic burden of road construction upon the state became increasingly evident (and exacerbated by the costs of dealing with increasing direct action protest). Organizations such as FoE and CPRE have consequently developed a much stronger relationship with members of Treasury.

Paragovernmental activity and recent developments in taxation and Royal Commission proposals notwithstanding, the United Kingdom lingers well behind Germany and Norway in the ecological modernization stakes. It remains to be seen whether effective inclusion in the state's core via that route is possible. However, these developments do suggest that ecological modernization is possible in a context that has become passively inclusive—not just in actively inclusive Norway and passively exclusive Germany. Moreover, the arrival of social movement activism in the 1990s suggests that the kind of oppositional politics necessary for a strong version of ecological modernization to gain ground is present. However, so far the movement activism is mostly in the anti-roads area, where the state still shows its actively exclusive face. The ecological modernization policy momentum is entirely in the area of pollution control, where there is no oppositional movement activity. Without that activity, radical proposals such as the 1998 report of the Royal Commission on Environmental Pollution are unlikely to be implemented. That social movement activity is possible on pollution issues is illustrated by the anti-toxics and environmental justice movements in the United States, which have no British counterpart. This absence in the United Kingdom is puzzling in light of the risk-society thesis—according to which one would expect movement mobilization in the vicinity of risks imposed by pollution. The risk that attained greatest prominence in the 1990s came with BSE/Mad Cow disease, but this became framed as a health and food supply issue, not an environmental one, and it inspired little movement activity. Risks associated with genetically modified organisms and their release into the environment saw more in the way of movement action (e.g., destruction of crops), and the 2001 Foot and Mouth disease outbreak brought loud criticisms from both environmentalists and farmers. Again, all of these issues are distant from the anti-pollution policy area where ecological modernization has been stirring, but the broad issue of industrialized food production—where both the government and science have taken a beating in public opinion—may offer a unique route to ecological modernization policies.

Still, the continuing influence of the partial and ineffective inclusion that long characterized the United Kingdom remains detrimental when it comes to anything like strong ecological modernization. The shock of active exclusion that characterized the 1980s was perhaps unfortunate in that it demonstrated to environmental groups that a return to this traditional alternative was not so bad after all. At any rate, our conclusion about the destructive impact of active exclusion on the prospects for ecological modernization remains a firm one. It remains to be seen whether the United Kingdom can escape this destructive legacy.

Germany: stronger ecological modernization and the professionalization of sub-politics?

We have argued in previous chapters that a passively exclusive state is conducive to a vibrant civil society. Confronted with this situation, a passively exclusive state will, we argue, be the most likely of our four kinds of states to move from weak towards strong ecological modernization, a move that looks improbable in actively inclusive Norway. This very development may in turn change the character of the state, softening its exclusive aspect. However, because even mainstream groups are less likely to suffer co-optation than their counterparts in more inclusive settings, the movement as a whole is likely to retain resistance alongside engagement—the kind of dual strategy we discussed in Chapter 6.

With the 1998 election of an SPD/Green coalition government, German environmental NGOs had to revise upwards their assessment of the prospects for pursuing their ends through policy initiatives based on ecological modernization. Environmental NGOs enjoy better access to public officials as a result of the Greens' participation in parliament and government, although activists say that opportunities for access to and influence in the SPD/Green federal government have remained far behind expectations (Musiol, interview, 1999). Moreover, social movement activity had by 1998 declined from its peak a decade or so earlier. As the Green Party's parliamentary spokes-woman for the environment explains:

> Sixty-eighters—members of the peace and anti-nuclear movements—are getting older. These people no longer attend protest marches. Young people today tend to be apolitical and have a different relationship to the state. And the state itself has changed for the better so that radical opposition is no longer necessary. Politically motivated young people are more likely to seek change through conventional political channels such as interest groups and parties (Hustedt, interview, 1999).

More than in most countries, the discourses of sustainable development and ecological modernization have come to dominate environmental politics in Germany. At issue is an 'ecological transformation of industrial society' not its abolition. In a comparative light, Germany's environmental policy record in the 1980s and 1990s looks positive. Stricter government regulations have dramatically reduced acute water, air, and soil pollution (Jänicke and Weidner, 1997b), though green taxes have so far made a smaller impact than might be expected. Nuclear energy—once the chief cause of environmental mobilization—has suffered serious setbacks, culminating in the SPD/Green government's plan for gradual phase-out.[8] This plan, which envisions that few if any plants will be taken off the grid immediately, is a compromise that aims to balance the state's economic and legitimation needs, but it is hard to see how even this could have happened without the prior new social movement activity (Hunold, 2001). The nuclear issue is the most graphic and significant kind of risk society issue that Germany has confronted. However, issues of environmental risk have been especially salient since the 1970s, perhaps having much to do with the fact that Germany is a heavily industrialized, densely populated country in the heart of Europe, subject to risks generated beyond its borders as well as within them. Beck's work on the risk society, though presented in generally applicable terms, owes much to his German location.

The environmental movement's confrontational strategies of the late 1970s and 1980s had their successes, as we showed in Chapter 6. Not least of these was inducing the state to adopt ecological modernization as both discourse and policy practice in the mid-1980s, when environmentalists were still excluded. Still, confrontation has now receded, along with acceptance of environmental policy tools that rely on market-based ecological incentives (Zittel, 1996). The pragmatists who control the Green Party today reject ecoradical goals and embrace the aim of ecologically modernizing the country's social-market economy. Environmental advocacy organizations willing to support the ecological modernization project have, after a long struggle, gained access to the state, whose passively exclusive character they have diluted. But radical groups such as BBU still face passive exclusion. Their role in sub-political activity has been largely supplanted by a burgeoning sector of for-profit environmental consultancies. Public agencies and firms interested in ecologically sustainable development increasingly bypass traditional environmental NGOs altogether in favour of these for-profit consultancies for information and advice.

The pattern of movement history, then, is one of radical beginnings followed by the formation of a political party that has become a permanent presence on the German electoral landscape, together with the creation of environmental advocacy groups, some of which have close ties to the Greens but all of which have kept an independent organizational and political identity. These environmental NGOs do not have quite the discursive and democratic vitality of the social movements of the 1970s and 1980s. Given that the Greens were in opposition until 1998 (and lacked an effective parliamentary presence 1990–4) and there were limited opportunities for environmental NGO inclusion in environmental policymaking, professionalized sub-politics became a mainstay of environmental NGO activity alongside the state's modest ecological modernization program.

Unlike some among the US environmental justice movement and radical wilderness defenders, German environmental NGOs active in oppositional civil society do not reject the country's consensus-seeking, rationalist discourse and policy style. The discourse of German environmental politics, state-centred or not, fits the give-and-take between expertise and counter-expertise described in Beck's model of ecological democracy. As noted above, some environmentalists' hopes of special treatment by the SPD/Green government have been disappointed. But in the long run this could actually help them pursue a viable dual strategy vis-à-vis a left-of-centre/green government for whom the environment is not as central an issue as environmentalists would like. Few activists we interviewed believed that co-optation was going to be a serious problem for environmental NGOs.

Mainstream environmental NGOs too regard ecological modernization as the best means of securing the role of environmental values in politics and economy even though environmental issues no longer occupy the top of the political agenda, taking a back seat to economic policy. Environmental NGOs believe their ability to shape public policy against the wishes of business associations and labour unions waxes and wanes in direct proportion to public concern for environmental issues. The dominance of business in environmental policy debates derives from the fact that individual firms must implement many of the policies demanded by environmental NGOs and enacted by state officials. In that context, it makes sense for environmental NGOs wishing to be seen as 'reasonable' partners in policy making to embrace conceptions of environmentalism that seek to bridge the gap between business and environmental protection. A veteran environmental scientist and Social Democrat MP summarizes the challenge as follows:

> The situation is completely different today: local environmental calamities and protests and the spirit of unruliness are gone. But environmental calamities reach all areas of life. Keeping them politically alive requires scientific proof that there is no necessary contradiction between ecology and the economy. If you can piggyback the issue of the environment onto the dominant issue of the economy, the environment will have a chance again. For all intents and purposes, it is no longer possible to argue for the environment against the opposition of business (von Weizsäcker, interview, 1999).

Business too has changed. Even industrial managers, particularly younger ones who work outside major corporations, include sustainable development and climate protection among their top ethical priorities, albeit subject to the constraints of capitalist growth (Hustedt, interview, 1999). Consequently, the movement's earlier battle with industry in some sectors has turned into a 'critical dialogue' rooted in the shared discourse of sustainable development. Yet Germany's big businesses, particularly utilities and chemical companies, have not budged very much from their negative view of environmentalists, and vice versa. Although both NABU and the Institute for Applied Ecology have accepted eco-sponsoring agreements, and the Institute for Applied Ecology has signed eco-auditing contracts with large firms (Musiol, interview, 1999), many environmentalists believe the greatest potential for making an impact lies with medium-sized firms, where managers have greater freedom over setting company policy than do managers in publicly held corporations beholden to the stock market as well as politically conservative managers and supervisory boards. This is not to say that environmentalists and managers share the same conception of sustainable development. In fact, acceptance of sustainable development by major industrial associations is seldom more than superficial (Streese, interview, 1999). However, that some corporations are willing to discuss environmental measures and have accepted sustainability as their guiding principle indicates a cultural shift in power relations in favour of environmentalists and gives further credence to the ecological modernization thesis.

The kind of programme accepted by Green politicians and NGOs alike is a far cry from that championed by the citizens' initiatives of the 1970s. Still, pragmatically inclined environmental NGOs claim the state and certain sectors of the economy have become less resistant to environmental change. Nuclear power is a dramatic case in point. The anti-nuclear movement and the economic imperative long conflicted. But following forty years of federal support for nuclear energy, after the 1998 election environmentalists suddenly had the federal government on their side on this issue—up to a point. The SPD/Green government's commitment to end the nuclear energy programme signalled the persistent political influence of the antinuclear movement. However, it is economics ministry officials and nuclear industry representatives who have negotiated the terms of ending the programme, with environmentalists and the environment ministry apparently sidelined. Thus passive exclusion persisted, and so did protest. Radicals such as the BBU still operate outside policy making. As discussed in Chapter 5, the anti-nuclear movement has shown that it can still mobilize large numbers of activists in a separate public sphere, even after the elevation of the Greens in 1998. When numerous environmental groups rejected the Greens' acquiescence on the shipment of reprocessed nuclear wastes, non-violent protests continued.

Though now restricted mostly to the anti-nuclear movement, the oppositional sphere remains comparatively large and vital. The 'strong' aspect of ecological modernization in Germany is confirmed by the fact that sub-political activities have not been killed off, although they have been thoroughly professionalized. One of the movement's responses to the state's exclusionary strategy of the 1970s and early 1980s was to establish environmental policy institutions outside the state. The Working Group of Ecological Research Institutes alone comprises approximately eighty institutes (Hey and Brendle, 1994: 133). The Institute for Applied Ecology, founded in 1977, sought to meet the demand for scientific and technical data that could be used to support plaintiffs challenging environmentally questionable industrial facilities in the courts. Additional independent ecological research institutes were established in subsequent decades. Examples include the Wuppertal Institute for Climate, Environment, and Energy and the Potsdam Institute for Climate Research. These two research organizations receive part of their funding from the federal government. While they do not share the movement origins of the Institute for Applied Ecology, they contribute information and proposals to environmental policy debates. Politically, they counterbalance the various state agencies charged with gathering and interpreting environmental data for the public bureaucracy.

The greatest potential for assuming paragovernmental powers lies in technical areas of environmental policy that require expertise rather than mass mobilization (Brendle, interview, 1999). Eco-auditing is a good example. Advising companies on retooling their operations according to ecological criteria is one of the bread-and-butter issues of the Institute for Applied Ecology. In keeping with the theme of ecological modernization, moreover, Germany's independent ecological research institutes have pioneered market-based environmental policy tools that would shift more regulatory tasks to the private sector. Setting industrial norms and standards—a task historically left to private-sector associations—is a further case in point. Here government has been more than happy to outsource highly technical deliberations and decisions to nongovernmental experts, including environmental NGO representatives.

There remain obstacles to stronger ecological modernization, prominent among which is the German tradition of administrative secrecy in government, which has been slow to bow to European pressure to improve public access. The Federal Freedom of Access to Environmental Information Act of 1994 came into force eighteen months after the deadline set by the 1990 EU directive had passed (Gebers, 1996: 97). The Act empowered citizens to obtain environmental information held by administrative agencies at all levels of government. Previously citizens without a legally valid claim based on property rights could not obtain such data.

The Act, however, has failed to dislodge an entrenched pattern of administrative secrecy. Most of the country's 445 municipal and county environmental agencies remained less than forthcoming with information concerning the quality of drinking water and the existence of industrial soil contamination. Many also charged steep fees for information (Ahrens and Stoller, 1995). In a 1999 ruling, the European Court of Justice struck down most of the Act's provisions, although it found no grounds for the EU Commission's complaint that German fees for access to environmental information effectively discouraged freedom-of-information requests. However, the Court ruled that German citizens' access to environmental data remained inadequate by EU standards.

This situation may be about to change. The SPD/Green coalition signed the 1998 Aarhus UN Convention, which seeks to narrow the grounds on which administrators may withhold environmental data. Some environmental activists believe the new convention will help to eliminate the worst elements of what Thomas Lenius (BUND) calls Germany's 'Prussian administrative tradition' (Pötter, interview, 1998).

Germany has, then, experienced stronger ecological modernization than the other three countries in this study, however short it may fall of Christoff's ideal type. Along with the salience of risk issues, this experience in turn has enabled more effective connection of environmental movement interests to core state imperatives than in the UK, USA, and even Norway. These developments were facilitated by the passively exclusive character of the German state, which provided the space and impetus for the development of a green public sphere. Ironically, strong ecological modernization is advanced as exclusion diminishes with the entry of environmental NGOs into the state and the Greens into government. How long will this combination of circumstances persist? Our structural and historical analysis could be deployed to suggest that this state of affairs is unstable, because it is conditional on the recent experience and memory of an autonomous green public sphere. If so, Germany may lapse into passive inclusion of the American sort, with an associated weakening of ecological modernization. Alternatively, it could develop a kind of environmental corporatism that includes a green elite, but passively excludes others. The latter would not necessarily be a bad outcome, possibly constituting a further turn of the historical spiral in which passive exclusion means a revitalized public sphere—which in turn might go the way of the social movements of the 1970s and 1980s. But that is to look too far into the future.

Conclusion

If we array our four countries according to the degree they have achieved strong ecological modernization and associated sub-politics, and by implication the degree of connection this reveals between environmental movement interests and core state imperatives, then Germany is in front. Germany has reaped the benefits of a history of passive exclusion, though how long it can continue to do so remains unclear. Norway may be ranked second, but only on the dimension of ecological modernization, which it is pursuing in weak form. But nothing stronger can be envisaged for Norway, and no sub-politics is apparent. Ecological modernization now plays some role in public policy making in the UK, which together with some paragovernmental activity would place the UK behind Germany and Norway but in front of the USA. In the USA, national politics still exhibits an old-fashioned stand-off between economy and environment. The USA featured by far the strongest connection between environmental values and core state imperatives around 1970, but that peak has not been attained since, nor does it look to return in the near future.

At first sight it might seem paradoxical, but on reflection understandable, that in the 1990s we see a resurgence of environmental activism in the ecological modernization laggards, the USA and UK. This resurgence occurs despite the passively inclusive character of the US state that ought to absorb and neutralize radicalism, and despite the recovery of the UK from the active exclusion of the Thatcher era. These developments themselves indicate the waning of structural determinism: state structure may be less important than it once was in determining the form taken by social movements. This in turn indicates heightened reflexivity. That is, activists can look to the history of inclusion and see what it has and, more to point, has not accomplished, and draw lessons for their own orientation to the state.

Notes

1. The basic literature on ecological modernization includes Weale (1992), Hajer (1995, 1996), Christoff (1996), and the special issue of *Environmental Politics* 9(1) (Spring 2000) on 'Ecological Modernisation Around the World'. Theorization of the concept can be traced to the works of German social scientists Joseph Huber and Martin Jänicke in the early 1980s (see Spaargaren, 2000: 46–50).

2. Our three European states are bound by the European Union's Directive 90/313 on freedom of access to environmental information passed in 1990. Although Norway is not a member of the EU, it does belong to the European Environmental Area and has agreed to incorporate existing and future environmental directives in its national laws (Andersen and Liefferink, 1997: 31).

3. Source: OECD Database on Environmentally Related Taxes, www.oecd.org/env/policies/taxes/index.htm.

4. From our point of view, Hajer's terminology is unfortunate (as well as unnecessary), because, as we will show for Germany, corporatism can actually be instrumental to strong ecological modernization.

5. From the ecological modernization theory side, this connection is accepted by Spaargaren, 2000: 62–4.

6. Sociologically, this is because today's leaders of German mainstream groups were themselves anti-statist activists in the 1970s and early 1980s—unlike many of their US counterparts most of them have not been shaped by the calculating professional ethic imparted by elite law, business, and public policy schools. But the divide is less sharply defined in Germany also because passive exclusion has given Germany's mainstream environmental NGOs far fewer opportunities for

inclusion in the state, thus preserving a sense of opposition and exclusion among all environmental activists—at least until the 1998 election.

7. The Environment Council's origins in the conservation movement can be traced much earlier, although in its new form it focuses on liaison and education between the movement, business community and government. Forum for the Future, led by former Green Party figures Jonathon Porritt and Sarah Parkin, seeks to develop constructive policy solutions as a consultancy to business and government.

8. The SPD called for ending the country's reliance on nuclear energy as early as 1988, but the environmentalist wing of the party is in the minority. Prior to Chernobyl, the SPD was closely allied with nuclear industry unions and sharply critical of the anti-nuclear movement.

References

Adler, Allan R. (ed.). 1997. *Litigation Under the Federal Open Government Laws*, 20th edn. Washington, DC: American Civil Liberties Union.

Ahrens, Ralf, and Detlef Stoller. 1995. 'Geheimniskrämer in Amtsstuben.' *Tageszeitung*, 28 July, p. 7.

Andersen, Mikael Skou, and Duncan Liefferink (eds.). 1997. *European Environmental Policy: The Pioneers*. Manchester: Manchester University Press.

Anderson, Terry L., and Donald R. Leal. 2001. *Free Market Environmentalism*. New York: Palgrave.

Andrews, Richard N. L. 1997. 'United States.' 25–44 in Martin Jänicke and Helmut Weidner (eds.), *National Environmental Policies: A Comparative Study of Capacity-Building*. Berlin: Springer.

Beck, Hanno, Brian Dunkel, and Gawain Kripke. 1998. *Citizens' Guide to Environmental Tax Shifting*. Washington, DC: Friends of the Earth.

Beck, Ulrich. 1992. *Risk Society: Towards a New Modernity*. London: Sage.

—— 1999. *World Risk Society*. Cambridge: Polity.

Brick, Philip, Donald Snow, and Sarah Van De Wetering (eds.). 2000. *Across the Great Divide: Explorations in Collaborative Conservation and the American West*. Washington, DC: Island Press.

Brown, Phil, and Edwin J. Mikkelsen. 1990. *No Safe Place: Toxic Waste, Leukemia, and Community Action*. Berkeley: University of California Press.

Bryner, Gary C. 2000. 'The United States: "Sorry-Not Our Problem".' 273–302 in William M. Lafferty and James Meadowcroft (eds.), *Implementing Sustainable Development: Strategies and Initiatives in High Consumption Societies*. Oxford: Oxford University Press.

Byrne, Paul. 1997. *Social Movements in Britain*. London: Routledge.

Christoff, Peter. 1996. 'Ecological Modernisation, Ecological Modernities.' *Environmental Politics*, 5: 476–500.

Daly, Herman E. 1996. *Beyond Growth: The Economics of Sustainable Development*. Boston: Beacon.

Davis, S. H. 1996. *Public Involvement in Environmental Decision-Making: Some Reflections on the Western European Experience*. Washington, DC: World Bank.

Dryzek, John S. 1990. 'Designs for Environmental Discourse: The Greening of the

Administrative State?' 97–111 in Robert Paehlke and Douglas Torgerson (eds.), *Managing Leviathan: Environmental Politics and the Administrative State*. Peterborough, Ont.: Broadview.

—— 1997. *The Politics of the Earth: Environmental Discourses*. Oxford: Oxford University Press.

European Environment Agency. 2000*a*. *Environment in the European Union at the Turn of the Century*. Copenhagen: European Environment Agency.

—— 2000*b*. *Environmental Taxes: Recent Developments in Tools for Integration*. Copenhagen: European Environment Agency.

Fischer, Frank. 1995. 'Hazardous Waste Policy, Community Movements and the Politics of Nimby: Participatory Risk Assessment in the USA and Canada.' 165–82 in Frank Fischer and Michael Black (eds.), *Greening Environmental Policy: The Politics of a Sustainable Future*. London: Paul Chapman.

—— 2000. *Citizens, Experts, and the Environment: The Politics of Local Knowledge*. Durham, NC: Duke University Press.

Gebers, Betty. 1996. 'Germany.' 95–110 in Ralph E. Hallo (ed.), *Access to Environmental Information in Europe*. Dordrecht: Kluwer.

Gonzalez, George A. 2001. 'Democratic Ethics and Ecological Modernization: The Formulation of California's Automobile Emission Standards.' *Public Integrity*, 3: 325–44.

Grove-White, Robin. 1997. 'Brent Spar Rewrote the Rules.' *New Statesman*, 20 June: 17–19.

Hajer, Maarten A. 1995. *The Politics of Environmental Discourse: Ecological Modernization and the Policy Process*. Oxford: Oxford University Press.

—— 1996. 'Ecological Modernisation as Cultural Politics.' 246–68 in Scott Lash, Bron Szerszynski, and Brian Wynne (eds.), *Risk, Environment, and Modernity: Towards a New Ecology*. London: Sage.

—— and Frank Fischer. 1999. 'Beyond Global Discourse: The Rediscovery of Culture in Environmental Politics.' 1–20 in Frank Fischer and Maarten A. Hajer (eds.), *Living with Nature: Environmental Politics as Cultural Discourse*. Oxford: Oxford University Press.

Hallo, Ralph E. (ed.). 1996. *Access to Environmental Information in Europe: The Implementation and Implications of Directive 90/313/EEC*. London: Kluwer Law International.

Hey, Christian, and Uwe Brendle. 1994. *Umweltverbände und EG. Strategien, politische Kulturen und Organisationsformen*. Opladen: Westdeutscher Verlag.

Hunold, Christian. 2001. 'Corporatism, Pluralism, and Democracy: Toward a Deliberative Theory of Bureaucratic Accountability.' *Governance* 14: 151–68.

Hutcheson, Ron. 2001. 'Bush Stands Firm on Missile Defense.' *Philadelphia Inquirer*, 20 July.

Jacobs, Michael. 1997. 'Introduction: The New Politics of the Environment.' 1–17 in Michael Jacobs (ed.), *Greening the Millennium? The New Politics of the Environment*. Oxford: Blackwell.

Jänicke, Martin. 1997. 'The Political System's Capacity for Environmental Policy.' 1–14 in Martin Jänicke and Helmut Weidner (eds.), *National Environmental Policies: A Comparative Study of Capacity-Building*. Berlin: Springer.

—— 2000. *Ecological Modernization: Innovation and Diffusion of Policy and Technology*. Berlin: Forschungsstelle für Umweltpolitik.

—— and Helmut Weidner (eds.). 1997*a*. *National Environmental Policies: A Comparative Study of Capacity-Building*. Berlin: Springer.

—— 1997*b*. 'Germany.' 133–55 in Martin Jänicke and Helmut Weidner (eds.), *National Environmental Policies: A Comparative Study of Capacity-Building.* Berlin: Springer.

Jansen, Alf-Inge, and Oddgeir Osland. 1996. 'Norway.' 179–256 in Peter Munk Christiansen (ed.), *Governing the Environment: Politics, Policy and Organization in the Nordic Countries.* Copenhagen: Nordic Council of Ministers.

Kasa, Sjur. 2000. 'Policy Networks as a Barrier to Green Tax Reform: The Case of CO_2-Taxes in Norway.' *Environmental Politics*, 9 (4): 104–22.

Kemp, René. 1997. *Environmental Policy and Technical Change: A Comparison of the Technological Impact of Policy Instruments.* Cheltenham: Edward Elgar.

Killingland, Tore. 1996. 'Norway.' 309–16 in Ralph E. Hallo (ed.), *Access to Environmental Information in Europe.* Dordrecht: Kluwer.

Langhelle, Oluf. 2000. 'Norway: Reluctantly Carrying the Torch.' 174–208 in William M. Lafferty and James Meadowcroft (eds.), *Implementing Sustainable Development: Strategies and Initiatives in High Consumption Societies.* Oxford: Oxford University Press.

Lovei, Magda, and Charles Weiss, Jr. 1998. *Environmental Management and Institutions in OECD Countries: Lessons from Experience.* Washington, DC: World Bank.

Lovins, Amory B., and L. Hunter Lovins. 1999. *Natural Capitalism: Creating the Next Industrial Revolution.* Boston: Little, Brown.

Luke, Timothy. 1999. 'Eco-managerialism: Environmental Studies as Power/Knowledge Formation.' 103–20 in Frank Fischer and Maarten A. Hajer (eds.), *Living with Nature: Environmental Politics as Cultural Discourse.* Oxford: Oxford University Press.

Midttun, Atle, and Svein Kamfjord. 1999. 'Energy and Environmental Governance under Ecological Modernization: A Comparative Analysis of Nordic Countries.' *Public Administration*, 77: 873–95.

Mol, Arthur P. J. 1996. 'Ecological Modernisation and Institutional Reflexivity: Environmental Reform in the Late Modern Age.' *Environmental Politics*, 5: 302–23.

—— and Gert Spaargaren. 2000. 'Ecological Modernisation Theory in Debate: A Review.' *Environmental Politics*, 9: 17–49.

O'Brien, Mary. 2001. *Making Better Environmental Decisions.* Cambridge, Mass.: MIT Press.

Reitan, Marit. 1998. 'Ecological Modernisation and "Realpolitik": Ideas, Interests and Institutions.' *Environmental Politics*, 7: 1–26.

Roderick, Peter. 1996. 'United Kingdom.' 249–76 in Ralph E. Hallo (ed.), *Access to Environmental Information in Europe.* Dordrecht: Kluwer.

Rose, Chris. 1993. 'Beyond the Struggle for Proof: Factors Changing the Environment Movement.' *Environmental Values*, 2: 285–98.

Sabel, Charles, Archon Fung, and Bradley Karkkainen. 2000. 'Beyond Backyard Environmentalism.' *Boston Review*, online http://bostonreview.mit.edu/BR24.5/sabel.html

Sarewitz, Daniel. 1996. *Frontiers of Illusion: Science, Technology, and the Politics of Progress.* Philadelphia: Temple University Press.

Sclove, Richard. 1995. *Democracy and Technology.* New York: Guilford.

—— 2001. 'Is There Really a Link Between Neo-Corporatism and Environmental Performance? Updated Evidence and New Data for the 1980s and 1990s.' *British Journal of Political Science*, 31: 686–92.

Spaargaren, Gert. 2000. 'Ecological Modernization Theory and the Changing Discourse on Environment and Modernity.' 41–73 in Gert Spaargaren, Arthur P. J. Mol, and Frederick Buttel (eds.), *Environment and Global Modernity*. London: Sage.

Street, John. 1992. *Politics and Technology*. New York: Guilford.

Sverdrup, Liv Astrid. 1997. 'Norway's Institutional Response to Sustainable Development.' *Comparative Politics*, 6: 54–82.

Voisey, Heather, and Tim O'Riordan. 1997. 'Governing Institutions for Sustainable Development: The United Kingdom's National Level Approach.' *Environmental Politics*, 6: 24–53.

Wapner, Paul. 1996. *Environmental Activism and World Civic Politics*. Albany, NY: State University of New York Press.

—— 1997. 'United Kingdom.' 89–108 in Martin Jänicke and Helmut Weidner (eds.), *National Environmental Policies: A Comparative Study of Capacity-Building*. Berlin: Springer.

—— 2001. 'Can We Democratize Decisions on Risk and the Environment?' *Government and Opposition*, 36: 355–78.

Weber, Edward. 1998. *Pluralism by the Rules: Conflict and Cooperation in Environmental Regulation*. Washington, DC: Georgetown University Press.

Weiland, Sabine. 2001. 'Models of Integrating the Environment into Society: Ecological Modernisation and the Case of Chemicals Control in Britain, Germany, and the US.' Paper presented at Keele University Summer School on Environmental Politics and Policy, 10–21 Sept.

Welsh, Michael M. 2000. 'Toward a Theory of Discursive Environmental Policy Making: The Case of Public Range Management.' Unpublished Ph.D. dissertation, University of Oregon.

Wildavsky, Aaron. 1995. *But Is It True? A Citizen's Guide to Environmental Health and Safety Issues*. Cambridge, Mass.: Harvard University Press.

World Bank. 1998. *Environmental Management and Institutions in OECD Countries: Lessons from Experience*. World Bank Technical Paper No. 391. Washington, DC: World Bank.

Young, Stephen C. 2000. 'The United Kingdom: From Political Containment to Integrated Thinking.' 245–72 in William M. Lafferty and James Meadowcroft (eds.), *Implementing Sustainable Development: Strategies and Initiatives in High Consumption Societies*. Oxford: Oxford University Press.

Zittel, Thomas. 1996. Marktwirtschaftliche Instrumente in der Umweltpolitik. Opladen: Leske & Budrich.

Part Three

Greening lifecycles and lifestyles

Chapter 14

Gert Spaargaren and Maurie J. Cohen

GREENING LIFECYCLES AND LIFESTYLES: SOCIOTECHNICAL INNOVATIONS IN CONSUMPTION AND PRODUCTION AS CORE CONCERNS OF ECOLOGICAL MODERNISATION THEORY

Introduction: the green revolution in consumption and production

THE TRANSITION TO AN ecological society characterised by ecologically modernised systems of consumption and production has been considered from a variety of perspectives. For instance, environmental scientists and engineers have conducted life-cycle analyses (LCA) and deployed other 'clean production' technologies to induce improvements at the upstream end of product supply chains (see e.g. Hunt 1996; Thomas et al. 2003; Udo de Haes et al. 2004). Other specialists have sought to facilitate sustainability transitions through a more comprehensive (re)organisation of production processes (Vezzoli 2003; Tukker and Tischner 2006). A key liability of these approaches is that they have largely disregarded consumers as potentially vital participants in the conceptualisation and management of environmental change. Only in recent years have environmental social scientists begun to articulate a broader shift 'from production to consumption' that gives explicit attention to the multifarious and seemingly incongruous practices of consumers (Cohen and Murphy 2001; Princen 2002; Jackson 2006).

The challenge that this turn toward consumption raises for reflexive Ecological Modernisation Theory (EMT) centres on how to navigate between the dark green romantic dismissal of modernity and the naïve endorsement of market-driven, liberal eco-technotopias. Sociological theories that emphasise the importance of agency, governance and the cultural and institutional dimensions of environmental change are likely to be of crucial importance for scholars working on realist agendas for alternative (eco)modernities. The aim of Part Three of this volume, organised around the theme of greening life cycles and lifestyles, is to demonstrate that EMT has moved beyond a straightforward emphasis on eco-technological innovation and now focuses as well on the central roles of agency, politics and institutions. At the same time, the original emphasis on upstream actors and processes has been complemented with studies that meaningfully combine both upstream and downstream actors and processes. As a result, EMT in its current formulation can be said to address the greening of both consumption and production (Mol and Spaargaren 2004; Jensen and Gram-Hanssen 2008; see also Carolan 2004; Hobson 2003).

The next section begins with an examination of EMT's initial attention on technological environmental innovation and this review is followed by a discussion of three more recent applications of the theory that emphasise particular aspects of sustainable consumption: infrastructures of consumption, political consumerism and sociotechnical changes in consumer practices. After a sketch of the unfolding international politics of sustainable consumption and production in the fourth section, we conclude the chapter with a discussion of the different weight that various ecological modernisation scholars attach to consumption relative to production.

The upstream orientation of environmental technology debates

Sociotechnical innovation has been a primary focus for EMT scholars since the earliest developments of this theoretical perspective during the 1980s. Partly as a critical response to the romantic 'small-is-beautiful' view common within calls for appropriate technologies, the German sociologist Joseph Huber (1982) formulated an initial version of EMT around elaborate and ambitious innovations predicated upon, for example, genetic modification and advanced information and communications technologies. Despite their obvious differences regarding the role of technology in tempering ecological harms, both conceptual approaches maintain that social systems and their environmental performances are shaped by technology in fundamental and decisive ways. Because of the crucial role that it assigns to (eco)technological innovation and development, EMT has typically been classified as belonging to the 'industrial society' school within the social sciences that has scholars like Daniel Bell and Alvin Toffler as its most prominent proponents (Badham 1986; Cohen 1997). The prominence within EMT of technological innovation and its impact on society, moreover, generates some overlap with prevalent economic theories of innovation as well as with perspectives common in the sociology of science and technology (Geels 2005b; Perz 2007).

Although this technological emphasis has been a relatively consistent feature of EMT, the ways in which innovation processes have been conceptualised have varied over time and contributors to this literature have employed different terms and concepts. In the earlier period extending through the early 1990s, debate hinged on the replacement of end-of-pipe equipment (e.g., filters, scrubbers, and other *post hoc* treatment techniques) with more process-integrated, preventive technologies. Such 'clean technology studies' continue to constitute a recognisable and indispensable facet within the EMT body of literature (Cramer and Schot 1990; Schot 1992; Cramer 2006; see also Sonnenfeld 2002). The article by Joseph Murphy and Andy Gouldson (2000) that appears at the start of Part Three is a classic example of the application of this type of EMT approach in empirical research. This analysis of integrated pollution control (IPC) policies in the United Kingdom is situated primarily at the company level and the authors discuss the conditions under which different generations of environmental technologies came to be implemented and the role of environmental regulations in influencing these decisions. Consistent with most clean technology studies carried out at the time, the investigation concentrates exclusively on production and limits itself to environmental policy making within a domestic context.

The search for preventive environmental technologies, as well as commitments to the 'new politics of pollution' that Weale (1992) identified at approximately the same time, has not however been limited to companies or to individual production facilities. Inter-firm collaborations based on industrial ecology techniques – including by-product exchanges and waste cascading – have also become increasingly common in some manufacturing applications (Lowe 1997; Desrochers 2004; Heeres et al. 2004). In other areas, integrated (or green) supply-chain management (ICM) is being used to improve articulation of the vertical relations among collaborating firms (Van Koppen and Mol 2002). For example, powerful actors in the retail food sector have demonstrated an ability to effectively use ICM to systematise environmental management across nearly the entire length of the supply chain (Oosterveer et al. 2007). In these studies of the organisational dimensions of industrial ecology, attention on hierarchical power relations and cross-firm interactions is complemented by a focus on the horizontal (and spatial) relations among the participating companies that are themselves embedded in different production systems (see Seuring 2004; Ashton 2008). Some of this work has gained prominence over the past decade, with two specialised English-language academic journals (*Journal of Industrial Ecology* and *Progress in Industrial Ecology*) and the launch of several pilot projects around the world. Attention within the field of industrial ecology more generally though remains heavily centred on the scientific measurement of material flows within the production sphere (Cohen and Howard 2006; Salmi and Toppinen 2007). The article by Kris van Koppen and Arthur Mol (2002) in Part Three employs EMT and the triad-network approach to discuss some of the latent institutional dimensions of these IE studies. The authors argue that a stronger emphasis on agency and policy is necessary to understand the improvements in environmental efficiencies that result from these collaborative organisational synergies. Although the industrial district in the Danish city of Kalundburg regularly serves as the classic example of such interactions, Van Koppen and Mol discuss a new industrial park in Vietnam to demonstrate the applicability of IE theory and methods in a non-European context.

EMT-inspired studies on technology-driven environmental reform have always faced fierce criticism from dark green critics who contend that any reformist programme based on the pursuit of eco-efficiency and resource productivity sidesteps the real global crisis of human survival. This line of criticism emphasises that a fundamental restructuring of capitalism is necessary and putative ecological modernisation improvements will always be overwhelmed by inevitable rebound effects (Jänicke 2008; see also Polimeni et al. 2008). What is needed instead – so the contrarian green critics argue – is a radical reorientation of social and political priorities focused on (voluntary) simplicity, restraint and sufficiency (Ullrich 1979; Princen et al. 2002; Sachs and Santarium 2005).

Following Damian White (2002) in this respect, we argue that the strong polarisation which has emerged in the environmental social sciences about the role of (first generation) technology in environmental change does not help us much further. Processes of environmental change in reflexive modernity are too complex to be judged using binary schemes and codes like optimism versus pessimism, catastrophe versus cornucopia, and radical transformation versus status quo. The academic debate on sustainable development should instead focus on the conditions and possibilities for a 'viable ecological modernist project' that includes not only the technological manipulation of material flows, but also the 're-legitimisation of the public sphere' and the pursuit of 'credible industrial policies' and 'intelligent urban planning' (White 2002).

As the study of sustainable sociotechnical transitions has expanded, emphasis has turned toward the management of multi-level, multi-actor processes of ecological modernisation.[1] A series of so called 'transition studies' serves as an illustration of this shift towards a broader, social analysis of technological change (Schot and Geels 2007). In this work, the role of far-reaching innovation is still quite evident, but the focus on technological improvement is generally combined with analyses of the institutional dimensions of environmental change. Recent research considers different societal sectors (e.g., energy, agriculture, housing) and seeks to identify (non-linear) patterns of change that could encourage leapfrogging and switchover processes for sustainable development (Tukker 2005; Tukker et al. 2008). Key questions centre on how ensembles of niche-based innovation might begin to link up to create possibilities for system change and catalyse the emergence of more sustainable regimes for, say, generating energy or producing food. Scholars have also sought to identify large-scale factors (the 'landscape' in the language of this literature) that might facilitate or hinder regime transitions. Transformation toward carbon-neutral energy provision has, perhaps not surprisingly, become one of the most active fields for the application of this perspective (Verbong and Geels 2007).

On the basis of this overview of EMT studies of environmental technology and sociotechnical innovations we can identify a number of different tendencies that have became manifest over the past three decades. First, there has been during this timeframe general agreement that a shift occurred away from mitigative, end-of-pipe technologies that for the most part could be implemented without altering organisational structures or product chains. Recent years have seen the emergence of more preventive, proactive strategies that entail a (leapfrogging) restructuring of existing modes of consumption and production.[2] Second, the institutional and managerial dimensions of technologically induced changes have gradually been assigned a more important role in analyses of eco-transition processes. Linear and single actor-oriented investigations that have concentrated on one policy level (e.g., the nation) are being supplanted by systemic, multi-actor studies that consider different scalar levels and the interactions that take place across them, as our discussion of transition studies has aimed to illustrate (Geels 2005a; Schot and Geels 2007; Voss et al. 2006).

Since the late-1990s, a third major trend emerged challenging existing notions of environmental technology for their continued emphasis on providers and technological systems (e.g., on 'upstream' activities). The article by Spaargaren (2003) in Part Three exemplifies this trend. He argues that downstream actors in production-consumption chains – consumer groups in particular – have typically been neglected (or at least underestimated) as important co-drivers of environmental change by both scientists and policy makers. To compensate for the disregard of a consumer focus, and to complement the existing body of knowledge on (eco)technological transitions, recent studies have concentrated on end-users (e.g., householders, automobile drivers, tourists) and their lifestyle patterns (Spaargaren and Van Vliet 2000; Spaargaren 2003; Spaargaren et al. 2007). To develop this perspective on consumers, technology and everyday life an international research network was established at the University of Lancaster during the 1990s. Particularly important to this project was the stimulus provided by sociologist Elizabeth Shove. Research carried out on 'infrastructures of consumption' was especially valuable in advancing the field and building bridges to span the gap between producer- and consumer-focused approaches of technological transitions (Van Vliet et al. 2005). Shove herself has systematically developed several of the concepts at the heart of contemporary transition studies and assessed their significance for analysing technologies of everyday life (see Shove 2003; Shove and Walker 2007). By putting the sociotechnical transition of ordinary social practices at centre stage, this emergent body of work has already helped to introduce a consumption perspective to the field of transition studies and to reinforce the shift from production to consumption as discussed in more detail in the next section.

From production to consumption: a consumerist turn in EMT studies?

The central focus of EMT is on production-consumption chains (or cycles) and this implies, at least in the strict conceptual sense, that consumers (or other end-users) are implicitly included in the framework. Despite this tacit recognition of the relevance of consumption, actual practice has been quite different. The longstanding tendency in both environmental science and politics has been to ignore consumption and its adverse impacts or to sidestep such issues by conceptualising the utilisation of goods and services primarily in terms of individual motives and responsibilities. The main argument from policy makers for neglecting consumption has been based on the claim that it is too difficult and inefficient to seek changes in the behaviour of consumers and strategies aimed at producers provide greater leverage for improvement.[3] The production processes of large industrial firms have typically been viewed as more predictable and manageable than the diverse and often contradictory lifestyles of individual consumers. Policy initiatives to green consumption have therefore been subsumed under efforts to enhance the sustainability of production and distribution structures because they can be pursued more or less 'behind the backs' of consumers. In those rare cases where policy makers have sought to engage with consumption directly, they have tended to do so through interventions based on public information and education. Social marketing campaigns have stressed the importance of conserving residential energy and water, encouraging the use of paper shopping bags instead of their plastic alternatives, and promoting public transportation (and bicycles) as substitutes for private automobiles (Jackson 2006).

Within EMT studies, a third strategy with which to address consumption is suggested for the space between the systemic and the individualist approaches (Micheletti 2003; Spaargaren et al. 2006). Strategies to quietly green consumption without the direct participation of consumers are likely to be inadequate because the effective redesign of production systems requires the commitment of end-users. Initiatives that try to work around them frequently result in inappropriate uptake because the sustainability innovations do not mesh with extant consumer practices. Moreover, programmes that simply impart information about green products or advise consumers about the adverse impacts of their actions will ultimately prove insufficient in motivating consequential changes. Such directives need to be combined with attractive alternatives that are within the realistic grasp of end-users. In other words, ecologically favourable products and services must fit into the complex dynamics of everyday lifestyles and household dynamics. Within this 'contextual approach', consumers can be transformed into change agents without the need to lapse into isolated, individualistic strategies (Wallenborn 2007; Jackson 2006).

Like the debates over environmental technology described above, the discussions surrounding sustainable consumption will likely over time come to be characterised by customary social scientific divisions. At present, because active consideration of this issue is still relatively new, scholars subscribing to the relevant perspectives have not yet fully elaborated the respective approaches and books published to date demonstrate considerable empirical and theoretical diversity (Cohen and Murphy 2001; Jackson 2006; Zaccaï 2007; Van Vliet et al. 2005). Nonetheless, if we leave aside the individualistic approaches (most commonly associated with economics and social psychology), the different contextual understandings of sustainable consumption that are emerging can be subdivided into three broad groups.

First, we can identify a set of studies that constitutes an 'infrastructural' school of thought to sustainable consumption and is focused on the networks that provide households with energy, water, electricity and other basic services. Although the focus is on customary environmental flows, this research bears little resemblance to the work on domestic conservation that reached its apogee during the 1970s and 1980s. Prudent energy use and waste separation belong to the 'first generation of household environmental behaviours' that have become largely accepted and normalised by most people in affluent countries. Contemporary studies on infrastructures of consumption deal with new issues that gained prominence in Europe during the 1980s due to the privatisation and liberalisation of public services (Guy and Marvin 1996; Graham and Marvin 2001; Guy et al. 2001; Chappels et al. 2000; Van Vliet 2002; Spaargaren 2003; Southerton et al. 2004; Van Vliet et al. 2005). This scholarship centres on the differentiation and fragmentation of provisioning networks, the (lack of) transparency and communication between providers and end-users, the emergence of new (green) supply choices, the empowerment of domestic consumers, the changing power relations between actors on both sides of the utility meter, and new ways of service co-provisioning (and self-provisioning). The importance of these studies for EMT relates less to the analysis of sociotechnical systems than it does to the new institutional forms and relationships that are being built around these innovations. While the most interesting developments are taking place in and around projects that involve newly constructed housing, it is widely recognised that the major challenge for sustainable housing will be in applying already familiar sociotechnical systems (e.g., rainwater catchment, photovoltaic panels) to existing housing stock and to other elements of the supply infrastructure (Green and Vergragt 2002; Smith 2007; Farhar and Coburn 2008; Brown and Vergragt 2008).

Second, political scientists and other social scientists have begun to identify new expressions of 'political consumerism' embedded in campaigns to green global production-consumption chains (Micheletti 2003; Oosterveer 2007; Boström and Klintmann 2008). The aim of these efforts is to elaborate novel roles for consumers that enable them to use their purchasing practices as a source of power for promoting sustainable transitions. Characteristic for the political consumerism perspective is the notion of chain inversion: to better understand the complex dynamics governing networks and chains of production and consumption, the influence of organised citizen-consumers is taken as the main starting point of the analysis. Proponents of this perspective argue that under certain circumstances the political and ethical values of consumers can become important drivers in the ecological modernisation of consumption and production. Determination of how to most efficaciously organise this activity provides the primary focus for the empirical research carried out by scholars associated with this school of thought. The work of Michelle Micheletti (2003) has been prominent in setting this agenda, with boycotts, and especially 'buycotts', coming under scrutiny as emblematic of these new modes of political commitment and engagement.[4]

These activities are an alternative to traditional types of political engagement and they lower the barrier for entry into politics. The myriad ways in which ordinary people can use their buying power do not typically require special training, bureaucratic competence, party membership or elaborate institutional logistics. Political consumerism is a form of (sub)politics that is accessible, volatile, voluntary and closer to everyday life than the customary repertoire of practices that constitute more conventional political participation (Beck et al. 1994). As the recent work of Boström and Klintmann (2008) shows, the organised use of (environmental) information flows that travel through production-consumption chains can be very useful for purposes of ecological modernisation. The massive increase in labelling and certification schemes over the past decade, especially in Europe (but increasingly in most advanced and even some developing countries), is evidence that an entirely new generation of instruments is emerging with the potential to heighten consumer commitment and involvement in environmental change. Peter Oosterveer (2007) has also demonstrated the prominence and importance of labeling and highlighted how these tools can be interpreted as an evolving response to the challenges of building consumer trust in global provisioning chains.

Consumer groups and environmental nongovernmental organizations (ENGOs) have crucial roles to play in mobilising consumers to become effective agents of political consumerism (Kong et al. 2002; O'Rourke 2005; Cohen 2006a; Church and Lorek 2007). By lending support to labelling schemes and other assurance programmes, producers can effectively foster and (re)establish trust among consumers. This marshalling of grassroots actions can put pressure on providers at the upstream end of production-consumption chains without the need for official regulation or governmental participation (Woods and Schneider 2006). Also at the international level, there are indications that ENGOs can be effective facilitators of trust even among remote actors (Tysiachniouk 2006).

Finally, contextual studies in sustainable consumption have focused on domestic technologies and everyday household practices. As mentioned above, this perspective came to be constituted around a series of summer and winter schools at Lancaster University during the 1990s. Over the course of many meetings, researchers primarily from Europe and the United States shared experiences about home use of refrigerators and freezers, bathroom equipment (e.g., low-flow showers, power showers) and 'organic kitchens'. The participants in this network emphasised the importance of historical and comparative studies of ordinary practices like bathing, lighting, washing, cooking, heating and cooling to grasp the dynamics of (environmental) change at the level of everyday-life consumption routines (Pantzar 1992; Røpke 1999; Shove 2003, 2006; Shove and Warde 2001; Southerton et al. 2004; Van Vliet et al. 2005; Hegger 2007; Gram Hanssen 2008; Quitzau and Røpke 2008).

To develop more sophisticated policy interventions to green these daily routines it is necessary to acknowledge and deconstruct the interdependencies between the technologies and the norms governing social practices. Only by recognising the deep connections that link, say, heating and cooling technologies with prevailing notions of what constitutes a healthy, comfortable indoor climate can we begin to identify possibilities for more sustainable alternatives. This work underscores the fact that (sustainability) transitions in everyday life entail the co-evolution of technology, norms and social practices (see Shove 2003).[5] The contribution by Spaargaren in Part Three makes an intentional and explicit connection between EMT and the social practices of everyday life. This article argues that we cannot understand the structuration of social practices (Giddens 1984) without reference to the constant flows of environmental technologies, products and norms that have entered and reshaped our daily lives since the mid-1980s. The greening of everyday social practices is happening in front of our eyes and this proximity makes these developments more tangible, particularly in comparison to the parallel and interdependent global processes that are occurring across industries.[6]

Although each of these three approaches embodies its own specific emphasis, there is significant overlap with respect to their theoretical roots and research orientations. The various schools of sustainable consumption scholarship also share a common interest in combining investigations of sociotechnical innovations with institutional and social dynamics of consumption and production. Researchers working in these areas refuse to subscribe to standard social scientific separations between micro- and macro-levels of analysis, but instead seek to identify the interactions and interdependencies among processes working at different scales from the global down to the local. They moreover all make specific contributions to our understanding of the roles that human actors play as change agents in societal transitions toward more sustainable consumption practices, lifestyles and politics. The next section describes how an international politics of sustainable consumption has developed since the early 1990s.

The Marrakech process: developing the global SCP agenda

The sustainability dimensions of consumption first came to be conceptualised in political terms in 1992 at the Río Earth Summit. Agenda 21, the conference action plan, pronounces that '[a]ll countries should strive to promote sustainable consumption patterns' and encourages developed countries to assume primary responsibility for leading this change (Robins and Roberts 1998; Cohen 2001; Manoochehri 2002). This novel focus on consumption – as opposed to the traditional emphasis on production discussed above – generated a powerful wave of controversy that was fueled by the negative reaction it provoked from the United States and several other affluent countries. During the aftermath of the Río conference, the Nordic Council of Ministers and the Organisation for Economic Cooperation and Development (OECD) played constructive roles to temper passions and to flesh out the conceptual substance of a nascent sustainable consumption policy programme (see, e.g., Nordic Council of Ministers 1995; OECD 1998). During this period, however, the primary approach, at least among governmental institutions, was to redefine sustainable consumption in terms of more familiar (and less volatile) strategies and to subsume it under the umbrella of cleaner production, life cycle analysis and integrated product management.

Dissatisfaction with this conceptual approach, especially among developing countries and some ENGOs, gave consumption issues visible prominence at the 2002 Johannesburg Summit and generated a resultant commitment to produce a Ten-Year Framework of Programmes on Sustainable Consumption and Production under the joint auspices of the United Nations Environment Programme (UNEP) and the United Nations Department of Economic and Social Affairs (UNDESA) (Barber 2003). The formulation of this 'framework' – what has come to be known as the Marrakech Process – got off the ground in 2003 with a sequence of international experts meetings first in Morocco and two years later in Costa Rica (Clark 2007). In tandem, a series of regional stakeholder dialogues was initiated in Latin America and the Caribbean, Africa, Europe and the Asia-Pacific region to address questions of more localised relevance.[7]

A notable characteristic of the Marrakech Process is the tension that exists regarding how the concept of sustainable consumption should be defined and further developed. Virtually all proponents contend that, like sustainable development before it, the aims of sustainable consumption are too expansive to warrant a strict definition and efforts to do so will necessarily limit its effectiveness. While this purposeful ambiguity has allowed people with widely divergent interests to gather around the same table, it has led to considerable confusion over the terms of the debate and the content of the discussion (Tukker et al. 2006). As applied to affluent countries, sustainable consumption is generally meant to imply decreases in consumers' utilisation of energy and materials (and hence reduction in per capita ecological footprint).[8] For developing countries (and some ENGOs), the notion of sustainable consumption seeks to draw attention to chronic and persistent 'underconsumption'. The aim therefore is to provide the resources to enable the majority of the planet's population to increase consumption so as to more adequately satisfy basic needs.[9]

A lack of agreement on definitional parameters and policy objectives has however not been the only factor that has limited progress advancing an international agenda on sustainable consumption.[10] The UN system is manifestly hierarchical and UNEP does not enjoy a very high level of stature among its peer agencies. The problem is further compounded by chronic budgetary problems, and the fact that aside from UNEP virtually all of the constituent organisations of the UN are charged in one way or another with facilitating customary objectives of economic growth. Sound environmental management is seen to stand in the way of this objective and any efforts that could be interpreted as dampening consumer demand are viewed as utterly inconsistent with the overarching aims of the international system. Without support from other key actors, UNEP has been forced to build its initiatives around the voluntarism of a handful of sympathetic national governments. These projects have centred on the preparation of reports, training manuals and other informational and educational materials. Life cycle analysis, government procurement and product-service systems have been especially favoured approaches for these projects.

From this short overview of the international politics of sustainable consumption and production we can conclude that systematic policies pertaining to lifestyles and consumption are not yet visible on the global level though there is a considerable amount of organisational activity currently taking place outside of public view. Like in the early days of production-oriented EMT, there is a pressing need for political modernisation to invent new policy arrangements and innovative forms of governance to overcome the problems of 'state-failure' with respect to (national) consumption policies (Jänicke 1993; see also Jänicke 2008). The commitment to negotiate a common framework on the greening of consumption certainly represents an important first step, but there is presently less agreement between actors across the North–South divide regarding consumption than there is on the ecological modernisation of production where the task has been pursued largely as a technical exercise. This situation is largely due to unresolved political and economic controversies about the necessity and efficacy of specific policies to facilitate sustainable consumption. As we describe in the final section, these debates about consumption have also not yet been settled within EMT.

From consumption (back) to production?

The overall structure of our argument has been that the original emphasis within EMT on (clean) production and technological innovation at the upstream, producer end of production-consumption chains has gradually been complemented by a supplementary focus on consumption and everyday-life technologies. The consequences for EMT have been that additional perspectives and new theoretical emphases have developed since the eco-rationalities of consumption and lifestyles are quite different from the eco-rationalities that govern companies and producer-based policies. It is our contention that EMT must – and indeed already has started to – renew itself by confronting and accommodating the two most significant challenges that have emerged in the social sciences over the past twenty years: the globalisation of consumption and production and the 'consumerist turn' (Featherstone 1991).

'Modernising' EMT turns out to be necessary because these two developments have fundamentally altered the character of production–consumption chains and networks (Redclift and Benton 1994; Spaargaren et al. 2000). The major institutions responsible for managing consumption and production under conditions of 'simple modernity' – from the end of World War II until the mid-1980s – have been radically reformed and have gained a new dynamic in the present phase of 'reflexive modernity' (Beck et al. 1994; Beck 2005). As a social theory, EMT has been forced to (co)evolve together with its main objects of theoretical and empirical interest: production-consumption networks and chains. EMT, in fact, has contributed to this process of reform by stressing the need to include eco-rationalities in the rules governing the organisation of production and consumption. When interpreted along these lines, theoretical renewal is a sign of strength, of a theory being alive and relevant because social actors are subjecting it to frequent use, discussion and adaptation.

The shifting emphases within EMT have also given rise to theoretical debates within the community of scholars working with the theory. With respect to the impact of globalisation, especially the role of nation-states and their seemingly diminished power to influence the behaviour of market actors and civil society institutions, has become a subject of fierce debate (see Part Two of this volume). Regarding the consumerist turn, we have illustrated in the preceding sections that there is an evolving effort to understand the significance of a consumer orientation in environmental science and politics. While there seems to be consensus that this consumer orientation has gained momentum in recent years and is now becoming more widespread, it is not necessarily the case that this trend is equally welcomed by all EMT scholars. The contribution by Joseph Huber (2004) that concludes Part Three can be read as an explicit argument against a downstream focus on the end-users of products and services. He contends that the consumerist turn – what we have characterised as the shift since the mid-1980s from exclusively upstream sustainability policies toward initiatives designed to combine both upstream and downstream strategies – is compromising the ability of EMT to concentrate on the most critical impacts of contemporary decision making. Huber asserts that the higher up one is able to engage with chains and networks, the more 'effective' and consequential the resultant interventions will be from a sustainability point of view. Huber estimates that the environmental impacts of the end-user phase are generally less than 30 percent of overall effects and this relatively low proportion does not justify efforts to actively interfere with consumers, automobile drivers, home-owners and tourists.

On the basis of this observation, we can conclude that debates both about and within EMT do not evolve in a linear way, with 'new' assertions simply supplanting 'old' arguments. Through processes of continual discussion and revision, and by re-interpreting classic texts in the light of novel challenges and societal conditions, the field of EMT studies has retained its vibrancy and under such conditions has demon-strated an ability to contribute to a more sustainable future.

Notes

1. Although transition theory originated primarily in the Netherlands, it has expanded geographically over the past five years with a number of related studies taking place in Germany, the United Kingdom and throughout the Nordic countries. Some recent key references include Geels (2005a, 2005b), Geels and Schot (2007), Schot and Geels (2007) and Verbong and Geels (2007).

2. Examples of preventive technologies with the potential to restructure existing modes of consumption and production are renewable energy collectors (e.g. wind, solar and biomass), energy-producing greenhouses that collect and store heat during the summer and release it during the winter, zero-emission homes and energy co-producing households.

3. This view obviously does not stand up to scrutiny as experiences during wartime and in cases of other extreme emergencies provide ample evidence that the modification of consumption practices is well within the institutional capacity of most governments. See, for example, Zweiniger-Bargielowska (2000).

4. Buycotts and boycotts are 'political forms of engagement' with the potential to encourage environmental change. Boycotting products is implemented with the aim of putting pressure on producers to respond to consumer concerns since the former will generally be sensitive to threats of lost market share. With buycotts, reforms on the producer side are rewarded by positive action from consumers. In other words, consumers will buy the (greener) alternative themselves and help organise (new) markets for green products by mobilising other consumers. See Micheletti (2003) and Spaargaren and Mol (2008) for a more elaborate discussion on the role of citizen-consumers in environmental change.

5. Although almost all of the empirical studies that Shove (2003) discusses have direct relevance for environmental theory and policy, she does not enter directly into debates on EMT in the context of everyday-life consumption routines.

6. See Spaargaren et al. (2007) for useful empirical examples of the greening of consumption practices in the domains of food, housing, mobility, clothing and personal care, and leisure. Cohen (2006b) seeks to conceptualise sustainable consumption in terms of household finance and consumer economics.

7. See the various reports and other documents available at the website that supports the Marrakech Process at http://esa.un.org/marrakechprocess

8. As part of their commitment to the Johannesburg proceedings, several countries (e.g. Finland, Germany and the United Kingdom) have produced national sustainable consumption plans. For instructive reviews, see Seyfang (2004) and Spangenberg (2004).

9. These differences on how to interpret the policy objectives of an 'international sustainable consumption agenda' are grounded in the fact that on a global level consumption is massively inequitable. Clark (2007) reports that the wealthiest quintile of the world's population is responsible for 86% of total private consumption expenditures and the poorest quintile for approximately 1%. It merits noting that proponents of the 'basic needs' interpretation of sustainable consumption are also cognizant that large consumption disparities continue to exist within otherwise affluent countries.

10. Some critics of sustainable consumption have also charged that efforts in affluent countries to encourage the use of ecolabels, the relocalisation of consumption

and other similar strategies are really prompted by a desire to erect a new generation of trade barriers against the importation of goods from developing countries (see Bush and Oosterveer 2007).

References

Ashton, W. (2008) 'Understanding the organization of industrial ecosystems: a social network approach', *Journal of Industrial Ecology* **12** (1), pp. 34–51.

Badham, R. (1986) *Theories of Industrial Society*. London: Croom Helm.

Barber, J. (2003) 'Production, consumption, and the World Summit on Sustainable Development', *Environment, Development, and Sustainability* **5** (1–2), pp. 63–93.

Beck, U. (2005) *Power in the Global Age: A New Global Political Economy*. Cambridge: Polity.

Beck, U., Giddens, A. and Lash, S. (1994) *Reflexive Modernisation: Politics, Tradition, and Aesthetics in the Modern Social Order*. Cambridge: Polity.

Boström, M. and Klintmann, M. (2008) *Eco-Standards, Product Labelling and Green Consumerism*. London: Palgrave.

Brown, H. and Vergragt, P. (2008) 'Bounded socio-technical experiments as agents of systemic change: the case of a zero-energy residential building', *Technological Forecasting and Social Change* **75** (1), pp. 107–130.

Bush, S. and Oosterveer, P. (2007) 'The missing link: intersecting governance and trade in the space of place and the space of flows', *Sociologia Ruralis* **47** (4), pp. 384–399.

Carolan, M. (2004) 'Ecological modernisation theory: what about consumption?', *Society and Natural Resources* **17** (3), pp. 247–260.

Chappells, H., Klintman, M., Linden, A.-L., Shove, E., Spaargaren, G. and Vliet, B. van (2000) *Domestic Consumption, Utility Services, and the Environment*. Final Domus Report. Wageningen: Wageningen University.

Church, C. and Lorek, S. (2007) 'Linking policy and practice in sustainable production and consumption: an assessment of the role of NGOs', *International Journal of Innovation and Sustainable Development* **2** (2), pp. 230–240.

Clark, G. (2007) 'Evolution of the global sustainable consumption and production policy and the United Nations Environment Programme's (UNEP) supporting activities', *Journal of Cleaner Production* **15** (6), pp. 492–498.

Cohen, M. (1997) 'Risk society and ecological modernisation: alternative visions for postindustrial nations', *Futures* **29** (2), pp. 105–119.

Cohen, M. (2001) 'The emergent environmental policy discourse on sustainable consumption', in M. Cohen and J. Murphy (eds) *Exploring Sustainable Consumption: Environmental Policy and the Social Sciences*. London: Elsevier, pp. 21–37.

Cohen, M. (2006a) 'Sustainable consumption research as democratic expertise', *Journal of Consumer Policy* **29** (1), pp. 67–77.

Cohen, M. (2006b) 'Consumer credit, household financial management, and sustainable consumption', *International Journal of Consumer Studies* **31** (1), pp. 57–65.

Cohen, M. and Howard, J. (2006) 'Success and its price: the institutional and political foundations of industrial ecology', *Journal of Industrial Ecology* **10** (1), pp. 79–88.

Cohen, M. and Murphy, J. (eds) (2001) *Exploring Sustainable Consumption:. Environmental Policy and the Social Sciences*. London: Elsevier.

Cramer, J. (2006) *Duurzaam ondernemen: van defensief naar innovatief.* Rotterdam: Inaugu-rele Rede.

Cramer, J. and Schot, J. (1990) *Problemen rond innovatie en diffusie van milieutechnologie. Een onderzoeksprogrammeringsstudie verricht vanuit een technologiedynamica perspectief.* Rijswijk: RMNO.

Derochers, P. (2004) 'Industrial symbiosis: the case for market coordination', *Journal of Cleaner Production* 12 (8–10), pp. 1099–1110.

Farhar, B. and Coburn, T. (2008) 'A new market paradigm for zero-energy homes: a comparative case study', *Environment* 50 (1), pp. 19–32.

Featherstone, M. (1991) *Consumer Culture and Postmodernism.* London: Sage.

Geels, F. (2005a) 'Processes and patterns in transitions and system innovations: refining the co-evolutionary multi-level perspective', *Technological Forecasting and Social Change* 72 (6), pp. 681–696.

Geels, F. (2005b) *Technological Transitions and System Innovation: A Co-Evolutionary and Socio-Technical Analysis.* Cheltenham: Edward Elgar.

Giddens, A. (1984) *The Constitution of Society.* Cambridge: Polity.

Graham, S. and Marvin, S. (2001) *Splintering Urbanism.* London: Routledge.

Gram-Hanssen, K. (2008) 'Consuming technologies – developing routines', *Journal of Cleaner Production* 16 (11), pp. 1181–1189.

Green, K. and Vergragt, P. (2002) 'Towards sustainable households: a methodology for developing sustainable technological and social innovations', *Futures* 34 (5), pp. 381–400.

Guy, S. and Marvin, S. (1996) 'Transforming urban infrastructure provision: the emer-ging logic of demand-side management', *Policy Studies* 17 (2), pp. 137–147.

Guy S, Marvin, S. and Moss, T. (eds) (2001) *Urban Infrastructure in Transition; Networks, Buildings, Plans.* London: Earthscan.

Heeres, R., Vermeulen, W. and Walle, F. de (2004) 'Eco-industrial park initiatives in the USA and the Netherlands: first lessons', *Journal of Cleaner Production* 12 (8–10), pp. 985–995.

Hegger, D. (2007) *Greening Sanitary Systems: An End-User Perspectiv.* Dissertation. Wage-ningen: Wageningen University.

Hobson, K. (2003) 'Consumption, environmental sustainability, and human geography in Australia: a missing research agenda', *Australian Geographical Studies* 41 (2), pp. 148–155.

Huber, J. (1982) *Die verlorene Unschuld der Ökologie. Neue Technologien und superindustrielle Entwicklung.* Frankfurt am Main: Fisher.

Huber, J. (2004) *New Technologies and Environmental Innovation.* Cheltenham: Edward Elgar.

Hunt, R. (1996) 'LCA – how it came about: personal reflections on the origin and the development of LCA in the USA', *International Journal of Life Cycle Assessment* 1 (1), pp. 4–7.

Jackson, T. (2006) 'Challenges for sustainable consumption policy', in T. Jackson (ed.) *The Earthscan Reader in Sustainable Consumption.* London: Earthscan, pp. 109–129.

Jackson, T. (ed.) (2006) *The Earthscan Reader in Sustainable Consumption.* London: Earthscan.

Jänicke, M. (1993) 'Über ökologische und politische Modernisierungen', *Zeitschrift für Umweltpolitik und Umweltrecht* 2, pp. 159–175.

Jänicke, M. (2008) 'Ecological modernisation: new perspectives', *Journal of Cleaner Production* 16 (5), pp. 557–565.

Jensen, J. and Gram-Hanssen, K. (2008) 'Ecological modernization of sustainable buildings: a Danish perspective', *Building Research and Information* **36** (2), pp. 146–158.

Kong, N., Salzmann, O., Steger, U. and Ionescu-Somers, A. (2002) 'Moving business/industry towards sustainable consumption: the role of NGOs', *European Management Journal* **20** (2), pp. 109–127.

Lowe, E. (1997) 'Creating by-product exchanges: strategies for eco-industrial parks', *Journal of Cleaner Production* **5** (1–2), pp. 57–65.

Manoochehri, J. (2002) 'Post-Rio "sustainable consumption": establishing coherence and a common platform', *Development* **45** (3), pp. 47–53.

Micheletti, M. (2003) *Political Virtue and Shopping; Individuals, Consumerism, and Collective Action*. London: Palgrave.

Mol, A. and Spaargaren, G. (2004) 'Ecological modernisation and consumption: a reply', *Society and Natural Resources* **17** (3), pp. 261–265.

Murphy, J. and Gouldson, A. (2000) 'Environmental policy and industrial innovation: integrating environment and economy through ecological modernization', *Geoforum* **31** (1), pp. 33–44.

Nordic Council of Ministers (1995) *Sustainable Patterns of Consumption and Production: Reports from the Seminar on Instruments to Promote Sustainable Patterns of Consumption and Production*. Copenhagen: Nordic Council of Ministers.

O'Rourke, D. (2005) 'Market movements: nongovernmental organization strategies to influence global production and consumption', *Journal of Industrial Ecology* **9** (1–2), pp. 115–128.

Oosterveer, P. (2007) *Global Governance of Food Production and Consumption: Issues and Challenges*. Cheltenham: Edward Elgar.

Oosterveer P., Guivant, J. and Spaargaren, G. (2007) 'Shopping for green food in globalizing supermarkets: sustainability at the consumption junction', in J. Pretty, A. Ball, T. Benton, J. Guivant, D. Lee, D. Orr, M. Pfeffer and H. Ward (eds) *Sage Handbook on Environment and Society*. London: Sage, pp. 411–428.

Organisation for Economic Cooperation and Development (OECD) (1998) *Toward Sustainable Consumption Patterns: A Progress Report on Member Countries*. Paris: OECD.

Pantzar, M. (1992) 'The growth of product variety–a myth?', *Journal of Consumption Studies and Home Economics* **16** (4), pp. 345–362.

Perz, S. (2007) 'Reformulating modernisation-based environmental social theories: challenges on the road to an interdisciplinary environmental science', *Society and Natural Resources* **20** (5), pp. 415–430.

Polimeni, J., Mayumi, K., Giampietro, M. and Alcott, B. (2008) *Jevons' Paradox and the Myth of Resource Efficiency Improvements*. London: Earthscan.

Princen, T., Maniates, M. and Conca, K. (eds) (2002) *Confronting Consumption*. Cambridge: MIT Press.

Quitzau, M. and Røpke, I. (2008) 'The construction of normal expectations: consumption drivers for the Danish bathroom boom', *Journal of Industrial Ecology* **12** (2), pp. 186–206.

Redclift, M. and Benton, T. (eds) (1994) *Social Theory and the Global Environment*. London: Routledge.

Robins, N. and Roberts, S. (1998) 'Making sense of sustainable consumption', *Development* **41** (1), pp. 28–36.

Røpke, I. (1999) 'The dynamics of willingness to consume', *Ecological Economics* **28** (3), pp. 399–420.

Sachs, W. and Santarius, T. (2005) *Fair Future: begrenzte Ressourcen und globale Gerechtigkeit – ein Report*. Munich: Beck.

Salmi, O. and Toppinen, A. (2007) 'Embedding science in politics: "complex utilization" and industrial ecology as models of natural resource use', *Journal of Industrial Ecology* **11** (3), pp. 93–111.

Schot, J. (1992) 'Constructive technology assessment and technology dynamics: the case of clean technologies', *Science, Technology, and Human Values* **17** (1), pp. 36–56.

Schot, J. and Geels, F. (2007) 'Niches in evolutionary theories of technical change: a critical survey of the literature', *Journal of Evolutionary Economics* **17** (5), pp. 605–622.

Seuring, S. (2004) 'Industrial ecology, life cycles, and supply chains: differences and interrelations', *Business Strategy and the Environment* **13** (5), pp. 306–319.

Seyfang, G. (2004) 'Consuming values and contested cultures: a critical analysis of the UK strategy for sustainable consumption and production', *Review of Social Economy* **62** (3), pp. 323–338.

Shove, E. (2003) *Comfort, Cleanliness and Convenience: The Social Organization of Normality*. Oxford: Berg.

Shove, E. (2006) 'Efficiency and consumption: technology and practice', in T. Jackson (ed.) *The Earthscan Reader in Sustainable Consumption*. London: Earthscan, pp. 293–305.

Shove, E. and Warde, A. (2001) 'Inconspicious consumption: the sociology of consumption, lifestyles, and the environment', in R. Dunlap, F. Buttel, P. Dickens, and G. Gijswijt (eds) *Sociological Theory and the Environment: Classical Foundations, Contemporary Insights*. Lanham, MD: Rowman and Littlefield, pp. 230–241.

Shove, E. and Walker, G. (2007) 'CAUTION! Transitions ahead: politics, practice, and sustainable transition management', *Environment and Planning, A* **39** (4), pp. 763–770.

Smith, A. (2007) 'Translating sustainabilities between green niches and socio-technical regimes', *Technology Analysis and Strategic Management* **19** (4), pp. 427–450.

Sonnenfeld, D. (2002) 'Social movements and ecological modernisation: the transformation of pulp and paper manufacturing', *Development and Change* **33** (1), pp. 1–27.

Southerton D., Chappells, H. and Vliet, B. van (2004) *Sustainable Consumption: The Implications of Changing Infrastructures of Provision*. Cheltenham: Edward Elgar.

Spaargaren, G. (2003) 'Sustainable consumption: a theoretical and environmental policy perspective', *Society and Natural Resources* **16** (8), pp. 687–701.

Spaargaren, G. and Mol, A. (2008) 'Greening global consumption: redefining politics and authority', *Global Environmental Change: Human and Policy Dimensions* **18** (3), pp. 350–359.

Spaargaren, G., Vliet, B.J.M. van (2000) 'Lifestyles, Consumption and the Environment: The Ecological Modernisation of Domestic Consumption', *Environmental Politics* **9** (1), pp. 50–77.

Spaargaren, G., Martens, S. and Beckers, T. (2006) 'Sustainable technologies and everyday life', in P. Verbeek and A. Slob (eds) *User Behaviour and Technology Development: Shaping Sustainable Relations Between Consumers and Technologies*. Dordrecht: Springer, pp. 107–118.

Spaargaren, G., Mol, A. and F. Buttel (2000) *Environment and Global Modernity*. London: Sage.

Spaargaren, G, Mommaas, H., S. Burg, S. van den, Maas, L., Drissen, E., Dagevos, H., Bargeman, B., Putman, L., Nijhuis, J., Verbeek, D. and Sargant, E. (2007) *More Sustainable Lifestyles and Consumption Patterns: A Theoretical Perspective for the Analysis of Transition Processes within Consumption Domains*. Wageningen: Environmental Policy.

Spangenberg, J. (2004) 'A great step further, but still more to go', *Environment* **46** (8), pp. 42–45.

Thomas, V., Theis, T., Lifset, R., Grasso, D., Kim, B., Koshland, C. and Pfahl, R. (2003) 'Industrial ecology: policy potential and research needs', *Environmental Engineering Science* **20** (1), pp. 1–9.

Tukker, A. (2005) 'Leapfrogging into the future: developing for sustainability', *International Journal of Innovation and Sustainable Development* **1** (1–2), pp. 65–84.

Tukker, A. and Tischner, U. (eds) (2006) *New Business for Old Europe: Product Services, Sustainability, and Competitiveness*. Sheffield: Greenleaf.

Tukker, A., Cohen, M., Zoysa, U. de, Hertwich, E., Hofstetter, P., Inaba, A. and Lorek, S. (2006) 'Oslo Declaration on Sustainable Consumption', *Journal of Industrial Ecology* **10** (1), pp. 9–14.

Tukker, A., Emmert, S., Charter, M., Vezzoli, C., Stø, E., Munch Andersen, M., Geerken, T., Tischner, U., Lahlou, S. (2008) 'Fostering change to sustainable consumption and production: an evidence based view', *Journal of Cleaner Production* **16** (11), pp. 1218–1225.

Tysiachniouk, M. (2006) 'NGOs between globalization and localization: the role of global processes in mobilizing public participation in forest settlements', *Journal of Sociology and Social Anthropology* **9**, pp. 113–158.

Udo de Haes, H., Heijungs, R., Suh, S. and Huppes, G. (2004) 'Three strategies to overcome the limitations of life-cycle assessment', *Journal of Industrial Ecology* **8** (3), pp. 19–32.

Ullrich, O. (1979) *Weltniveau: In der Sackgasse der Industriegesellschaft*. Berlin: Rotbuch Verlag.

Van Koppen, C. S. A. (Kris) and Mol, A. (2002) 'Ecological modernisation of industrial systems', in P. Lens, L. Hulshoff Pol, P. Wilderer, and T. Asano (eds) *Water Recycling and Resource Recovery in Industry: Analysis, Technologies, and Implementation*. Wageningen: IWA Publishing, pp. 132–158.

Van Vliet, B. (2002) *Greening the Grid: The Ecological Modernisation of Network-bound Systems*. Dissertation. Wageningen: Wageningen University.

Van Vliet, B., Chappells, H. and Shove, E. (2005) *Infrastructures of Consumption: Environmental Innovation in the Utility Industries*. London: Earthscan.

Verbong, G. and Geels, F. (2007) 'The ongoing energy transition: lessons from a socio-technical, multi-level analysis of the Dutch electricity system (1960–2004)', *Energy Policy* **35** (2), pp. 1025–1037.

Vezzoli, C. (2003) 'A new generation of designers: perspectives for education and training in the field of sustainable design–experiences and projects at the Politecnico di Milano University', *Journal of Cleaner Production* **11** (1), pp. 1–9.

Voss, H.-P., Bauknecht, D. and Kemp, R. (eds) (2006) *Reflexive Governance for Sustainable Development*. Cheltenham: Edward Elgar.

Wallenborn, G. (2007) 'How to attribute power to consumers? When epistemology and politics converge', in E. Zaccaï (ed.) *Sustainable Consumption, Ecology, and Fair Trade*. London: Routledge, pp. 57–71.

Weale, A. (1992) *The New Politics of Pollution*. Manchester: Manchester University Press.

White, D. (2002) 'A green industrial revolution? Sustainable technological innovation in a global age', *Environmental Politics* **11** (2), pp. 1–26.

Woods, D. and Schneider, R. (2006) 'Toxic Dude: the Dell Campaign', in T. Smith, D. Pellow, and D. Sonnenfeld (eds) *Challenging the Chip: Labor Rights and Environmental Justice in the Global Electronics Industry*. Philadelphia: Temple University Press, pp. 285–297.

Zaccaï, E. (ed.) (2007) *Sustainable Consumption, Ecology, and Fair Trade*. London: Routledge.

Zweiniger-Bargielowska, I. (2000) *Austerity in Britain: Rationing, Controls, and Consumption, 1939–1955*. Oxford: Oxford University Press.

Chapter 15

Joseph Murphy and Andrew Gouldson

ENVIRONMENTAL POLICY AND INDUSTRIAL INNOVATION: INTEGRATING ENVIRONMENT AND ECONOMY THROUGH ECOLOGICAL MODERNISATION

Introduction

THE THEORY OF ECOLOGICAL modernisation has become a focus for academic debate in recent years. As part of an essentially sociological discussion on the nature of modernity, ecological modernisation has been identified as one of the ways in which late-modern society is responding to its increased awareness of, and anxiety about, the ecological risks associated with industrialism (see for example Huber, 1985; Giddens, 1991; Beck, 1991, 1995; Lash et al., 1996; Christoff, 1996). Also, the theory has been developed both as a way of analysing emergent policy discourses (see Hajer, 1995) and as a theoretical basis from which various policy prescriptions can be brought forward to encourage a shift toward more environmentally benign modes of industrial development (see Huber, 1985; Jänicke et al., 1989; Simonis, 1989a; Mol, 1995; Gouldson and Murphy, 1997). Thus, the theory of ecological modernisation has been used in both a descriptive and prescriptive way (Mol, 1997).

In each case ecological modernisation is centrally concerned with the relationship between industrial development and the environment. Consequently, it is concerned with the capacity of modern industrial societies to recognise and respond to existing and emergent environmental problems. In this respect, the prescriptive use of ecological modernisation as a programme of policy reform has attempted to establish a widespread appeal by arguing that certain forms of policy intervention can simultaneously result in both economic and environmental benefits. At the micro-economic scale this is particularly the case where policies serve to promote the development and application of new and innovative clean (or cleaner) technologies and techniques. However, despite the emphasis that ecological modernisation places on the potential for policies to stimulate innovation, existing work in this area has yet to be informed by a detailed understanding of the innovation process.

This paper reviews the nature of industrial innovation and its outputs. It highlights the significance of barriers to innovation and argues that these barriers justify regulatory intervention. More particularly, it suggests that regulation can help companies to overcome the short-term barriers to innovation that commonly prevent them from moving beyond control technologies to consider clean technologies, from complementing technological change with organisational change and from exploring the strategic as well as operational opportunities for improvement.

This argument is then explored empirically, based on research conducted during 1996 and 1997. This work examined the impact of regulation on industrial innovation by considering the implementation and impact of integrated pollution control (IPC) as introduced in England and Wales by the 1990 Environmental Protection Act. The research suggests that IPC is helping to stimulate forms of innovation that improve both the environmental and the economic performance of regulated companies in the short- to medium-term. This is particularly the case where sustained interaction between the regulator and regulated companies enhances the capacity of the latter to explore the various opportunities for environmental improvement. However, the research also indicates that because IPC is not associated with challenging targets for environmental improvement in the medium- to long-term it fails to establish the imperative for environmental improvement. As a result environmental issues have yet to be adopted as a strategic concern in regulated companies. One outcome of this is that the opportunities for more radical forms of environmental improvement are unlikely to be exploited.

Based on this theoretical and empirical discussion this paper shows that the claim made by ecological modernisation theory is supported, at least in some respects, by an understanding of the nature of innovation. Regulation can be used to drive the process of industrial innovation with environmental and economic gains realised as a result. This is most likely to be the case where regulation successfully helps to build the capacity of regulated companies to address environmental problems whilst also establishing the imperative for such action to be taken, particularly with the aim of influencing strategic decision making. However, in the case of IPC, given that a continual focus on incremental rather than radical change is likely to encounter diminishing returns, both environmentally and economically, the common elements it shares with ecological modernisation are likely to exist only in the short- to medium-term. This is the case because in the longer-term the barriers to radical innovation, which IPC fails to address, will result in structural constraints that prevent environmental and economic gains from being realised simultaneously.

Ecological modernisation, industrial restructuring and policy reform

In *The Refinement of Production*, Mol (1995) states that Joseph Huber should be acknowledged as the founding father of ecological modernisation theory because of his theoretical contributions to the environment and society debate from the 1980s onwards. In this work Huber (1982, 1984, 1985) began to promote the idea that environmental problems could be addressed by entrepreneurs and industrialists through 'super-industrialisation'. For Huber super-industrialisation involves addressing the environmental impacts of industrial society largely through the transformation of industrial production based on the development and application of advanced technologies. Huber offers such an approach as a desirable solution to environmental problems. In his own words:

> the dirty and ugly industrial caterpillar will transform into a[n] ecological butterfly (Huber, 1985, p. 20 as quoted by Mol, 1995, p. 37).

Therefore, to some extent the ideas of Huber resonate with the wider discussion of reflexive modernity and the risk society put forward by authors such as Giddens (1991), Beck (1991, 1995) and Lash et al. (1996). These authors have suggested that modern industrial society is becoming less secure about its future and that it is responding by seeking to change the nature of industrial development. However, they also argue that modern society is increasingly questioning the faith that it has traditionally placed in science, technology and the institutions of government, and here this work differs from the ecological modernisation debate significantly.

Since Huber initially developed the theory of ecological modernisation the associated ideas have been explored from a number of different academic perspectives. Initially Jänicke (1985), Jänicke et al. (1988, 1989) and Simonis (1989a, b) moved beyond the focus on industrial innovation and highlighted the macroeconomic restructuring of advanced industrial economies as another essential component of ecological modernisation. They suggested that technological change was important along with a shift in the sectoral composition of industrial economies, away from heavy industry toward less resource intensive and environmentally burdensome sectors such as tourism and financial services. Following this Weale (1992) made a major contribution to the debate in his book *The New Politics of Pollution*. This work had a more practical policy focus and firmly identified ecological modernisation as a government led programme of action based on strategic planning and innovative policy instruments with industry as the centre of attention. Weale identified that government action which is consistent with the theory of ecological modernisation will be based on the belief that there is no necessary conflict between environmental protection and economic growth. Indeed, more important still, it will assume that addressing environmental problems can make industry more efficient and may actually generate future growth.

More recently, with the core elements of ecological modernisation established, Hajer and Mol have transferred the idea from politics to sociology and explored it in detail from two different perspectives. Hajer's (1995, 1996) approach has been to view ecological modernisation as a policy discourse which is composed of a set of storylines. These storylines concern the link between environment and economy and are mobilised by members of discourse coalitions to promote specific interests. Thus ecological modernisation is not viewed as something that is real but instead it is examined as a set of attractive ideas which are adopted and communicated by policy elites. In contrast Mol (1995) has explored ecological modernisation as an identifiable phenomena. More specifically he has treated ecological modernisation as an example of institutional reflexivity in the Dutch chemicals industry as a result of environmental pressures.

As a result of this work, the theory of ecological modernisation has increasingly become established as a relatively clear set of ideas, although as Mol (1997) points out it is important to distinguish between the descriptive use of the concept and its prescriptive deployment. With respect to the prescriptive dimension it views environmental problems as resulting from industrialism. It proposes that these problems can be addressed within the framework of modernity, largely through the actions of the state and industry, with the result of encouraging further economic growth. On the part of the state the need is for active engagement with the aim of promoting ecological modernisation and managing environmental impacts. This will require strategic planning and the promotion of structural change at the macro-economic level, particularly involving emphasis on less resource intensive and environmentally burdensome means of generating wealth. This is likely to require a range of innovative policy instruments and approaches to replace the traditional understanding of the regulation of industry, particularly through the incentivisation of environmental improvement. In industry itself ecological modernisation involves the invention and diffusion of new technologies and techniques of production which will simultaneously reveal environmental and economic benefits. Critically, innovative and strict environmental regulation by government is seen as a way of driving the innovation process in industry to this end. However, as the review of ecological modernisation literature shows, despite some of the excellent work in this area, the potential for regulation to drive innovation in this way has not been examined in detail in association with this theory.

Environment, economics and innovation

Innovation is normally taken to include all stages of new economic activity including the "search for and discovery, experimentation, development, imitation and adoption of new products, new processes and new organisational set-ups" (Dosi, 1988, p. 222). Innovations may be either radical, which involve discontinuous change and the introduction of new technologies and techniques, or incremental, which involve gradual improvement of existing technologies and techniques. While radical innovations often rely on incremental improvement for their success, incremental innovations must eventually encounter diminishing marginal returns as they encounter both economic and technical limits. Therefore, the periodic introduction of radical or discontinuous change is a pre-requisite for successful phases of incremental innovation.

It is apparent that innovations are not adopted on the basis of their isolated characteristics, such as cost or quality, but also on the extent of their compatibility with existing systems and structures. New innovations depend upon a system or network of relations without which their adoption is impossible (Kemp, 1993). As a consequence, new technologies and techniques must be introduced into systems which have often been developed for, and adapted to, older technologies and techniques. Consequently the introduction of a new innovation may require far-reaching changes to ensure compatibility with the system as a whole. Considerable resistance and inertia may be apparent as a result. For example, firms which have successfully mastered the operation of old technologies and techniques commonly face difficulties in overcoming the limits of existing skills and knowledge and in acquiring the new skills and knowledge needed to successfully apply new technologies and techniques (OECD, 1992). Depending on the ability of an innovation to influence existing systems, an innovation which requires only incremental changes is more likely to be adopted than one which requires more radical change. Consequently, the dynamic nature of innovation tends to favour incremental rather than radical innovation and as a consequence change normally takes place within particular trajectories in an evolutionary way.

However, the ability of a new technology or technique to influence existing systems and structures varies over time, particularly as a result of the dynamic and self-reinforcing impact of scale and learning effects (Soete and Arundel, 1995; Kemp, 1993). These effects mean that new innovations commonly realise improvements in quality and reductions in cost as their production expands and as experience with their application accumulates. Thus, the effectiveness and efficiency of a new innovation increases as it is diffused so that rather than encountering diminishing returns, innovations commonly experience a period of increasing returns as they are adopted (OECD, 1992). While these effects appear to be beneficial, and as their impacts increase, for a period both the supply-side and the demand-side are likely to become increasingly locked into the particular trajectory of the selected innovation. By comparison, whatever their apparent potential, those innovations which have not benefited from the dynamic scale and learning effects, are, or are perceived to be, expensive, inadequate, incompatible, unproven, unknown and without the necessary support mechanisms. In essence, therefore, potentially beneficial innovations may not be selected due to their initial inefficiency, although, paradoxically, the only way that they can become efficient is by being selected (OECD, 1992). The rationality of actors choosing whether to develop or use a new innovation is thus bounded by the nature of the existing system.

It is apparent from this brief assessment of the innovation process that there is commonly an in-built bias towards incremental rather than radical change. This bias stems particularly from a desire not to disrupt existing systems but it is also linked to other barriers to innovation associated with, for example, managerial capacity and access to capital. However, although the preference for fine-tuning existing technologies and techniques before introducing new ones may appear to be rational in the short- to medium-term, it is evident that if explored in isolation incremental innovations generally encounter diminishing marginal returns so that the costs of sustained improvement eventually escalate. While new technologies and techniques have the potential to allow sustained improvements in performance for a longer period and in the absence of escalating costs, in the short- to medium-term they tend not to be adopted. This is the case partly because they have yet to benefit from the scale and learning effects that would reduce their cost and improve their quality.

The opportunities for environmental improvement in industry

The generic characteristics of the innovation process are reflected in the various options available to companies seeking to improve their environmental performance.

In such instances, companies typically face a number of choices related to those technological and organisational options that can incrementally improve the environmental performance of existing operations and to the strategic options that offer the potential for more radical change.

In relation to the technological opportunities for environmental improvement, companies generally face a choice between control and clean (or at least cleaner) technologies. Control technologies are end-of-pipe additions to production processes which capture and/or treat a waste emission in order to limit its impact on the environment. By contrast, clean technologies are general processes or products which fulfil a non-environmental objective as their primary purpose but which integrate environmental considerations into their design to avoid or reduce their impact on the environment. Thus, control technologies can be seen as a reactive response to emissions while clean technologies are more anticipatory in their nature.

Because of their reactive nature, control technologies do not normally require a significant redesign of the processes or products with which they are associated. Also, as a distinct market for control technologies exists, suppliers are easily identified and a generic analysis of their economic and environmental performance is relatively straightforward. Consequently, they are often relatively easy to purchase and install when compared to clean technologies. In contrast, and as a consequence of their integrated nature, the compatibility of clean technologies with existing systems is usually more limited. This means that it is often expensive to retro-fit clean technologies into existing processes. Furthermore, as they are designed to be integrated into existing processes on a case-by-case basis, a generic analysis of their economic and environmental performance is normally more complicated.

Collectively, these characteristics have tended to support the development and application of control technologies, particularly as a consequence of their short- to medium-term expedience. This initial bias towards control technologies has subsequently been compounded by the scale and learning effects that have allowed their price and performance to be improved. In many instances, clean technologies have yet to benefit from a similar process because they have been less widely adopted. This tends to be the case despite the fact that clean technologies are generally considered to be more economically efficient and environmentally effective in the medium- to long-term as they can improve resource efficiency, thereby increasing economic efficiency whilst also addressing environmental impacts at source (OECD, 1987; CEC, 1993; DTI/DoE, 1994).

As well as investments in new technologies, companies can also improve their environmental performance through organisational change and the introduction of new managerial techniques. Organisational innovations of this nature can be adopted on their own or in combination with technological changes. In each instance organisational change can be associated with a range of direct and indirect benefits. For example, relatively simple managerial or organisational innovations may negate the need for potentially costly investments in technology. Where technological investments are necessary, managerial or organisational innovations can help to secure the direct and indirect benefits associated with the new technology, particularly by facilitating further incremental improvement. The application of a management system (traditionally concerning quality but more recently involving health, safety and environment) is one of the most important organisational innovations in business due to the way that it can pull a disparate organisational system into an integrated and organised whole. Potentially such an organisational innovation can not only bring rapid economic and environmental gains but it can also have a positive impact on the environment into which new technologies must be introduced, thereby enhancing the potential for clean technologies to be integrated into existing systems.

Finally, in association with technological and organisational change in the short- to medium-term, companies can explore the potential for improving their environmental performance by integrating environmental concerns into their strategic as well as their operational management processes. While incremental improvement to existing processes through technological or organisational change can secure some benefits, incremental improvements alone must eventually encounter diminishing marginal returns. However, these can be avoided if opportunities for more radical change are considered at a time when more elements of the production process can be altered. This is most likely to occur if companies integrate environmental concerns into their strategic decision making processes. In this way companies can explore the opportunities for radical change, the benefits of which can be further strengthened by subsequent phases of incremental improvement through the technological and organisational options outlined above. However, it is likely that the barriers to innovation of this nature will be more pronounced. This is the case because radical changes will require greater managerial and financial inputs to develop production approaches which are more likely to be relatively untried and untested.

Therefore it is apparent that a variety of options are available to companies who are seeking to improve their environmental performance. These options are related to the nature of industrial innovation and they can be presented as a simple analytical framework with three dimensions:

- the technological dimension (control/clean technologies);
- the organisational dimension (technologies/management techniques);
- the strategic dimension (operational/strategic focus – incremental/radical change).

However, although this framework usefully summarises the previous discussion it is important to note that in reality such discrete categories do not exist. For example, the classification of something as a control or clean technology may not always be clear. Also, as outlined above, the framework does not acknowledge the barriers to innovation which tend to become more pronounced as more disruptive or more radical changes are sought. Thus, the barriers to innovation increase as companies move beyond control technologies to consider clean technologies, as they complement technological change with organisational change and as they explore the strategic as well as the operational opportunities for improvement. These barriers to innovation reduce the likelihood that new technologies and techniques will be developed and adopted. Consequently, they tend to increase the short-term path-dependencies generated particularly by the impacts of the scale and learning effects that shape the development of industrial activity. However, given the longer-term potential of new technologies and techniques, it is appropriate to examine the role that regulation can play in helping companies to overcome these barriers to innovation.

Environmental regulation and industrial innovation in England and Wales

In the empirical part of this paper the link between environmental regulation and industrial innovation is explored. Therefore the discussion examines one of the central aspects of ecological modernisation, namely the potential for regulation to promote improvement in the economic and environmental performance of industry. The empirical analysis focuses on industrial environmental regulation in England and Wales in the form of the IPC regulations introduced in 1990 by the Environmental Protection Act. Consequently it also focuses on processes covered by this legislation – largely involving the bulk chemicals and speciality chemicals sectors but also including power stations, ferrous and non-ferrous metal smelting works and cement kilns. The choice of IPC as a focus is justified because it covers the most environmentally significant industrial operations in the UK. Also, systems of IPC are presumed to represent the 'state of the art' in industrial environmental regulation (see particularly Haigh and Irwin, 1990) and the UK's approach to regulation in this area is similar to that which will be applied throughout Europe as a result of the European Union's Integrated Pollution Prevention and Control Directive. Thus the potential exists for lessons to be learnt from the UK which may have wider significance.

The analysis draws extensively on research conducted during 1996 and 1997. It is based on data collected during approximately 40 semi-structured interviews with pollution inspectors and the environmental managers of regulated companies. During this research the intention was to develop an understanding of the experience and process of regulation, rather than to record actual changes in emissions from regulated sites, or in detail to investigate specific pieces of technology. Broadly the work combined top-down and bottom-up approaches to implementation research (see Sabatier, 1986). From the top-down perspective it focused particularly on the role of the individual 'street-level bureaucrats' responsible for establishing the specific demands of regulation through their interaction with regulated companies (Lipsky, 1980). From the bottom-up perspective the analysis examined the conditions for innovation in regulated companies and the influence of regulation as one of a number of factors which might motivate such activity.

The issues covered in these semi-structured interviews were initially identified through a literature review and subsequently refined through early pilot interviews. Thus, the top–down analysis of implementation focused firstly on the nature of the broader policy framework and latterly on the regulatory structures and styles associated with IPC. In relation to the structures of regulation the analysis assessed the significance of different legal, institutional and resource structures. In relation to the styles of regulation the analysis considered the importance of different design, delivery and enforcement styles. The bottom-up analysis focused particularly on the conditions for innovation in regulated companies, the nature of any innovative activities that had taken place and the significance of IPC as one of a wide range of possible influences stimulating such innovation. The qualitative data collected in these interviews was subsequently transcribed from tape and coded according to the innovation categories as developed above.

By design the main objective of IPC is to prevent, minimise and, as the last option, render harmless emissions of prescribed substances from prescribed industrial processes. It applies to those processes and substances that represent the greatest immediate threat to the environment or human health. The operators of these processes have had to apply for IPC authorisations at specific times from 1991, with the final round of authorisations being applied for at the start of 1996. To achieve its aims, IPC employs two important principles. First, the 'best practicable environmental option' (BPEO) for emissions should determine their destination. In other words, following prevention and minimisation, the most appropriate media (air, water or land) for a given pollutant should be chosen. Second, the 'best available techniques not entailing excessive cost' (BATNEEC) should be applied to regulated processes. Consequently, IPC focuses on the production process, it seeks to promote the avoidance rather than the treatment of pollution and it attempts to encourage the wider uptake of the most effective technologies and techniques, within economic constraints, that are determined on a case-by-case (or sector-by-sector) basis. From its inception in 1990, responsibility for the implementation of IPC was delegated to Her Majesty's Inspectorate of Pollution (HMIP). HMIP had been established in 1987 as a combination of a number of pre-existing central government pollution inspectorates and was responsible for implementing IPC regulations between 1990 and 1996 when it became part of the Environment Agency in a move toward further integration of regulatory structures.

From this brief description it is clear that from the mid-1980s the institutional and legal structures associated with industrial environmental regulation in the UK began to be integrated. However, apart from the praise that has accompanied these structural changes the implementation of IPC has caused some concern (Jordan, 1993; Ghazi and Grant, 1995; ENDS, 1996; Smith, 1997). Three broad issues are worth mentioning before implementation and impact of IPC is discussed in detail. First, it has been suggested that the resources allocated by central government for the implementation of IPC have not been sufficient to allow effective regulation. According to various authors an insufficient number of pollution inspectors has been particularly problematic. Second, concern has been voiced, particularly by industry, that IPC is expensive legislation to implement. One reason for this is that because the staff recruited to implement IPC typically have 10 years or more prior industrial experience they receive salaries which are significantly above what would normally be associated with the role of a pollution inspector for regulating a comparatively small number of processes. Combined with this is the statutory requirement for the costs of IPC to be recovered from regulated companies.

Third, and finally, it has also been argued that regulatory capture has been associated with the implementation of IPC. The industrial backgrounds of the inspectors, the consensus based and conciliatory relationship between the regulator and the regulated, and particularly the asymmetry of knowledge between these two have been highlighted as significant in this respect. And, as a result, the view has emerged that the imperatives for environmental improvement introduced by IPC are weak and that industry sets both the pace and the direction of environmental improvement (Smith, 1997). With these concerns in mind it is now possible to examine the impact of IPC against the three dimensions of innovation outlined above, drawing on the interviews conducted.

The implementation and impact of IPC

With respect to the technological and organisational impacts of IPC both inspectors and industry perceive it to be evenly divided between changes that are primarily technological in nature and those that are primarily organisational. However, the balance between technological and organisational change demanded or encouraged by the regulator and pursued by industry is heavily influenced by the nature of the regulated plant in question. Broadly, emphasis is commonly placed on the introduction of new technologies for new plant and for plant approaching a major upgrade, whilst during normal operations emphasis is placed on developing organisational improvements.

Where IPC conditions require technological solutions inspectors suggest that end-of-pipe control options, as opposed to cleaner and more integrated solutions, are still the first to be proposed by industry. The explanation for this centres largely on the short-term expedience of end-of-pipe control technologies, which may be the result of a variety of pressures as described by the environmental manager of a large chemical plant:

> Yes, end-of-pipe solutions still creep in. They are easy to plan for and produce quick solutions. It shows that you're doing something and you can cost for it. Of course, in the longer term they may not be desirable.

However, the position of the regulator on this issue is clear, as one IPC inspector explained:

> We would always encourage companies to improve the way they are carrying on the process to meet whatever emission standards we require. But if they can't do that, then they may have to fit abatement technologies. But the preferred way is to have them minimise at source and have the minimum of abatement equipment necessary.

Although in the early stages of IPC the preferred short-term response of industry to regulation was the installation of end-of-pipe technologies, industry increasingly accepts that the application of integrated cleaner technologies may be preferable, both economically and environmentally, at least in the medium- to long-term. An example of this increasing realisation is provided by the following quote from the site environmental manager of a multinational chemical company:

> Well, we have deliberately moved away from end-of-pipe whenever we can. Even as recently as 1992 we were moving quite swiftly toward an end-of-pipe effluent treatment plant, but it is not an environmental

solution, particularly for our type of effluent and our location. It would just turn organic waste into sewage sludge and make it an environmental problem in some other media. We very deliberately and very publicly decided that wasn't the solution and instead would try waste minimisation. We got full backing from the Inspector. Unfortunately it means we do not have an overnight success, and now we will have to manage maybe 30 or 40 small projects with all their problems. But in the long run it is a better option.

These examples confirm the economic and environmental implications associated with the selection of different forms of technological innovation outlined above, namely that although integrated clean technology responses may be less expedient in the short-term, in the medium- to long-term they commonly offer a range of economic and environmental benefits relative to responses based on the application of control technologies. Economically the range of relative and in some cases absolute benefits associated with investments in clean technology can be significant. The benefits can either relate to tangible reductions in cost or to less tangible improvements in product quality as illustrated by the following extract from an interview with the group environmental manager of a multinational company:

> We were producing one product, and not much of it, but we had to do a recrystallisation from methanol. This was not very good, there were significant wastes associated, the yields were low and there were significant losses to atmosphere. We have been looking around for ways to obviate the use of methanol and in the last 2 or 3 yr we have found a way. We have now gone on to a new process using water based technology. The yield is significantly higher, there is virtually zero waste coming out of the process, we are not using methanol at all so we do not have the same emissions to air. Also, it has enabled us to instead of having an oven dry product to go for a spray dry product which our customers love.

Consequently IPC has clearly played a role in encouraging firms to examine the potential for the application of integrated clean technologies before automatically accepting the need for an end-of-pipe response to environmental problems. Those firms that have developed and applied clean technologies as a response to IPC regulations have commonly achieved the necessary improvements in environmental performance at a lower cost than those which would be incurred through investments in end-of-pipe technologies. In some cases the improvements in process efficiency or product quality that have been realised as a consequence of the investment in clean technologies have secured economic benefits which have been in excess of associated costs. As a result many of the companies regulated by IPC are satisfied with the overall costs of compliance as for the moment they are recovering costs or securing benefits as a result of the changes made.

As well as these technological impacts, IPC has also stimulated a range of organisational impacts, involving changes in management or working practices.

Technological and organisational changes have commonly been required in combination to ensure that the technological responses to regulation are successfully integrated into existing systems. However, despite their combined importance, it is still
possible to identify explicitly organisational innovations which have occurred as a
result of the IPC regime. These are equally important, and in some cases more
important, than the technological solutions upon which attention commonly
focuses.

In general IPC has improved awareness of environmental issues in regulated
companies, largely as a result of the interaction between the company and the
regulator. Evidence suggests that a critical issue in this respect is the experience of
the inspector, including not only their formal knowledge and expertise but also
their tacit understanding of the issues affecting the environmental performance of
an industrial facility. Although the inspector may not know more about a particular
process than the operator, IPC has facilitated the transfer of information and understanding to and between regulated companies. Clearly the background and experience of the inspectors who implement IPC reduces the asymmetries of information
to some degree, as does the intensive approach to implementation as each inspector
typically regulates around 10 industrial processes in similar industrial sectors.
Through their background and experience, and through regular interaction with a
relatively small number of processes and operators, the regulators understanding
not necessarily of the processes themselves but of the efficacy of the various
improvement options (particularly those applied by similar processes in the sector)
becomes significant.

As regulated companies have interacted with the regulator and assimilate
information and understanding, their ability to search for and apply new technologies and techniques has increased. In many instances the need to search for and
apply new technologies and techniques has been sustained by the regular presence
of an inspector. Their willingness to do so has been increased by a heightened
awareness of the potential benefits of some improvement options. One of the
most significant outcomes of this is that regulated companies have been increasingly
willing and able to respond to IPC by changing the organisation and management
of their operations rather than by developing or installing new technologies.
This change in emphasis from the technological to the organisational has been
encouraged by the regulator. In the words of one inspector:

> I think we have been getting the operators to improve their techniques
> in the soft sense rather than in a technological sense. Also to look at
> minimisation and look at prevention, look back through their data on
> how the process builds up, improve their training and management
> structures, improve their software techniques.

From the industry perspective placing the emphasis as much on the techniques
of environmental management as on the technologies has had a considerable
impact. In most cases regulated companies, with the help of the regulator, have
identified areas for environmental improvement which entail almost no investment
and improve both environmental and economic performance. The environmental
manager of one large manufacturing facility stated that:

> We've had success in terms of environmental improvements from sitting down and thinking and improving existing processes, by reducing the amount of product changes . . . People are prepared to sit and think about what they do and there has been success because of it.

Another environmental manager explained the impact of IPC as follows:

> To be fair, when we started off the whole object was to comply with the legislation. As it turned out that has almost become secondary because we gained so much from it . . . We have saved a lot of money from environmental measures. In fact very few of the environmental measures we have put in, apart from one, have cost us money. The pay back period has been remarkable, some have cost us nothing and we have saved thousands of pounds. The staff have got little projects going all over the site now. Most of them are very simple operating conditions. Instead of putting that drainage down there if it goes down there we can use it again. If we turn that pipe around it won't go into the river and we can reuse it. Regulation was the kick start for it really.

Thus, in many instances regulated companies have not had to invest in relatively expensive technologies in order to improve their environmental performance and comply with IPC. Many improvements have been derived from managerial and organisational changes, for instance from changes to operating instructions, instrumentation, maintenance and general working practices. However, in many cases regulated companies acknowledge that, despite their economic and environmental benefits, relatively simple organisational changes would not have taken place without the initial stimulus of IPC. While the imperatives for environmental improvement were rarely cited as a significant influence on industrial behaviour, the sustained process of interaction with a relatively knowledgeable inspector was. In this respect, the regular presence of an inspector on site was seen to maintain the search for improvement options, to transfer information and understanding particularly related to the experiences of other similar processes and initiatives and to raise the capacity of companies to respond to IPC in a relatively effective and efficient way.

However, although IPC has clearly stimulated both technological and organisational change in regulated companies, it has yet to have a significant influence on the strategic decision making agenda in those companies and thus the evidence of radical change is scarce. In part, this is due to the fact that by nature the IPC regulations seek to enact a process (i.e., to ensure that companies apply BATNEEC) rather to achieve particular environmental objectives. Thus, the regulatory agency itself has lacked a strategic focus with regard to the objectives of IPC, although this has been corrected to a degree recently by the introduction of various strategies and plans to guide the activities of the Environment Agency. This lack of strategic direction has had a tangible impact on regulated companies as the regulator has not introduced medium- to long-term targets for improvement for regulated companies. Without a comprehensive framework of targets individual inspectors regularly identified that they have encountered difficulties in ensuring that the environment influences the strategic agenda of industry. This was explained by one inspector as follows:

> Our focus is operational as it should be. But it should also be strategic shouldn't it. But for that to happen we need a strategy to work within and I'm not sure we are getting that from government. It certainly isn't clear to me what role IPC plays within a wider approach to managing the environment. It isn't clear what IPC is ultimately aiming at.

In this instance a clear link is being made between IPC and the broader policy framework as it is established in the UK in documents such as This Common Inheritance (HMG, 1990a,b) and Sustainable Development: The UK's Strategy (HMG, 1994). However, for new processes or for processes that are being significantly redesigned, the regulator is becoming increasingly involved at the design stage to ensure that the opportunities for environmental improvement are exploited effectively when there is flexibility to explore the various options. In such instances the costs of the additional inputs into the design stage are typically more than offset by significant improvements in the resource efficiency and therefore the environmental and economic performance of the process once it becomes operational.

Discussion and analysis

In the above discussion it was shown that although ecological modernisation has been examined from a number of important perspectives the debate has not yet been informed by a detailed understanding of the innovation process and its outputs. This is significant because one of the most important tenets of ecological modernisation concerns the ability of regulation to encourage innovations which simultaneously reduce the impact of industry on the environment and make it more economically competitive. The subsequent review of innovation theory highlighted a number of important conceptual and theoretical issues, particularly relating to the barriers to innovation that commonly prevent the development and application of new technologies and techniques, whatever their longer-term potential. It also indicated that the impact of regulation could be assessed in terms of its influence on the technological, organisational and strategic dimensions of innovation in regulated companies. As a conclusion to the theoretical discussion it was suggested that one justification for environmental regulation could be its use in helping companies to overcome the barriers to innovation.

The empirical section examined the ability of industrial environmental regulation in the form of IPC to stimulate innovation. More particularly it assessed the ability of IPC to encourage companies to explore the economic and environmental potential of cleaner technologies, of organisational changes and of the integration of environmental concerns into strategic decision making. The strategic dimension is particularly important because the discussion of innovation theory suggested that achieving both environmental and economic benefits from innovation over the longer-term is only possible if, periodically, there is a radical change which then allows for further incremental improvements to be made.

In relation to the influence of industrial environmental regulation on the development and application of new technologies, IPC appears to be playing a role in persuading regulated companies to consider clean technology options before simply opting for control solutions, despite the lack of imperatives for environmental improvement established by authors such as Smith (1997). In part this is a consequence of the design of IPC, namely that it is integrated, process focused and anticipatory legislation which seeks to encourage prevention before control. However, it is also a consequence of the approach to implementation which depends upon expert inspectors that deliver the legislation in a relatively intensive and co-operative way through a close working relationship with regulated companies. Although it is not suggested that these inspectors necessarily know more about regulated processes than the operators they are still able to improve the capacity of regulated companies to respond to the demands of regulation in a more effective and efficient way. In essence the capacity for innovation in regulated companies is enhanced as a result of their relationship with the regulator. This is the case because the inspector can transfer information to and between companies, often has more knowledge of alternatives and solutions, or may simply encourage the company to think more broadly about its environmental impacts and ways to address them.

In relation to the organisational impacts of IPC, it is apparent that the legislation has many features which allow it to promote managerial and organisational solutions to environmental problems. In many cases these changes negate the need to invest in either control or clean technologies. This is particularly the case with respect to the adoption of environmental management systems. Perhaps more importantly, however, these organisational changes appear to be creating conditions in regulated companies that are conducive to innovation, particularly where new technologies and techniques have to be integrated into existing systems. The ability of IPC to promote organisational change is linked only partially to the design of the legislation itself. Once again it is also determined by the expertise of the staff involved and the opportunities presented to them to interact with regulated companies on a regular basis.

Finally, in relation to the impact of IPC on strategic decision making in regulated companies it is apparent that it is not associated with clear and consistent medium to long-term targets for regulated companies to work towards it in a strategic way. As a result, IPC so far has not consistently promoted the integration of environmental concerns into the strategic decision making agendas of regulated companies. One important outcome of this is that regulated companies are more likely to make incremental changes to existing approaches than they are to explore the potential of more radical change. However, although this focus on operational issues has resulted in improved performance to some degree, eventually diminishing marginal environmental and economic returns are likely to be encountered as the opportunities for incremental improvement of existing processes are exploited.

Thus, the results of this research suggests that the strengths of IPC stem from the sustained interaction with a relatively expert regulator involved in the building of environmental capacity in regulated firms, rather than from the imposition of imperatives for environmental improvement. The associated beneficial impacts are on the whole felt in the short- to medium-term, particularly where cleaner technologies and organisational changes have been considered. This has commonly resulted in improvements in both the economic and the environmental perform-ance of regulated companies (quantitative economic and environmental data to support this conclusion is given in HMIP and Allied Colloids (1996) and HMIP and Business in the Environment (1996)). However, the weakness of IPC is that it is not delivered within a framework of demanding targets and thus the environment has not been established as a strategic concern in industry. Consequently the imperative for improvement has not been clearly established. By drawing on the theory of innovation the problem with this becomes clear. As the opportunities for incre-mental improvements to existing production processes decline, escalating costs are likely to be associated with further environmental gains. Therefore, although it has promoted some combined economic and environmental improvements in the short- to medium-term, in the absence of a regulator that can impose challenging imperatives for environmental improvement it seems unlikely that IPC will pro-mote the radical innovations that must be associated with ecological modernisation in the longer-term.

With this conclusion in mind it is useful to reflect again on some of the concerns raised about IPC by other authors. With respect to the costs of IPC it is clear that the financial gains made by industry as a result of the legislation have commonly not been included when the burden of the regulation has been calcu-lated. In the past groups such as the Chemical Industry Association have been particularly at fault here. Related to this evidence presented by Smith (1997) and supported by Gouldson and Murphy (1998) does show that the resources allocated for the implementation of IPC were not sufficient, particularly with respect to the numbers of pollution inspectors. This can be explained with reference to the broader political context of the creation of HMIP, the implementation of IPC and then the development of the Environment Agency. All this was overseen by a government that was committed to rationalising the involvement of the state in the market and who assumed that environmental regulation was a threat to com-petitiveness (see Weale, 1992). The evidence presented here does not support this conclusion.

Finally, with respect to the subject of regulatory capture – which it is claimed occurs as a result of the industrial background of inspectors, the consensus and conciliatory nature of their relationship with companies, and the asymmetry of knowledge between them and the regulated – the picture is complicated. It is clear that much work, and particularly that of Smith (1997), has criticised the implemen-tation of IPC because it has failed to establish the imperative for environmental improvement. This has been explained as an example of regulatory capture. How-ever, this has been done without also considering the capacity building function of regulation. Approaching regulation from an innovation perspective requires that the network of relationships and the transfer of knowledge throughout this network is considered and as a result capacity building is brought into focus as an important subject. Thus the analysis presented in this paper can be seen as a useful addition and contrast to other interpretations of the implementation of IPC. Although it does not deny the importance of establishing the imperative for environmental improvement, particularly in the longer-term, it also emphasises the value of cap-acity building through regulation, something that occurs through a consensus based and cooperative relationship between the regulator and the regulated.

Conclusion

Ecological modernisation theory describes a way of addressing some of the environmental problems associated with industrialism whilst at the same time improving economic competitiveness. The discussion of innovation above has shown that clean technologies, organisational change and in the longer-term the development and wider adoption of radical innovations is the only way in practice to realise the goal of ecological modernisation. However, the barriers to innovation of this kind in industry are considerable. In fact, standard operating conditions in industry often promote control technologies, limited organisational change and a focus on operational issues. Consequently, as suggested by ecological modernisation theory, there is a role for environmental regulation in forcing or facilitating the adoption of innovations which, despite being environmentally and economically beneficial in the medium- to long-term, are unlikely to be adopted. However, the potential for regulation in this area is two-fold, it can establish the imperative for improvement whilst at the same time improving the capacity of industry to respond to that imperative.

From this perspective the implementation of IPC in England and Wales is an interesting case. The strengths of IPC are that it is process focused and anticipatory legislation which emphasises pollution prevention before control. It involves the application of qualitative principles which allow the legislation to be fine-tuned to acknowledge the specifities of individual industrial sites. It has also been implemented to date by relatively expert pollution inspectors working in a conciliatory and consensus building way with regulated companies, where this has been possible. These elements have allowed the pollution inspectors to enhance the capacities of regulated companies through the transfer of knowledge and to some extent to establish the imperative for action on environmental issues.

However, the analysis has also highlighted the main weaknesses of IPC. The legislation is not delivered within a strict framework of targets, which in combination with the influence that industry can exert through negotiation with the regulator, reduces the extent to which imperatives for environmental improvement are introduced. In particular the legislation has failed to establish environmental issues as a strategic concern of industry and one outcome of this is likely to be that the potential of the legislation to promote radical innovations is not maximised. Thus, although the design and implementation of IPC in its current form has a number of significant strengths, from the perspective of its impact on innovation it cannot be considered as an example of ecologically modern regulation. Linked to this is the problem that the implementation of IPC in the UK can be characterised as mainly, if not exclusively regulating to build the environmental capacity of firms. Equivalent attention has not been given to the equally important issue of regulating to build public confidence and trust.

More broadly it is useful to conclude by raising some doubts concerning ecological modernisation as a useful guide over anything other than the short- to medium-term. It has been suggested that even if environmental regulation can establish the imperative for action at the strategic level in industry, and that this does result periodically in radical innovations, there may in the end be structural limits which make it impossible to continually realise combined economic and environmental improvements as a result of innovation. This argument raises doubts about the technocentric and overly optimistic nature of the core ecological modernisation ideas and more research is needed in this area. However, taken together with other concerns, such as the way that the theory divorces social justice issues from environmental issues and its focus exclusively on the environmental problems of advanced industrial countries, there must be serious doubts about ecological modernisation as a 'solution' to environmental problems over the longer-term.

Acknowledgements

We would like to thank Jenny Robinson and two anonymous referees for their helpful comments on an earlier draft of this paper. The research reported in this paper was carried out with the support of the UK Economic and Social Research Council under grant no. L323253015 'Research and Development in Environmental Technology: Assessing the Role of EU Regulatory Frameworks'.

References

Beck, U., 1991. *Risk Society: Towards a New Modernity*. Sage, London.

Beck, U., 1995. *Ecological Politics in an Age of Risk*. Polity, Cambridge.

CEC – Commission of the European Communities, 1993. *Growth, Competitiveness and Employment*, Office for Official Publications of the European Communities, Luxembourg.

Christoff, P., 1996. Ecological modernisation, ecological modernities. *Environmental Politics* 5 (3), 476–500.

Dosi, G., 1988. The nature of the innovative process. In: Dosi, G., Freeman, C., Nelson, R., Silverberg, G., Soete, L. (Eds.), *Technical Change and Economic Theory*, Pinter Publishers, London, pp. 221–237.

DTI/DoE – Department of Trade and Industry/Department of the Environment, 1994. *The UK Environmental Industry: Succeeding in the Changing Global Market*. Her Majesty's Stationery Office, London.

ENDS – Environmental Data Services, 1996. *HMIP urged to find incentives for industry to improve*. The ENDS Report 252, January, 34.

Ghazi, P., Grant, I., 1995. Polluters left to report own spills. *The Observer*, September 17, 8.

Giddens, A., 1991. *The Consequences of Modernity*. Polity Press, Cambridge.

Gouldson, A., Murphy, J., 1997. Ecological modernisation: economic restructuring and the environment. *The Political Quarterly* 68 (5), 74–86.

Gouldson, A., Murphy, J., 1998. *Regulatory Realities: The Implementation and Impact of Industrial Environmental Policy*. Earthscan, London.

Haigh, N., Irwin, F. (Eds.), 1990. *Integrated Pollution Control in Europe and North America*. Institute for European Environmental Policy, London.

Hajer, M., 1995. *The Politics of Environmental Discourse: Ecological modernisation and the Policy Process*. Oxford University Press, Oxford.

Hajer, M., 1996. Ecological modernisation as cultural politics. In: Lash, S., Szerszynski,

B., Wynne, B. (Eds.), *Risk, Environment and Modernity: Towards a New Ecology*, Sage Publications, London, pp. 246–268.

HMG – Her Majesty's Government, 1990a. *This Common Inheritance*. HMSO, London.

HMG – Her Majesty's Government, 1990b. *This Common Inheritance: A Summary of the white paper on the environment*. HMSO, London.

HMG – Her Majesty's Government, 1994. *Sustainable Development: The UK Strategy*. HMSO, London.

HMIP – Her Majesty's Inspectorate of Pollution – and Allied Colloids, 1996. *3Es Project Concluding Report*. HMIP, London.

HMIP – Her Majesty's Inspectorate of Pollution and Business in the Environment. 1996. *Profiting From Pollution Prevention: the 3Es Methodology*. HMIP, London.

Huber, J., 1982. *Die Verlorene Unschuld der Ökologie: Neue Technologien und Superindustrielle Entwicklung*. (The Lost Innocence of Ecology: New Technologies and Superindustrialized Development.) Fisher Verlag, Frankfurt am Main.

Huber, J., 1984. *Die Zwei Gesichter der Arbeit: Ungenutzte Möglichkeiten der Dualwirtschaft*. (The Two Faces of Labour: Unused Possibilities of the Dual Economy.) Fisher Verlag, Frankfurt am Main.

Huber, J., 1985. *Die Regenbogengesellschaft: Ökologie und Sozialpolitik*. (The Rainbow Society: Ecology and Social Politics.) Fisher Verlag, Frankfurt am Main.

Jänicke, M., 1985. *Preventive environmental policy as ecological modernisation and structural policy*, Discussion Paper IIUG dp 85–2, Internationales Institut Für Umwelt und Gesellschaft, Wissenschaftszentrum Berlin Für Sozialforschung (WZB).

Jänicke, M., Mönch, H., Ranneberg, T., Simnois, U., 1988. *Economic structure and environmental impact: empirical evidence on thirtyone countries in east and west*, Working Paper FS II 88–402, Internationales Institut Für Umwelt und Gesellschaft, Wissenschaftszentrum Berlin Für Sozialforschung (WZB).

Jänicke, M., Monch, H., Ranneburg, T., Simonis, U., 1989. Economic structure and environmental impacts: east west comparisons. *The Environmentalist* 9(3).

Jordan, A., 1993. Integrated pollution control and the evolving style and structure of environmental regulation in the UK. *Environmental Politics* 2 (3), 405–427.

Kemp, R., 1993. An economic analysis of cleaner technology: theory and evidence. In: Fischer, K., Schot, J. (Eds.), *Environmental Strategies for Industry: International Perspectives on Research Needs and Policy Implications*, Island Press, Washington, pp. 79–116.

Lash, S., Szerszynski, B., Wynne, B., 1996. *Risk, Environment and Modernity: Towards A New Ecology*. Sage Publications, London.

Lipsky, M., 1980. *Street Level Bureaucracy: Dilemmas of the Individual in Public Services*. Russell Sage Foundation, New York.

Mol, A., 1995. The Refinement of Production: Ecological Modernization Theory and the Chemical Industry. Van Arkel, Utrecht.

Mol, A., 1997. Ecological modernization: industrial transformations and environmental reform. In: Redclift, M., Woodgate, G. (Eds.), *The International Handbook of Environmental Sociology*, Edward Elgar, Glos, pp. 138–149.

OECD – Organisation for Economic Cooperation and Development, 1987. *The Promotion and Diffusion of Clean Technologies*. OECD, Paris.

OECD – Organisation for Economic Cooperation and Development. 1992. *Technology and the Economy: The Key Relationship*. OECD, Paris.

Sabatier, P., 1986. Top-down and bottom-up approaches to implementation research: a critical analysis and suggested synthesis. *Journal of Public Policy* 6, 21–48.

Simonis, U., 1989a. Ecological modernization of industrial society: three strategic elements. *International Social Science Journal* 121, 347–361.

Simonis, U., 1989b. *Industrial restructuring for sustainable development: three points of departure*, Working Paper FS II 89–401, Internationales Institut Für Umwelt und Gesellschaft, Wissenschaftszentrum Berlin Für Sozialforschung (WZB).

Smith, A., 1997. *Integrated Pollution Control*. Ashgate, Aldershot.

Soete, L., Arundel, A., 1995. European innovation policy for environmentally sustainable development: application of a systems model of technical change. *Journal of Public Policy* 2 (2), 285–385.

Weale, A., 1992. *The New Politics of Pollution*. Manchester University Press, Manchester.

Chapter 16

Kris van Koppen and Arthur P.J. Mol

ECOLOGICAL MODERNIZATION OF INDUSTRIAL ECOSYSTEMS

Introduction

FEW WOULD DOUBT THAT industrial ecology is rapidly developing into a significant field of study around the world. If the *Journal of Industrial Ecology*, published by the MIT Press since 1997, is taken as a point of reference it becomes clear that increasing numbers of case studies parallel various more theoretical informed elaborations on topics as diverse as industrial metabolism, Factor X, dematerialization, design for environment, eco-industrial parks, technological transition and industrial symbiosis. Increasingly, industrial ecology incorporates longer standing academic traditions such as those in life cycle analysis and design, environmental management systems, cleaner production and ecoefficiency. By that industrial ecology becomes for some a common denominator of the wide area of studies that focus on the interrelationship between industrial production (increasingly extended to consumption) and the natural environment. For others, however, the idea and concept of industrial ecology shows especially its usefulness in more restricted studies on (eco)industrial estates.

No matter how wide or narrow one interpretes industrial ecology, most scholars would not object to the observation that industrial ecology studies, and the various related academic traditions, have proved to be especially strong in engineering and environmental sciences. Although a recent textbook claims that "important disciplines contributing to industrial ecology include physical and biological sciences, engineering, economics, law, anthropology, policy studies and business studies" (Allenby 1999), and most would agree on the necessity of all these disciplines for industrial ecology, the practice of industrial ecology studies proves otherwise. Of central attention to industrial ecology scholars and studies are the substance flows between the natural environment and various units of the production-consumption system, as well as the technological changes necessary to bring the two more in 'harmony'. Mapping out the relations within the industrial production system and between that system and the environment is primarily describing and analyzing the 'material' interactions, the "additions and withdrawals" (Schnaiberg 1980). In making such analyses of material flows and suggesting primarily technological solutions, the social, institutional, organizational and policy dimensions that parallel the substance flows remain often undertheorised. As Jackson and Clift (1998) rightly state in their own words:

> (I)ndustrial ecology requires what sociologists would call a 'theory of agency'. That is, it needs to have some idea who the actors in the industrial ecology are, and what motivates their actions. (. . .) Indeed, the whole success of the industrial ecology metaphor depends crucially on its assumptions about agency.

In line with this observation, we aim in this contribution to extend the more technological/material flow dimensions of industrial ecology with an actor and institutional perspective. Such a perspective or theory provides us the tools to analyse which actors and institutions (can) play a vital role in putting the concept of industrial ecology to work. In this contribution, we will develop a more actor and institutional perspective on industrial ecology by starting from an ecological modernization perspective, and operationalising this social theory via network models of interdependent actors. The proof of the pudding of such a perspective is of course in the eating: what does such a perspective offers us in analysing concrete situations of greening clusters of industries? We will use our actor and institutional perspective to analyse the roles of various actors in different stages and configurations of eco-industrial parks and draw some conclusions on the future agenda of both industrial park environmental management and industrial ecology studies.

Industrial ecology as a concept of industrial transformation

The emergence of industrial ecology

Ideas of industrial ecology have been around for some 30 years, but until the 1990s they had little resonance in society – with the notable exception of Japan, where early industrial ecology studies had significant impacts on industrial policy (Erkman 1997). The more recent flowering of the concept in academic and business circles goes back to Frosch and Gallopoulos' article in the *Scientific American* special issue on 'Managing the Earth' (1989). Inspired by this article Tibbs (1991), a consultant of Arthur D. Little Inc., wrote 'Industrial ecology: A New Environmental Agenda for Industry', which spread the idea further among industry. This set the stage for an increasing number of publications. Côté *et al.* (1994), Graedel and Allenby (1995), Lowe *et al.* (1995; 1997), and Ayres and Ayres (1996) are just a few names that are often encountered in Industrial Ecology literature. A growing stream of articles on industrial ecology has appeared in several environmental and economic journals, among which, since 1997, the *Journal of Industrial Ecology*.

The core idea of industrial ecology is to study the industrial system from an ecosystem angle. This perspective basically involves two starting points. First, the industrial system itself should be interpreted and analysed as a particular system with a distribution of materials, energy and information flows (not unlike the ecosystem). Second, the industrial system relies on resources and services provided by the biosphere. Both the flows within the industrial system and between the system and the biosphere have to be optimised from a closed-loop perspective, as is exemplified by natural ecosystems. Several authors take the ecological analogy further still, and seek to apply principles from biological processes to industrial processes. Examples are the use of solar power as sole energy source; the recycling of rest materials; the local concentration of toxic substance generation and use; and the application of decentralised, self-organising processes (see references in Lowe 1997). But, as most industrial ecologists would agree, as helpful and inspiring such analogies may be, they cannot be extended to all aspects of industrial processes (e.g. Lowe 1997; Boons and Baas 1997).

As a review of industrial ecology literature learns, many of the environmental management concepts that have been developed independently in the last decades increasingly become subsumed under the umbrella of industrial ecology: pollution prevention, environmental management systems, new approaches in environmental policy, life cycle assessment and life cycle management, design for the environment, and others. It is this potential for an integrative approach that is characteristic for industrial ecology and explains some of the popularity of the concept. Notwithstanding the diversity in underlying concepts, there is a fair consensus on the demarcation of industrial ecology and the related concept of industrial ecosystems.

Industrial ecosystems are characterised as communities of companies and other organisations that collaborate to achieve enhanced environmental and economic performance – in other words, greater eco-efficiency. Another characteristic is the spatial setting. In the case of *eco-industrial parks*, the communities are mainly situated within an industrial estate or an industrial area. In other cases of industrial ecosystems, the geographical setting is wider, covering a municipality, or a region.

Industrial ecology is regarded as a broader concept, encompassing industrial ecosystems as well as approaches that pursue environmental improvement along the lines of life cycle management, design for environment (or eco-design), and service economy (e.g. Gertler 1995; Lowe 1997). Thus, industrial ecology studies industrial systems in two main settings: the setting of industrial ecosystems and the setting of product chains, or – as economists would have it – value chains. In the ecosystem setting, eco-industrial parks are the main direction of industrial ecology research; in the product chain setting the main direction is that of dematerialisation and service economy (Erkman 1997).

Industrial metabolism, a concept pioneered by Ayres (1989), focuses on the analysis of material flows in systems of production and consumption. It therefore may be regarded as one of the tools of industrial ecology.

In reviewing the industrial ecology literature it should not only be concluded that eco-industrial parks have received most attention of industrial ecologists. It also becomes clear that most of industrial ecology's more practical efforts have been dedicated to eco-industrial parks. As Lowe *et al.* (1997) state: "The concepts of industrial ecosystems and eco-industrial parks (EIPs) have embodied the industrial ecology approach in very concrete terms. The story of Kalundborg, an industrial ecosystem identified in Denmark, has circled the planet as a sign that industry can coexist with nature in a more benign manner." In developing and illustrating an actor and institutional perspective within the industrial ecology tradition, this chapter also starts with a focus on eco-industrial parks. As we will see, however, when tracing the variation in developments of eco-industrial parks, the product chain perspective will eventually emerge again.

Eco-industrial parks

The Fieldbook for the Development of Eco-Industrial Parks (Lowe *et al.* 1995) defines an eco-industrial park as follows:

> A community of manufacturing and service businesses seeking enhanced environmental and economic performance through collaboration in managing environmental and resource issues. By working together, the community of businesses seeks a collective benefit that is greater than the sum of the individual benefits each company would realise if it optimised its individual performance only. The goal of an EIP is to improve the economic performance of the participating companies while minimizing their environmental impact. Components of this approach include new or retrofitted design of park infrastructure and plants, pollution prevention, energy efficiency, and inter-company partnering. Through collaboration, this community of companies becomes an 'industrial ecosystem'.

Inspired by industrial ecology, or independently of it, a wide variety of initiatives have developed that to some extent meet this description. Examples of eco-industrial parks that involve collaboration of several industries on different environmental aspects are:

- The "Industrial Park as an Ecosystem" project in Burnside Industrial Park, Nova Scotia, Canada is underway since 1992, and has among others resulted in an infrastructure for rest materials exchange (Côté *et al.* 1994).
- The INES (INdustrial EcoSystems) project was started in 1994 in the port of Rotterdam region, the Netherlands. Several potential options for environmental improvement were identified, among others in the areas of energy reuse and sustainable water use (INES 1997). Currently, a start has been made with implementing the most promising options, and a project organisation, INES Mainport, is established to further explore and disseminate industrial ecology potentials (INES 2001).
- The Houston Ship Channel is a large petrochemical complex, on each side of the channel to Houston, USA. Since long, industries in this area have exchanged by-products of their production activities, for instance chemical substances stripped in the purification of natural gas. Different from the projects above, which were initiated by industrial ecology researchers, this collaboration was mainly driven by economic motives (Lowe *et al.* 1997).

In addition to these examples, there is a large number of smaller projects with only a few companies involved and just one or two environmental aspects taken into consideration. Côté and Cohen-Rosenthal (1998) list several initiatives in the USA and Canada; for the Netherlands an overview is presented in Ministerie van Economische Zaken (1998). A well-known international activity is the Zero Emissions Research Initiative (ZERI), which has initiated several projects that combine different industrial processes to minimise environmental impacts. Projects where organic waste from beer brewing is reused for purposes of bread production, mushroom farming and fish farming, for example, are initiated in Namibia, Canada, Germany and Japan (ZERI 2001). Also worth mentioning are project studies on potential EIPs, such as the Brownsville/Matamaros study (Martin *et al.* 1996). A substantial part of the EIP examples quoted in literature belongs in fact to this kind of project studies.

By far the most well-known and elaborate example of an eco-industrial park is Kalundborg (see below). Kalundborg's outstanding case of industrial ecology, however, not only illustrates the possibility of realising an eco-industrial park, but also the difficulty of it. A decade of worldwide industrial ecology initiatives has brought us no other example that can compete. Why is that so? Several factors play a role, and a major one concerns the social and institutional aspects of eco-industrial parks. As Côté and Cohen-Rosenthal (1998) epitomise: "The lesson of Kalundborg is not found in mapping its pipes but in the unfolding of the existing relationships. What makes Kalundborg a model is that its participants allowed and continue to encourage interaction, not that it had a particularly spectacular technical breakthrough."

Industrial ecology in need for social theory

While the initial and more conventional industrial ecology investigations emphasise the physical relations between industrial 'organisms', several recent contributions call for attention to the social and organisational relations between industrial and non-industrial 'actors' in the development of EIPs (e.g. Lowe 1997; Boons and Baas 1997; Ruth 1998; Côté and Cohen-Rosenthal 1998; Jackson and Clift 1998). Most of these authors agree that especially the socio-institutional and actor dimensions have remained undertheorised in industrial ecology studies up till now. If anything, the role of the so-called 'social context' is often given credit in industrial ecology studies by a 'stakeholder' analysis: a poorly differentiated and often arbitrary listing of actors that in one way or the other relate to industries. Such stakeholder analyses often lack theoretical background, analytical clarity and a systematic analysis of the social dynamics, mechanisms and processes via which these actors interact in processes of environmental reform.

If we want to bring industrial ecology perspectives from the design table more into practice, it seems essential to further develop an actor and socio-institutional perspective of these kind of industrial transformations. In contributing to such a perspective and analytical framework at least two central issues are at stake. First, the co-operation of companies within a perspective of company strategy needs to be understood and clarified. The benefits of industrial ecology, and thus the motives for firms to step in, are often presented in terms of simultaneous environmental improvements and financial savings. Company environmental strategy, however, is more complex and dynamic than simple calculations of material and financial flows. Second, in addition to company strategies we need to obtain a better analytical understanding on the embeddedness of individual companies in their social environment, in order to clarify how and to what extent the interrelations between the eco-industrial park companies and their social environment do or do not trigger industrial ecology collaboration. It has increasingly been acknowledged that other actors than the 'businesses' of the Fieldbook definition have entered the stage of industrial ecology. Governments, residents, environmental organisations, universities, consumers, credit and insurance companies, and so on can no longer be perceived as merely contextual actors at the periphery of industrial ecology activities. Understanding their role, position and interdependencies seems essential in clarifying the successes and failures of eco-industrial parks.

In contributing to the development of what Jackson and Clift (1998) have called a 'theory of agency', but that we will call an actor and institutional perspective on industrial ecology, we follow three steps. First, a social theory that is according to us a useful overall framework for a socio-institutional and actor perspective on environment-informed industrial transformation is presented. Secondly, that theory will be operationalised using network perspectives on social continuity and change, much in line with and building upon initial ideas made by Côté and Cohen-Rosenthal (1998) and Boon and Baas (1997) on industrial networks. Thirdly, a categorisation on company environmental management strategy will be developed.

Agents and institutions in industrial transformation

Ecological modernization as social theory

One of the most well-known contemporary social theories which take industrial transformations among its core objects of reflection is ecological modernization theory (Spaargaren and Mol 1992; Mol 1995; Spaargaren 1997; Mol and Sonnenfeld 2000). Ecological modernization theory emerged in the late 1980s in reaction to the then dominant neo-Marxist, de-industrialist and neo-liberal perspectives on industry-environment relations. The first two theoretical perspectives denied that any significant environmental reform was possible under the present conditions of industrial capitalism. The latter perspective claimed that industrial capitalism would by its internal dynamic and logic automatically solve any environmental problem. By contrast, ecological modernization theory outlined the core dynamics, mechanisms, institutions and actors that could move the unsustainable industrial system into more environmentally sound directions. In the environmental reform of industrial systems, technological and economic institutions and actors play an important role, next to more conventional actors, such as states and environmental NGOs. A short introduction to the theoretical premises of ecological modernization will be given below.

Basic premise in ecological modernisation theory is the centripetal movement of ecological interests, ideas and considerations in social practices and institutional developments, which results in the constant ecological restructuring of modern societies. Ecological restructuring refers to the ecology-inspired and environment-induced processes of transformation and reform going on in the central institutions of modern society. Institutional restructuring should, of course, not be interpreted as a new phenomenon in modern societies, but rather as a continuous process that has accelerated in the phase that is often labelled late, reflexive or global modernity (cf. Beck *et al.* 1994). The present phase differs from the pre-1980s, however, in the increasing importance of environmental considerations in these institutional transformation processes in industrial societies.

Within ecological modernisation theory, this is conceptualised on an analytical level as the growing autonomy or independence of an ecological perspective and an ecological rationality vis-a-vis other perspectives and rationalities (cf. Mol 1995 and 1996; Spaargaren 1997). Sociological theories during at least the last century have often analysed the emergence of relatively independent economic, political and socio-cultural rationalities, logics and perspectives in the maturation of modern societies out of traditional societies (cf. Polanyi 1957; Giddens 1984). Different domains or systems in society (e.g. production and consumption; the political system; etc.) are governed by these distinct logics, distinct 'rules', distinct rationalities and distinct institutions. For instance, only from the emergence of capitalism the economic domain could be distinguished as a separate domain, with its own institutions, its own logic of profit making, its own rules of the economic game, etc. By claiming that a new phase is starting in the modernisation of (western) societies, ecological modernisation theory conceptualises the emergence of a new ecological perspective and rationality out of the existing ones. Within policies/politics and culture/ideologies some notable environment-informed transformations took already place in the 1970s and early 1980s. The construction of governmental organisations, departments and laws dealing with specifically environmental issues date from that era. While in the 1970s a separate green ideology – materialised in for instance environmental NGOs and environmental periodicals – started to emerge, it is especially in the 1980s that this ideology became more and more independent from and could no longer be interpreted in terms of the old political ideologies of socialism, liberalism and conservatism, as for instance Cotgrove (1982), Spaargaren and Mol (1992) and Giddens (1994) have emphasised.

But the crucial transformation, which makes the notion of growing autonomy of an ecological perspective, logic and rationality especially relevant, is of even more recent origin. After that the ecological perspective and rationality have become relatively independent in the political and socio-ideological domains in the 1970s and 1980s, this process of growing independence extends to the economic domain of production and consumption. That economic domain was up till than primarily governed by an economic logic and rationality. And as – according to most scholars – this economic domain proves especially consequential to 'the ecological question', this last step is the most decisive one. The consequence will be that – slowly but steadily – in the economic domain processes of production and consumption will be and are increasingly analysed and judged as well as designed and organised from an economic and an (relatively independent) ecological point of view. Some first profound institutional changes started to appear from the late 1980s onward that show the growing independence of an ecological rationality and logic in the economic domain of production and consumption. Among these changes are the emergence of environmental management systems and cleaner production, the introduction of economic valuations of environmental goods via among others eco-taxes, the use of environment-inspired liability and insurances, the increasing importance of environmental goals among public and private utility enterprises such as natural resource saving and recycling, the inclusion of the environment in the design of new products, and the articulation of environmental considerations in economic supply and demand. One easily recognises the various concepts and strategies that are put together under the heading of industrial ecology.

Two additional remarks are essential for a full understanding. First, a full and equal integration of environmental interests with other interests is only possible after that an ecological rationality, logic and perspective can be and is distinguished from other logics, rationalities and perspectives. When ecological perspectives and rationalities remain unarticulated, any integration will result in a subservience. Second, the fact that these transformations toward ecological independence should be analysed as *institutional* changes includes their semi-permanent character. Although the process of ecology-induced transformation should not be interpreted as linear and irreversible, the changes have some permanency and are difficult to reverse.

A (triad-)network approach to institutional change

Environment-informed industrial transformations, such as those analysed by eco-logical modernization theory and put on the agenda by industrial ecology, do not unfold automatically. To understand how individual companies as well as industrial estates take environmental considerations on board, we need to be more specific on the role various (industrial and non-industrial) actors play in distinct settings. Network models that relate industrial firms to the societal, economic and policy environment are useful in this. Some industrial ecology scholars have made signifi-cant steps in this direction especially with respect to what we will label industrial or economic networks (e.g. Boon and Baas 1997; Côté and Cohen-Rosenthal 1998). But Jackson and Clift (1998) rightly conclude that a theory of agency that is solely build on economic actors governed by economic rules and resources is unlikely to fulfil the need for a more sociological account of industrial ecology. We need to broaden our network model beyond a mere economic perspective.

These network models are useful for several reasons. First, they form an intermediary that links the rather abstract theoretical notions of ecological modern-ization with empirical developments in social systems of production and consump-tion. A second reason is of a more theoretical nature. On the 'continuum' of action theory and system theory, network models occupy a middle position and offer a more fruitful theoretical perspective than either of the two other theories. Both action theory or agency theory as Jackson and Clift (1998) call it and system theory have their limitations as conceptual models, as among others Giddens (1984) has emphasised in the development of his structuration theory. Brown (1992), when evaluating different strands of industrial sociology, underlines the importance of overcoming the dualism of action vs. structure, and concludes that a structuration theory perspective might be helpful to bring these different theoretical perspectives together. Network models, while focusing on institutional analysis, can be interpreted in line with a structuration perspective. Thirdly, the empirical scope of network models corresponds well with the main aim of industrial ecology and industrial transformation perspectives, which is to identify (and contribute to) institutional changes in relation to improvement of the environmental performance. Network models are able to encompass an institutional analysis, including the contribution of capable agents in bringing about such institutional changes in industrial systems.

In other words, network models have the advantage of combining both the structural properties of institutions and the interactions between actors constructing a network. Networks can be characterised as social systems in which actors engage in more or less permanent, institutionalised interactions. There appear several analytical network models in the literature, also with respect to environment-informed transformations in economic sectors. Most of them are rather constructed on an ad hoc basis, following directly from one or a limited number of empirical studies. A triad-network model (Mol 1995) helps us bridging the perspective of industrial ecology and eco-industrial park development on the one hand with the theory of ecological modernization on the other. As indicated above, the ecological modernization theory focuses on the growing independence, 'emancipation' or empowerment of the ecological perspective or sphere from the three basic analytical spheres or perspectives in modern society: political, economic and societal. The triad-network model continues in this line as it is a conceptual model for analysing the extent to which the ecological 'perspective' penetrates and transforms the social practices predominately governed by each of the three basic perspectives in modern society. It thus enables us to analyse to what extent the ecological perspective (and thus ecological interests, arguments, considerations, etc.) 'makes a difference' independently from the other three perspectives in practices such as industrial production.

In order to be able to do so the triad-network model combines these three analytical perspectives with three networks around given social practices: an economic, a policy and a societal network. Each of the three interdependent networks constitutes thus a combination of a specific analytical perspective, distinctive institutional arrangements and a restricted number of interacting (collective) actors, which are considered to be most important regarding that perspective. Applied for industrial transformation and industrial ecology, the following networks can be distinguished:

- Within policy networks, interactions and institutional arrangements between state organisations and industry are primarily governed by political-administrative rules and resources. Other interactions and institutional arrangements are also inspired by and can be analysed from this perspective, although they will then prove to be more peripheral. Policy network studies analyse the interdependencies between these actors, the 'rules of the game' which put these policy networks to work, the resources dependencies (regarding power, money, knowledge, information etc.) between the various actors and agents dominant in these policy networks, the common or diverging world view along which communication and joint strategies are developed or not, etc. There exists a considerable amount of literature (e.g. in neo-corporatism and policy community studies) that provides evidence of the usefulness of such analyses of the transformations and continuities in these interdependent networks actors (e.g. Grant *et al.* 1988; Marsh and Rhodes 1992; Smith 1993; Mol 1995).

• Economic networks basically focus on economic interactions via economic rules and resources between economic agents in and around the industrial park, chain or sector that form the object and unit of analysis. Although, the intellectual background of economic network analyses are mainly to be found in industrial organisation theory, institutional economics and organisational sociology, the basic concepts differ only partially from those in policy network analyses. Economic network studies analyse the relationships between the firms, the network structures in terms of power and resource dependencies, and the economic processes of continuity and transformation. They look at: (i) the vertical interactions from raw material producer up until the final consumer and beyond; (ii) the horizontal relations between competitors and on the level of the industrial branch association providing some collective interest representation; and (iii) regional relations and interactions in restricted geographical areas. The relations with conventional industrial ecology analyses of the material flows in geographical areas or production–consumption chains (e.g. LCA) are evident, but in the network studies the emphasis is rather on the non-material dimensions of the park/chain/ network (economic relations, power, information monopoly and exchange, knowledge, control, ownership, etc.). Håkansson and Johanson (1993) seem to apply a structuration theoretical approach in analysing industrial network and their constraints and opportunities in behaviour and change. Håkansson (1988), Martinelli (1991) and Grabher (1993) are other relevant volumes that provide valuable conceptual tools for analysing these economic networks in detail.

• Thirdly and finally, societal networks aim at identifying relations between the industrial park, chain or sector in question and civil society organisations and arrangements associated with what is usually called 'the life world', both directly and indirectly via state agencies. It is the rich tradition of (new) social movement research that provides the conceptual tools to analyse the interaction patterns, and their continuity and transformation, between on the one hand environmental and consumer organisations and on the other industrial firms.

It should be remembered that these three (interdependent) networks are only distinguished in order to conceptualise different mechanisms of and perspectives on how the social environment of industrial parks engages in industrial ecology initiatives. In reality, the actors and institutional arrangements found in the three different networks closely interact. Together they help us in understanding why industrial ecology initiatives in industrial parks are taken, whether and how these initiatives are successfully institutionalised and how such environmental reforms change the existing structures and arrangements. But these interdependent networks and mechanisms do not work in a similar way, nor with equal 'force' and not with comparable outcome. How in the end the 'social environment' influences industrial park cooperation on environmental issues is also determined by the individual companies and their intentions and strategies. This brings us to the last step in our analytical framework.

Stages of company environmental strategy

To complete our framework for analysing the social dynamics behind industrial ecology, a closer look has to be taken at company environmental strategies. The choice to collaborate in an eco-industrial park is basically or ultimately a matter of company environmental strategy. In the process of industrial transformation, as it is analysed by ecological modernisation theory, ecological rationality will play an increasing role in the companies strategy, affecting different aspects of the organisation. But not all companies are equally moved by, interested in or forced to adapt their company strategy along environmental lines. In taking these differences into account, a three-stage model of company environmental management has been developed by Van Koppen and Hagelaar (1996). This model (Table 16.1) identifies three distinct stages in company environmental management by looking at five main internal aspects (ambition, knowledge and information, organisation, technology, and budget allocation) and two aspects of the company's external relations (environmental risks and environmental opportunities). The model describes the environmental strategy development of a company as a dynamic process that depends on these different factors. Three more or less stable environmental strategy configurations (three stages) are identified: a crisis-oriented strategy, a process-oriented strategy, and a chain-oriented strategy.

These configurations may be interpreted as stages in the ecological modernisation of a company's strategy. This is not to suggest that all companies neatly fit into the model's categories, nor that the model predicts the development of companies in a mechanistic, evolutionary way. Company environmental management is a widely varying, company-specific process, in which individual actors can bring about decisive changes for better or for worse. The model draws attention to the different factors that may influence these changes, and to the patterns of coherence across these factors. Aiming at a green identity, for instance, is highly feasible when market opportunities exist and environmental risks are under control. For a company that has large inherent environmental risks – e.g. an oil refinery or a coal-fired power plant – a process-oriented strategy may be a more adequate approach. As will be elaborated in the rest of this chapter, combining such insights with network perspectives provides new understanding in the divergence of industrial ecology in eco-industrial parks.

Eco-industrial park configurations

Overview

In this section, the different stages in industrial management strategies on the environment will be applied on eco-industrial park development, in combination

Table 16.1 Configurations of environmental management (translated from Van Koppen and Hagelaar 1998)

Company characteristics	Crisis-oriented strategy	Process-oriented strategy	Chain-oriented strategy
Internal:			
Ambition Knowledge and information	Compliance Knowledge is directed to only a few (prescribed) aspects; little horizontal or vertical information exchange	Eco-efficiency Knowledge is directed to production process; information exchange on operational and tactical level	Green identity Knowledge is directed at product chain; information exchange up to the strategic management level
Technology	End-of-pipe-technology, directed at cleaning and filtering	Process-integrated technology, directed at prevention	Process- and product innovations from a product life cycle perspective
Organisation	Environmental tasks are focused and isolated	Environmental management system	Environment-oriented organisational networks encompassing marketing, R&D, suppliers and customers
Budget	Financial budget for environmental investments is limited	Financial budget for environmental investments (with 1–4 years pay-back period)	Budget for strategic environmental investments
External:			
Environmental risks	Environmental risks are serious and inherent to the production process	Environmental risks are limited or convertible	Environmental risks are not (no longer) a major constraint
Environmental opportunities	Hardly any environmental opportunities	Indirect environmental opportunities (e.g. corporate image)	Direct environmental opportunities (e.g. green market)

with the three network perspectives. At least three main configurations of EIPs can be distinguished, corresponding with the environmental strategies that prevail among the companies occupying the industrial park. In each of these configurations, specific roles emerge for actors and institutions in the different networks. The configurations are summarised in Table 16.2.

Table 16.2 Networks in three ideal-typical eco-industrial park configurations

EIP configuration	Policy network	Industrial network	Societal network
Efficient compliance and 'gratis effect' (crisis-oriented environmental strategy prevails)	Prescribing combined treatment, tight, but well-adapted permit regulation, facilitating pollution prevention	Rest material exchange, exchange of information on environmental control, mutual advisory support	NGO's pressing for environmental regulation, and raising funds for stimulating industry
Industrial symbiosis (process-oriented environmental strategy prevails)	Shifting to (tradable) performance requirements, site-wide permitting, adjusting economic conditions (energy and waste processing costs)	Exchanging and sharing energy and material flows, sharing production and transport facilities, exchanging information for inter-firm linkages	NGO's supporting industrial initiatives with expertise, funding and publicity, and pressing government to establish EIPs
Nodes of environmental innovation (chain-oriented environmental strategy prevails)	Control at distance, facilitating environmental planning as a joint effort of government, industry and societal groups	Joint planning and management of estates, joint development of green technology and green products, natural landscaping, EIP labelling	Cooperation of industry and societal groups in product development and environmental planning

Eco-industrial parks: efficient compliance and 'gratis effect'

In this configuration, companies in industrial parks have predominantly a crisis-oriented environmental strategy (or no environmental strategy at all). Companies are oriented towards compliance of the most pressing environmental requirements, needs and interest, often articulated by state environmental authorities. In contrast to the common belief that industrial ecology forms the final stage in an evolution of environmental reform, also in this setting of 'crisis-oriented firms' modes of industrial ecology can be found. In such a setting the actors and networks that bring about environmental transformation in firms and estates have specific characteristics, as will be described below.

Especially in developing countries strong and sophisticated informal and formal systems of rest material flow reuse can be identified. Rest or by-products of industrial processes are either sold to formal industries or are collected by informal companies. Both groups transform them into valuable products that are often sold on the domestic market. It is these economic networks around by-products that form an unplanned and often unorganised industrial ecosystem: an environmental gratis effect. The term 'gratis effect' is borrowed from the German political scientist Martin Jänicke (1986), who analysed that successes in environmental reforms are sometimes hardly related to environmental intentions but rather the unintentional side-effects of other developments. Environmental reforms can then be considered as a free-rider on economic or other developments. In the case of industrial ecosystems in developing countries, the major drivers behind this effect are economic mechanisms, as the economic value of 'waste' rather than state regulation or public pressure forms the key motive to enter into industrial ecology arrangements.

Policy networks are important in bringing about environmental improvements in these firms with predominantly a crisis-oriented environmental strategy, basic-ally by pushing for compliance of existing laws and regulation. Often the key environmental actors in policy networks are weak and understaffed, which limits more facilitating and negotiating modes of environmental governance. Usually, the unplanned industrial ecology system is put in jeopardy by more effective state regulation and enforcement as well as further technological and economic development, each removing the main drivers behind the existing recycling mechanisms such as low wages, a market for cheap products, capital extensive production processes, and under-priced natural resources and pollution. For the informal sector, a further regulation of waste management introduces higher investment costs for recycling and re-use practices. This undermines the informal industrial ecology arrangements, often without having new modes and arrange-ments of industrial ecology present.

Therefore, these 'gratis effects' should be taken into account in the development and transformation of industrial estates. In addition to informal and unplanned waste recycling activities, there are also opportunities for more active and planned industrial ecosystem arrangements. Examples are collective waste (water) collection and treatments systems, collective monitoring and enforcement, and collective environmental management and advisory systems. Actors within both the policy and economic networks can be active in organising these common facilities, depending on the division of tasks and responsibilities between estate management authorities and public regulating authorities.

Community and NGO pressure form a trigger for environmental authorities to be serious on regulation and enforcement. The media is a crucial instrument for the latter category of actors. Especially in countries with limited traditions and possibilities of environmental NGOs and a crisis-oriented environmental manage-ment orientation (e.g. China and Vietnam), community complaints constitute a crucial driver for industrial transformation (Wang 2000; Phung Thuy Phuong 2002). Such complaints as well as NGO pressure often target the waste streams not yet included in industrial ecology arrangements, adding thus to environmental improvements without jeopardising existing (in)formal industrial cooperation on the environment.

Case: industrial estates in Vietnam

In Vietnam some 60 industrial estates form the nucleus around which the industrialisation process of this country proceeds since the early 1990s. The rapid industrialisation (with annual growth figures above 10%) goes along with many environmental ills, especially around the geographical concentrations of industries.

Severe pollution of air and water occurs, despite that a large amount of industrial by-products that would have been classified as waste in the OECD countries are re-used and recycled as resource inputs in smaller firms that produce mainly for the domestic market (Phung Thuy Phuong 2002, Nguyen Phuc Quoc 1999). No environmentally sound collection and treatment systems exist in the industrial estates for final wastes. Provincial and national environmental authorities that are responsible for environmental control and enforcement, are severely understaffed, while monitoring systems are deficient. Complaint systems of communities form an alternative monitoring system for these authorities, and media attention to these complaints puts additional pressure on authorities to act: community driven regulation (O'Rourke 2000; Woltjer 2001). While foreign direct investments bring to Vietnam some of the more advanced environmental management strategies of multinational companies that have their headquarters in OECD countries and produce for the world market, the economic networks and actors these foreign companies engage with are different from the ones domestic companies engage with. This limits the transfer of both knowledge and practices of industrial ecology. The more advanced and front-running industrial estates have started to construct common waste (water) collection and treatment facilities, common environmental advisory and auditing consultancies, and new monitoring systems (cf. Bien Hoa II industrial zone in Dong Nai province). The industrial state authorities and the state environmental authorities are also starting to reformulate their division of tasks in terms of control, monitoring, auditing, advice, extension and technological diffusion in the field of environment (Phung Thuy Phuong 2002).

Eco-industrial parks: industrial symbiosis

Most of the EIPs that are frequently cited in the literature, such as Kalundborg in Denmark, INES in the Netherlands, and Burnside in Canada, are examples of industrial symbiosis. In industrial symbiosis, greater eco-efficiency is achieved by inter-company exchange of materials and energy. Due to the fame of these examples, EIPs sometimes become almost synonymous to industrial symbiosis, although they constitute only one of the possible EIP configurations. The industrial symbiosis perspective of EIPs typically applies to companies in a process-oriented stage. To achieve environmental optimization on the park level, companies need to have knowledge of pollution prevention and process-integrated technology change. They need to have insight in rest material and energy flows and the environmental costs attached to these flows. To fully realise the potentials of industrial symbiosis, a substantial budget for investments may be needed (Martin *et al.* 1996). Industrial symbiosis also requires organisational skills. Horizontal communication is needed, particularly between the environmental staff and process engineering staff, as well as vertical communication between these staff departments and the top management.

Industrial symbiosis is primarily based on networks of economic actors. To establish these networks, a common orientation on environmental improvement and mutual trust are of vital importance. In most reported cases of industrial symbiosis, long processes of co-operation and trust-building in other domains are at the root of succesfull environmental collaboration. For a good understanding of industrial symbiosis, it is important not to overestimate its financial and environmental benefits. In most cases, the benefits should be regarded as an extra, but modest improvement on top of what pollution prevention efforts within the company can achieve (for Kalundborg, see below; for Brownsville/Matamaros, see Martin *et al.* 1996).

The predominant economic and policy conditions do not always favour industrial symbiosis. The economic situation often demands flexible responses of companies across global production chains. Markets often value innovative products and human resources much higher than rest materials and energy. For these reasons, corporate managers may regard industrial symbiosis as something of the past rather than the future. For example, a senior environmental manager of Unilever argued that industrial ecology was characteristic of the 20th century production system of their company. Such a symbiosis would be too rigid and locally restricted to meet Unilever's current corporate strategy, which is character-ised by large-scale production chains, a focus on core business, and an emphasis on cost-effectiveness and flexible response to consumer demands. He considered the instruments of pollution prevention and life cycle management to be more adequate to this strategy (Dutilh 1996).

An important role of policy networks, therefore, is to create more favourable conditions for company collaboration. Local governments can collaborate with industries to facilitate energy and material exchange, e.g. by subsidizing infra-structure facilities at the park and by enhancing its accessibility through (environ-mentally friendly) transport facilities. Another important role of government concerns regulation. Governmental regulation can be an important driver for col-laboration, for instance when it increases the costs of waste disposal or the prices of materials and energy. But in other cases, governmental regulation disencourages industrial symbiosis, since it is directed to single companies and it tends to be very restricted towards the use of waste for re-use purposes (Lowe 1997; Martin *et al.* 1996). A company that receives and reuses waste materials from other companies will frequently face restrictions rather than incentives from authorities. Experi-ments with tradable permits or collective permits – also referred to as 'bubble concept', or site-wide permits – can help to overcome such set-backs. A park-wide environmental management system, certified against ISO 14001, might serve as an instrument to enhance the self-regulatory role of park management.

In contrast to the former EIP configuration the role of environmental organisa-tions under these circumstances focuses on advocating and stimulating the estab-lishment of EIPs. For instance, in the Netherlands an environmental NGO has published a brochure with suggestions for business collaboration (De Vries 1998), and a regional environmental organisation is represented in the strategic advisory board of the INES Mainport organisation (INES 2001).

Case: industrial symbiosis in Kalundborg

The cluster of industrial activities that goes around the world as *the* case of indus-trial symbiosis is situated in an industrial area near Kalundborg, a small city of about 20,000 inhabitants in Denmark. Many descriptions of the Kalundborg case circulate in literature, with different figures on the environmental and economic benefits (among others dependent on which activities are taken into consideration). Our description and figures of Kalundborg focus on inter-firm activities and are based on Christensen (1994), Evans (1995), Gertler (1995), and Jacobsen (2001).

The five major actors in the symbiosis are: Asnaesvaerket, a coal-fired electrical power station; Statoil, a large oil refinery; Gyproc, a plaster board factory; Novo Nordisk, a large producer of enzymes, insulin and penicillin; and the city of Kalundborg. Exchange of energy and material flows between these actors started already in the early 1970s, with Gyproc using gas from Asnaesvaerket. Further collaboration was established in the 1980s and 1990s, resulting in an industrial ecosystem with several inter-firm linkages between the five actors as well as to external actors (see Figure 16.1 for the situation in the mid-1990s). More recently the gas delivery from Statoil to Gyproc has terminated and a new actor, a soil remediation company, has entered the stage.

Among the environmental benefits of these linkages (figures of 1994) are savings on the use of coal (30,000 tons/year or 2% of total use) and water (600,000 tons/year or 20% of total use), reduction of the emissions of CO_2 (130,000 tons/year or 3% of total emission) and SO_2 (3700 tons/year or 13% of total emission), and waste reduction as well as materials savings from the reuse of fly ash, sulphur, gypsum, and nutrients (N and P). The total investments for these linkages (in 1994) were approximately 60 million USD, generating annual revenues of 10 million USD.

In sum, Kalundborg has proven that industrial symbiosis can be organised on an enduring, economically sound basis, and can bring substantial benefits to the environment. What factors made this success possible? Obviously, many of the linkages are possible because of the geographical proximity of the actors: especially those involving transport of water and heat. Some linkages, however, stretch far beyond Kalundborg, such as the transfer of fly ash to cement industries and of sulphur to the fertiliser industry. Just as important as physical proximity is the social proximity. The fact that Kalundborg is a small community and managers of firms often knew one another from various contacts has contributed to building

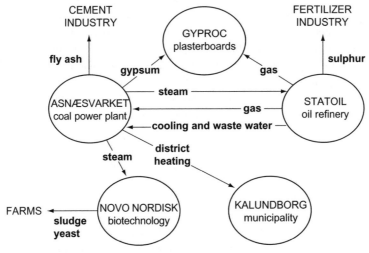

Figure 16.1 The Kalundborg industrial symbiosis (around 1994)

mutual knowledge and trust. Employees of different companies were allowed to communicate directly in case of problems.

The policy network, too, has been influential to the success of industrial symbiosis. Although the early linkages were made out of economic considerations, regulation has been an important driver of the further development of the symbiosis. Water reuse, for instance, was stimulated by water pollution regulations, and the gypsum exchange was an indirect result of stricter requirements on SO_2 emission to the atmosphere. An important aspect, in this regard, is that the environmental requirements were set as performance standards rather than technology standards, allowing companies to create their own strategies of emission prevention and control.

Societal networks have undoubtedly been of influence too. In fact, several of the managers involved have a strong personal motivation for environmental concerns, and together acted as an environmental network for disseminating the lessons of Kalundborg over the world. Also, the connections with the local community, via a Community Environmental Committee, have been important to the realisation of the symbiosis.

Eco-industrial parks: nodes of environmental innovation

In moving towards a chain-oriented strategy, companies will look beyond the boundaries of their industrial park or geographical region, to encompass the chains of production and consumption they contribute to and depend upon. This broader outlook may lead to another perspective on EIPs, in which the two directions of industrial ecology, the industrial ecosystem approach and the life cycle approach, meet. Exchange of energy and rest materials, and facility sharing remain of interest to companies, but only within a much broader view of closing material loops. That recycling loops can extend over long distances is, for instance, shown by Xerox in Europe. This company has established a comprehensive system for the recycling of copy machines, in which parts are collected all over Europe and revised and re-assembled in one facility in the Netherlands (Xerox 1996). Dismantling and recycling of end-of-life vehicles, glass and plastics recycling systems, and paper recycling are just a few other examples of reuse systems that extend over the production chain. From such a broader point of view, the EIP is only one link in the wider production and consumption system, and closing material loops and saving energy are analysed and optimised at higher scale levels. This does not mean that EIPs are no longer important to industrial transformation. New innovative roles for EIPs can emerge; three of them are highlighted below:

- *Exchange and co-generation of strategic knowledge for environmental development.* As has become clear from industrial network studies, one of the main assets of regional company networks is knowledge exchange and collective innovation. EIPs may provide a fertile ground for environmental innovation (Côté and Cohen-Rosenthal 1998). For such a purpose, in particular knowledge-intensive environment-oriented companies need to cluster.
- *Creation of regional niches for environmental innovation.* EIPs can be a focal point of innovative experiments with new modes of transport, service economy,

and sustainable consumption. In other words, they are an important potential instrument in what has been called 'strategic niche management': the creation of social and economic spaces for environmental innovation. The networking of the EIP with the surrounding region is crucial here, since the scope of innovation encompasses both production and consumption. Elements of such a perspective are found, for example, in the Port of Cape Charles Sustainable Technologies Park (see cases).

- *Natural landscaping of industrial parks.* One of the challenges of EIPs in an advanced stage is to integrate elements of nature conservation into the park design. Landscaping of industrial parks is desirable from an angle of bio-diversity, environmental quality, and historical or cultural identity of the region. Obviously, it is difficult and costly to realise, and in general industries are not interested. For companies with a chain-oriented strategy, however, landscaping makes also sense commercially. A high quality, green setting of their facilities and offices, within a park that is renowned as an ecological forerunner, will help them establishing a green identity, and thus improve their competitiveness on green markets.

Knowledge, rather than gypsum or steam, will flow between the businesses of this type of EIPs. Industrial networking therefore will be primarily geared to the exchange and generation of environmental expertise in a broad sense. The role of the policy and societal networks should mainly be conceived in terms of co-operation and mutual support. Particularly, where the creation of strategic niches for environmental innovation and the integration of industry and landscape in regional planning are concerned, participatory ways of decision making, involving industries, local governments, ngo's and residents can be useful.

Cases: the Cape Charles model and the PALME label

As yet, there is no full-fledged example of a 'chain-oriented' EIP. But there are examples that move towards this stage, as illustrated below.

The Cape Charles Sustainable Technology Park is one of the EIP demonstration projects in the USA. According to its principles, established in 1995, this EIP aims not only to support, attract, and create ecologically compatible enterprises, but also to stimulate sustainable development of the community, and to serve as a model for environmentally sound coastal development. It includes many aspects mentioned above. It will be designed in a way that preserves and advances the historical and cultural landscape of the region, as well as the natural values of the site. Its development is embedded in the Port area community, and the planning process included, among others, a community design charrette. Dissemination of information and technologies is an important aspect. In several ways, the project seeks collaboration between companies and citizens of Cape Charles, for instance in water saving and solar energy projects (Lowe *et al.* 1997). The companies locating in this area are typically green, knowledge-intensive firms, e.g. a solar energy company, an environmental consulting firm, and a German wind farm developing company (Cape Charles 2001).

The PALME eco-label for industrial parks. In France, a consulting group has developed a 'Programme d'actions labelisé pour la maîtrise de l'environnement' (PALME). EIPs can submit to this program, and so obtain the PALME label. The requirements are very demanding and include a landscaping plan and maintenance of the natural flora and fauna; an advisory service for clean technologies: a liaison mechanism with relevant local authorities; as well as plans and provisions for regulatory compliance, management of energy, waste and emissions, and 'clean construction'. PALME focuses on park management rather then physical flows (Côté and Cohen-Rosenthal 1998). The PALME label seems particularly adapted to a chain-oriented strategy, because it sets a high standard and offers companies the opportunity to show that they act to that standard.

Conclusions and outlook

This chapter started from the observation that a better understanding of the role of actors and institutions seems essential for both the advancement of industrial ecology as a scientific project and the advancement of industrial ecology in the practice of eco-industrial parks. In contributing to such a perspective ecological modernization theory was introduced. The role of actors and institutions in the ecological modernization of industrial ecosystems was operationalised at two levels. At the level of the social environment of companies, actors and institutions can be conceptualised as networks that take on board and institutionalise ecological requirements and demands in their relations and interactions with companies. Three such networks perspectives were distinguished: policy networks, industrial networks, and societal networks. At the level of the company three strategies in dealing with growing ecological requirements were identified: ecological rationality as external threat (crisis-oriented strategy), ecological rationality integrated in internal production processes (process-oriented strategy), and ecological rationality extended towards product chains and company identity (chain-oriented strategy).

In applying these actor and institutional perspectives to eco-industrial parks three ideal-typical eco-industrial park configurations could be distinguished, each corresponding with specific actions and strategies of both the companies that make up the eco-industrial park and the external industrial, policy and societal networks that interact with these companies. Each configuration provides insight in the roles different actors and institutions can play in making an eco-industrial park work.

The general conclusion from analysis should be that industrial ecology perspectives in general, and eco-industrial parks in particular, can contribute significantly to increased environmental performance. But to move industrial ecology ideas from the drawing table into the practice of existing industrial parks a better understanding is needed of the actors and institutions that make up the configurations companies are being part of. The findings in this chapter suggest that physical exchange of 'waste' is not the only and perhaps not the most promising industrial ecology option. Interesting options for improvement are also to be found in more social areas, and that not only counts for the last type of configuration. In the first configuration type (efficient compliance and 'gratis effect') a collective approach building on existing networks can be most successful, and in addition to that, more effective environmental policies integrated in social and economic policies. In the second configuration type (industrial symbiosis) innovations in permitting and economic policy instruments seem promising. In the last configuration type (nodes of environmental innovation) there are good perspectives for exchange and joint development of knowledge, and also for regional environmental planning across the boundaries of the park. These examples are certainly not the only strategies to be followed in developing eco-industrial parks in practice, but they provide enough evidence of the necessity to expand the industrial ecology approaches towards the social sciences, both in their further theoretical development and in their practical application.

References

Ayres, R.U. (1989) Industrial metabolism. (eds. J.H. Ausubel and H.E. Sladovich, *Technol. environ.* pp. 625–636, National Academic Press, Washington DC.

Ayres, R.U. and Ayres, L. (1996) *Industrial ecology: towards closing the materials cycle.* Edward Elgar, London.

Beck, U., A. Giddens, A. and Lash, L. (1994) *Reflexive Modernisation. Politics, tradition and aesthetics in the modern social order.* Polity Press, Cambridge.

Boons, F.A.A. and Baas, L.W. (1997) Types of industrial ecology: the problem of coordination. *J. Cleaner Prod.* **5**(1–2), 79–86.

Brown, R.K. (1992) *Understanding Industrial Organisations. Theoretical Perspectives in Industrial Sociology*, Routledge, London/New York.

Cape Charles (2001) *Cape Charles Sustainable Technology Park.* Website: www.sustainablepark.com (2001-11-06).

Christensen, J. (1994) *From the drawing board to production.* Hand-outs of a presentation at the Industrial Ecology Workshop on February 10, 1994, Toronto.

Côté, R.P., Ellison, R., Grant, J., Hall, J., Klynstra, P., Martin, M. and Wade, P. (1994) *Designing and operating industrial parks as ecosystems*, Dalhousie University, Halifax.

Côté, R.P. and Cohen-Rosenthal, E. (1998) Designing eco-industrial parks: a synthesis of some experiences. *J. Cleaner Prod.* **6**, 181–188.

Cotgrove, S. (1982) *Catastrophe or Cornucopia. The Environment, Politics and the Future*, John Wiley, Chichester.

De Vries, A. (1998) *39 Ideeën voor een nieuw duurzaam bedrijventerrein*, Milieufederatie Noord-Holland, Zaandam.

Dutilh, C.E. (1996) Lezing op 16 april 1996. In *Congresverslag Industrial Ecology*, (ed. J.L.A. Geurts *et al.*), pp. 9–10, Katholieke Universiteit Brabant, Tilburg.

Erkman, S. (1997) Industrial ecology: an historical view. *J. Cleaner Prod.* **5**(1–2), 1–10.

Evans, L. (1995) Lessons from Kalundborg. *Business environ.* **6**(2), 5.

Frosch, R.A. and Gallopoulos, N.E. (1989) Strategies for manufacturing. *Scientific American* 1989 (September), pp. 144–152.

Gertler, N. (1995) *Industrial ecosystems: developing sustainable industrial structures*, Massachusetts Institute of Technology (MSc Thesis), Boston.

Graedel, T.E. and Allenby, B.R. (1995) *Industrial ecology*, Prentice Hall, Englewood Cliffs.

Giddens, A. (1984) *The Constitution of Society*, Polity Press, Cambridge.

Giddens, A. (1994) *Beyond Left and Right. The Future of Radical Politics*, Polity Press, Cambridge.

Grabher, G. (ed.) (1993) *The Embedded Firm. On the Socioeconomics of Industrial Networks*, Routlegde, London.

Grant, W., Paterson, W. and Whitston, C. (1988) *Government and the Chemical Industry. A Comparative Study of Britain and West Germany*, Clarendon, Oxford.

Håkansson, H. (1988) Evolution processes in industrial networks. In *Industrial Networks. A New View of Reality* (eds. B. Axelsson and G. Easton), Routledge, London.

Håkansson, H. and Johanson, J. (1993) The network as a governance structure: inter-firm cooperation beyond markets and hierarchies. In *The Embedded Firm. On the Socioeconomics of Indutrial Networks* (ed. G. Grabher), Routledge, London.

INES (1997) *INES INdustrieel EcoSysteem. Eindrapport*, Stichting Europoort/Botlekbelangen, Rozenburg.

INES (2001) *Industriële ecologie voor het Rotterdamse haven en industriegebied*. Website: www.inesmainport.nl (2001-11-06).

Jackson, T. and Clift, R. (1998) Where's the profit in industrial ecology? *J. Industrial Ecol.* **2**(1), 3–5.

Jacobsen, N.B. (2001) Eco-industrial networking – a case study of Kalundborg industrial symbiosis. Cornell University, Ithaca.

Jänicke, M. (1986) *Staatsversagen. Die Ohnmacht der Politik in die Industriegesellschaft*. Piper, München.

Lowe, E.A. (1997) Creating by-product resource exchanges: strategies for eco-industrial parks. *J. Cleaner Prod.* **5**(1–2), 57–65.

Lowe, E., Moran, S.R. and Holmes, D. (1995) *A Fieldbook for the Development of Eco-Industrial Parks*. Prepared for the U.S. Environmental Protection Agency, Oakland, Indigo Development.

Lowe, E.A., Warren, J.L. and Moran, S.R. (1997) *Discovering industrial ecology. An executive briefing and sourcebook*, Battelle Press, Columbus.

Marsh, D. and Rhodes, R.A.W. (eds.) (1992) *Policy Networks in British Government*, Clarendon, Oxford.

Martin, S.A. *et al.* (1996) *Eco-Industrial Parks: A Case Study and Analysis of Economic, Environmental, Technical, and Regulatory Issues*. Final report prepared for U.S. Environmental Protection Agency, Research Triangle Institute, Research Triangle Park.

Martinelli, A. (ed.) (1991) *International Markets and Global Firms. A Comparative Study of Organized Business in the Chemical Industry*, Sage, London.

Ministerie van Economische Zaken (1998) *Duurzame bedrijventerreinen. Handreiking voor het management van bedrijven en overheid*. Ministerie van Economische Zaken, Den Haag.

Mol, A.P.J. (1995) *The Refinement of Production. Ecological Modernisation Theory and the Chemical Industry*. Jan van Arkel/International Books, Utrecht.

Mol, A.P.J. (1996) Ecological modernisation and institutional reflexivity. Environmental reform in the late modern age. *Environ. Politics* **5**(2), 302–323.

Mol, A.P.J. and Sonnefeld, D. (eds.) (2000) *Ecological Modernisation around the world: Perspectives and Critical Debates*, Frank Cass and Co, Ilford.

Nguyen Phuc Quoc (1999) *Industrial Solid Waste Prevention and Reduction. Circular Metabolism in Rubber Products and Footwear Manufacturing Industries*. Wageningen University (MSc thesis; REFINE MSc thesis series no. 4), Wageningen.

O'Rourke, D.J. (2000) *Community Driven Regulation: The Political Economy of Pollution in Vietnam, Berkeley*, University of California (dissertation), Berkeley.

Phung, Thuy Phuong (2002) *Ecological Modernization of Industrial Zones in Vietnam.* Wageningen University (dissertation [. . .]), Wageningen.

Polanyi, K. (1957 [1944]) *The great transformation*, Beacon Press, Boston.

Ruth, M. (1998) Mensch and mesh: perspectives on industrial ecology. *J. Industrial Ecol.* **2**(2), 13–22.

Smith, M.J. (1993) *Pressure, Power and Policy. State Autonomy and Policy Networks in Britain and the United States*, Harvester Wheatsheaf, New York.

Spaargaren, G. (1997) *The Ecological Modernisation of Production and Consumption. Essays in Environmental Sociology*, Wageningen Agricultural University (dissertation), Wageningen.

Spaargaren, G. and Mol, A.P.J. (1992) Sociology, environment and modernity. Ecological modernisation as a theory of social change. *Society and Natural Resources* **5**, 323–344.

Tibbs, H. (1991) *Industrial ecology, An Environmental Agenda for Industry*. Arthur D. Little, Inc. An updated version of the paper was published in Whole Earth Review, Winter 1992, no. 77, pp. 4–16.

Van Koppen, C.S.A. and Hagelaar, J.L.F. (1998) Milieuzorg als strategische keuze. Van bedrijfsspecifieke situatie naar milieuzorgsystematiek. *Bedrijfskunde* **70**(1), 45–51.

Wang, X. (2000) *Information Strategy in China's Environmental Management – A case study in Nanjing's industrial water pollution treatment*. Wageningen University, department of Social Sciences (MSc thesis), Wageningen.

Woltjer, L. (2001) *Coping with Industrial Zones. An Analysis of Community Influence on the Environmental Performance of Industrial Zones in Southern Vietnam*. Wageningen University (MSc thesis), Wageningen.

Xerox, (1996) *EMAS-Milieuverklaring. Periode 1 januari 1996 tot 1 juli 1997*. Xerox Manufacturing B.V., Venray.

ZERI (2001) Zeri systems. Website: www.zeri.org/systems.htm (2001-11-06).

Chapter 17

Gert Spaargaren

SUSTAINABLE CONSUMPTION: A THEORETICAL AND ENVIRONMENTAL POLICY PERSPECTIVE

BY BRINGING CONSUMPTION ISSUES to the fore, this article reflects the general trend to attach greater importance to the role of citizen-consumers in shaping and reproducing some of the core institutions of production and consumption. At the present time, the opinions and behaviors of citizen-consumers matter increasingly for companies, policymakers, and social movements. The shifting emphasis from production to consumption and the concomitant weight attached to the dynamics of civil society are expressed by stating that there has been a consumerist turn within sociology (Featherstone 1983). This article makes the central argument that environmental sociologists need to conceptualize sustainable consumption behavior, lifestyles, and daily routines in such a way as to avoid the pitfalls of many of the so called micro-approaches that have been developed to date. We argue for a contextual approach to sustainable consumption and for that purpose try to develop a conceptual model that combines a focus on the central role of human agency with proper treatment of the equally important role of social structure. Structuration theory as developed by Anthony Giddens (1984; 1991) provides the conceptual key to a solution of the classical actor-structure dilemma by putting forward *social practices* as the proper unit of analysis for researchers and policymakers. We use the social practices model, first to analyze the process of diffusion of environmental innovations in utility sectors, and second to discuss the emergence of a variety of lifestyles (lifestyle groups) in the field of the sustainable consumption of energy, water, and waste services. In the third section of the article, we describe the general policy discourse on consumer-oriented environmental politics and look at some of the most relevant policy experiments in the field of sustainable consumption in the Netherlands in order to develop the future agenda for research and policy-making on sustainable consumption. We end by arguing that an international network of social scientists working in the field of sustainable consumption can help significantly in developing a comparative perspective to sustainable consumption in a globalizing world.

A contextual approach to consumption: the social practices model

After a long period of abstract criticisms on consumer culture, general sociology has now taken up the study of consumption in a serious way. Despite this, pride of place within the emerging sociology of consumption (Warde 1990; Otnes 1988) has not yet been given to environmental issues. The first edited volume on sustainable consumption in this field was issued only recently (Cohen and Murphy 2001).

Within environmental sociology, the study of consumption behavior and lifestyles has for a long time been the domain of empirical researchers who in many cases are theoretically inspired by a variant of the social psychological model of human behavior. They use variants of the model as it was introduced into the social sciences by Fishbein and Ajzen (1975). When applied within the field of environmental studies,[1] this so called Attitude-Behavior model uses the individual attitude or norm to predict the concrete future behavior of individuals. This behavior is conceived of in terms of a great number of separate items or issues that are used as indicators for a more encompassing environmental consciousness or awareness. Among the symbolic representations of a high environmental consciousness and the corresponding environmentally correct behaviors were the intended use of nonphosphate detergents; the promise of avoiding throwaway plastic bags and packages while shopping; and the tendency to use bicycles instead of cars for short-distance mobility.

The sociological model as presented in Figure 17.1 differs from the sociopsychological Attitude-Behavior model in some crucial respects. First, at the center of the model one finds not the individual attitude or norm but rather the actual behavioral practices, situated in time and space, that an individual shares with other human agents. Second, the model does not focus on specific, isolated behavioral items but rather looks into the possibilities for designated groups of actors to reduce the overall environmental impacts of their normal daily routines involving clothing, food, shelter, travel, sport, and leisure. Third, the model analyzes the

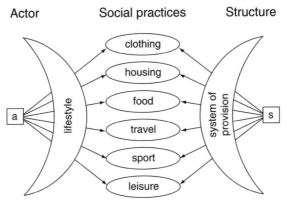

Figure 17.1 The social practices model

process of reducing the environmental impacts of consumption in distinct domains of social life in terms of the deliberate achievements of knowledgeable and capable agents who *make use of the possibilities offered to them in the context of specific systems of provision*. By adding this focus on systems of provision (Fine and Leopold 1993; Shove 1997; Shove and Chappells 1999), social structures are no longer treated as external variables but are brought into the center of the analysis.

There are critical changes in the results when the social practices model is to replace the Attitude-Behavior model as the central framework for analyses in environmental research and policymaking. The sections that follow briefly discuss three consequences of working with this model.

The end of the individual as the (only) relevant unit of analyses

First, the social practices model implies the end of the individual as the central unit of analysis. Human agency is analyzed not in terms of the isolated individual or the "homo clausus" (Elias 1971) but rather in terms of the twin concepts of lifestyles and social practices. The lifestyle of each individual is constructed from a series of building blocks—corresponding to the set of social practices an individual invokes when pursuing his or her daily life. The lifestyle of an individual human agent is defined by Giddens as the set of social practices that an individual embraces, together with the storytelling that goes along with it (Giddens 1991, 81). When interpreted in this way, it becomes clear that the concept of a green or sustainable lifestyle is different from the concept of an environment-friendly attitude since it cannot be measured using only one dimension or scale.[2] Green lifestyles are composed of lifestyle segments or sectors that may vary considerably among themselves with respect to the contribution they make to the net environmental impact of the lifestyle of the individual human agent. Even individuals who state that it is their intention to put to work as often and strictly as possible some environmental criteria they embrace as part of the foundational principles of their lifestyle will act against these rules at certain times and under some circumstances at some sectors or segments of their lifestyle. Some people deliberately insulate specific practices or lifestyle segments from the environmental considerations they accept and apply as legitimate rules most of the time, and for most other segments of their lifestyles. In Dutch environmental discourse this mechanism is recognized and has come to be represented by the image of a typical Harley-Davidson motorcyclist who also proves to be an active member of a so called eco-team or GAP team[3] in his neighborhood. This image is used in information campaigns by environmental policymakers to illustrate the fact that a person does not have to be written off for climate change politics once he or she apparently acts against the established rules for sustainable behavior in one or two domains of daily life (VROM 2000). The probably considerable CO_2 emissions of the Harley-Davidson can, for example, be compensated for by a set of rather strict energy policies applied in the domain of domestic consumption.

Environmental policy goals to be defined from a lifeworld perspective

A second consequence of working with the social practices model is the need to reformulate most of the existing targets in environmental policymaking. Within the present policy framework, the central aims are formulated in an exclusively technical language, which is addressed mainly to institutional actors in the sphere of industrial production. When daily routines like clothing, food, shelter, travel, sport, and leisure are taken as a starting point for policymaking, these policy definitions must be reformulated and specified from a life-world perspective in order to be recognized and taken on board by concrete (lifestyle) groups of citizen–consumers.

As we discuss later in more detail, key actors in Dutch environmental policy circles are taking up the challenge to develop sets of environmental indicators for all the major social practices. Until now, however, these sets of indicators continue to rely on the technical rationale that is so prominent in environmental policymaking.[4] In order to make the policy targets in the different social practices or consumption domains recognizable and perceptible for citizen–consumers, the indicators have to be reformated according to the rationale of the life-world itself. Only when a social dimension is added to the technical criteria, citizen–consumers—in the context of enacting their daily routines—will be able to arrive quickly, correctly and routinely at the position in which they "know" about the relevant environmental aspects of their routines. To help them accomplish this, we would argue for the need to develop so-called *environmental heuristics* for each social practice. Environmental heuristics can be regarded as rules of thumb to be used by citizen-consumers in determining "how to go on" in a more sustainable way in the context of the time/space-bound daily routines they are involved in. These heuristics connect the predominantly technical rationale of environmentalism to the social rationale of the life-world and at the same time reduce the complexity of sustainable consumption in such a way that it fits the practical logic of daily life. Some examples of more or less well-known heuristics are the principle of waste prevention and separation ("keep them apart"), the use of eco-labels ("buy green"), and the use of the modal-shift notion in transport ("park and travel").

Contextualizing individual responsibility for environmental change

The third consequence has to do with the issue of individual responsibility for environmental social change. In the social practices approach, the responsibility of the individual towards environmental change is analyzed in direct relation with social structure. The individual possibilities and the disposition of the individual towards environmental social change are analyzed in conjunction with the levels and modes of green provisioning. When there is a high level—both in quantitative and qualitative respects—of green provisioning, people are more or less brought into a position in which the greening of their corresponding lifestyle segment becomes a feasible option. By emphasizing the role of systems of provision, the enabling aspects of social structure are emphasized.

When looking at the present state of affairs in the Netherlands, it is obvious that within some social practices the levels of green provisioning are much higher than in the case of others. This is due to the fact that the ecological modernization of production and consumption does not develop at the same pace and in the same way through the different sectors of production and the corresponding systems of provision. For example, in many European countries there are nowadays many more green options available in the food chain (Schuttelaar & Partners 2000) as compared to the travel industry (Bargeman 2001; Beckers et al. 2000).

If we contextualize the norms and environmental behaviors of individual human actors, we not only move away from overly individualistic accounts of environmental change, but at the same time open up a new research agenda for environmental sociology in studying environmental change from a life-world perspective. Some elements of this new research agenda are discussed in the next section.

Sustainable consumption in context: the example of utility provision

That levels of provisioning for sustainable alternatives differ among social practices is well recognized within the environmental sciences. The kinds of choices that are made available to citizen-consumers, as well as the classical properties like quality and prices of products and services, have also been investigated to a considerable extent (Vringer and Blok 1993; Blok and Vringer 1995). Much less attention has been paid to the social conditions and circumstances that frame the green or sustainable alternatives. Environmental sociologists can make a vital contribution of this point by investigating the social relations that go together with the socio-technical innovations implied in the ecological modernization of consumption. It must be clarified whether, on the one hand, the greening of social practices like "feeding" or "dwelling" makes citizen-consumers more dependent on expert systems that are hard to get to and impossible to control, or, on the other, whether the sustainable versions enhance the authority of consumers, who are thereby able to exert democratic control over the major actors involved in providing the green alternatives.[5]

The changing context for sustainable consumption and more specifically the changing social relations between providers and citizen-consumers formed the central questions for the Domus research project (1997–2000), which was funded by the European Union and conducted by an international team of researchers from the United Kingdom, Sweden, and the Netherlands, using a comparative research methodology (Chappells et al. 20001; van Vliet et al. 2000; Raman et al. 2000). We briefly discuss both the theoretical and empirical aspects of this research project, because it can help to elucidate the ways in which social context can be given a proper place in the analyses of sustainable consumption.

An infrastructural and historical perspective

When the social practice of "inhabiting a house" is taken as a starting point, the possibilities for householders to green their consumption can be said to be determined to a large extent by the green alternatives made available in the field of energy, water, and waste. The Domus team investigated the kind of green alternatives that some of the major providers in different European countries had endorsed from about the mid-1980s onward. To analyze the packages of green alternatives offered by the providers of energy, water, and waste services, the system of provision perspective—also referred to as the "infrastructural perspective"—was used as an analytical tool. Within this infrastructural perspective, special emphasis is laid on the ways in which modes of design, production, and distribution at the providers side of the chain do or do not correspond with certain mode of access, use, and disposal at the consumer end of the chain.

From the empirical research it became evident first that power relations between providers and consumers are multisided phenomena and second that these relations go through a rapid and rather profound process of change. To make sense of this variety and to establish the overall direction of the changes involved, a historical perspective had to be introduced.

In Europe, the process of householders "serving and being served" (Otnes 1988) by utility-like organizations used to be rather predictable and one-dimensional during the so-called Fordist period from about World War II up until the 1970s. In most European countries, householders were connected to networks that were most of the time monopolistic in nature and controlled by national governments. Householders had no choice other than buying—against a fixed tariff—the commodities and services they needed in order to be able to sustain their domestic social life. Today, in contrast, we witness major, long-term changes in utility provision, changes that will alter the very nature of the provision-consumption process and that are comparable to the kind of changes Ruth Cowan analyzes in her impressive book *More Work for Mothers* (1983). But although Cowan describes the ways in which the industrialization of households in the United States resulted in a new set of power relations and technical interdependencies between networks of utility providers on the one hand and householders on the other, we are nowadays witnessing the dissolution of this "modern" form of serving and being served, to be replaced by a late-modern variation of utility provisioning. In the late-modern condition of the global network society we must, as argued by Simon Guy and Simon Marvin (Guy and Marvin 1996; Guy et al. 1997; 2001) in a careful and detailed way, analyze the provision and consumption of utility goods and services against the background of *splintering, fragmented*, or *differentiating* networks, linking providers and consumers in many different ways. Figure 17.2 presents a very simplified representation of the premodern, modern, and late-modern configurations in utility service provision and consumption, also illustrating the different positions of householders vis-à-vis provider networks.

The social dimension of environmental innovation

With the help of the infrastructural and historical perspective, it could first be established that in the field of domestic consumption of water, energy, and waste services, a substantial number of innovations were introduced under the heading of sustainability in several European countries from the mid-1980s onward.[6] Second, the Domus researchers were able to demonstrate that these sociotechnical innovations did vary considerably with respect to the social relations that go along with them. When combining both results, one can conclude that the ecological modernization within networks of utility provision is to a great extent carried by and dependent on technological innovations, but these innovations are socially variable in the sense that social relations are not determined by sustainable technologies.

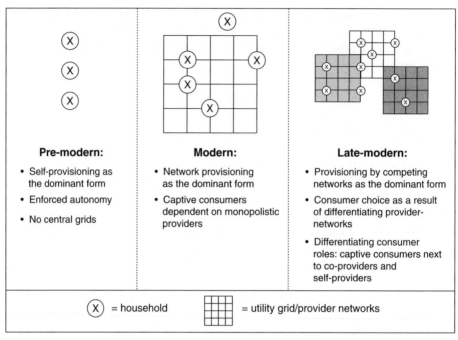

Figure 17.2 Utility provisioning and changing household-grid-relations

Perhaps this may sound obvious to sociologists; it is not so for many environmental engineers who are working in the expert systems involved in the renewal of utility provision. In the example given next, green electricity provision with the help of photovoltaic (PV) cell technology is discussed as an example to illustrate this main point.

PV technology in the built environment: two case studies

PV solar panels are among the most widely recognized and used symbolic representations of sustainable energy provision and consumption in the Netherlands, and over the past 10 years the application and integration of PV technology in the built environment has become one of the main goals of energy policies. However, the ways in which the PV panels are integrated in living areas, and especially the social conditions of their application and use, are very different indeed.

In the Amsterdam area, there has been a major pilot project for the large-scale integration of PV panels in the built environment by roofing hundreds of family dwellings[7] with PV panels. The panels are owned by the utility company, and the electricity generated by these panels is plugged directly into the central electricity grid, without the householders knowing about the actual energy performance of the panels, let alone that they were able to use the green electricity generated on the roof of their own dwelling. In short, householders were not regarded as—nor offered the possibility to become—functional *stakeholders* (www.zon-pv.nl).

It was possible to compare this situation with a technically identical situation in the eastern part of the country, where the municipalities of Amersfoort and Apeldoorn decided to work together with the (privatized) utility companies to offer—on the basis of attractive financial conditions—PV roofs to citizens who had just bought a house in a newly built area. Citizen-consumers could either rent or buy the PV-paneled roof from the utility company. There was also a double metering system installed, monitoring both the self-produced solar energy and the amount of energy that householders bought from or redelivered to the central grid, with this grid functioning as a backup system for the domestic PV systems. One can imagine the commitment citizens-consumers in Amersfoort and Apeldoorn will develop with respect to their "own," visible, servicing PV system to be qualitatively and quantitatively different when compared to the Amsterdam situation. From the outside, though, the dwellings look identical.

From this and many similar empirical cases we were able to conclude that the social relations that accompany the application of sustainable technologies at the domestic level show remarkable and relevant differences and that these differences cannot be explained by or reduced to differences in the technologies themselves (Spaargaren 2000).[8] As a consequence, we conclude that the greening of domestic consumption cannot be properly understood by looking into the technology in an isolated way. The development, diffusion, and application of sustainable technical devices must be analyzed against the background of the broader societal dynamics in utility provisioning and consumption (van Vliet 2002). Environmental innovations are shaped by the forces of liberalization and privatization, but also contribute to these developments in a specific way, as we discuss in the next section.

Liberalization, privatization, and the emergence of "green choices"

Although the pace of change will be different in different sectors and societies,[9] the general direction of change within the European Union is clear. In Europe, there is a process of *differentiation* of utility provision and consumption evolving, which seems to have no precedent in history. This process of differentiation implies that more options are offered to citizen-consumers in these sectors than before. Green options are emerging first and foremost with respect to the resource base of domestic energy provision; the quality and/or reusability of domestic drinking water, the nature of domestic sewage systems; and the schemes for handling domestic solid waste. This differentiation in the Resource Base (Rs) of domestic consumption is displayed in the column at the right side of Figure 17.3. However, the process of differentiation does not halt here. Differentiation pertains to the social or power relations between providers and citizen–consumers as well. Due to the liberalization of electricity markets, citizen–consumers are given the choice to select their own provider and also to have a say in the social conditions surrounding the provision of domestic utility services. In the Netherlands, the case of green electricity schemes again might serve as an example, now that Dutch householders have been free to choose their provider of green electricity since July 2001. With the liberation of formerly captive consumers, a process is set in motion in which providers have to compete for end users, including households. In the context of this competition, green products, services, and strategies start to be taken on as strategic resources by utility companies. The differentiation between providers

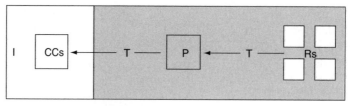

Step 1: resource-differentiation

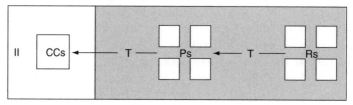

Step 2: (also) providers-differentiation

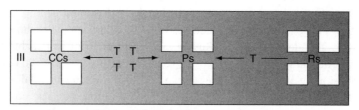

Step 3: (also) consumers-differentiation

Figure 17.3 Differentiation in provisioning and consumption within utility sectors

(Ps) and with respect to provider–consumer relations is expressed in the middle column of Figure 17.3.

The most interesting aspect of the differentiation process however is to be found in the left-hand column of Figure 17.3. Here it is envisaged that, due to the differentiation process in utility provision, in the end there will also emerge different lifestyle groups of citizen–consumers (CCs). These lifestyle groups differentiate among each other by the different ways in which they engage themselves with sustainable utility consumption and (co)provision. Some providers—both companies and governments—begin to recognize this process and start to make use of lifestyle and distinction aspects to promote more sustainable patterns of domestic consumption.[10] In Sweden, for example, one electricity company contributed to the lifestyle patterning of householders by introducing different colors, not for the different resource bases themselves but for different lifestyle groups of citizen–consumers. The red, blue, yellow, or green lifestyle groups are characterized by a specific use pattern of electricity, which to a certain extent is said to result from their lifestyles (Summerton 2000).

In sum, we might conclude from the discussion in this section that in the European Union a process of differentiation in utility sectors can be detected that results—among other things—in more green options for distinct groups of citizen-consumers. Such choices refer not only to the technical dimension of utility provision and consumption, but to the social dimension as well. Consequently, based on the dynamics of differentiation and the resulting green choices, lifestyle groups could come forward that distinguish themselves by the specific ways in which they do—or do not—make use of more sustainable alternatives to green different segments of their lifestyles.

Policymaking and research on sustainable consumption in the Netherlands

The (non)handling of sociotechnical innovations by emerging lifestyle groups of citizen-consumers has the potential to become an important subject for future research and policymaking. Research should clarify when and how, under what lifestyle conditions, different groups of citizen–consumers make use of the socio-technical innovations made available to them in the various consumption domains. At this moment the research agenda on the politics of sustainable consumption is just barely in the making, and most of the work still needs to be done. As a contribution to the debate, we introduce and discuss three pilot projects that were recently conducted in the Netherlands in an effort to develop new, consumer-oriented environmental politics aimed at reducing the environmental impacts of consumption. The pilot projects were designed and implemented in close collaboration between scientific researchers and policymakers and intended to deliver some essential building blocks for consumer-oriented environmental politics in the future. We evaluate the pilot projects with the help of the social practices model as it was developed in the second section, and look into the kind of questions that the social practices model itself might generate for future research on the politics of sustainable consumption.

The "domain explorations"

By conducting a number of so called "domain explorations" (Domeinverkenningen), researchers tried to establish the precise environmental profile of a series of social practices that were judged to be relevant in the context of consumer-oriented environmental politics (Schuttelaar & Partners 2000; CREM 2000; TNO 1999). Food, "clothing," "housing," "leisure," "personal care," and "sport" were assessed with respect to their environmental impacts as well as their potential for reducing these impacts. When the nature of the environmental impacts within the different consumption domains would be established, the government would be capable of organizing in the next phase of the process a consultation with the main actors involved in the various consumption domains. For the social practice of food or cooking, these so-called chain platforms—bringing together all the major actors in the food chain—were already organized in 2001.[11]

The innovative element of the domain explorations refers to the effort to translate the environmental goals, as set the national level, into a series of daily routines. However, the job of translating was only halfway completed, since the domain explorations still hold on to the technical vocabulary that appears to be characteristic of environmental policymaking in general. In terms of the social practices model, one could conclude that the domain explorations delivered the raw materials from which the environmental heuristics for the different consumption domains should be processed.

"Citizens and the environment"—experiments

A second series of policy pilots, conducted under the label "citizens and the environment experiments" (Burger en Milieu-experimenten), stressed the need to take into account the life-world rationality that citizen-consumers use to enact their daily consumption routines (B&A Groep 2000). To illustrate the pressing need to evaluate environmental policymaking along these lines, researchers went out to visit shopping malls, pubs, and child care centers to observe "on the spot" the (environmentally relevant) behaviors of these "situated" groups of citizen-consumers, and to invite them to a focus-group meeting to be organized at a later moment. During these meetings, which were also attended by environmental policymakers, citizen–consumers could bring to the fore their views on the (mis)-match of formal environmental policies on the one hand and the rationality of their everyday life on the other. From these "citizen and the environment" experiments it was revealed first that environmental policies in general were lacking a sensitivity to the problems of everyday life, and second, that for most of the policy goals, most of the time, most of the citizens are not able to establish the link with everyday life by themselves.

The "future perspective" project

Assuming that the environmental goals with respect to all the relevant consumption domains (clothing, feeding, traveling, etc.) are clear and also accepted by citizen-consumers, would it then actually be possible to substantially reduce the environmental impacts of daily routines in modern industrialized societies like the Netherlands? In other words, are there enough sociotechnical innovations (made) available to citizen–consumers that they can make use of to green their lifestyles without the need to abandon the existing high quality levels of modern consumption? To explore these possibilities, a third research experiment was conducted in concerted action by policymakers and environmental researchers. In the "Future Perspective" project (Project Perspectief), a small number of householders accepted the challenge to substantially reduce, within a circumscribed period of time, the environmental impacts of their daily consumption routines. In exchange for their willingness to participate, citizen–consumers were provided with the relevant resources (money, knowledge[12]) by the government. From this research project it was learned that the possibilities for sustainable consumption are indeed different within different domains of daily consumption. Some routines are more difficult to change than others, partly because the emotions invested in them are different, but definitely and substantively also because of the fact that the level of provisioning of green alternatives turned out to be different for the distinguished social practices.[13]

The various research and policy experiments just described help to develop the future agenda for consumer-oriented research and policymaking. The social practices model can be seen as an attempt to combine some of the essential building blocks into an integrated strategy, emphasizing the need to:

- Establish sets of *environmental heuristics for all the major social practices* that can be said to be relevant for environmental policymaking. These environmental heuristics inevitably involve a mixture of the technical rationality that dominates the provider-networks on the one hand, and the life-world rationalities used by citizen–consumers to organize their daily lives on the other. The "domain-explorations" as well as the "citizen and environment" experiments provide valuable inputs for the development of these heuristics. From the "citizen and environment" experiments it can be concluded that the consultation of circumscribed groups of citizen–consumers at different phases of the process is a prerequisite.
- Identify and analyze in some detail the actual and potential "routes for innovation" within the selected social practices. The level of sustainable provisioning in terms of both the quantity and the quality of the sustainable alternatives offered to specific lifestyle groups of citizen–consumers must be determined. With the help of the system of provision perspective, these innovation routes can and should be specified in terms of the modes of design, production, distribution, access, use, and disposal that prevail in the selected social practices or consumption domains. To understand why some sociotechnical innovations are being picked up by specific groups of citizen–consumers and others not, lock-in factors or "slots" have to be indentified and analyzed, as they can occur at all the relevant sections of the production–consumption chains.[14]
- Provide a description and analyses of the different lifestyle groups that emerge from the *stylized and symbolized use* that is being made of the environmental innovations at the level of the particular social practices. This implies not only that new conceptual tools have to be developed to discuss classes or "tribes" (Bauman 1983) of sustainable consumers, but also that new research methodologies and forms of data collection are needed.[15]
- (Re)define the role of both Governmental and Non-Governmental actors in consumer-oriented environmental policymaking at different levels. These roles will be different for the different social practices and also depend on the phase of the process and the tasks at hand.[16] The urge to take into account participatory forms of policymaking should be obvious in this respect.

This agenda may turn out to be overly ambitious, and it will take a considerable amount of time and research to develop it into a body of knowledge that is scientifically solid and relevant to policymaking. However, the goal of turning environmental policymaking and policymakers a little into the direction of citizen–consumers makes it worth the effort.

Epilogue

While our analysis of sustainable consumption draws upon empirical research from European countries only, we are well aware of some of its eurocentric biases and the problems that can occur when applying the analysis in other parts of the global network society (Mol 2001). For example, the greening of food practices of consumers in rich countries could very well result in new inequalities within global food chains, and ecotourism has been recognized already for its social distributional impacts. To protect the emerging field of the study of sustainable consumption from small-mindedness, we would argue for a transnational, comparative approach (Wilhite 1997). The further development of international networks of researchers on sustainable consumption[17] should be welcomed in this respect.

Notes

1. We are aware of the fact that many authors have developed and refined the original formulation of the Attitude–Behavior model since its first introduction, without altering, however, the basic assumptions on which the model rests. For a recent overview, see the discussion by Peter Ester in Beckers et al. (1999).
2. For an interesting piece of research which tries to combine both perspectives, see Thøgersen and Ölander 2001.
3. Eco-teams are part of the international Global Action Plan (GAP). The eco-team brings together for several times a small number of people who are interested in reducing the overall environmental impacts of their domestic consumption (Staats and Harland 1995).
4. See also section on policymaking and research on sustainable consumption in the Netherlands, where the so-called domain explorations (Domeinverkenningen) are being introduced and discussed.
5. A case in point to illustrate the power relations discussed here is the way in which the actual use of utility flows is measured or monitored. Metering technologies can be applied in such a way that they serve the surveillance of end users by providers or, alternatively, to make the expert system before the meter more transparent to the end user. See also van den Burg et al. (2001) and Marvin et al. (1999).
6. For a detailed description of over 150 sociotechnical innovations in the three countries from the Domus project, see Raman et al. (2000).
7. In Amersfoort, for example, there were 500 houses included, covered with 12,000 m^2 of PV solar panels.
8. Van Mierlo looked into the factors that could help explain the differences between the Amsterdam and the Amersfoort projects and arrived at the conclusion that the different dynamics between the providers networks involved contributed most to the explanation (van Mierlo 1997).
9. These uneven developments are described in a series of "national reports" on the historical development of the utility sectors in Sweden, the Netherlands, and the United Kingdom. For example, while the United Kingdom is a front-runner in privatization, the Netherlands and Sweden have in general a more developed environmental history in utility provision. Also in almost every country the energy sectors turn out to be ahead of the water sectors with respect to privatization. And last but not least: All privatization efforts are fueled by EU policies to a considerable extent, as well as most of the environmental measures which go along with these changes (Chappells et al. 2000).
10. The Dutch government, for example, agreed to liberalizing first only the green electricity schemes, while also offering relief from existing eco-taxes on

domestic energy to those householders who selected such a scheme. As a result, green electricity markets are no longer regarded as insignificant niche markets, with 1 million users (total population 16 million) of green electricity 1 year after liberalization of the market (www.greenprice.nl) in the Netherlands and with revenues from eco-taxes in many countries of the EU mounting from 6 up to 10% of total taxes. In the Netherlands in 2001 eco-taxes mounted to 15 billion euros, representing 15% of total tax revenues (Bos 2001).

11.   The chain platforms were inspired by the rather successful forms of horizontal policymaking as applied in the 1980s with respect to different target groups in the sphere of production. These forms of joint environmental policymaking (JEP) resulted in a considerable number of voluntary agreements between government and industry and were regarded as rather effective from an environmental policy-making point of view (for further information on JEP, see Mol et al. 2000).

12.   Twelve families received a 10% increase in income and were regularly assisted by a "coach" who helped to organize the monitoring of the environmental impacts on a daily bases and who could also advise the families about possible alternative courses of action (CEA 1999). A software program was made available to the families to check the environmental impacts of the products they purchased.

13.   The consumption pattern of the participating households were again evaluated a year and a half after the formal ending of the project. In turned out that, for example, the new established routines in the area of food were judged by the participants to be more easy to sustain than those in the area of leisure/tourism (Rescon 2000).

14.   As will be manifest from the language used in this respect, we think environmental sociologists can profit very much from the works on the innovation and diffusion of technology as conducted by Pinch and Bijker (1987), Schot (1992; 1998), Kemp (1994), Rotmans et al. (2001), Shove and Southerton (1998), and many others.

15.   Research methodologies and data sets as developed by Socio-Consult in the context of the Bourdieu inspired "motivaction project" might serve as an example here (Motivaction 1999).

16.   In Spaargaren (2001) we provide a discussion on the different phases of the process in terms of the policy cycle model as it was introduced in the Netherlands by Winsemius (1986).

17.   The Lancaster-based ESF network on "consumption, everyday life, and the environment" could be mentioned as an example, although less developed countries are under-represented.

References

B & A Groep. 2000. *Burger en Milieu. Verslag van een verkenning naar de potentie en meerwaarde van 'burger en milieu'.* Hoofdrapport. Den Haag: VROM.

Bargeman B. 2001. *Kieskeurig Nederland.* Dissertation. Tilburg: Tilburg University.

Bauman, Z. 1983. Industrialism, consumerism and power. *Theory Culture Soc.* (3):32–43.

Beckers, T. A. M., P. Ester, and G. Spaargaren. 1999. *Verklaringen van duurzame consumptie: Een speurtocht naar nieuwe aanknopingspunten voor milieubeleid.* Publicatiereeks Milieustrategie 1999/10. The Hague: VROM.

Beckers, T. A. M., G. Spaargaren, and B. Bargeman. 2000. *Van gedragspraktijk naar beleidspraktijk: een analytisch instrument voor een consument-georiënteerd milieubeleid*, 78. The Hague: Ministerie van VROM, 2000. Publicatiereeks milieustrategie 2000; 8.

Blok, K., and K. Vringer. 1995. *Energie-intensiteit van levensstijlen*. Utrecht: Vakgroep Natuurwetenschap en Samenleving.

Bos, W. 2001. Public lecture by Secretary of State Wouter Bos on the significance of Green Taxes on behalf of the ministry of Financial Affairs in the Netherlands. Wageningen.

CEA. 1999. *Minder energiegebruik door een andere leefstijl?* Eindrapportage Project Perspectief. The Hague: VROM.

Chappells, H., M. Klintman, A. L. Lindèn, E. Shove, G. Spaargaren, and B. van Vliet. 2000. *Domestic consumption, utility services and the environment*, p. 185. Final Domus report. Wageningen: Wageningen University.

Cohen, M. J., and J. Murphy, eds. 2001. *Exploring sustainable consumption: Environmental policy and the social sciences*. Amsterdam: Pergamon.

Cowan. R. S. 1983. *More work for mothers*. New York: Basic Books.

CREM. 2000. *Domeinverkenning recreëren. Milieuanalyse recreatie en toerisme in Nederland*. Amsterdam: CREM.

Elias, N. 1971. *Wat is sociologie?* Amsterdam: Aula.

Featherstone, M. 1983. Consumer culture: An introduction. *Theory Culture Society* (3):4–9.

Fine, B., and E. Leopold. 1993. *The world of consumption*. London: Routledge.

Fishbein, M., and I. Ajzen. 1975. *Belief, attitude, intention and behaviour*. Reading, MA: Addison-Wesley.

Giddens, A. 1984. *The constitution of society*. Cambridge: Polity Press.

Giddens, A. 1991. *Modernity and self-identity*. Cambridge: Polity Press.

Guy, S., and S. Marvin. 1996. Transforming urban infrastructure provision—The emerging logic of demand side management. *Policy Stud.* 17:137–147.

Guy, S., S. Graham, and S. Marvin. 1997. Splintering networks: Cities and technical networks in 1990s Britain. *Urban Stud.* 34(2).

Guy, S, S. Marvin, and T. Moss, eds. 2001. *Urban infrastructure in transition: Networks, buildings, plans*. London: Earthscan.

Kemp, R. 1994. Technology and the transition to environmental sustainability; The problem of technological regime shifts. *Futures* 26(10):1023–1046.

Marvin, S., and H. Chappels. 1999. Pathways of smart metering development: Shaping environmental innovation. *Computers, Environment and Urban Systems* 23: 109–126.

Mol, A. P. J. 2001. *Globalization and environmental reform: The ecological modernization of the global economy*. Cambridge, MA: MIT Press.

Mol, A. P. J., V. Lauber, and J. D. Liefferink, eds. 2000. *The voluntary approach to environmental policy: Joint environmental policy-making in Europe*. Oxford: Oxford University Press.

Motivaction. 1999. *Milieubelevingsgroepen in Nederland: een kwantitatief onderzoek naar drijfveren, profielen en mogelijkheden tot communicatie*. Amsterdam: Socioconsult.

Otnes, P., ed. 1988. *The sociology of consumption*. NJ: Humanities Press.

Pinch, T., and W. Bijker, eds. 1987. *The social construction of technological systems*. Cambridge, MA: MIT Press.

Raman, S., H. Chappells, M. Klintman, and B. van Vliet. 2000. *Inventory of environmental innovations in domestic utilities. The Netherlands, Britain and Sweden*. Universities of Lancaster, Wageningen, and Lund.

Rescon. 2000. *Het project perspectief anderhalf jaar later. De stand van zaken bij 11 huishoudens*. Utrecht: Novem.

Rotmans, J., R. Kemp, and M. van Asselt. 2001. More evolution than revolution: Transition management in public policy. *Foresight* 3(1).

Schot, J. W. 1992. Constructive technology assessment and technology dynamics: The case of clean technologies. *Science Technol. Hum. Values* 17(1): 36–56.

Schot, J. W. 1998. The usefulness of evolutionary models for explaining innovation. The case of the Netherlands in the nineteenth century. *Hist. Technol.* 14:173–200.

Schuttelaar & Partners. 2000. *Domeinverkenning Voeden. Ingrediënten voor een gezond milieu*. The Hague.

Shove, E. 1997. *Notes on comfort, cleanliness and convenience*. Paper for the ESF workshop on Consumption, Everyday Life and Sustainability, Lancaster.

Shove, E., and H. Chappels. 1999. *DSM working paper*. Lancaster: Domus.

Shove, E., and D. Southerton. 1998. *Frozen in time: Convenience and the environment*. Paper for the second ESF workshop on Consumption, Everyday Life and Sustainability, Lancaster.

Spaargaren, G. 2000. Ecological modernization theory and domestic consumption. *J. Environ. Policy Plan.* 2(4):323–335.

Spaargaren, G. 2001. *Milieuverandering en het alledaagse leven*. Inaugural address, Tilburg University, Tilburg.

Staats, H. J., and P. Harland. 1995. *The ecoteam program in the Netherlands*. Leiden: Leiden University.

Summerton, J. 2000. *Out of anonymity: Shaping new identities for electrons, utilities and consumers in Swedish electricity in the 1990's*. Contribution to the winter workshop on Infrastructures of Consumption and the Environment, Wageningen.

Thøgersen, J., and F. Ölander. 2001. *Spillover of environment-friendly consumer behaviour*. Paper for the 5th Nordic Environmental Research Conference, Aarhus.

TNO. 1999. *Duurzame consumptie: Verkenning Kleding*. Delft: TNO/STB.

Van den Burg, S. W. K., G. Spaargaren, and A. P. J. Mol. 2001. *Environmental monitoring: Opportunities and threats of a consumer-oriented approach*. Paper presented at the ISA conference on New Natures, New Cultures, New Technologies, Cambridge University.

Van Mierlo, B. C. 1997. *De totstandkoming van twee grote Pilotprojecten met zonnecellen in nieuwbouwwijken. Een vergelijking tussen Amsterdam en Amersfoort*. Amsterdam: IVAM.

Van Vliet, B. J. M. 2002. *Greening the grid*. Wageningen: Wageningen University (PhD thesis).

Van Vliet, B., R. Wüstenhagen, and H. Chappells. 2000. *New provider–consumer relations in electricity provision. Green electricity schemes in the UK, the Netherlands, Switzerland and Germany*. Paper for the Business Strategy and the Environment Conference, Leeds.

Vringer, K., and K. Blok. 1993. *The direct and indirect energy requirement of households in the Netherlands*. Utrecht: Vakgroep Natuurwetenschap en Samenleving.

VROM. 2000. *De warme golfstroom: heroriëntatie op communicatie over milieu*. The Hague: Centrale Directie Communicatie DGM.

Warde, A. 1990. Introduction to the sociology of consumption. *Sociology.* 24(1):1–4.

Wilhite, H. 1997. *Cultural aspects of consumption*. Paper for the ESF workshop on Consumption, Everyday Life and Sustainability, Lancaster.

Winsemius, P. 1986. *Gast in eigen huis: Beschouwingen over milieumanagement*. Alphen aan de Rijn: Samsom.

Chapter 18

Joseph Huber

UPSTREAMING ENVIRONMENTAL ACTION

THIS BOOK* IS ABOUT key environmental innovations, i.e. new technologies, products and practices that are of particular importance to solving urgent environmental problems.

One of the challenges of our time still is how to achieve ecologically sustainable living standards with a decent level of affluence for all of the 6–8 billions of people on earth. The approach of technology life cycle analysis explains why a viable answer to the problem needs to be conceived of in terms of ecological modernisation, based on scientific research and technology, controlled by market economies and apt administration, and aimed at structurally changing the industrial metabolism so as to re-embed it into nature's metabolism. Hence the author's long-standing interest in environmentally benign new technologies, referred to hereafter as technological environmental innovations, or TEIs for short.

TEIs are more than just environmental technology such as exhaust-air catalytic converters, or filters in chimneys or sewage water purification plants. This kind of environmental technology is called end-of-pipe technology, applied within the context of add-on measures. They represent a downstream approach in that they come in after the point where some damage or pollution has already occurred, without changing the upstream source of the ecological perturbations in question. Foreseeably, add-on measures will continue to be important in many cases. The far more important portion of TEIs, however, are those new technologies which prevent environmental pollution and deterioration from happening in that they come further upstream in the manufacturing chain and in a technology's life cycle, and change the source of perturbation or avoid it altogether.

TEIs are not necessarily being developed for environmental reasons only. They represent a new generation of innovative technologies that fulfil ecological criteria as much as technical criteria of efficiency, operational safety and reliability. And TEIs have of course to match economic criteria such as price and profitability

---

* [Huber (2004)].

sooner rather than later as they move along their learning curve. Examples of such TEIs include the following:

- replacement of fossil fuels with clean-burn hydrogen, the use of which does not require additional end-of-pipe purification of emissions
- substitution of clean electrochemical fuel cells for pollutant furnaces and combustion engines
- fuelless energy such as photovoltaics and further regenerative energies which make use of sun radiation, geothermal heat, or wind and water currents
- transgenic biochemistry which makes use of enzymes and microorganisms especially designed and bred for various production tasks, thus replacing the conventional high-temperature high-pressure chemistry that is dangerous and poses a heavy burden on the environment
- sophisticated low-hazard specialty chemicals which are, besides other product properties, biodegradable, non-persistent, non-accumulative and non-toxic
- new materials which are simultaneously ultra-light and ultra-strong, thus saving larger volumes of conventional materials and energy
- micromachines and nanotechnology which relieve pressure on resources and sinks compared to larger conventional machines and chemical production
- substituting sonar, photonic and microfluidic analyses for cumbersome conventional methods involving many hazardous ingredients, and thus considerably improving quality and performance of production
- circulatory production processes in which water, auxiliary substances, metals, bulk minerals and fibres are recycled at an optimum rate
- last but not least, overcoming the ecologically devastating practices of today's over-intensified and inappropriately chemicalised agriculture by introducing sound ecological practices in combination with high-tech precision farming and, again, modern biotechnology which makes use of transgenic organisms.

The appendix contains a list of examples from eleven realms of TEIs. It might be advisable at this point to have a look through this list which will give an overall impression of what the empirical basis of this book [Huber 2004] is about.

To clarify the message straight away: TEIs may help to reduce the quantities of resources and sinks used, be they measured as specific environmental intensity per unit output, or as average consumption per capita, or even in absolute volumes. Overriding priority, however, is given to improving the qualities and to changing the structures of the industrial metabolism. Rather than doing less of something, TEIs are designed to do it cleaner and better by implementing new structures rather than trying to increase eco-productivity of a suboptimal structure which has long been in place. TEIs are about using new and different technologies rather than using old technologies differently.

The attitude does not arise out of taking 'creative destruction' as an ideology, but is derived from the fact that successful systems innovations which represent new technological paradigms and which break new ground tend to have a much bigger learning-curve potential than mature generations of technologies which are approaching the end of their learning curve. TEIs, to give an example, certainly include recycling of organic solvents. Preferable, though, is to replace organic solvents, particularly in open-air applications, with water-based agents or super-critical carbon dioxide. This does not exclude those special cases in which organic solvents remain an unbeatable auxiliary for the time being. In these cases it may in fact be preferable to use organic solvents in a leakage-tight closed-loop process.

An environmental strategy based on TEIs certainly cares about ecological *sufficiency*, i.e. keeping the use of resources and the burdens on sinks within critical limits. This, to be sure, presupposes the availability of some valid knowledge about these limits, not just vague precautionary fears. Given these limits, TEIs will also have to attend to ecological *efficiency*, though this is not an end in itself. If, for example, the ecological problem arises from burning carbon or plutonium, burning it more efficiently does not make much sense. Instead, different fuels or completely different technologies of power generation will have to be developed and diffused. In this sense, an environmental strategy based on TEIs will above all care about the ecological *consistency* of the industrial metabolism as illustrated by the TEI approaches mentioned above.

The idea of eco-consistent TEIs is 'fitting in with' in order to avoid anthropogenic destabilisation of geo- and ecodynamics. This is achieved by contriving technological ways and means of maintaining an industrial metabolism that is effectively compatible with nature's metabolism at an optimum level of efficiency, even at large volumes of materials turnover. This contrasts with an oversimplistic notion of 'dematerialisation' that lumps together totally different things, makes no difference between clean and dirty energy, hazardous and harmless substances, and develops no understanding of the major or minor ecological sensitivity of different kinds of materials.

To give just two examples. One is the automobile as the supposed 'environmental enemy #1'. The automobile does indeed have a long list of environmental problems associated with it, ranging from road safety and noise, to land utilisation, soil sealing, and further to the consumption of raw materials on a large scale and to a complex problematic of airborne emissions. All of these problems, however, have proven to be accessible to technological regimes which can keep the environmental impact to acceptable levels, with the exception of CO_2 and some other airborne emissions caused by gasoline burned in internal-combustion engines. So, in principle, there is not too big a problem with being an automobile society, something which most people consider to be an achievement. But there is a major problem with metabolically inconsistent vehicle propulsion, a problem that could already have been solved for the most part if industrial elites had not been so sluggish in developing the necessary awareness and will to deal with that bottleneck.

A second example is energy demand. Intensive use of energy is the physical basis and one of the most significant indicators of industrial development. High and still growing levels of per capita consumption of energy are normally not based on inefficient energy consumption. Efficiency increases right from the beginning of a technology's life cycle, and growth is in turn based on increasing efficiency.

So any environmental strategy aimed at 'using less energy', in the sense of bringing down the overall level of energy demand, would have detrimental effects on the ongoing evolution of modern society. What we need instead is a strategy of clean energy which fits in with nature's metabolism even on a large scale in the giga and tera range. This approach certainly entails using energy as efficiently as possible. But the real purpose of increased efficiency along the learning curve of a system, in technical systems as much as in natural organisms, is never to get by on less but to stabilise further growth and development. As can be seen from most of the cases represented in the TEI database of this book [Huber 2004], important eco-consistent TEIs tend by themselves to come equipped with much higher eco-efficiencies than previous like technologies have. Mere increases in efficiency, by contrast, tend to occur during the later life cycle stages of mature technologies. The conclusion is obvious: Rather than calling on green saving commissioners, TEIs are calling for green inventors, innovators and investors.

Not long ago, green-minded people used to be at odds with the idea of the greening of industry. Thinking of technological environmental innovation, and expecting science, technology, industry and finance to take a leading role in that greening seemed to be a contradiction in itself. As a Neapolitan colleague once put it, it seemed like entrusting the Mafia with the task of combating crime. Today, green ideas have been diffused throughout society, big science and industry included. The environmental movement has assimilated itself to the general science and technology paradigm such as it is dominating the knowledge base of modern society. Pinning hopes on technology is no longer automatically deemed 'techno-cratic'. Technology has actually turned out to be the key component of any viable response to ecological challenges.

*   *   *

Environmental policy has much to gain from two basically simple insights. First, solving environmental problems means looking for metabolically consistent and efficient technological solutions, thus TEIs. Second, these solutions are to be found upstream rather than downstream in the manufacturing chain and a technology's life cycle.

Certain scholars of economics, sociology and humanities tend to belittle the role of technology, be that for pretensions to the supremacy of their own discipline, or out of fear of succumbing to a 'technocratic' ideology, while actually missing the pivotal function of technology in modern society which is obvious to everybody else. Some of the classics have clearly recognised the role of technology, among them Karl Marx, Joseph Schumpeter, Jacques Ellul, Lewis Mumford, Daniel Bell, all of them anything but apologists for technology. Social sciences must incorporate a social theory of technology and technosystems development. Otherwise they cannot fully cope with their task of explaining social systems and the place of humans therein. This is all the more true with regard to the environment and human society's relationship with nature.

Ecology refers to the physical exchanges which a population of a given number and a given level of instrumental capacities realises within its living space, i.e. the locations where a population lives and the entire space which is covered by its activities. In particular it includes the natural resources a population makes available and the environmental sinks it employs to get rid of resulting phase-out products. It also includes cooperation and competition with other populations concerning the utilisation of resources and sinks within that living space. Ecology thus deals with the metabolism between a population and its natural surroundings within the geo- and biosphere, how in this process the environment is being transformed, and how populations and their environment co-evolve.

Environmental problem solving needs to be technological in the first place because environmental problems are perturbations of ecological systems of which humankind is part by way of physical operations carried out through its instrumental capacities, i.e. the technology of the time. The origin of ecological perturbations may be geogenic or anthropogenic, or both, as is the case with global climate change. Anthropogenic perturbations are immediately caused by the technological means employed to fulfil certain operative functions. That is why resolving the problem unavoidably includes restructuring or substitution of the technology in place. Also with geogenic perturbations, some kind of technological solution will be needed to face the challenge. Typically these are measures aimed at shielding humans (as well as animals and plants which are considered to be useful) from natural influences, and protecting land and settlement structures.

Ecology is a natural-science discipline which refers to something physical. 'Ecology of freedom' or 'ecology of mind' are unsuitable terms which confuse metabolic operation (exchange of physical substances) with communication (exchange of ideas). It is certainly true that ecological effects which are physically caused by human operations originate ultimately in the value base of humans, in their worldview and certain cultural concepts such as utilitarianism in modern society. In the first instance, however, ecological effects come about through physical operations which modify or transform the state of the environment, for example, by gathering, hunting, building, manufacturing, by extracting, processing and using physical things in the realm of res extensa.

Higher animals and particularly humans have extended the natural instruments of their organism by developing artificial methods and instruments, embedded in more complex social organisation: tools, machines, infrastructures and cooperative arrangements, in brief, technology which serves special operative purposes. Technology thus is the immediate ecological factor in human society besides the fact of humans' sheer biological existence.

The metabolism effectuated by labour and technology needs to be analysed in terms of engineering and natural sciences such as physics, chemistry and biology. At that level ecology cannot be understood in terms of the social sciences, including economics. For example, postulating that 'the problem of anthropogenic climate change is greed' or 'economic growth pressure' or 'distorted prices' or 'compensatory consumption' or 'godlessness' or 'bad government' and 'poor lawmaking', may rightly hint to certain factors which could play a role in a more complex analysis, but it misses the ecological fact that, at the physical source, the problem is an inappropriate way of utilising resources and sinks through technological regimes which are apparently less highly developed than is commonly believed.

In modern society, the metabolism between humans and their environment has taken on the forms of industrial metabolism (Ayres 1993, Ayres/Ayres 1996, Ayres/Simonis 1994). The resulting relationship of industrial society with its environment represents an industrial ecology (Lowe 1993, Allenby/Cooper 1994, Socolow et al. 1994, Frosch 1996). It results in an ecological transformation (Bennett 1976) of the earth's ecosystems unknown in scope and scale to traditional societies. Industrial man is currently causing a per-capita turnover of materials that is 5–10 times the quota of traditional agrarians, in the same way as the agrarian quota in its time was 5–10 times that of primitive hunter-gatherers. Industrialised contemporaries live on 1,320 kg a day, while their stone-age ancestors are said to have lived on just 35 kg (Fischer-Kowalski 1997). Facing the responsibilities ensuing from this opens up the perspective of earth systems management or eco-systems management (Allenby 1999). Today's still rather unfocussed environmental policies represent preliminary steps towards such comprehensive sustainability regimes. Their effective tool will be technology because intended as well as unintended ecological effects are caused by nothing other than powerful technologies in a context of industrialised production.

There is no 'post-industrial' society today and none can be envisaged in the foreseeable future. What is called the service economy, or the information economy, knowledge society, high-tech society or scientific civilisation, remain different aspects of industrial society progressing on its evolutive path. Expecting environmental ease from the sectoral changeover to the service economy has been one of the vain ideas prevalent in environmental discourses. The service society is based on industrial production in ever larger volumes. The supposed paperless office of the future quickly became another chapter in the book of yesterday's tomorrows. Another such futile expectation is that of substituting communications technology for transport of passengers and goods, when in real industrial history transport and communications have always developed in tandem. Industrial productivity is the basis upon which services build up. As a consequence, the service society is much more resources- and energy-intensive than earlier industrial society used to be, in the same way as that earlier industrial society was based on ever more productive agriculture and proved to be much more agro-intensive than any societal formation before.

We can, of course, subdivide the ongoing development of industrial society into different epochs. For example, there are recurrent epochs of technological systems innovation, which introduce new key technologies and so lay the foundations for new industries and markets. These are the so-called long waves – also named, after two of the early researchers in this field, Schumpeter cycles or Kondratievs – occurring within a window of opportunity that seems to open up for about one or two decades every 40–60 years. Thus far they have been mechanisation in the decades around 1800, railroads around 1850, electrification and chemistry around 1900, mass motorisation and telecommunications around 1950, and computerisation, the now somewhat aged new economy, in the past decades.

Another such subdivision that could prove to be useful is that of industrial ages, to be measured in centuries rather than decades. So far, two of these could be identified: industrial age I from the beginnings of the industrial revolution more than 200 years ago until the 1960–80s, and industrial age II which has been in the ascendant since around the middle of the 20th century. This can be seen as another long-term stage in the transsecular life cycle of ongoing modernisation of nations in the world system. The defining feature which distinguishes the two is the basis of reference for modernisation. While industrialisation in age I took place in a context of traditional practices of agrarian, crafts and household production which it replaced step by step, industrial age II is now proceeding by modernising the now older modern structures that were created during industrial age I.

Contemporary sociology has chosen to call that process of self-reference 'reflexive modernisation' (Beck et al. 1994). Aficionados of business buzz-words would say industry is re-inventing itself, for example, by substituting the 21st century's specialty steels and a broad range of new materials for the 19th century's cast iron which in turn replaced the wood and stone materials of earlier centuries; and by substituting the 21st century's clean fuels and fuelless energy for the 20th century's oil and plutonium, and the 19th century's coal that replaced the wood, straw and peat of traditional society.

As industrial workers replaced feudal peasants and craftsmen, the industrial workforce of the 19th and 20th centuries is now being replaced by skilled specialists and scientifically educated experts. Production has increasingly become based on scientific methods and advanced technology. The greater part of employment and economic turnover may indeed occur in services, though most of these have taken on scientific characteristics themselves. Apart from this, many services remain directly linked to industrialised production in agriculture, basic industries, manufacturing and crafts. In advanced industrial society each activity is becoming technology-based or technology-intensive, as all technologies are increasingly based on scientific methods of research and development.

There has been little controversy over the expectation that 'post-industrial' society would be high-tech. Hopefully there will soon be no more controversy either on the idea that high technology will be used to environmentally re-embed society. During the last decades of the 20th century, industrial age I has increasingly been coming to its end. At the same time, the high-tech society or industrial age II has been arriving rather than still coming.

*   *   *

The purpose of technology is to enable and empower human operations. Technology is the working medium in the operative subsystem of industrial society. In that function, technology is socially embedded in many ways. To put it in a slightly different way, it is functionally interrelated, co-related with, and co-directional, sometimes counter-directional, to other societal subsystems such as the economic system, the ordinative system of law and administration, the political system, the knowledge system, the value base and further aspects of a society's culture.

Starting from such a systemic–evolutive basis of analysis, we cannot fail to understand that no technical change can occur without concomitant changes in the knowledge base and co-directional inspirations from the value base. Conversely, no sustained change in social ideas can occur without co-directional restructuring of social practices, particularly patterns of work and production, i.e. society's technosystem and co-related institutional and economic control systems. In this sense, a transition to ecologically readapted technologies and practices in, say, agriculture requires that those involved gain certain ecological insights, change their mindset accordingly, acquire more sophisticated knowledge, take decisions on political goals, with whom to cooperate, with whom to compete and whom to fight, how to regulate what needs to be done, and to what ends to invest available money.

Formative factors such as awareness, consciousness, values and knowledge should not be set up in opposition to effectuative factors such as law, financial means and technology, because all of these belong to functionally different subsystems that need to co-evolve. They certainly display individual features, specific 'laws' and constraints of their own, but they do not have an autonomous existence detached from the entire system. Within the whole system, the different subsystems act as a selective and conditioning context to each other, resulting in system integration and functional control.

Functional interrelatedness signifies that the formative subsystems of culture and politics and the effectuative subsystems of law and administration, economy, and production, are all interdependent. For its evolvement, each subsystem needs corresponding evolvements or conditions in other subsystems. Functional inter-relatedness thus clearly contradicts any hypothesis of 'factor substitution'. For example, money and capital cannot substitute for work, technology and resources. Substitutional alchemy has been a feature of neo-classical economics, including environmental economics, which was brought back down to earth by ecological economics (Krishnan et al. 1995, Jansson et al. 1994). In much the same sense, environmental awareness cannot substitute for green technology, neither can good consumer behaviour substitute for sustainable production practices, nor administrative procedures and financial policy instruments for those technologies which need to be invented, developed and used. Regulation and financial means can selectively control, but never create and place themselves in lieu of what they control. And seen from the particular angle of ecology, i.e. the geo- and biospheric metabolism between humans and nature, whatever we may believe, think, wish, decide, order and spend money on, nothing will effectively change unless we change technology.

* * *

According to the empirical findings underlying this book [Huber 2004], techno-logical environmental innovations (TEIs) can in general be characterised as being upstream rather than downstream, i.e. upstream in the manufacturing chain or product chain respectively, as well as upstream in the life cycle of a technology.

In the environmental literature, the term 'life cycle analysis' is often used interchangeably with 'product chain analysis' or 'eco-balances' that try to gauge the environmental impact of a product from first input by extraction of raw materials to last output by being definitely phased out as waste, the analysis in this sense stretching 'from cradle to grave'. What they represent more precisely, however, is analysis of the vertical manufacturing or product chain. By contrast, another mean-ing of life cycle analysis, the one preferred here, refers to an innovation life cycle or technology life cycle, i.e. to the evolutive existence of a kind of product or technology, to which the metaphor 'from cradle to grave', i.e. from being born to dying, would apply much better. Life cycle analysis then describes the creative process by which a new technology or product comes into existence, by which it is furthered and diffused throughout a population of adopters, until it finally reaches a stage of retention, or decline and intended phase-out or unintended die-off.

In this sense, technological life cycle analysis includes the stages of structura-tion of an innovation, from invention or discovery, via successive steps of develop-ment to the unfolding of that innovation, followed by later stages of maturation and increasing structural stability before eventually becoming too firmly established to remain adaptive enough for further evolvement. This process of the structural unfolding of an innovation is related to a process of societal diffusion from emergence to take-off and niche saturation. In this process, innovative promoters of a novel item are opposed by the established defenders of previous like items. In terms of evolution theory, the confrontation between challengers and defenders, which can in most cases not be avoided, is a genuine struggle for life, a struggle to gain and maintain a living space, a 'niche' in a society's minds and markets, a position in the divisional structure of society, and to serve a function in its subsystems, which in the case of TEIs are operative production functions that meet ecological requirements and human needs. In the process of unfolding of an innovation life cycle, producers and users of a novel thing have as important a role to play as the inventive originators, and then communicators, investors and regulators who are present throughout the process of innovation.

The discussion of metabolically consistent TEIs thus relies on a coherent framework of innovation life cycle analysis as outlined in part II of this book [Huber 2004]. Even though innovation life cycle analysis builds upon the wealth of empirical research and theorising that has been done in this field during recent years, part II does not represent an all-encompassing recapitulation of innovation and diffusion theories. Rather, it represents a rearrangement and streamlining of particular categories with regard to their specific potential to contribute to the understanding of TEIs, combined with the introduction of a number of new categories and considerations.

If this book [Huber 2004] were an academic dissertation, part II would have come first, and part I second, because part II lays down the general model and hypotheses of innovation life cycle analysis as the basis on which the more specific discussions of TEIs in part I are drawing. Wherever certain terms or statements in part I may need more explanation in terms of innovation life cycle analysis, the corresponding references to chapters in part II are made in brackets. To scholars of processes of technological innovation and societal diffusion of innovations, part II might be worth reading in its own right.

As will be discussed in more detail, it is important here to understand that with technologies, as with living organisms, the key features of a novel thing and its life course are determined upstream rather than downstream, for instance with regard to an organism, in its genetic code and in its early days of growth, experiencing and learning; and similarly with regard to technologies, in their conceptualisation and design, in the early stages of research and development, i.e. the early stages of structuration and diffusion. What remains to be determined during later stages consists of incremental changes and modifications of minor importance, after the point of inflection of a learning curve has been passed. Most of the environmental pressure which is caused by producing and using a certain kind of product or technology is determined right at the beginning with the conceptualisation and design of that product or technology. Once in place, there is not much left which can be done with it, aside from some improvements in later new-generation variants of that product, some percentage points of materials and energy savings in the factory, and a few percentage points by being a good consumer. But a truly significant difference can only be made upstream in the life cycle, by changing path on the basis of new technological systems, new products and new practices.

*   *   *

Beyond and in addition to 'upstream' in the sense of occurring during early stages of a learning curve, the second meaning is upstream in the vertical manufacturing chain (the product chain as shown in Figure 18.1). The product chain includes the following steps which lead from upstream to downstream:

- Starting from extraction of raw materials, i.e. denaturation of resources in mining, quarrying, agriculture, forestry and fishery
- via processing those materials in successive steps in order to obtain food, fuel, working materials and auxiliary agents
- to producing intermediate products, production machinery and infrastructure, all of which flow into or contribute to
- the production of buildings, vehicles and other short- and long-lived, simple and complex end products, which are in turn going into
- final use or consumption, which is just another form of production, thus the next step in the product chain, in that it uses physical things in order to obtain certain results (as any production consumes resources and sinks, and any such consumption results in metabolic products)
- until being recycled back to the foot of the entire chain as used material or waste, thus on the one hand becoming secondary raw material or fuel for re-entry into the product chain upwards, or on the other being phased out of human production and released back into natural cycles as re-natured or near-nature materials.

As a rule of thumb one can say the more products and production processes are placed chain-upwards the more important the potential of their environmental

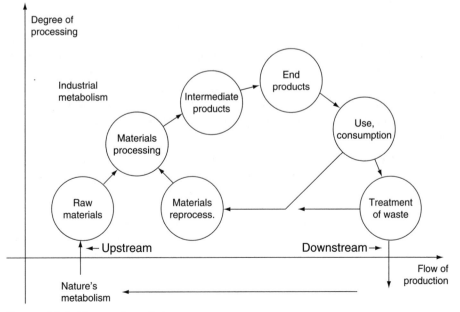

Figure 18.1 The product chain

impact is. This is particularly the case with regard to the difference between all of the steps of extracting and successively (re)processing materials, which regularly cause large environmental impact, and the final assembly or otherwise finishing of final products, which causes comparatively less. This insight has implicitly been present in concepts of gauging environmental impact such as the ecological footprint by Rees and Wackernagel (1994) or the materials rucksack by Schmidt-Bleek (1994).

In car-making, for example, the production steps which have high environmental impact are the production of metals, rubber, plastics and textiles from the extraction and processing of the raw materials to their shaping, surface working and coating. The rest, which includes many steps of transport and successive finishing and assembly, is tainted with much less environmental impact. So production steps concerning the original extraction and processing of materials, fuels and food, and the related generation of working energy, seem to be more important targets of TEIs than later steps downwards in the product chain related to the finishing of end products and to final use or consumption.

There is, however, an important exception to this rule: long-lived complex energy apparatus such as motors in cars, jet engines, power plants, heating systems and electric appliances the use of which consumes large quantities of fuel. Food and fuels have a metabolic rate of almost 100 per cent. This makes a big difference to materials which keep their physical structure when used, even if in the production process of materials these are also partly transformed in their physico-chemical structure. Metabolic products of intensive food and feed consumption are highly problematic as we know from environmental and hygienic problems of agriculture and sanitation, although some of that impact has been brought under control. The same holds true for intensive consumption of fuels, particularly carbon and nuclear fuels.

If there are important, unresolved environmental problems to be found in the final assembly, use or consumption of end products, they indeed have to do with the fuels and the energy apparatus involved in using those products. This is most obvious with regard to houses, vehicles and all kinds of electric appliances used in homes and offices. For example, building houses and roads involves vast flows of construction materials and rubble. Most of this, however, with the exception of the production of cement, does not pose a severe environmental problem. By far the most important environmental impact of a building is rooted in its energy design, i.e. in heating and air conditioning.

Energy remains of the utmost importance because there is an energy input into each step along the entire product chain. As a consequence of this another upstream orientation can be formulated: energy-related aspects of producing and using products should be viewed as upstream of any other aspect of product design and production process. This is valid at least for the present age of carbon and nuclear energy. Clean and safe energy must be a priority target in developing TEIs because many things can be done in favour of metabolically consistent and efficient products and processes on the precondition that there is clean and safe energy available.

* * *

A paradigm shift from downstream to upstream in the vertical product chain and technologies' life cycles implies a parallel shift in the emphasis of environmental action and policy.

First, action has to be refocused onto those industrial operations where large environmental impact actually occurs – in energy, raw materials, agriculture, chemistry and base industries, partly also in building and vehicles, not, however, in distribution, use and consumption.

Second, the key actor groups that have to be mobilised are technology developers, product designers and producers rather than users and consumers. With a focus on products and production, one would not be spending too much time with services, user behaviour and consumer demand. According to the environmental paradox of consumer society, it may be true that environmental effects are ultimately caused by attitudes and the demand of final users, but most of the ecological pressure of producing and using a product is normally not caused end-of-chain in use or consumption, but in the more basic steps of the manufacturing chain, and, seen in a perspective of life cycle analysis, by the conceptual make-up, the technological principles and design of that product. Approaches such as 'sustainable consumption' or 'sustainable household' are undoubtedly useful to some extent. But due to their end-of-chain approach they cannot be particularly effective in changing the industrial metabolism.

Third, demand will continue to play an important role, though the decisive part of it is demand by the manufacturers of end products such as buildings, vehicles, appliances and other more complex goods, and also demand by large retailers such as mail order firms, rather than demand by end users and consumers. Demand is a selective factor in the diffusion of innovations. Demand itself, however, cannot innovate, it cannot 'buy into existence' things which are not on the market yet, with one exception, which is the demand by key manufacturers and large retailers. They are in the position to effectively influence innovation processes and to implement supply chain management, i.e. act as defining chain managers, because they are able to directly exert influence on suppliers along the product chain and have a defining influence on the design and redesign of products. Innovative chain management is none of a user's or consumer's nor of a government's business.

Fourth, government's business is to implement policies which foster TEIs. Upstreaming environmental policy would induce a shift of emphasis from regulation to innovation. Environmental policy will thus also become a policy of technology development, or will have to cooperate systematically with technological R&D. In consequence there would have to be a shift in the pattern of environmental policy from bureaucratic command-and-control to coordination of national innovation networks, European at the EU level, including suitable and well balanced financial support by granting regular research funds and seed money, as well as providing venture capital and introductory aids in appropriate ways.

What has to stay, though, is to set strict environmental performance standards. This remains by far the most effective control instrument for environment and innovation alike – which is not astonishing given the fact that environmental performance standards are of a parametric technical character by themselves.

[. . .]

Appendix: systematics of technological environmental innovations

Energy

Safer nuclear

    Pebble bed reactor

    CAESAR (Clean and environmentally safe reactor; just uranium-238)

    Transmutation of nuclear waste to shorten half-life to a few decades

    Nuclear fusion (unclear)

    Cold fusion (unclear)

Cleaner carbon

    Beneficiation of coal (artificial ageing, desulphuration)

    Synfuels (synthesis gas, liquefied coal)

    Pressurised-fluidised bed combustion

    Co-generation of heat and power, e.g. district heat and power stations

    Combined cycle power plants (steam turbine plus hot gas turbine)

    Integrated gasifier combined cycle IGCC, and

    CO_2 sequestration in central power stations, in connection with steam reformation (production of H_2)

New fuels

    Methane hydrate

    Biofuels

        Fuel crops

        Biodiesel

        Combustion of biomass

        Biogas from fermenters

    Silane (unclear)

    Hydrogen production

        Steam reformation

        Electrolysis

        Gravitational electrolysis

        Solar hydrogen

        Brown's gas

    Hydrogen storage

        Cryotanks

        Absorption into sodium hydride (Borax)

        Absorption into metal hydride

        Nanocubes

Fuel Cells

    SOFC (solid oxide fuel cell) i.e. high temp, large, stationary

    MCFC (melted carbonate fuel cell) i.e. high temp, large, stationary

    PAFC (phosphoric acid fuel cell) i.e. medium temp, medium-sized, stationary

    PEM-FC (proton-exchange membrane, or polymer-electrolyte membrane fuel cell) i.e. relatively low temp, small and miniature, stationary and mobile

    DMFC (direct methanol fuel cells) i.e. low temp, small and miniature, mobile and portable

Flow cell batteries (large plants for interim power storage)

Latest generation batteries and accus

Micro-electro-mechanical systems (MEMS)
    Free piston micro engines
    Miniaturised gas turbines
    Nuclear MEMS

Photovoltaics. Next generation solar cells made of
    Ultra-thin-layer silicon
    Special metals (copper-iridium-gallium, cadmium-telluride)
    Flexible organic polymers with inorganic nanorods

Solar thermal
    Solar thermal heat
    Concentrating power
            Parabolic trough plants
            Parabolic dish reflectors
            Solar towers

Wind
    Up-current towers
    Wind turbines, stand-alone and in wind farms
            Offshore wind farms
            Lightweight wind turbines with hinged blades

Hydroelectric power
    Running-water power stations (river barrages)
    Storage power stations (large dams)
    Tidal power
            Tidal power plants
            Tidal underwater turbines (watermills)
    Underwater-current turbines
    Wave power
            Power buoys
            Limpet (land-installed marine-powered energy transformer)
            Energy ship (hybrid of wind and hydropower)
            Ocean thermal energy conversion (OTEC)

Geothermal
    Heat in combination with heat pumps (about 100 m depth)
    Combined heat and power plants (hot water from 4–5,000 m depth)

Distributed power generation (micropower)
Integrated two-ways-flow grid management

Natural Resources

Cascadic retention management of water and groundwater
    Sustainable groundwater regime
    Drainage management of surface waters
    Flood prevention regime (dams and restoration of flooding zones)

Percolation management of rainwater and purified effluent water

Rainmaking in arid regions
Decentral water supply with small plants
Drinking water by desalination of sea water
Low-impact mining (coal, metals)
Low-impact mining (oil)
>
Single shaft recovery
Multi-directional wells, aided by seismic imaging of oil fields
Multi-phase pumping

Mining biotechnology: phytomining, bioleaching
Sustainable forestry
>
Sustainable forestry regimes adapted to local conditions
Licensed logging and certified wood industry

Fisheries
>
Scaling down of open sea fisheries
Aquaculture
>>
Fish farming, naturally bred and transgenic varieties
Offshore ranching, naturally bred and transgenic varieties

Agriculture

Organic farming
Ecological modernisation of conventional agroindustry
>
Fermentation of manure (biogas plants) in intensive animal farming
External auditing and brand licensing in animal and field farming
Adoption of a number of green regime rules in animal farming concerning, e.g., animal medication, animal density, animal transport
Phytase (enzymatic additive to feedingstuffs)
Biological pest control
Integrated pest management
Clean Weeder in combination with new way of growing rice
Precision farming in the field
>>
Sensor-controlled automated rake
Camera-assisted weed-detection
Tractor tyres with variable pressure
Computer-aided and satellite-based systems (global positioning) on and off the field which control precise dosage of seeds, fertiliser, pesticides, water, exactly to the square meter

Closed-loop greenhouses
Closed-loop agrofactories on several floors
Crop and animal identity verification (produce tracking)
Industry crops (natural fibres and high-content crops as feedstocks for industrial production)

Transgenic crops and animals
>
Conventional cross-breeding enhanced by genetic engineering, such as genetic markers

DNA fingerprint technology
Automated genetic screening
GM crops tolerant of pesticides and herbicides
GM crops tolerant of climate stress and hostile environments
GM crops resistant to special pests and immune to special disease
GM high-yield crops of high quality and content-value
Apomixis-stimulated seeds, i.e. self-replicating ('self-cloning') genderless crops
Molecular farming in crops (nutrients, pharmaceuticals and chemical specialty substances)
Molecular farming in animals (secretion of useful extractable substances in the milk of female animals)
Transgenic animals with optimised useful properties
EnviroPig (more complete absorption of phosphate nutrients)
Medaka fish and salmon with higher level of growth hormones
Transgenic spider silk

Chemistry and Chemicals

Biofeedstocks (vegetable oil, fat, starch, sugar, cellulose, fibres, etc.), natural and transgenic
Phytochemistry partly replacing carbo- and petrochemistry
Biotechnological processes
Microbial biosynthesis
Microbial photosynthesis (photolytic production of hydrogen)
Biocatalytic processes
GM-enhanced fermentation
Enzymatic catalysis

Biosensors
DNA-chip technology
Enzymatic
Microbial
Biooptical
Benign substitution of harmless or low-impact substances for
Heavy metals
Persistent organic pollutants
Halogenated compounds
Organic solvents
Concentrated acids or alkaline solutions

Low-impact chemistry (inherently safe products, short-range chemistry, green chemistry)

New Materials

Secondary materials
Secondary raw materials from reprocessing of waste materials
Cycleware
Biotic compound materials (made of natural fibres and plant proteins)

New plastics and synthetic fibres
New metal alloys
New ceramics
Crystal breeding
New composite materials
> Composite materials on the basis of carbon fibres, e.g. carbon fibre reinforced
> plastics, carbon fibres reinforced by synthetic resin
> Metal-like materials such as organic metals (e.g. Polyanilin), aerogels, metallic
> glass or transparent glass-like metals
> Polytronics, i.e. semiconducting materials made of organic polymers
> and metals

Nanomaterials
> Carbon nanotubes (as working material in electronics, chemistry and
> medicine)
> Nanosensors
> Nanofluids
> Thermoelectric materials (direct conversion of heat into electricity, or vice
> versa)
> Artificial zeolites (nano-porous minerals)

Materials Processing

Nanotechnical surface treatment
> Self-protecting, self-healing, self-cleaning surfaces through protein/enzyme
> nanocoatings and micro-rough bristly surfaces
> Light-refracting nanostructured surfaces (colours without pigments)
> Ultra-thin extra-strong coatings by colloidal (electrostatic) self-assembly
> of layers
> Ultra-thin extra-strong coating by thermal gaseous spraying
> Inductive decoating of surfaces
> Plasma treatment of surfaces (among others, plasma treatment of plastics,
> vacuum-sputtering of solar-thermal surfaces, dry decontamination)

Dry powder coating technology
Regenerative combustion
Clean-burn technology on the basis of porous burners (also flameless oxidation)
Advanced membrane technology
High-performance lasers
Sonic devices, high and low frequency
Powder metallurgy, foamed metals, dry metal-working
Sulphur- and chlorine-free pulping
Biotechnological processing of biotic materials (e.g. in pulping and paper making,
leather tanning, textiles manufacturing, food processing)

Building

Re-densification of urban space, spatial re-integration of operative functions
Multi-floor buildings in closed street blocks/carées preferable to high-rise buildings,
as both are preferable to small stand-alone houses

Building materials
    Design for reusability of construction elements and recyclability of building
        materials
    Elimination of hazardous substances in building materials
    High-quality building materials made of mineral and wooden cycleware
    Polymer matrix composites
    Foamed minerals

Energy design
    North–south alignment of new buildings
    Daylighting instead of artificial lighting
    Automated lighting control
    Transparent heat insulation (windows, façade glass)
    Sandwich walls
    High-performance insulation materials
    Thermo paints
    Photovoltaic panels roof-top or façade-integrated
    Solar thermal collectors roof-top, maybe in combination with
        Calorific boilers
        Solar air and ventilation system
        Heat recirculation in pipes beneath the floor
        Heat pumps
        Geothermal heating and cooling

Water design
    Roof-top turf
    Minimisation of soil sealing

Facility management and contracting

Vehicles

Optimum modal split of transport
Logistic traffic optimisation
Soft car driving
Internal-combustion engines
    Diesel engines
        Diesel soot microfilters
        Plasma diesel-exhaust cleaning
        Diesel common rail injection
        Elsbeth motors
    Incremental improvements of internal combustion in Otto motors
        Three-way exhaust catalyst
        Lean-burn motor
        Fast starter-generator
        Pressure wave turbocharger/supercharger
        Direct fuel injection
        Cranked connecting rod
        Variable valve system

> Camless engine
> Cartronics (interconnecting and fine-tuning controls of components)
> Mechatronics (similar, including substitution of electromechanical actuators for conventional mechanics)

Hybrid propulsion (combination of two different propulsion systems in one car, e.g. electric motor combined with internal combustion)
Hydrogen-fuelled and fuel cell-powered cars

> Hydrogen-fuelled internal-combustion motor car
> Hybrid car combining hydrogen combustion with electric motor
> E-cars powered by fuel cells (various designs of FCs, fuel, and fuel storage)

Materials

> Product stewardship. Take-back of disused cars
> Recycling of car parts and materials
> Design for disassembly and recyclability
> Increased share of aluminium and new compound materials
> Increased share of cycleware and biotic materials

FC-powered ships and submarines
FC-powered propeller machines
Hydrogen-fuelled jet planes
Fuel-efficient redesign of aircraft components, including new materials
Airship cargolifter
Micro-electro-mechanical propulsion systems in small, unmanned airplanes
Magnetic levitation train (Maglev)

Utility Goods: Office and Household Appliances

Teaming-up in use
Leasing/renting instead of owning
Modular design
Power-efficient office and household appliances and devices

> Power-efficient refrigerators, washing machines, TV sets, etc.
> Power-efficient light bulbs and fluorescent tubes
> LEPs (light emitting polymers) or LEDs (light emitting diodes)

New refrigeration technology

> Closed-cycle air refrigeration
> Thermo-acoustic refrigeration
> Magnetic refrigeration

Preference for high-quality pure materials
Design for disassembly and recyclability
Increased share of cycleware and biotic materials
Environment-oriented chain management

Materials Reprocessing and Waste Management

Refurbishing and reuse of machines, plants, and product parts
Industrial symbiosis (combined processes, i.e. cascadic re-entry of waste products or byproducts in subsequent production processes)

Dry-stabilate process (one-bin principle in municipal waste collection and separation with integrated recovery of valuable content materials)
Automated detection and separation in sorting plants
> High-pressure waterbeam crushing
> Sonic and optical (colour reflection) sensors detecting materials
> Software-and-actuator systems controlling mechanic separation

Disassembly of complex products in semi- or fully-automated plants
Materials recycling
> In-process recycling (within a factory or production line)
> In-process recovery of valuable and hazardous waste content
> Macrostructural recovery and recycling (secondary raw materials) particularly metals such as steel, aluminium, platinum, gold, silver, copper, and building materials, plastics, textile fibres and organic base materials (e.g. amino acids)

Biological treatment of biotic final wastes
> Composting in closed plants with recovery of biogas
> Co-fermentation of various wastes in biogas plants

Thermal treatment of final waste and hazardous waste
> Incineration plants with heat and power co-generation
> Waste incineration in industrial blast furnaces
> Solar and mechanical drying of wet sludges for incineration
> Smouldering-burning process (pyrolysis)
> Thermoselect process (pyrolysis)

Closing down of all landfills, except a few for slag and ash and selected mineral construction waste, as far as these are not recyclable yet

Emissions Control

Airborne emissions
> CO_2 sequestration (as under energy)
> Car exhaust treatment (as under vehicles)
> Electric filtration of airborne particles
> Aerosol micro filtration
> Dry desulphuration (in contrast to conventional wet desulphuration)
> Catalytic removal of NO_x
> Removal of VOCs and CVOCs by
>> Catalysts
>> Flameless oxidation
>> UV oxidation
>> Biocatalytic waste air treatment with microbes
> Advanced membrane-type filtration of gases

Sewage water purification
> 3-phase-sewage water purification plant with integrated ammonia-stripping
> Advanced membrane-type filtration of effluent
>> Ceramics membranes and cellulose-stainless steel membranes
>> Ultrafiltration with various diaphragms

Gel stripping of high-molecular substances
Electrolysis of brine (replacing disinfection of water with chlorine)
Biocatalytic purification
New breeding generations of bacteria (GM)
Combining bacteria with coke
Microfiltration through biomembranes
Sequential reactor plants
Plant-growth purification stations
Small purification units in decentral applications
Catalytic
Electrolytic
Microfiltration through biomembrane

Site remediation
Enshrining contaminated soil in a concrete coffin
Thermomechanical treatment in rotary skilns, off- and on-site
Chemical decontamination
Use of neutralising agents
Bauxsol™ (red mud)
Phytoremediation
Microbial bioremediation
Various cases of GM-enhanced bacteria
High-pressure aeration with biological and mineral additives
Restoration of damaged habitats, i.e. recultivation and reforestation of dug-over, abandoned or desertificated land, e.g. newly developed sites, traffic ways, shut-down mining areas and slag heaps, fields devastated by agroindustrial use, formerly forested areas

Environmental Measuring and Monitoring

Spectroscopic detection and measurement
Ion mobility spectrogram synthesiser
Mobile case-sized mass spectrometer

Chromatographic detection and measurement
Chemical fingerprinting (isotope markers)

Laser-based metrology
Laser-based quality/consistency probing
Laser-based sewage water meter and analyser
LIPAN (laser induced plasma analyser)

Optical, photo-acoustic and sonar methods
32-channel array detector (UV light)
Photo-acoustic sensors
Combined sonar and radar remote detection of pollutants

Membrane-type dialytical analysis
Electrolytic analyses (nitrate monitoring)
Electrical analyses (measuring resistance, e.g. mercury monitoring)
DNA-chip technology, optical biosensors, enzymatic and cell biosensors

Natural bioindicators (lichens, indicator plants, small organisms)
Ultra-fine soot monitoring
Wireless environmental sensor networks (interconnected motes, i.e. remote sensors and radio transceivers)
Satellite-based earth observation (environmental monitoring)

References

Allenby, Braden R. (1999), 'Earth Systems Engineering. The Role of Industrial Ecology in an Engineered World', *Journal of Industrial Ecology*, **2** (3), 73–93.

Allenby, Braden R./Cooper, William E. (1994), 'Understanding Industrial Ecology from a Biological Systems Perspective', *Total Quality Environmental Management*, Spring 1994, 343–354.

Ayres, Robert U. (1993), 'Industrial Metabolism. Closing the Materials Cycle', in Jackson, T. (ed.), 165–188.

Ayres, Robert U. and Leslie W. (1996), *Industrial Ecology. Towards Closing the Materials Cycle*, Cheltenham: Edward Elgar.

Ayres, Robert U./Simonis, Udo Ernst (eds) (1994), *Industrial Metabolism. Restructuring for Sustainable Development*, Tokyo: United Nations University Press.

Beck, Ulrich/Giddens, Anthony/Lash, Scott (1994), *Reflexive Modernization. Politics, Tradition and Aesthetics in the Modern Social Order*, Cambridge: Polity Press.

Bennett, John William (1976), *The Ecological Transition. Cultural Anthropology and Human Adaptation*, New York: Pergamon Press.

Fischer-Kowalski, Marina (1997), 'Society's Metabolism', in Redclift/Woodgate (eds), 119–137.

Frosch, Robert A. (1996), 'Toward the End of Waste. Reflections on a New Ecology of Industry', *The Liberation of the Environment, Daedalus*, special issue, **125** (3), 199–212.

Hubert, Joseph (2004), *New Technologies and Environmental Innovation*, Cheltenham, UK/ Northampton, MA: Edward Elgar.

Jansson, AnnMari/Hammer, Monica/Folke, Carl/Costanza, Robert (eds) (1994), *Investing in Natural Capital. The Ecological Economics Approach to Sustainability*, Washington: Island Press.

Krishnan, Rajaram/Harris, Jonathan M./Goodwin, Neva R. (eds) (1995), *A Survey of Ecological Economics*, Washington, D.C./Covelo, Ca.: Island Press.

Lowe, Ernest (1993), 'Industrial Ecology. An Organising Framework for Environmental Management', *Total Quality Environmental Management*, Autumn 1993, 73–85.

Rees, William E./Wackernagel, Mathis (1994), 'Ecological Footprints and Appropriated Carrying Capacity. Measuring the Natural Capital Requirements of the Human Economy', in Jansson et al. (eds), 362–391.

Schmidt-Bleek, Friedrich (1994), *Wieviel Umwelt braucht der Mensch? MIPS, das Maß für ökologisches Wirtschaften*, Berlin: Birkhäuser.

Socolow Robert/Andrews, Clinton/Berkhout, Frans/Thomas, Valerie (eds) (1994), *Industrial Ecology and Global Change*, Cambridge University Press.

Part Four

Environmental reform in Asian and other emerging economies

Chapter 19

David A. Sonnenfeld and Michael T. Rock

ECOLOGICAL MODERNISATION IN ASIAN AND OTHER EMERGING ECONOMIES

INTEREST IN ECOLOGICAL MODERNISATION Theory (EMT) is substantial and growing around the world, including in rapidly industrialising Asian and other emerging economies. In China, Vietnam, South Africa and elsewhere, academics, government policymakers, business leaders, citizen activists and others are exploring this school of thought's usefulness for facilitating environmental reform in widely varying cultures, political and institutional contexts and biophysical conditions. Because global sustainability is so closely linked to the success of institutional environmental transformation in emerging economies, an evaluation of the applicability of EMT in such locations is critically important. Richly detailed and geographically diverse, the readings in Part Four of this volume help us understand the dynamics and effectiveness of environmental reform in emerging economies by examining the actors, institutions and conditions conducive to such reform; the diverse paths such reforms take; and the challenging relationship between global and local inequalities and environmental reform, among other dimensions. Together, these studies demonstrate that ecological modernisation theory has much to offer around the globe.

This essay provides a brief overview and review of ecological modernisation and related scholarship on environmental reform in Asian and other emerging economies. Beginning with a thumbnail sketch of the political economic and ecological context of rapid industrialisation in this era, it continues with a survey of studies on environmental reform in such economies, and finishes by drawing conclusions and suggesting future directions for research in this critically important arena. The readings included in Part Four are introduced and contextualised along the way.

The political ecology of late industrialisation

With the end of the Fordist model of economic development in the North in the 1970s, global capital sought out new, emerging markets and manufacturing bases, finding them in the Four Tigers (Hong Kong, Taiwan, South Korea and Singapore) of East Asia and the bureaucratic authoritarian industrialising regimes (BAIRs) of Argentina, Chile, Mexico and Brazil in Latin America. Rapid, large-scale, industrial development flourished in these authoritarian economies. Initially, industrial production and foreign direct investment by Northern multinationals served local markets, but when the returns to import substitution slowed, developmentalist governments turned to the export of manufactures. This happened first in the Four Tigers in East Asia beginning in the late 1960s. Subsequently, additional waves of export-oriented industrial development occurred in the so-called Mini-Dragons (Malaysia, Thailand, Indonesia, and later Vietnam) of Southeast Asia, and the newly liberalised economies of Latin America. With the end of apartheid, South Africa began to realise its potential as an important economy in southern Africa. China entered the fray after 1978, and so did India in the 1990s – both by liberalising domestic markets and opening their economies to trade and foreign investment.

Intensified economic, institutional and cultural globalisation and the shift in emerging economies toward the export of manufactures to OECD countries have had complex effects on the natural environment and on institutional reforms in emerging economies (Mol 2001). In each of the economies mentioned above, the rapid growth of export-oriented industrial production initially occurred in the absence of much concern for the environment. This lack of concern amounted to a 'grow first, clean up later' environmental strategy that led to substantial increases in pollution and serious human health problems. Subsequently, substantial evidence, much of it assembled by EMT analysts, was produced suggesting that the global-local linkages attending export-oriented industrial development strategies have also had positive environmental effects on firms and industries in emerging economies. These positive or 'win-win' effects occurred as a consequence of product and process regulations in Northern markets (such as the European Union's Waste Electrical and Electronic Equipment [WEEE] and Reduction of Hazardous Substances [RoHS] directives – see Geiser and Tickner 2006); as reactions of global finance, insurance firms, and stock markets to bad environmental news for firms located in emerging markets; through the import of cleaner technologies (see Sonnenfeld 1999); through multinational corporations' global firm-wide environmental goals (see Rock and Angel 2005); and through linkages between nongovernmental organisations (NGOs) in the North and South (see Sonnenfeld 2002). By the beginning of the third millennium, EMT analysts had demonstrated that institutional environmental reforms were taking place in many emerging economies. Environmental ministries and basic laws governing pollution were established in most emerging Asian economies between the early 1970s and the mid-1990s (Rock 2002a; Sonnenfeld and Mol 2006a, 2006b), with similar developments occurring in other emerging economies. At the same time, local communities and NGOs were increasingly involved in pressing firms to reduce pollution, compensate victims and adopt new, cleaner technologies and practices.

As Giddens (1990; 2007), Castells (1996/2000), Frank et al. (2000), Mol

(2001) and others have pointed out, the environment became an interest of global concern particularly after the United Nations Conference on Environment and Development (UNCED) in 1992. Ironically, such heightened concern for the global environment was accompanied by continued and increased environmental destruction and devastation in emerging economies. This combination was not lost on critics of globalisation, capitalism and EMT. Doubting that environmental improvements are occurring in the emerging economies, critics ask whether *any* environmental improvements (in the extraction and utilisation of raw materials, the efficiency of manufacturing and lessening of pollution or the accumulation of various forms of waste) in the emerging economies are large enough to offset the huge increases in the scale of production and consumption in those economies; whether population growth will cancel out any positive environmental reforms in those countries; and whether any form of global capitalism is sustainable. Critics also hold that most environmental reforms in emerging economies serve simply to 'greenwash' fundamentally destructive institutions and their practices. Such fundamental critiques will not be resolved here – see the conclusion of this volume for further discussion of these and other debates.

This essay suggests, rather, after Buttel (2000, included in this volume), that ecological modernisation theory is the study of actually-existing environmental reform processes at every level of societies around the world, including in newly emerging economies. Scholarship by EMT analysts and others indicates that environmental reforms *are* being developed and instituted in these economies, from a variety of vantage points, and with varying effectiveness. EMT analysts have asked: How are such reforms being developed? Which social forces are engaged in and necessary for their development and implementation? Why do environmental reforms vary in effectiveness from place to place? Such research is as critically important for institutions, organisations and individuals engaged in environmental policymaking in emerging economies as it is for students of environmental reform. Spurred by environmental destruction, rallied by environmental improvements, given a sense of urgency by escalating and persistent media attention, few things today are more timely than understanding what works where in emerging economies to lighten human impact on the natural environment.

Exploring environmental reform in emerging economies

At least three distinct 'schools' in the study of environmental reform in emerging economies have developed over the last several decades, including: (i) scholars using EMT as an analytical and policymaking framework; (ii) kindred analysts using approaches similar to but different from EMT; and (iii) a more critical group of academics who question the limits and efficacy of environmental reform efforts in emerging economies.

Ecological modernisation-oriented studies

Ecological modernisation-oriented studies of emerging economies have a number of distinguishing characteristics. First, in terms of scale of analysis, many have examined the adoption of environmental policies, technologies and practices at the level of the facility, firm or industry (see Frijns et al. 1997; Sonnenfeld 2000, included in this section; Tran et al. 2003, included in this section; see also Mol and van Buuren 2003); others analysed environmental policy innovations at the state or national level (Frijns et al. 2000; Lang 2002; Yang 2005; Studer et al. 2006; China Centre for Modernization research 2007; see also Rinkevicius 2000; Gille 2000). Second, in their geographic focus, many of these studies have examined newly industrialising economies in east and south-east Asia (see Sonnenfeld 2000, in this section; Studer et al. 2006; Mol 2006; and Gouldson et al. 2008); while others focused on the transitional economies in central and eastern Europe (Rinkevicius 2000; Gille 2000; Andersen 2002). Initially, emerging African and Latin American economies received limited attention from EMT scholars (but see Frijns et al. 1997 and Mol 2001), however more recent scholarship includes those continents as well, as evidenced by the work of Oelofse et al. (2006) on South Africa, in this section; Wilson (2002) on Lake Victoria fisheries; Jepson et al. (2005) and Milanez and Bührs (2007) on Brazil; and other new studies from Africa around the idea of modernised mixtures as cited in the concluding chapter to this volume. Methodologically, most of the ecological modernisation-oriented studies of emerging economies utilised qualitative, field-research based approaches.

Ecological modernisation-oriented studies on Asian and other emerging economies analysed how governments, firms, industry associations, communities and markets interacted in a variety of ways leading firms to reduce pollution. Sometimes, reductions in pollution were sufficient to permit significant improvements in one or more ambient environmental indicators (Mol 2006, in this section). Other times, reductions in plant- and firm-level pollution were minimal or too small to counteract the scale effects of increased production (Oelofse et al. 2006, in this section; and Studer et al. 2006). In yet other instances, environmental institutions were too weak to affect the polluting behaviour of firms and industries in particular economies (Tran et al. 2003, in this section).

From these and related studies, three major factors affecting ecological modernisation in Asian and other emerging economies may be identified: (i) The failure of states to effectively address environmental problems using traditional 'command and control' regulatory approaches takes different forms in emerging or transitional economies than in the more advanced economies, due to the very different relationships between the former and global and local markets, and civic associations. (ii) Locally owned, small- and medium-sized enterprises play a crucially different role in the emerging economies, in contrast to the strong influence of transnational corporations in organising the greening of consumption and production in more advanced economies. And (iii) domestic NGOs and consumer organisations have a very different role in pressuring for environmental change in emerging than more advanced economies. All three factors are crucial to ecological modernisation in both types of economies; in Asian and other emerging economies they sometimes take very different forms, however.

Kindred studies of environmental transformation

A second, kindred group of scholars asks closely related questions about environmental reform in emerging economies and utilises similar methodologies, but bases its work on different theoretical frameworks and disciplinary traditions. Reform-oriented scholars from political science, sociology, economics and business studies, human geography, and other fields examine the institutionalisation of environmental policymaking in government agencies; adoption of environmental policies, technologies, and practices in the private sector; and the role of consumers, local communities, NGOs, and other third parties in environmental policymaking and practice in east and south-east Asia and elsewhere. Among the earliest such scholars, André Dua and Daniel Esty (1997), drawing from the experience of negotiating environmental side agreements to the North American Free Trade Agreement (NAFTA), explored the feasibility of utilising market mechanisms to strengthen environmental controls in east and south-east Asia, through the Asia-Pacific Economic Cooperation (APEC) framework. Yok-shiu Lee and Alvin So (1999), writing from socially and politically vibrant, *fin de siècle* Hong Kong, focused on social movements' influence on environmental policy reform across east and south-east Asia (see also Parnwell and Bryant 1996; Hirsch and Warren 1998; Sonnenfeld 1998a, 1998b, 2002; Hsiao and Liu 2002; Smith et al. 2006).

Michael Rock and David Angel have been engaged in the long-term study of institutional environmental change in east and south-east Asia from the mid-1990s. Their influential work examines the integration of environmental reforms into industrial policy institutions in the region; the relationship between trade and investment openness and domestic market liberalisation to local firms' ability to adopt newer, cleaner technologies and to reduce environmental intensities; the effect of new product and process regulations in the advanced, OECD economies on production processes in Asia; and the use of global, firm-wide environmental standards by OECD-based multinationals to improve environmental performance in developing economies while getting local suppliers to do the same. See Angel and Rock (2000), Rock (2002a), Rock (2002b, included in this section), and Rock and Angel (2005).

About the same time, Peter Evans and others at the University of California at Berkeley focused on dynamics of urban institutional environmental capacity-building and reform in Latin America, Asia and Central Europe (see Evans 2002; O'Rourke 2004; Gille 2007; see also Pezzoli 1998). Noting that 'three-fourths of those joining the world's population during the next century will live in Third World cities', Evans et al. examine local, national, and global dimensions of urban sustainability in the 21st century. Consistent with EMT analysts and others, they find ecological degradation 'a continuing threat' in the world's cities, but also 'a sense of hopeful possibility' in creative, civic-based initiatives to make these cities more liveable (Evans 2002, pp. 1, 27–28). In another interesting study, Ronie Garcia-Johnson (2000) examines the influence of U.S.-based chemical corporations in 'Exporting Environmentalism' in the form of voluntary corporate environmental policies and practices into the developing world, especially Brazil and Mexico. She finds that the chemical industry's 'Responsible Care' initiative had significant international impacts as local firms adopted the industry's voluntary standards.

However, these impacts varied by country (more effective in Mexico than Brazil) and firm, and in interaction with local NGOs and communities.[1]

More critical studies of environment reform

A third group of scholars has taken a more critical approach to the study of environmental reform in emerging economies. With reference to Europe's periphery, especially Ireland, David Pepper (1997, 1999) finds 'inconsistencies and practical failings in both "weak" and "strong" ecological modernisation's theorisation of sustainable development' (1999, p. 27). In particular, he finds that development interests tend to trump environmental concerns in the name of regional economic integration. One consequence of this is uneven development, particularly for rural areas. Oluf Langhelle (2000, included in this section) finds ecological modernisation to be a necessary, but insufficient basis for sustainable development. The latter, he argues, requires addressing issues of social inequality, resource scarcity, and possibly even 'ecological collapse' (p. 318). Both Pepper and Langhelle draw on early EMT scholarship, particularly that of Hajer and others, that emphasized ecological modernisation as an élite *policy discourse* in Europe. More geographically diverse and empirically grounded, a collection edited by Peter Utting (2002) examines corporate environmentalism and the greening of industry in Mexico (by David Barkin), Central America, Brazil (by Ricardo Carrere), Malaysia (by Martin Perry and Sanjiv Singh) and South Africa (by Jonathon Hanks), among others. These authors, like EMT scholars, find both 'incipient progress' and reason for disappointment (pp. 268–275).

Current research clusters on ecological modernisation

In the last decade, scholarship on environmental reform in emerging economies informed by Ecological Modernisation Theory has become more comprehensive, sophisticated and geographically diverse. Currently, four clusters of ecological modernisation research focusing on developing and newly industrialising countries can be identified.

A cluster of scholars in Germany, including Martin Jänicke at the Environmental Policy Research Centre (FFU) at the Freie Universität Berlin, Helmut Weidner at the Social Science Research Center of Berlin (WZB), and others, led a series of studies comparing the institutionalisation of environmental reform around the world. One of a number of resulting publications by Weidner and Jänicke (2002) includes analyses of ecological modernisation in Brazil, Bulgaria, Costa Rica, India, Morocco, Mexico, Poland, Taiwan, Vietnam and other countries. In a nice synopsis, Weidner (2002) suggests that, in their studies, 'the globalisation of environmental policies and proponents counteracts ecologically ignorant economic interests and fosters diffusion of environmental innovations'. He emphasises the importance of environmental actors' strategy, will, and a 'mixed strategy of cooperation and conflict . . . to win new friends in all sectors of society and to prevail over the powerful interest groups rooted deeply in most ecologically obstructive sectors' (pp. 1340, 1365).

Researchers associated with the Environmental Policy Group, Wageningen University, the Netherlands, have been singularly successful in engaging in long-term, advanced studies of environmental reform in east and south-east Asia. Research has focused on China and various south-east Asian countries, including Vietnam, Thailand and Malaysia, and has run the gamut from micro, firm-level, to meso- and macro-level industry, national, regional studies and beyond. Special attention has been given to agro-industrial, food processing, pulp and paper, and electronics industries, among others; and to the analysis of the opportunities and limitations for ecological modernisation in rapidly industrialising, economically transitional economies with a wide variety of political forms. These studies have found actors in most countries and sectors actively and seriously engaged in technological and institutional environmental transformation. Such dynamics are typically stronger in sectors more closely linked to global trade and international political networks, but existed as well in more domestically-oriented settings. The rapid development of environmental actors, institutions, policies and actions in east and south-east Asia in the last two decades has been stunning.[2]

A dynamic, new centre of EMT scholarship has emerged, at the Kadoorie Institute, University of Hong Kong. As a world city linking China with global financial markets, trading partners, and a wide variety of civic actors, Hong Kong is a fascinating location for studies of the promise and challenges of institutional environmental reform. In a series of articles, Peter Hills, Richard Welford and associates explore a rich variety of the city's environmental 'policy learning', looking especially at the respective roles of business and public policymaking (Hills et al. 2004; Studer et al. 2006; Mantel et al. 2007; and Gouldson et al. 2008). In one of their more recent works, following Jänicke and Weidner's work, they find in Hong Kong, as well, 'that [environmental policy] learning has been limited to those areas where there is a pressure for change, where opposition can be appeased or circumvented and where changes can be easily accommodated within existing institutional structures' (Gouldson et al. 2008).

And in South Africa, researchers at the University of KwaZulu-Natal have taken up the discursive strand of EMT, exploring its relevance in South Africa, the largest and most dynamic economy in Southern Africa (Oelofse et al. 2006, included in this section). Given that nation's history of apartheid, exploring the applicability of ecological modernisation approaches to questions of social and environmental inequality and justice are central. With a new, environmentally progressive constitution and implementing legislation, South Africa has focused considerable resources on building environmental capacity, with individual and collective actors not only influencing environmental outcomes, but also having opportunities to (re)structure processes and institutions in more equitable and sustainable ways. The researchers find that innovative, deliberative policymaking approaches are broadening the social basis of environmental reform in that country, providing a foundation for shifting from weaker to stronger forms of ecological modernisation.

In the last decade, additional centres of ecological modernisation research have emerged beyond OECD countries. In Brazil, the ANPASS network of environmental social scientists is increasingly including studies on environmental reform from an ecological modernisation perspective (for instance on food and retail: Guivant et al. 2003; Oosterveer et al. 2007). Across the Pacific, the China Centre for Modernization Research (2007) of the Chinese Academy of Sciences is a focal point for 'translating' western ecological modernisation theory to Chinese circumstances, while the Research Centre for Eco-Environmental Sciences of the same academy and the Department of Environmental Sciences at Tsinghua University apply ecological modernisation perspectives in their empirical studies on energy, water, circular economy and rural environmental protection (cf. Liu, et al. 2004; Zhong and Mol 2008; Han et al. 2008).

Conclusions and future directions

Given huge variation in the character of emerging economies (from vast nations to city-states, economic dynamos to quiet backwaters, democracies to dictatorships, and everything in between), ecological modernisation and kindred approaches have not attempted to develop 'one size fits all' grand theories of environmental reform. Rather, they have aimed to provide broadly encompassing, theoretically and practically informed frameworks for empirical research on conditions, processes and outcomes of institutional environmental transformation wherever it may (or may not) be taking place. Increasingly tied together by hyper-globalisation of finance, trade, culture, communication and transportation (among others), ecological modernisation theory has moved from the modern world of *nationally* based states, enterprises, civic associations and environmental policies, to the late- or post-modern world of *global* markets, supply-chains, technologies, institutions and other actors. A key component of late modernity, as Anthony Giddens (1990, 2007) has argued, is the emergence of global environmental concern, policymaking and practice, affecting every level of human societies and organisation, in more and more corners of the earth.

As evidenced by the readings in this section, ecological modernisation-oriented and kindred scholars have been vitally involved in exploring and explaining the character of the environmental reform process in emerging economies. They have demonstrated how environmental reform interacts with globalization and democratization, in varying ways in different economies, political, institutional and cultural contexts. Such dynamics have led sometimes to significant environmental improvement (Gouldson et al. 2008; Mol 2006; Rock 2002b, in this section), at other times, not (Tran et al. 2003, in this section). Future scholarship on environmental reform in emerging economies informed by EMT and related approaches will be carried out at a wide variety of scales, from the individual or household; to the local, firm or municipal level; to sub-regional, national, transboundary, supranational and even global levels. Some Asian economies, especially China and India, will continue to be central to such efforts in the 21st century. EMT-informed scholarship must also strengthen and further develop its examination of the institutionalisation of environmental reform in both more and less developed Africa, Latin America, central Asia and elsewhere.

A central question remaining in the study of environmental reform today is whether institutional and political transformations at various levels of scale will be large, effective and rapid enough to overcome the effects of increasing economic activity and population growth in Asian and other emerging economies. Conditions of rising global energy demand and fuel and food prices, shortages of potable water and increased urbanisation and pollution make answering this question critical. Such developments portend greater attention not only to alternative environmental products and processes, but also, and even more so, to conditions conducive to the adoption of institutional, political and technological environmental innovations. Ecological modernisation-oriented and related scholarship provide an empirically rich, theoretically, methodologically and geographically diverse, and steadily accumulating body of evidence on what kinds of environmental reforms work, under what internal and external conditions, and what policymakers and advocates can do to improve the likelihood of more sustainable development in Asian and other emerging economies.

Notes

1. Such studies complement research on the displacement of environmental pollution by relocating polluting industries from the OECD countries to the developing world (cf. Leonard (1988), Jänicke (1995), Mani and Wheeler (1997) and Mol (2001)). The overall conclusion from these studies is that environmental pollution (or a stringent environmental regime) is no reason for industrial displacement, but relocation of industrial pollution can coincide with a (geographical) restructuring of the global economy.

2. On Vietnam, see Frijns et al. (2000), Tran et al. (2003), in this section, and other selections in Mol and Van Buuren (2003). On China, see Ho (2005, updated and elaborated in Ho 2008) on 'Asian-style leapfrogging', Mol (2006), in this section, Carter and Mol (2007) and Zhang et al. (2007). On south-east Asia, see Sonnenfeld (2000), in this section, on pulp and paper manufacturing, Oosterveer et al. (2006) on agro-industry and Foran and Sonnenfeld (2006) on electronics. The collection edited by Sonnenfeld and Mol (2006a) provides a broad, historical overview of environmental trends in east and south-east Asia, as well as case studies of environmental transformation in those regions. See also a series of related PhD dissertations from Wageningen University, available at: http://www.enp.wur.nl/UK/education/PhD+program/Finalized+PhD+projects+at+ENP/.

References

Andersen, M.S. (2002) 'Ecological modernization or subversion? The effect of Europeanization on Eastern Europe', *American Behavioral Scientist* **45**, 9, pp. 1394–1416.

Angel, D.P. and Rock, M.T. (eds) (2000) *Asia's Clean Revolution: Industry, Growth and the Environment*. Sheffield, UK: Greenleaf Publishing.

Buttel, F.H. (2000) 'Ecological modernization as social theory', *Geoforum* **31**, 1, pp. 57–65.

Carter, N.T. and Mol, A.P.J. (eds) (2007) *Environmental Governance in China*. London/New York: Routledge.

Castells, M. (1996/2000) *The Information Age: Economy, Society and Culture*, 3 vols. Oxford, UK/Malden, MA: Blackwell.

China Centre for Modernization Research (2007), *China Modernization Report 2007: Ecological Modernization Study*. Beijing: Beijing University Press, pp. 455 (in Chinese).

Dua, A. and Esty, D. (1997) *Sustaining the Asia Pacific Miracle: Environmental Protection and Economic Integration*. Washington, DC: Institute for International Economics.

Evans, P.B. (ed.) (2002) *Livable Cities? Urban Struggles for Livelihood and Sustainability*. Berkeley: University of California Press.

Foran, T. and Sonnenfeld, D.A. (2006) 'Corporate Social Responsibility in Thailand's Electronics Industry', in T. Smith, D.A. Sonnenfeld and D.N. Pellow (eds), *Challenging the Chip: Labor Rights and Environmental Justice in the Global Electronics Industry*. Philadelphia, PA: Temple University Press, pp. 70–82.

Frank, D.J., Hironaka, A. and Schofer, E. (2000) 'The nation-state and the national environment over the twentieth century', *American Sociological Review* **65**, 1, pp. 77–95.

Frijns, J., Kirai, P., Malombe, J. and Vliet, B. van (1997) *Pollution Control of Small-scale Metal Industries in Nairobi*, Department of Sociology. Wageningen University, the Netherlands.

Frijns, J., P.T. Phung and Mol, A.P.J. (2000) 'Ecological Modernisation Theory and Industrialising Economies: The Case of Viet Nam', in A.P.J. Mol and D.A. Sonnenfeld *op cit* (eds), *Ecological Modernisation Around the World: Critical Perspectives and Debates*, pp. 257–292.

Garcia-Johnson, R. (2000) '*Exporting Environmentalism: U.S. Multinational Chemical Corporations in Brazil and Mexico*'. Cambridge, MA: MIT Press.

Geiser, K. and Tickner, J. (2006) 'International Environmental Agreements and the Information Technology Industry', in T. Smith, D.A. Sonnenfeld and D.N. Pellow *op cit.* (eds), *Challenging the Chip: Labor Rights and Environmental Justice in the Global Electronics Industry*, pp. 260–272.

Giddens, A. (1990) *The Consequences of Modernity*. Cambridge: Polity Press.

Giddens, A. (2007) *Europe in the Global Age*. Cambridge: Polity Press.

Gille, Z. (2000) 'Legacy of Waste or Wasted Legacy? The End of Industrial Ecology in Post-Socialist Hungary', in A.P.J. Mol and D.A. Sonnenfeld *op cit.* (eds), *Ecological Modernisation Around the World: Critical Perspectives and Debates*, pp. 203–234.

Gille, Z. (2007) *From the Cult of Waste to the Trash Heap of History: The Politics of Waste in Socialist and Postsocialist Hungary*. Bloomington: Indiana Univ. Press.

Gouldson, A., Hills, P. and Welford, R. (2008) 'Ecological modernization and policy learning in Asia: lessons from Hong Kong', *Geoforum* **39**, pp. 313–330.

Guivant, J., Fernanda, M., Fonseca, de A.C., Sampaio, F., Ramos, V. and Scheiwezer, M. (2003) *Os supermercados e o consumo de frutas, legumes e verduras orgânicos certificados*, Relatório final de pesquisa, CNPq projeto 520874/01–3.

Han, J., Mol, A.P.J., Lu, Y. and Zhang, L. (2008) 'Small-scale bioenergy projects in rural China: lessons to be learnt', *Energy Policy* **36**, 6, pp. 2154–2162.

Hills, P., Lam, J. and Welford, R. (2004) 'Business, environmental reform and technological innovation in Hong Kong', *Business Strategy and the Environment* **13**, 4, pp. 223–234.

Hirsch, P. and Warren, C. (eds) (1998) *The Politics of Environment in Southeast Asia: Resources and Resistance*. London/New York: Routledge.

Ho, P. (ed.) (2005) Special issue on 'Greening Industries in Newly Industrialising

Countries: Asian-style leapfrogging', *International Journal of Environment and Sustainable Development* 4, 3, pp. 209–351.

Ho, P. (ed.) (2008) *Greening Industries in Newly Industrialising Countries: Asian-style Leapfrogging*. Hauppauge, NY: Nova Science Publishers.

Hsiao, H.M. and Liu, H.J. (2002) 'Collective Action toward a Sustainable City: Citizens' Movements and Environmental Politics in Taipei', in P. Evans (ed.), *Livable Cities?*, pp. 67–94.

Jänicke, M. et al. (1995) 'Green industrial policy and the future of 'dirty industries' '. Unpublished paper, Berlin: Free University.

Jepson, W.E., Brannstrom, C. and de Souza, R.S. (2005) 'A case of contested ecological modernisation: the governance of genetically modified crops in Brazil', *Environment and Planning C: Government and Policy* 23, 2, pp. 295–310.

Lang, G. (2002) 'Deforestation, floods and state reactions in China and Thailand', in A.P.J. Mol and F.H. Buttel (eds), *The Environmental State under Pressure*. Amsterdam/New York: Elsevier, pp. 195–220.

Langhelle, O. (2000) 'Why ecological modernisation and sustainable development should not be conflated', *Journal of Environmental Policy & Planning* 2, 4, pp. 303–322.

Lee, Y.F. and So, A.Y. (eds) (1999) *Asia's Environmental Movements: Comparative Perspectives*. Armonk, NY/London: M.E. Sharpe.

Leonard, H.J. (1988) 'Pollution and the struggle for the world product. Multinational corporations, environment and international comparative advantage'. Cambridge: Cambridge University Press.

Liu, Y., Mol, A.P.J. and Chen, J. (2004) 'Material Flow and Ecological Restructuring in China. The Case of Phosphorus', *Journal of Industrial Ecology* 8, 3, pp. 103–120.

Mani, M. and Wheeler, D. (1997) 'In Search of Pollution Havens? Dirty Industry in the World Economy, 1960–1995' World Bank Policy Research Department, Working Paper.

Mantel, S., Cheung, D., Welford, R. and Hills, P. (2007) 'Cooperation for Environmental Reform, Business-NGO Partnerships in Hong Kong', *Journal of Corporate Citizenship* 27 (Autumn), pp. 91–106.

Milanez, B. and Bührs, T. (2007) 'Marrying strands of ecological modernisation: A proposed framework', *Environmental Politics* 16, 4, pp. 565–583.

Mol, A.P.J. (2001) *Globalization and Environmental Reform: The Ecological Modernization of the Global Economy*. Cambridge, MA: MIT Press.

Mol, A.P.J. (2006) 'Environment and modernity in transitional China. Frontiers of ecological modernization', *Development and Change* 37, 1, pp. 29–56.

Mol, A.P.J. and Buuren, J.C.L. van (eds) (2003) *Greening Industrialization in Asian Transitional Economies: China and Vietnam*. Lanham, MD.: Lexington Books.

Mol, A.P.J. and Sonnenfeld, D.A. (eds) (2000) *Ecological Modernisation Around the World: Critical Perspectives and Debates*. London, UK/Portland, Ore.: Frank Cass/Routledge.

Oelofse, C., Scott, D., Oelofse, G. and Houghton, J. (2006) 'Shifts within ecological modernization in South Africa: deliberation, innovation and institutional opportunities', *Local Environment* 11, 1, pp. 61–78.

Oosterveer, P., Kamolsiripichaiporn, S. and Rasiah, R. (eds) (2006) Special issue on 'The Greening of Industry and Development in Southeast Asia', *Environment, Development, and Sustainability* 8, 2, pp. 217–227.

Oosterveer, P., Guivant, J. and Spaargaren, G. (2007) 'Shopping for green food in globalizing supermarkets: sustainability at the consumption junction', in J. Pretty, A. Ball, T. Benton, J. Guivant, D. Lee, D. Orr, M. Pfeffer and H. Ward (eds), *Sage Handbook on Environment and Society*. London: Sage, pp. 411–429.

O'Rourke, D. (2004) *Community-Driven Regulation: Balancing Development and Environment in Vietnam*. Cambridge, MA: MIT Press.

Parnwell, M.J.G. and Bryant, R.L. (eds) (1996) *Environmental Change in South-East Asia: People, Politics and Sustainable Development*. London/New York: Routledge.

Pepper, D. (1997) 'Sustainable Development and Ecological Modernization: A Radical Homocentric Perspective', *Sustainable Development* **6**, 1, pp. 1–7.

Pepper, D. (1999) 'Ecological Modernisation or the 'ideal model' of Sustainable Development? Questions Prompted at Europe's Periphery', *Environmental Politics* **8**, 4, pp. 1–34.

Pezzoli, K. (1998) *Human Settlements and Planning for Ecological Sustainability: The Case of Mexico Cit*. Cambridge, MA: MIT Press.

Rinkevicius, Leonardas (2000) 'The Ideology of Ecological Modernization in 'Double-Risk' Societies: a case study of Lithuanian environmental policy', in G. Spaargaren, A.P.J. Mol and F.H. Buttel (eds), *Environment and Global Modernity*. London: Sage Studies in International Sociology, pp. 163–186.

Rock, M.T. (2002a) *Pollution Control in East Asia: Lessons from Newly Industrializing Economies*. Washington, DC: Resources for the Future Press.

Rock, M.T. (2002b) 'Integrating environmental and economic policy making in China and Taiwan', *American Behavioral Scientist* **45**, 9, pp. 1435–55.

Rock, M.T. and Angel, D.P. (2005) *Industrial Transformation in the Developing World*. Oxford, UK: Oxford University Press.

Smith, T., Sonnenfeld, D.A. and Pellow, D.N. (eds) (2006) *Challenging the Chip: Labor Rights and Environmental Justice in the Global Electronics Industry*. Philadelphia: Temple University Press.

Sonnenfeld, D.A. (1998a) 'From Brown to Green? Late Industrialization, Social Conflict, and Adoption of Environmental Technologies in Thailand's Pulp and Paper Industry', *Organization & Environment* **11**, 1, pp. 59–87.

Sonnenfeld, D.A. (1998b) 'Social Movements, Environment, and Technology in Indonesia's Pulp and Paper Industry', *Asia Pacific Viewpoint* **39**, 1, pp. 95–110.

Sonnenfeld, D.A. (1999) 'Vikings and Tigers: Finland, Sweden and Adoption of Environmental Technologies in Southeast Asia's Pulp and Paper Industries', *Journal of World-Systems Research* **5**, 1, pp. 26–47.

Sonnenfeld, D.A. (2000) 'Contradictions of ecological modernisation: pulp and paper manufacturing in Southeast Asia', in A.P.J. Mol and D.A. Sonnenfeld (eds), *Ecological Modernisation Around the World: Critical Perspectives and Debates*, pp. 235–256.

Sonnenfeld, D.A. (2002) 'Social Movements and Ecological Modernization: The Transformation of Pulp and Paper Manufacturing', *Development and Change* **33**, 1, pp. 1–27.

Sonnenfeld, D.A. and Mol, A.P.J. (2006a) Special issue on 'Environmental Reform in Asia', *Journal of Environment and Development* **15**, 2.

Sonnenfeld, D.A. and Mol, A.P.J. (2006b) 'Environmental Reform in Asia: Comparisons, Challenges, Next Steps', *Journal of Environment and Development* **15**, 2, pp. 112–137.

Studer, S., Welford, R. and Hills, P. (2006) 'Engaging Hong Kong Businesses in Environmental Change: Drivers and Barriers', *Business Strategy and the Environment*, **15**, 6, pp. 416–431.

Tran, M.D.T., Phung, P.T., van Buuren, J.C.L. and Nguyen, V.T. (2003) 'Environmental management for industrial zones in Vietnam', in A.P.J. Mol and J.C.L. van Buuren (eds), *Greening Industrialization in Asian Transitional Economies: China and Vietnam*, pp. 39–58.

Utting, P. (ed.) (2002) *The Greening of Business in Developing Countries: Rhetoric, Reality and Prospects*. London/New York: Zed Books.

Weidner, H. (2002) 'Capacity Building for Ecological Modernization: Lessons from Cross-National Research'. *American Behavioral Scientist*, **45**, 9, pp. 1340–68.

Weidner, H. and Jänicke, M. (2002) *Capacity Building in National Environmental Policy: A Comparative Study of 17 Countries*. Berlin: Springer-Verlag.

Wilson, D.C. (2002) 'The global in the local: the environmental state and the management of the Nile perch fishery on Lake Victoria', in A.P.J. Mol and F.H. Buttel (eds), *The Environmental State under Pressure*. Amsterdam/New York: Elsevier, pp. 171–192.

Yang, L.F. (2005) 'Embedded autonomy and ecological modernisation in Taiwan', *International Journal of Environment and Sustainable Development*, **4**, 3, pp. 310–330.

Zhang, L., Mol, A.P.J. and Sonnenfeld, D.A. (2007) 'The Interpretation of Ecological Modernisation in China', *Environmental Politics* **16**, 4, pp. 659–668.

Zhong, L. and Mol, A.P.J. (2008) 'Participatory environmental governance in China: Public hearings on urban water tariff setting', *Journal of Environmental Management* **88**, 4, pp. 899–913.

Chapter 20

David A. Sonnenfeld

CONTRADICTIONS OF ECOLOGICAL MODERNISATION: PULP AND PAPER MANUFACTURING IN SOUTH-EAST ASIA

Introduction

'**ECOLOGICAL MODERNISATION**' **IS A** relatively new concept, coined in the early 1980s by German sociologist Joseph Huber (Spaargaren and Mol, 1992; Mol, 1995; Hajer, 1995). In simple form, it might be thought of as industrial restructuring with a green twist. Poetically, Huber suggests that 'The dirty and ugly industrial caterpillar transforms into an ecological butterfly' (Huber (1985) cited in Spaargaren and Mol (1992: 334)). Hajer (1995) defines ecological modernisation theory as 'the discourse that recognises the structural character of the environmental problematique but . . . assumes that existing institutions can internalise the care for the environment'.

As a theory of social change, ecological modernisation suggests we have entered a new industrial revolution, one of radical restructuring of production, consumption, state practices, and political discourses along ecological lines (Mol, 1995; Hajer, 1995; also Mol and Spaargaren, 2000). As a 'normative theory' or 'political program' (Spaargaren and Mol, 1992; Mol, 1995), ecological modernisation advocates resolving environmental problems through 'harmonizing ecology and economy' (Simonis, 1989), and through 'superindustrialisation' rather than de-industrialisation (Spaargaren and Mol, 1992).

To date, ecological modernisation theory has been considered applicable primarily for advanced industrial countries, due to prerequisites for green industrial restructuring, 'e.g. the existence of a welfare state, advanced technological development . . . a state regulated market economy . . . and . . . widespread environmental consciousness' (Mol, 1995: 54). Ecological modernisation theory may also be 'increasingly relevant for Newly Industrialising Countries' (*ibid.*: 55). The first empirical studies addressing this question are only now beginning to show results (*cf.* Frijns *et al.*, 1997; Mol and Frijns, 1999; Hengel, 1998).

Building on a study of Indonesian, Malaysian and Thai pulp and paper firms' adoption of environmental technologies in the early 1990s,[1] this study aims to contribute to a discussion of the applicability of ecological modernisation theory to newly industrialising countries (NICs). Concerns are raised for scholarship on ecological modernisation especially in regard to dematerialisation, North–South equity, and the role of small- and medium-sized enterprises (SMEs).

Operationalisation

A first step in examining the applicability of ecological modernisation theory to NICs is formally to define and operationalise the concept of ecological modernisation. Various authors have made such an effort. Early scholars focused on actual improvements in environmental performance. Paulus (1986), cited in Simonis (1989: 347), for example, suggests that 'Ecological modernization focuses on prevention, on innovation and structural change towards ecologically sustainable development . . . It relies on clean technology, recycling and renewable resources . . . .'

Jänicke et al. (1989: 100) suggest an important premise of ecological modernisation: 'that a reduction in the resource input of production will lead to a reduction in the amount of emissions and waste and also the costs of production'. Hajer (1995: 25–6) suggests three basic concepts emerging out of early work on ecological modernisation:

- 'Make environmental degradation calculable' (especially monetarily);
- 'Environmental protection is . . . a "positive-sum game" '; and
- 'Economic growth and the resolution of ecological problems can, in principle, be reconciled'.

From these and other works, it might be suggested that ecological modernisation has three immediate and two ultimate technological/material objectives: in the short term, *waste reduction and elimination, resource recovery and reuse*, and *dematerialisation;*[2] in the long term, *resource conservation* and *clean production.*[3]

Later scholars have focused on mechanisms and broader social dynamics of ecological modernisation. Mol (1995: 39), for example, suggests that ecological modernisation's material objectives obtain both through '*economising the ecology*', using vehicles such as monetary valuation of natural resources, levying of environmental taxes, and establishment of market incentives; and '*ecologizing the economy*', via re-engineering production, improving industrial co-processing ('industrial ecology'), and boosting superindustrialisation.[4]

Mol goes on (1995: 58; also Mol and Spaargaren, 2000) to examine institutional and social dimensions of environmental transformation, arguing that in ecological modernisation: the state shifts from top-down regulatory intervention to negotiation with industry; non-governmental organisations play a key role, including through direct interaction with industry; political and economic globalisation are supportive; and 'counterproductivity' ('small is beautiful', co-ops, etc.) policy approaches are eclipsed.

Together, such technological/material objectives, mechanisms and institutional dimensions of ecological modernisation provide a useful starting point for evaluating the applicability of the theory to particular cases, including NICs. This analysis addresses two questions: to what extent has pulp and paper manufacturing in Indonesia, Malaysia and Thailand been 'ecologically modernised' in recent years? How have these changes taken place?

Environmental reform

One of the world's ten largest industries, pulp and paper manufacturing has received world-wide attention for environmental pollution for decades. In the 1970s, much public and regulatory attention focused on air quality, including the industry's 'rotten egg' (sulphur dioxide (SO2)) smell. More recently, the industry has received attention for use of chlorine in pulping and bleaching, especially of wood. In the mid-1980s, chlorine used in paper production was positively identified as associated with the production of dioxin, a highly toxic chemical. Regulatory attention shifted to the presence of this chemical in wastewater streams and to water quality more generally. Beginning in the late 1980s, Greenpeace launched an international campaign for the elimination of chlorine in pulp and paper manufacturing.

In South-east Asia, prior to the late 1960s, most pulp and paper manufacturing was very small scale, producing printing and writing paper, newsprint, cigarette paper, and packaging materials, largely from agricultural wastes (bagasse, a residue of sugar cane processing; rice stalks) and recycled paper (especially old corrugated containers (OCC), often imported from East Asia and even North America) (Sonnenfeld, 1998c). Pulp and paper manufacturing expanded substantially in South-east Asia beginning in the late 1960s, an integral part of early industrial development in the region. Pollution became a major concern only with large-scale expansion of pulp manufacturing in South-east Asia beginning in the mid-1980s.

Protests in South-east Asia over pulp industry expansion and pollution in the late 1980s and early 1990s (Sonnenfeld, 1998a, 1998b) had a major impact however, leading to the establishment of new environment/technology regimes influencing adoption of newer, cleaner process technologies in the pulp and paper industry throughout the region (Sonnenfeld, 1996). I focus here especially on adoption of advanced elementally chlorine-free (ECF) pulping and bleaching technologies. Such developments may be analysed in terms of the actual technological/material improvements, and the social dynamics forcing those improvements in each country.

Technological improvements

Many environmental reforms in process and end-of-pipe technology were adopted by Indonesian, Malaysian and Thai pulp and paper firms, c. 1987–96, especially in the large-scale, export-oriented segments of the industry. During this period, new mills were built and brought on-line in South-East Asia utilising some of the most advanced technologies in the world, and some older mills were modified. The most acute problems remain with the oldest, smallest facilities, some government owned.

Perhaps most dramatic was the construction of six new bleached kraft pulp mills using elementally chlorine-free (ECF) pulping and bleaching technologies in Indonesia and Thailand from 1992–96 (see Table 20.1). These mills operate more efficiently and with less pollution per unit of output than mills in many advanced countries (Sonnenfeld, 1998b). During this period, no bleached kraft pulp mills were built without ECF technologies in these countries and Malaysia.

During the same period, companies improved environmental performance at older pulp mills in South-East Asia as well, through changing raw materials, upgrading pre-processing, modifying process technology, and improving waste treatment. Highlights include:

- Older Indonesian pulp mills added pre-treatment processes, reduced use of elemental chlorine, and improved wastewater treatment.
- When it was built in the late 1980s, Malaysia's only bleached kraft mill incorporated world-class technology including oxygen-activated sludge treatment. The mill progressively decreased use of elemental chlorine through the early 1990s.
- Thailand's first large-scale export pulp mill added oxygen delignification and upgraded wastewater treatment facilities. The country's largest pulp and paper group installed advanced wastewater treatment systems, and experimented with enzymes and bacterial pre-treatment for reducing elemental chlorine use.

The worst environmental problems exist among the smallest and oldest pulp and paper mills in the region, some government-owned. Efficiencies as well as pollution would dictate the closing of many of these operations. However, employment issues in high-unemployment economies have proved even more acute (cf. Hanafi, 1994). Even in this troubled segment, national and international research and assistance programmes worked to make environmental improvements.

How did the new mills come to incorporate the latest technologies? What motivated some of the older mills to modify their process and end-of-pipe technologies?

Social dynamics

Technological/material improvements in the manufacturing of pulp and paper in South-east Asia were the result of multiple social dynamics. At the local and country levels, community and environmental groups brought attention to companies' social and environmental practices; government agencies encouraged and sometimes forced companies to adopt cleaner production technologies; public and private sector research engineers creatively developed process modifications; and local environmental conditions came into play. These processes played out in both similar and dissimilar ways in each of the countries studied. In addition, global and regional dynamics contributed to the ecological modernisation of pulp production in South-east Asia.

Table 20.1 Correlates of environmental improvements at bleached pulp mills in Indonesia, Malaysia, and Thailand, c. 1987–96[5]

opns date	manufacturer	country	env. imp.	begin date	pre- proc.	upg. wwt	reduce Cl_2	ECF +	kraft	high capac.	strong regs.	strong mvts.	Nordic conn.	minority- owned	export- orient.
1968	Siam Pulp & Paper	Thailand	Y	?	Y	?	?	N	N	N	N	N	N	N	N
1981	Phoenix I	Thailand	Y	1989	Y	Y	Y	N	Y	N	Y	Y	Y	Y	Y
1984	Indah Kiat – Pulp Mill #1	Indonesia	Y	1987	Y	Y	Y	N	Y	Y	N	N	N	Y	Y
1988	Sabah Forest Industries	Malaysia	Y	1989	N	N	Y	N	Y	N	N	N	Y	Y	Y
1988	Inti Indorayon Utama	Indonesia	Y	1993	?	Y	–	Y	N	Y	N	N	Y	Y	Y
1991	Indah Kiat – Pulp Mill #2	Indonesia	Y	1992	Y	Y	Y	N	Y	Y	N	N	Y	Y	Y
1993	Siam Cellulose	Thailand	Y	1993	Y	Y	?	N	Y	N	Y	Y	Y	N	N
1994	Phoenix II	Thailand	–	–	–	–	–	Y	Y	Y	N	Y	Y	Y	Y
1994	Indah Kiat – Pulp Mill #8	Indonesia	–	–	–	–	–	Y	Y	Y	N	Y	Y	Y	Y
1994	Wira Karya Sakti	Indonesia	–	–	–	–	–	Y	Y	Y	N	Y	Y	Y	Y
1995	Riau Andalan Pulp & Paper	Indonesia	–	–	–	–	–	Y	Y	Y	?	Y	Y	Y	Y
1996	Advance Agro	Thailand	–	–	–	–	–	Y	Y	Y	N	Y	Y	Y	Y
1997	Kiani Kertas	Indonesia	–	–	–	–	–	Y	Y	Y	N	Y	Y	Y	Y

KEY:

opns date = date pulp operations started

env. imp. = implemented environmental improvements

beg. date = date environmental improvements began

pre-proc. = added environmental pre-processing improvements

upg. wwt = upgraded wastewater treatment

reduce Cl_2 = reduced use of elemental chlorine

ECF + = pulp produced without use of elemental (or any) chlorine

kraft = uses kraft chemical pulping process

high capac. = greater than 100,000 admt*/year of bleached pulp

*admt = air-dried metric ton

strong regs. = strong governmental environmental regulations/ enforcement

strong mvts. = strong community, environmental, or other social movements

Nordic conn. = relationship with Pöyry, Nordic firms, or Nordic aid agencies

Nordic aid = recipient of Finnish or Swedish foreign aid, credits, loans, etc.

minority owned = mill owned by ethnic/national minorities, or expatriates

export-oriented = mill designed for production/sales to global markets

Source: Sonnenfeld (1996).

Indonesia

As an important part of the country's export-oriented industrialisation strategy, Indonesia's pulp and paper industry grew rapidly in the late 1980s and early 1990s. New greenfield[6] mills were built on the 'outer' Indonesian islands of Sumatra and Borneo in locales with long traditions of subsistence agriculture. Pulpwood plantations, established by industrial timber estate concession holders as a precondition to government licensing of new mill projects, spread over hundreds of thousands of hectares, disrupted the lives of tens of thousands of people. Pulp mills piped their liquid wastes into rivers used for drinking, bathing, fishing, washing, irrigation, as well as by other industries. The adoption of environmental technologies in Indonesia's pulp industry was directly linked to popular protest against impacts of rural industrial development.

Minority residents in North Sumatra have been engaged in conflict for years with the Raja Garuda Mas (RGM) group, owners of PT Inti Indorayon Utama, Indonesia's first new export-oriented pulp mill, built with substantial government support. What began as an ethnically-charged local struggle over land and forest tenure became a conflict of national and international significance with the involvement of environmental groups and organisations, including the umbrella group, Indonesian Forum on the Environment (WALHI). A boiler explosion at Indorayon in November 1993, led to demonstrations in Medan and elsewhere, and a government statement that henceforth all new pulp mills built in Indonesia would have to be 'ECF or better'.[7]

Under the Suharto government,[8] while having limited success in getting RGM to improve operations at Indorayon, activists' campaigns against the company helped advance the adoption of green technology elsewhere in Indonesia, including RGM's next new pulp mill, PT Riau Andalan, built in the early 1990s in east central Sumatra. PT Riau Andalan was designed and built by the Jaakko Pöyry group, from Finland, and incorporated the latest pulping, bleaching and wastewater technologies.

The Sinar Mas group, Indonesia's leading pulp and paper producers,[9] took advantage of cyclical downturn in the international economy and global pulp industry in the early 1990s to purchase advanced new technology at a substantial discount. Its subsidiary, PT Indah Kiat Pulp and Paper Co. (IKPP), also in east-central Sumatra, had been a target of community protest over land, forest, and water resources. IKPP, the Indonesia Legal Aid Society (YLBH), WALHI, and BAPEDAL signed an historic memorandum of understanding in 1992, wherein the company agreed to clean up its operations and to assist development in surrounding communities (Sonnenfeld, 1996: Appendix E). IKPP utilised in-house engineering expertise, consulting engineering services, and international aid to improve process technologies at its older mill. At the same time, it also adopted new, cleaner technologies at its new mill, in part to help gain access to green export markets.

A continuing problem area in Indonesia relates to older, smaller pulp and paper mills, many located in densely populated areas on the island of Java, home to the great majority of the country's 200 million people. In East Java, for example, the provincial government, working together with various international agencies in attempting to clean up heavily used and polluted waterways, sought to shut down a number of smaller (5,000 tons per year), older, relatively inefficient, polluting pulp and paper mills. It was unable to do so, however, at least in the present, due to very high levels of unemployment in the area, and the impact such mill closures would have on the local economy (Hanafi, 1994).

Environmental activists in Indonesia had a co-operative working relationship with Emil Salim, Indonesia's first Minister of Population and Environment, and first head of the Bureau of Environmental Impact Management (BAPEDAL). Activists' relations with BAPEDAL cooled in the mid-1990s, as the agency's second chief, Sarwono Kusumatmadja, steered the agency towards a more voluntary-compliance based relationship with industry.[10] Activists had a positive relationship with the staff of the government-run Institute for Research and Development of the Cellu-lose Industry (IRDCI), which works primarily with state-owned and smaller, older mills.

International consulting engineer and technology supply firms played an important role in the adoption of environmental technology in Indonesia's pulp and paper industry. The Jaakko Pöyry group has advised almost every new pulp mill project in Indonesia. Technology suppliers received strong backing from their home governments, and benefited from close working relationship with consultants. Some suppliers became joint venture partners with Indonesian producers.

At least a half-dozen bilateral aid programmes also contributed to the adoption of environmental technologies in Indonesia's pulp and paper industry. Canadian and Australian agencies supported development of national environmental regula-tions and administration in Indonesia. Austraid (Australia) and other agencies con-tributed to PROKASIH, the Clean Rivers Project, under which the province of East Java developed perhaps the toughest local water environmental standards in Indone-sia. The United States Agency for International Development (USAID) supported a clean technology assistance programme, which included participation by Indone-sia's pulp and paper industry. Swedish and Japanese aid agencies supported environmental research at IRDCI.[11]

Malaysia

Sabah Forest Industries (SFI) was Malaysia's sole greenfield pulp mill during the period studied. Located in north-west Borneo, SFI is one of the most studied pulp mills in South-East Asia, with extensive baseline and follow-up environmental impact assessments conducted in association with the United Nations Environment Programme's (UNEP) Network on Industrial Environmental Management (NIEM) project (*cf.* Murtedza and Landner, 1993). Principal concerns articulated in those studies related to effects of the mill's effluent on fisheries in Brunei Bay. Even more consequential than the mill's discharge, however, was heavy siltation in the Bay due to soil erosion from upland logging (*ibid.*). It is not clear from Murtedza and Lander's study how much of that 'upland logging' was by SFI or its contractors.

During its start-up and initial years of operation, SFI was kept under strict environmental oversight by Malaysia's national government. SFI was required to conduct annual environmental audits and improve environmental performance. The close attention paid to SFI by Malaysia's national government may be partially explained by the mill's ownership by the state of Sabah. At the time, Sabah was one of only two states not led by the United Malays National Organisation (UMNO), Malaysia's ruling party.[12] SFI was caught in tension between local and national authorities on matters such as the granting of tariff protection and building of infrastructure.[13]

SFI used in-house research facilities to improve environmental performance, including reducing use of elemental chlorine and improving waste water processing. Technology and engineering supply firms, university researchers and the NIEM supplemented SFI's in-house efforts. The Swedish International Development Agency (SIDA) funded and provided training in Sweden for SFI engineers and machine operators.

Although there were no integrated pulp and paper mills in West Malaysia at the time of my fieldwork, there were a number of small and medium-sized paper mills utilising recycled paper as their primary raw material. Malaysia's weak environmental regulations governing the pulp and paper industry are meant at least in part to protect this small but economically important sector. In a recent white paper, for example, university researchers and government officials describe as 'valid' complaints from operators of small mills that they could not meet even existing weak environmental standards due to 'financial difficulty in installing efficient wastewater treatment plants as their production scales are too small' (Murtedza et al., 1995: 12).

Thailand

The pulp and paper industry in Thailand, too, has been at the centre of longstanding controversy. Small farmers, supported by NGOs, academics, urban professionals and the media, protested establishment of pulp plantations, industrial pollution, and loss of rural livelihood in Thailand. Two firms in the north-east region of the country received the greatest attention; one was perceived as foreign-owned, the other as minority-owned.

An industrial accident involving one of those firms had an important catalytic effect in public and governmental scrutiny of the industry. A massive fish kill in a nearby river was blamed on a mill operated by the Phoenix Pulp and Paper Co., joint venture of the 'European Overseas Development Corporation' (EODC)[14] and Ballarpur Industries (India). With an important election drawing near, the government responded to public demands for action, taking the unprecedented action of shutting down the Phoenix mill until it could upgrade its wastewater treatment system.

Phoenix management air-lifted a wastewater treatment system from Finland, got it operational in 30 days, and resumed operations. The company proceeded to add a second pulp line, supported with an interest-free Finnish loan. The new pulp mill incorporates the latest pulping and bleaching technology from Nordic supply firms. The construction project included upgraded pre-processing for the original mill.

With north-east Thailand beset by an extended drought, government agencies further required Phoenix to phase out all discharges into the nearby river. In response, the company developed a programme to use treated wastewater to irrigate pulpwood plantations in the vicinity of the mill. Optimistically, the company hoped to 'turn necessity into a virtue', by marketing its products to green markets internationally.[15]

The government's shutdown of the Phoenix mill rippled through the Thai pulp and paper industry. The Federation of Thai Industries' Pulp and Paper Club became involved in the USAID-sponsored Industrial Environmental Management (IEM) programme within weeks of the shutdown. Club members actively exchanged information about environmental technology, conducted environmental audits of each others' manufacturing facilities, and visited environmental technology suppliers in the USA.

The Soon Hua Seng group, a Sino-Thai business which had previously run into trouble establishing eucalyptus plantations, created a new subsidiary, Advance Agro Ltd., to develop a pulp mill, also in north-east Thailand. It hired Presko, a Thai-Finnish public relations firm, actively to monitor popular concern about the pulp industry in Thailand; and Jaakko Pöyry as designer and general contractor for new mill. The mill, which began operations in 1995, was designed with ECF pulping and bleaching processes and advanced wastewater treatment facilities.[16] In building a 'green' mill, Soon Hua Seng hoped to forestall getting into political trouble as it had before.

The Siam Pulp and Paper group, division of partially crown-owned Siam Cement, was less frequently the target of public and political pressure to make environmental improvements to its operations. Nevertheless, the group took strong environmental leadership in its operations and in the Thai Pulp and Paper Industry Association. The group upgraded wastewater treatment facilities at its largest pulp and paper production site, and experimented with using enzyme and bacterial pre-processing to reduce chemical use at its pulp mills.[17]

In contrast to the experience of the new, large-scale pulp producers, pollution by smaller, older mills often 'slip through cracks' of regulatory enforcement in Thailand. An official of the Pollution Control Department told me, for example, that smaller pulp and paper facilities were not a high enforcement priority for his office, because there were relatively few such producers, and he had to deal with many far more pressing problems.

* * *

In both Indonesia and Thailand, one pulp mill's troubled beginnings set the tone for the environmental reform of an entire industry. While having different immediate outcomes, the cases set in motion similar country-level dynamics. These can be portrayed graphically as in Figure 20.1. Essentially, these dynamics entail a combination of social movement (NGO), state regulation, and business factors driving the innovation process at the national level.

The core dynamics of adoption of environmental technologies as depicted in Figure 20.1 include the original, 'landmark' conflict; the establishment of new standards/levels of expectation for industry environmental performance; the encouragement of both firm and supplier innovation; and implementation/adoption of the new, cleaner production technologies. Key participants in these processes are local community groups, domestic and international business interests, non-governmental organisations, regulatory agencies, bi- and multi-lateral aid agencies, and 'green' consumers. Particular local environmental conditions (such as the poor character of Phoenix's site) also are a factor.

In all three countries, there were markedly different dynamics in the large-scale,

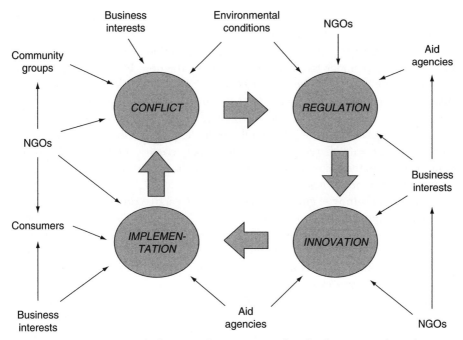

Figure 20.1 Dynamics of adoption of environmental technologies in pulp and paper manufacturing in Indonesia and Thailand, c. 1987–94[18]

Source: Sonnenfeld (1996).

often export-oriented pulp industry segment as compared to the segment of smaller, older producers. Although the former received great public and regulatory attention, the latter often 'slipped through the cracks' of both popular and regulatory oversight. Even when identified as causing serious pollution to public waterways, small and medium-sized pulp and paper mills typically were allowed to continue operating. In contrast, popular and government pressure was much greater on the larger, newer mills.

Global and regional dynamics

Global and regional institutions also contributed to environmental reform of (at least the large-scale, export oriented segment of) South-east Asian pulp manufacturing. Global market, finance, and technology considerations exerted influence on adoption of environmental technologies in the export-oriented, and financially dependent South-east Asian industries. International aid agencies and social movements played important roles as well.

On the one hand, various global factors motivated South-east Asian pulp producers to adopt the new, more environmental technologies. Environmental risk is factored in to the costs firms have to pay (in 'points' and interest) in world financial markets to issue corporate bonds. Maintaining access to global consumer markets also is important to export-oriented South-east Asian pulp producers. They pay attention to international eco-labelling efforts and aim to obtain eco-certification where possible. Phoenix, in Thailand, for example, was considering going after green markets in Japan and elsewhere with its non-wood (bamboo, kenaf), 'environmentally friendly' pulp.

On the other hand, global technology firms were highly motivated to sell their new technologies under favourable terms. When the new, ECF technologies were coming on the market in the early 1990s, Finland and Sweden were undergoing their worst economic (and political) crises in half a century; Europe and North America were in the midst of a major economic slump; and South-east Asia was one of the few regions of the world rapidly expanding. Technology supply firms and Nordic governments offered price discounts, trade credits, interest-free loans, and joint-venture partnerships to South-east Asian pulp producers to encourage them to adopt the new technologies.

Foreign consulting engineering firms played an active role in promoting adoption of environmental technology in Thailand's pulp and paper industry. Both new ECF mills in north-east Thailand were designed by Jaakko Pöyry and incorporate advanced, Nordic-sourced technology. H.A. Simons, a Canadian consulting engineering firm, was the major contractor for upgrading wastewater treatment facilities at the Siam Pulp and Paper group's mills.

International aid agencies, also, played an important role in the adoption of environmental technology in Thailand's pulp and paper industry. As noted above, the Finnish Agency for International Development (Finnaid) provided aid and trade credits for purchase of pulp manufacturing equipment, including at Phoenix Pulp and Paper in Thailand. USAID supported training in environmental management and waste minimisation in both Indonesia and Thailand.

Swedish assistance has been critical in carrying out the United Nations Environment Programme's Network on Industrial Environmental Management (NIEM). For more than seven years, the NIEM has conducted training, documentation, conferences, and workshops aimed at helping pulp industries in participating countries (China, India, Indonesia, Malaysia, the Philippines, Sri Lanka, Thailand, and Viet Nam) develop expertise and exchange information on environmental problems.[19]

International social movements played an indispensable role in the adoption of environmental technologies in South-east Asian pulp industries. The new, ECF technologies were developed in Sweden and Finland in response to European social movements' efforts to tighten environmental regulations and increase demand for green products. Transnational social movements helped disseminate environmental information to South-east Asian governmental, academic, and citizens' groups. Greenpeace, in particular, played a critical role in gathering information on the pulp industry and the environment, and making it available around the world. It consulted with pulp manufacturers, industry, governmental and university research institutes, regulatory agencies, and NGOs; and developed a good relationship with suppliers of environmental technology for the pulp industry, especially in Western Europe.

Expressed graphically, the global flow of environmental innovation in Southeast Asian pulp manufacturing might look like Figure 20.2. While having important country-level characteristics due to unique cultural, political, economic, and environmental factors; there is also an essential global dimension of the diffusion of technologies, regulatory standards, and social movement influence in the pulp and paper industry. South-east Asia, as a late industrialising region, was affected by and able to take advantage of these dynamics (Sonnenfeld, 1998b).

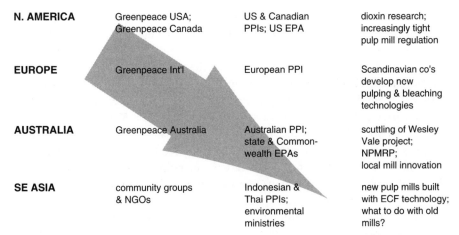

N. AMERICA	Greenpeace USA; Greenpeace Canada	US & Canadian PPIs; US EPA	dioxin research; increasingly tight pulp mill regulation
EUROPE	Greenpeace Int'l	European PPI	Scandinavian co's develop new pulping & bleaching technologies
AUSTRALIA	Greenpeace Australia	Australian PPI; state & Common- wealth EPAs	scuttling of Wesley Vale project; NPMRP; local mill innovation
SE ASIA	community groups & NGOs	Indonesian & Thai PPIs; environmental ministries	new pulp mills built with ECF technology; what to do with old mills?

Figure 20.2 Global flow of innovation in environmental pulping and bleaching technologies, c. 1985–94[20]

Source: Sonnenfeld (1996).

Social movement action in North America and Europe galvanised development of the new, ECF technologies. New environment/technology regulatory regimes also diffused from North to South, both directly through North-South regulatory agency cooperation, and indirectly, through the establishment of global environmental norms and expectations, including among corporate and individual consumers. Australia is included in Figure 20.2 due to its geographical proximity to South-east Asia, and the direct involvements of Australian regulatory agencies, social movements, and industry consultants in South-east Asia. This characterisation is especially pertinent to the newest mills.

Discussion

Based on this summary of research findings, to what extent can it be said that South-east Asian pulp and paper industries have 'ecologically modernised'? Let us return to the objectives, mechanisms, and institutional dimensions of ecological modernisation delineated above as a framework for addressing this question.

The first objective of ecological modernisation is *waste reduction and elimination*. South-east Asian pulp firms have made significant strides in reducing waste, at least in terms of the amount of waste per ton of product. The technologies utilised by the new pulp mills built and operating in South-east Asia are among the most efficient in the world. Modifications to reduce waste have been made to the previous generation of pulp mills as well. Where waste-reduction has least occurred is with the smaller, older pulp mills, some government-owned, which have been kept operating despite their inefficiencies and pollution. With regard to this criterion of ecological modernisation, South-east Asian pulp industries can be said to be well on the road to ecological modernisation, even while problems remain with small and medium-sized enterprises (SMEs).

The second objective of ecological modernisation is *resource recovery and reuse*. With regard to the pulp industry, several resources might be considered: water, chemicals, and fibrous raw materials. The new mills in South-east Asia show remarkable achievements in the reduction of the amount of water used per ton of pulp produced. Chemical recovery, as well, is highly advanced in the new mills. Historically, South-east Asian pulp industries were great *re-users* of fibre resources, utilising agricultural waste and wastepaper as raw materials. As the scale of production has increased, however, the industry has moved away from recycled inputs to greater reliance on virgin raw materials from native forests and tree plantations (Sonnenfeld, 1998c). In sum, South-east Asian pulp manufacturers are making substantial progress in recovering/reusing water and chemicals; fibre is another story, discussed more below.

Perhaps the biggest Achilles heel of South-east Asian pulp producers with regards to ecological modernisation is the criterion of *dematerialisation*. In ecological modernisation, dematerialisation is achieved through the substitution of high-technology for raw material inputs, or alternatively, the substitution of recycled or recovered waste for virgin raw materials. As South-east Asia's pulp industries expand and modernise, less water and fewer chemicals are used and less waste is produced per ton of product. At the same time, producers have adopted wholesale the industrial model promoted by their Northern technology suppliers and funding sources, one based on large increases in both absolute and relative use of virgin raw materials.

Of greatest immediate concern are the proliferating greenfield pulp mills, especially in Indonesia, each of which must be 'fed' a tremendous volume of virgin fibre to maintain full production. Typically, native forests provide the first round of raw materials in the form of 'mixed tropical hardwood' (MTH) chips, while extensive fast-growing plantations have been established to address future raw material needs. Government policies which awarded very large forest concessions at little or no cost to well-connected companies and individuals only compounded and accelerated this dynamic.

Resource conservation, one of the long-term objectives of ecological modernisation, thus remains in the distant future of South-east Asian pulp industries. As with the shorter-term goals of waste reduction and resource recovery, South-east Asian pulp industries fare much better with regard to the second longer-term objective of *clean production*. Perhaps nowhere is the contradiction between these two objectives more evident than in South-east Asian pulp industries, where there is imaginably a future of clean production while resource conservation is nowhere in sight with regard to virgin fibrous materials.

Through what mechanisms have the industries' accomplishments been achieved? A number of *market factors* were significant in the adoption of environmental technologies in South-east Asia's pulp industries. South-east Asian producers made good use of downturns in the global economy together with the reconfiguration of European trade in the wake of the dissolution of the USSR to negotiate rock-bottom prices for the new technologies. International environmental standards and policies were a minor but nonetheless present factor. Faced with making large investments in fixed capital, South-east Asian producers did not want to be locked out of current or future markets requiring International Standards

Organisation (ISO) certification, establishing environmentally preferred purchasing policies, or implementing eco-labelling.

Science and technology played a critical role both directly and indirectly. In South-east Asia, firm-based research and development laboratories, national industry research centres, and regional networking such as through the UNEP NIEM programme also contributed to the development, diffusion and modification of environmental pulp and paper industry technologies. South-east Asian producers also were able to take advantage of technological innovation in Finland and Sweden and other countries with major research and development efforts in pulping technology.

South-east Asian *states* played an interesting and important role in facilitating adoption of pulp industry environmental technologies. Though generally regarded as weak in their environmental regulatory efforts, in both Thailand and Indonesia, national environmental agencies were forced by popular pressure to intervene to halt production at pulp mills thought to be damaging the environment. They worked together with industry and NGOs, encouraging pulp producers to take preventative environmental measures so they could continue operations in a manner sensitive to community needs and less harmful to the natural environment. In Malaysia, national environmental officials pressed operators of Malaysia's only pulp mill at the time more than most businesses in the country.

Although regulatory standards were historically weak, government agencies, particularly in Indonesia, let industry know they would be setting progressively lower emission standards – giving industry added incentive to invest in cleaner technologies from the beginning. In Thailand, agencies established site-licensing and environmental impact assessment review procedures that strengthened more general and weak emission standards, and created a context for public input and oversight. In Indonesia and Malaysia, too, government agencies supplemented general environmental standards with site-specific licensing, audits, and environmental reviews. In this way, the pulp and paper industryy's big 'footprint' and high visibility made it an easy target for customised regulatory treatment.

The role of environmental and other *non-governmental organisations* (NGOs) was critical in the adoption of new environmental technologies in South-east Asian pulp production. Local community groups, often of ethnic and/or religious minorities, formed the 'first-line' of attack on the industry. National environmental, human rights, and alternative development NGOs took up the cause of local activists. Friendly journalists brought broad attention to the conflicts nationally and internationally. Together, local and national NGOs educated the public, forced regulators to act, negotiated with industry, and generally promoted an atmosphere in which producers were compelled to adopt new, environmental technologies.

Greenpeace International played a critical role in disseminating information about environmental risks associated with use of chlorine in pulp and paper manufacturing, spurring development of new technologies, and increasing regulatory scrutiny of the pulp industry. In South-east Asia, Greenpeace consulted with regulatory officials, NGOs, and even pulp companies interested in avoiding political hassles. Rather than be outside industrial and environmental planning, Greenpeace was very much 'inside' technology development (in Europe) and community oversight of industry (including in South-east Asia).

All of this took place in a context very much *interdependent with globalisation*. Technology, consumption, regulation, social movements all have global as well as local dimensions. None of this would have happened without the new technologies developed and available in the Nordic countries, green pulp and paper markets, converging global environmental regulatory standards, intergovernmental assistance programmes, and global interaction between social movements.

Lastly, no argument can be made that there has been a trend towards *de-industrialisation* in South-east Asia's pulp industry. There have been some calls for smaller-scale, appropriate technology in the global pulp and paper industry, for example the development of more mini-pulp mills to utilise agricultural waste rather than woody materials (Marchak, 1995; Smith, 1997; Sonnenfeld, 1998c). However, even the late 1990s 'Asian flu' (global financial crisis) has not reversed development of additional, large-scale pulp manufacturing capacity, utilising cleaner production technologies in South-East Asia (*cf.* APP 1998).

Using criteria drawn from the literature, then, it can be argued that a partial ecological modernisation *has* taken place in South-east Asian pulp industries, with a major exception in regard to the dematerialisation of pulp and paper production, and continuing problems with small and medium-sized pulp and paper firms. What implications do these findings have for ecological modernisation theory and its applicability to newly industrialising countries?

Conclusion

Clearly, ecological modernisation is on the agenda of newly industrialising countries. Cleaner production technologies are being used with beneficial effects relative to older technologies. Like their counterparts in advanced industrial countries, government environmental agencies in newly industrialising countries are moving toward collaborative relationships with producers. NGOs play a critical role both 'inside' and 'outside' environmental regulatory and management processes in South-east Asia as elsewhere.

Processes of ecological modernisation have specific characteristics in different types of economies and industrial sectors. In advanced economies, industrial restructuring involves modernising an ageing manufacturing base. With well-developed consumer markets and high wage and raw material costs, ecological modernisation in advanced economies combines improvements in environmental performance with productivity gains and manufacturing efficiencies. NICs, on the other hand, have the advantage of being able to use the latest, cleaner technologies from the onset of large-scale, modern manufacturing – 'leap-frogging', while benefiting from inexpensive raw materials and wages. Resource-extractive, export-oriented industries such as pulp and paper may be in a particularly strong position to benefit from such advantages. The marginal costs of adopting cleaner technology may thus be lower in the South, while the pull of international 'green' markets and standards may also be strong for export-oriented sectors.

The biggest problem for the applicability of ecological modernisation theory in NICs – and by extension for ecological modernisation theory more generally – may be in the area of dematerialisation. The cases examined here suggest that production is supermaterialising in the South, even if arguably dematerialising in the North. This is particularly troubling given the large areas of tropical rain forest being clearfelled to supply fibre and establish pulpwood plantations for this industry. This raises a critical question: Is ecological modernisation in advanced industrial societies dependent upon *increased* materialisation elsewhere? (Or more specifically, if global pulp and paper consumption continues to increase, where are those raw materials going to come from, if not native forests and timber plantations in the South?)[21]

A further concern is the applicability of ecological modernisation theory to small- and medium-sized enterprises, some of them government owned. Such firms play an important role in NICs in providing employment and serving domestic markets. In South-east Asia's pulp and paper industries, many SMEs are older, use poorer technology, and are more polluting.[22] While it may make environmental sense to phase out some or many of such firms, doing so would have high social costs. Rather, governments and international agencies need to devise incentives to encourage technology firms to develop more ecological, smaller-scale production technologies.

As ecological modernisation theory further develops, it must take into account the entire globe – not only as marketplace for new ecological ideas and technologies, but also as locus of shifting and increasing material production. Also, the theory must be broadened to include small- and medium-sized, as well as super-industrial, technologies and enterprises.

Notes

1. Data were collected from 1992–96 through fieldwork in those three countries, Singapore, Australia and the USA, and from published and unpublished secondary sources.
2. The 'reduction in the resource input' discussed by Jänicke *et al.* (1989).
3. Production without (un-recycled) waste.
4. Substitution of high-technology for material inputs.
5. Elementally chlorine-free (ECF) mills are indicated by a grey background.
6. That is, 'being built on undeveloped land especially when unpolluted' (Merriam-Webster, 1999).
7. RGM transferred ownership of Indorayon and Riau Andalan to its Singapore-based subsidiary, Asia Pacific Resources Investment Holdings Limited (APRIL) in 1995, in part to facilitate raising funds via the international corporate bound market.
8. Conflicts in North Sumatra regarding the Indorayon mill continued after Suharto's resignation, with the new government intervening more strongly than its predecessor. See Reuters (1998).
9. The group now operates its pulp and paper companies through Asia Pulp and Paper Co., like its competitor, APRIL, a Singapore-based holding company.
10. The relationship between environmental activists and government officials, along with many other things, is being renegotiated in the post-Suharto era.
11. For further information on the Indonesian case, see Sonnenfeld (1998a).

12. Underlying the opposition-led state government were cultural, religious, and ethnic differences between Sabah residents and West Malaysia (Vatikiotis, 1992; The Australian, 1994).

13. UMNO recaptured the state government in 1995.

14. Primary shareholder of which is USA-born expatriate, George Davison.

15. Phoenix continues to have trouble. It was shut down most recently in July 1998, again for pollution of the Nam Phong (The Nation, 1998).

16. In 1998, the Soon Hua Seng group entered into a three-way strategic alliance in its Advance Agro subsidiary, taking on the Enso Group (Finland) and Oji Paper Co. (Japan) as major partners and board members (Suwannakij, 1998).

17. For further information on the Thailand case, see Sonnenfeld (1998b).

18. Stronger influence is indicated by a solid line; weaker influence with dashes.

19. Interview, Mr Mark Radka, Co-ordinator, Network on Industrial Environmental Management, United Nations Environment Program/Regional Office for the Asia-Pacific, Bangkok, 6 August 1994. This project was particularly useful in assisting smaller and medium-sized enterprises (SMEs), some government-owned, with more limited access to financial and technical resources than the newer, larger, export-oriented pulp firms.

20. The four columns represent geographical region, social movement organisations, government and industry actors, and historical developments with regard to the 'greening' of pulp and paper technology, respectively. The arrow indicates the direction of diffusion of environmental transformation of industry: both horizontally, from social movements to government and industry; and vertically, from North to South.

21. Arthur Mol suggests that one way of addressing this question is to look at dematerialisation in pulp and paper manufacturing as a function of the amount (or proportion) of virgin raw material in the ultimate products. Thus increased paper recycling and/or use of agricultural wastes or other raw materials could result in dematerialisation, even if absolute quantities of paper produced continued to increase.

22. Industry structures vary; for example, in electronics, smaller supply firms may be obligated by contractual relationships with their customers to follow international environmental management standards and practices.

References

Asia Pulp and Paper (APP) (1998), 'Summary of APP's 1998/99 Expansion Plans' (Online). Available: http://www.asiapulppaper.com/expansion1.htm. Singapore (corporate information).

Australian, The (Sydney) (1994), 'Opposition Lonely Business in Malaysia', Jan., p.6.

Frijns, Jos, Paul Kirai, Joyce Malombe, and Bas van Vliet (1997), 'Pollution Control of Small-Scale Metal Industries of Nairobi', Department of Environmental Sociology, Wageningen Agricultural University, Wageningen, The Netherlands.

Hanafi Pratomo (1994), 'Impact of new government regulations on future existence of pulp mill', in Proceedings, 48th Appita Annual General Conference, Melbourne, Australia, 1–6 May 1994, Carlton, Australia: Australia-New Zealand Pulp and Paper Industry Technical Association.

Hajer, Maarten A. (1995), *The Politics of Environmental Discourse: Ecological Modernization and the Policy Process*, Oxford: Clarendon Press.

Hengel, Petra van der (1998), 'Textile Industry in Vietnam', M.Sc. thesis, Wageningen University, Environmental Sociology, The Netherlands, April.

Huber, Joseph (1985), *Die Regenbogengesellschaft. Ökologie and Sozialpolitik* (The Rainbow Society: Ecology and Social Policy), Frankfurt: Fisher.

Jänicke, Martin, Monch, Harald, Ranneberg, Thomas and Udo E. Simonis (1989), 'Structural Change and Environmental Impact', *Environmental Monitoring and Assessment*, Vol. 12, No. 2, pp. 99–114.

Marchak, Patricia (1995), *Logging the Globe*, Montreal: McGill-Queens University Press.

Merriam-Webster, Inc. (1999), *WWWebster Dictionary* (Online). Available: http://www.m-w.com/cgi-bin/dictionary. Accessed 14 Jan.

Mol, Arthur P.J. (1995), *The Refinement of Production: Ecological Modernization Theory and the Chemical Industry*, Utrecht: van Arkel.

Mol, Arthur P.J. and Jos Frijns (1999), 'Ecological restructuring in Industrial Vietnam: the Ho Chi Minh City region', *Asia-Pacific Development Journal*, Vol. 5, No. 2, pp. 117–138.

Mol, Arthur P.J. and Gert Spaargaren (2000), 'Ecological Modernisation in Debate: A Review', in Arthur P.J. Mol and David A. Sonnenfeld (eds), *Ecological Modernisation Around the World: Perspectives* and *Critical Debates*, London and Portland, OR: Frank Cass/Routledge, pp. 17–49.

Murtedza Mohamed and Lars Landner (1993), 'Sabah Forest Industries Sdn Bhd Pulp and Paper Mill: EIA Before Construction of Mill and Comparison with Actual Performance', in *Phase II Training Package 2, Volume II, Network for Industrial Environmental Management (NIEM)*. United Nations Environment Programme, Regional Office for Asia and the Pacific, Bangkok, March.

Murtedza Mohamed *et al.* (1995), 'Country Paper – Malaysia', paper presented at NIEM seminar on 'Regulatory Options for Fostering Improved Environmental Management in the Pulp and Paper Industry', Bangkok, 15–18 Nov.

Nation, The (1998), 'Phoenix Paper Mill To Be Closed' (Online). Available: http://203.146.51.4/nationnews/1998/199807/19980721/29181.html. 21 July.

Paulus, Stephan (1986), 'Economic Concepts for Industry-Related Environmental Policies', in *Proceedings – Forum on Industry and Environmentt*, New Delhi: Friedrich Ebert Foundation.

Reuters News Service (1998), 'APRIL Says Indonesia Comments Surprise', (Online). Available: http://biz.yahoo.com/rf/981007/x.html. 7 Oct.

Simonis, Udo E. (1989), 'Ecological Modernization of Industrial Society: Three Strategic Elements', *International Social Science Journal*, Vol. 41, No. 3 (Aug.), pp. 347–61.

Smith, Maureen (1997), *The U.S. Paper Industry and Sustainable Production: An Argument for Restructuring*. Cambridge, MA: MIT Press.

Sonnenfeld, David A. (1996), 'Greening the Tiger? Social Movements' Influence on the Adoption of Environmental Technologies in the Pulp and Paper Industries of Australia, Indonesia, and Thailand', Ph.D. thesis, Sociology, University of California, Santa Cruz.

Sonnenfeld, David A. (1998a), 'Social Movements, Environment, and Technology in Indonesia's Pulp and Paper Industry', *Asia Pacific Viewpoint*, Vol. 39, No. 1 (April), pp. 95–110.

Sonnenfeld, David A. (1998b), 'From Brown to Green? Late Industrialization, Social Conflict, and Adoption of Environmental Technologies in Thailand's Pulp and Paper Industry', *Organization and Environment*, Vol. 11, No. 1 (March), pp. 59–87.

Sonnenfeld, David A. (1998c), 'Logging versus Recycling: Problems of the Industrial Ecology of Pulp Manufacturing in South-East Asia', *Greener Management International*, No. 22 (Summer), pp. 108–22.

Spaargaren, Gert and Arthur P.J. Mol (1992), 'Sociology, Environment, and Modernity: Ecological Modernization as a Theory of Social Change', *Society and Natural Resources*, Vol. 5, No. 4 (Oct.–Dec.), pp. 323–44.

Suwannakij, Supunnabul (1998), 'Two foreign firms take 25.4% of AA', *The Nation* (Bangkok) (Online). Available: http://203.146.51.4/nationnews/1998/199809/19980929/32398.html. 29 Sept.

Vatikiotis, Michael (1992), 'Federal Excess: Sabahans Feel Exploited by Central Government', *Far Eastern Economic Review*, 18 June.

Chapter 21

Oluf Langhelle

WHY ECOLOGICAL MODERNIZATION AND SUSTAINABLE DEVELOPMENT SHOULD NOT BE CONFLATED

Introduction

THERE SEEMS TO BE widespread agreement that environmental policy has undergone substantial changes in the past 10–15 years (Weale, 1992; Hajer, 1995; Christensen, 1996; Christoff, 1996). Two 'paradigms', in particular, have been used to describe and explain these changes: ecological modernization and sustainable development. This paper argues that while ecological modernization and sustainable development are often conflated in the literature, there are, in fact, significant differences between these two ways of framing an approach to environmental policy.

Contrary to what seems to be the common perception, the argument put forward is that the different ways in which sustainable development and ecological modernization frame environmental problems have different implications for environmental policy. Although sustainable development and ecological modernization arguably lead to the same environmental policy in some areas, they do not necessarily do so in others. The two concepts have different frames of reference, and are directed towards different problems, which, in turn, leads to different goals and targets for environmental policy. Ecological modernization should, therefore, be seen as a necessary, but not necessarily sufficient, strategy for sustainable development, and the two concepts should not be conflated.

There are, no doubt, several similarities between the concept of sustainable development, as developed by the World Commission on Environment and Development in *Our Common Future* (WCED, 1987), and ecological modernization. For one thing, both are seen as primarily anthropocentric approaches. I shall return to other similarities below. The point here, however, is that this apparent similarity has led several commentators to the conclusion that *Our Common Future* (and sustainable development) is first and foremost an expression of ecological modernization. Weale (1992, p. 31), for instance, argues that the emergence of the new belief system called 'ecological modernization', most notably, is formulated in the Brundtland report. Hajer (1995, p. 26) makes the same point: 'The 1987 Brundtland Report *Our Common Future* can be seen as one of the paradigm statements of ecological modernization'. For both Weale and Hajer, therefore, it seems that *Our Common Future*, sustainable development and ecological modernization reflect the same belief system: ecological modernization.

Others, however, such as Dryzek (1997), Jänicke (1997) and Blowers (1998), seem to think of sustainable development and ecological modernization as overlapping, but not identical concepts. They disagree, nonetheless, as to which of the two perspectives has the most 'radical' policy implications. While both Weale and Hajer primarily use ecological modernization as a concept *describing* changes in the perception of environmental problems, the primary concern in this paper is the *prescriptive* aspects of these concepts. That is, the environmental policy that can be said to follow from either sustainable development or ecological modernization. It is, first and foremost, here that the differences between sustainable development and ecological modernization become crucial in light of Hajer's (1996, p. 247) argument that 'the framing of the problem also governs the debate on necessary changes'.

As pointed out by Jänicke (1997, p. 12), the 'leading paradigm of environmental policy actors' is seen as being increasingly important. Dryzek (1997, p. 5) expresses a similar view, and argues that 'the way we think about basic concepts concerning the environment can change quite dramatically over time, and this has consequences for the politics and policies that occur in regard to environmental issues'. In Dryzek's perspective, 'language matters', and the way 'we construct, interpret, discuss, and analyse environmental problems has all kinds of consequences' (Dryzek, 1997, p. 9). Hajer (1996, p. 257) also refers to this as 'the secondary discursive reality' of environmental politics, the 'layer of mediating principles that determines our understanding of ecological problems and implicitly directs our discussion on social change'.[1]

In this paper, therefore, the differences between the concepts of ecological modernization and sustainable development are seen in relation to the changes these concepts prescribe for environmental policy. The following questions are raised here. (1) To what extent can ecological modernization and sustainable development be said to overlap as paradigms for environmental policy? (2) What are the implications of sustainable development and/or ecological modernization for environmental policy? (3) Does it matter at all whether one views environmental policy from a sustainable development or an ecological modernization perspective? In my view, it does, and the rest of this paper is an attempt to substantiate this claim.

Sustainable development and ecological modernization as 'new' paradigms for environmental policy

Both sustainable development and ecological modernization are contested concepts. As Christoff (1996) points out, ecological modernization is used in different ways by different authors. Some use it to describe technological developments, others use it to define changes in environmental policy discourse. Others again seem to think of it as a new belief system. Mol & Spaargaren (1993) uses the term to cover a set of sociological theories about the development of modern industrialized society and a political programme favouring a particular set of policies. Christoff (1996) develops the concept of ecological modernization even further by introducing 'weak' and 'strong' versions of ecological modernization.

The same pluralism is present for the concept of sustainable development. There are endless lists of definitions (see Pearce *et al.*, 1989; Pezzey, 1992; Murcott, 1997) and a number of approaches to sustainable development. Several typologies of sustainable development have been developed. Dobson (1996, 1999) has developed a typology that now describes three broad 'ideal' conceptions of what he prefers to call 'conceptions of environmental sustainability'.[2] McManus (1996) identifies nine broad approaches to 'sustainability'. Others have made distinctions between very weak, weak, strong, and very strong conceptions of sustainable development (Pearce, 1993; Turner, 1993; Daly, 1996), and Baker *et al.* (1997) have developed what they call 'the ladder of sustainable development'.

Given the number of conceptions and approaches to both ecological modernization and sustainable development, any comparison between the two seems to be associated with difficulties. The most frequent link made between sustainable development and ecological modernization, however, runs through *Our Common Future* (WCED, 1987). Not only is the report seen as an expression of ecological modernization, but it is also said to represent 'the key statement of sustainable development'. According to Kirkby *et al.* (1995, p. 1), it marked the concept's political emergence and established the content and structure of the present debate.

In order to substantiate the claim that there are important differences between sustainable development and ecological modernization, it seems natural to concentrate on the WCED's understanding of the former term. If it can be shown that there are crucial differences between the conception of sustainable development in the Brundtland report and what is usually understood by the concept of ecological modernization, this would be sufficient evidence for the main argument in this article. I will, however, discuss different interpretations of *Our Common Future* and competing conceptions of ecological modernization and sustainable development in order to make clear the differences and similarities between them.

In the following, I will first give a short presentation of the two concepts in order to make a comparison. As there is no similar 'key statement' for the concept of ecological modernization (Weale, 1992), I will concentrate on the features that seem to be common in the literature on ecological modernization. Generally, there is a lack of clarity whether ecological modernization is used descriptively, analytically or normatively (Christoff, 1996), and these are not always easy to keep separate.[3] As the subject matter is the ensuing/associated consequences for environmental policy, I will, following Mol & Spaargaren (1993) and Weale (1993), concentrate not so much on ecological modernization as a set of sociological theories, but rather as a political programme favouring a particular set of policies.

Second, I will explore the relationship between the concepts of sustainable development and ecological modernization by looking at differences, similarities and implications for environmental policy. Finally, I will try to answer the above

questions concerning the relationship between sustainable development and ecological modernization and substantiate the conclusions presented above, that
sustainable development and ecological modernization, in fact, have different
implications for environmental policy.

The concept of ecological modernization

The concept of 'ecological modernization' originates from the works of Huber and
Jänicke. According to Spaargaren (1997), they can be regarded as the founding
fathers of the ecological modernization approach. As a political programme,
however, ecological modernization was originally intended as an interpretation of
the development of environmental policy in Germany and the Netherlands. Weale
(1992), referring to Germany, describes the 'ideology' of ecological modernization
as a denial of the validity of the assumptions underlying the pollution control
strategies of the 1970s. These strategies were, according to Weale, based on the
following assumptions:

> . . . that environmental problems could be dealt with adequately by a
> specialist branch of the machinery of government; that the character
> of environmental problems was well understood; that environmental
> problems could be handled discretely; that end-of pipe technologies
> were typically adequate; and that in the setting of pollution control
> standards a balance had to be struck between environmental protection
> and economic growth and development[4] (p. 75).

The strategies based on these assumptions soon proved to be incapable of solving
the environmental problems they were supposed to deal with. Instead, they resulted
in problem displacement, across time and space, rather than problem solving
(Weale, 1992, p. 76).

Nonetheless, the 'reconceptualization' of the relationship between economy
and the market represented a decisive break from the assumptions that informed
the first wave of environmental policy. The ideology of ecological modernization
challenged 'the fundamental assumption of the conventional wisdom, namely that
there was a zero-sum trade-off between economic prosperity and environmental
concern' (Weale, 1992, p. 31). Environmental protection, in this 'new' ideology, is
no longer seen as a burden upon the economy, but rather as a potential source of
future growth (Weale, 1992, p. 75).

Hajer (1995) gives a description of ecological modernization in accordance
with Weale's interpretation, and argues, in the same manner, that a decisive break
has taken place. In Hajer's perspective, however, ecological modernization is
presented not so much as a reaction to failures in environmental policy, but rather
as a reaction to the radical environmental movements of the 1970s:

> The historical argument, in brief, is that a new way of conceiving
> environmental problems has emerged since the late 1970s. This policy
> discourse of ecological modernization recognizes the ecological crisis as

evidence of a fundamental omission in the working of the institutions of modern society. Yet, unlike the radical environmental movements of the 1970s, it suggests that environmental problems can be solved in accordance with the workings of the main institutional arrangements of society. Environmental management is seen as a positive-sum game: pollution prevention pays (p. 3).

In its most general form, Hajer (1995, p. 25) defines ecological modernization as 'the discourse that recognizes the structural character of the environmental problematique but none the less assumes that existing political, economic, and social institutions can internalize the care for the environment'.

Dryzek (1997) argues that the core of ecological modernization is that there is 'money in it for business'. The following substantiates this. (1) 'Pollution is a sign of waste'; hence, less pollution means more efficient production. (2) Solving environmental problems in the future may turn out to be vastly more expensive than to prevent the problem from developing in the first place. (3) An unpolluted and aesthetically pleasing environment may give more productive, healthier and happier workers. (4) 'There is money to be made in selling green goods and services'. And (5), there is money to be made in 'making and selling pollution prevention and abatement products' (Dryzek, 1997, p. 142).

According to Hajer (1996), the 'paradigmatic examples of ecological modern-ization' are the following:

> . . . Japan's response to its notorious air pollution problem in the 1970s, the 'pollution prevention pays' schemes introduced by the American company 3M, and the U-turn made by the German govern-ment after the discovery of acid rain or *Waldsterben* in the early 1980s. Ecological modernisation started to emerge in Western countries and international organisations around 1980. Around 1984 it was generally recognised as a promising policy alternative, and with the global endorsement of the Brundtland report *Our Common Future* and the gen-eral acceptance of Agenda 21 at the United Nations Conference on Environment and Development held at Rio de Janeiro in June 1992 this approach can now be said to be the dominant in political debates on ecological affairs (p. 249).

Environmental politics is, therefore, now dominated by the discourse of ecological modernization, and seems, in addition, to encapsulate sustainable development (Hajer, 1996, p. 248).

The concept of sustainable development

According to Dryzek (1997, p. 123), however, it is not ecological modernization, but sustainable development around which 'the dominant global discourse of ecological concern' pivots. There are different opinions concerning the origin of the concept of sustainable development (see O'Riordan, 1993; Worster, 1993;

Jacob, 1996; McManus, 1996; Murcott, 1997). The 1980 World Conservation Strategy is often seen as one of the first to make use of the term,[5] but the earliest expression, to my knowledge, of something similar to sustainable development relates to work done within the World Council of Churches in the early 1970s. The following, which could have been a quotation from *Our Common Future*, is actually from a report made by a working group within the World Council of Churches in 1976:

> The twin issues around which the world's future revolves are justice and ecology. 'Justice' points to the necessity of correcting maldistribution of the products of the Earth and of bridging the gap between rich and poor countries. 'Ecology' points to humanity's dependence upon the Earth. Society must be so organized as to sustain the Earth so that a sufficient quality of material and cultural life for humanity may itself be sustained indefinitely. A sustainable society which is unjust can hardly be worth sustaining. A just society that is unsustainable is self-defeating. Humanity now has the responsibility to make a deliberate transition to a just and sustainable global society (in Abrecht, 1979).

Although this report speaks of a just and sustainable global society, and not sustainable development, social justice, ecology and the global dimension are also crucial parts of the framework of sustainable development. The definition of sustainable development in *Our Common Future* conceals, to some extent, all three dimensions and their (inter-)relationships. Sustainable development was defined by the WCED (1987, p. 43) as 'development that meets the needs of the present without compromising the ability of future generations to meet their own needs'.

Pearce (1993, p. 7) argues that defining sustainable development 'is really not a difficult issue'. The real problem lies 'in determining what has to be done to achieve it'. In one sense, this is true, but in another sense, it is wrong. The point of departure here is that how the problem is framed (which includes the way it is defined) also has implications for what is seen as necessary changes. This implies that the definition must be seen in the broader context of other concepts, conceptual and normative preconditions, and the implicit interrelations that shape the framework within the report (Verburg & Wiegel, 1997). Only by doing so can the dimensions of (the particular conception of) sustainable development in *Our Common Future* be identified.

The first step in such an analysis is to include the *two key concepts* that the definition of sustainable development is said to contain. These key concepts are often left out from quotations, but are of vital importance for understanding the concept of sustainable development:

- the concept of 'needs', in particular the essential needs of the world's poor, to which overriding priority should be given; and
- the idea of limitations imposed by the state of technology and social organization on the environment's ability to meet present and future needs (WCED, 1987, p. 43).

The satisfaction of human needs must, in light of both the definition and the first key concept, be seen as the primary objective of development (WCED, 1987, p. 43). Malnes (1990, p. 3) calls this the *goal of development* in *Our Common Future*. The qualification that this development must also be sustainable is a constraint placed on this goal, meaning that each generation is permitted to pursue its interests only in ways that do not undermine the ability of future generations to meet their own needs. Malnes (1990, p. 3) calls this the *proviso of sustainability*. As the sustainability constraint is a necessary condition for future need satisfaction, which is part of what sustainable development is supposed to secure, the proviso of sustainability becomes a necessary part of the goal of development, thus providing the inter-dependency of the concept. Moreover, as Malnes formulates it: 'the proviso is entailed by the very goal whose pursuit it constrains' (Malnes, 1990, p. 7).

Furthermore, social justice—understood as need satisfaction—is in this perspective *at the core* of sustainable development. The relationship between social justice and sustainable development, therefore, is not as Dobson (1999) argues, first and foremost 'empirical' or 'functional'. On the contrary, social justice is *the* primary development goal of sustainable development. Dobson (1999) is, of course, right in pointing out that *Our Common Future* strongly argues that there are 'empirical' and 'functional' relationships between social justice and sustainable development. Poverty is seen as a 'major cause and effect of global environmental problems' (WCED, 1987, p. 44), and the 'reduction of poverty itself' is seen as a 'precondition for environmentally sound development' (WCED, 1987, p. 69).

Yet the priority given to the world's poor is also *independent* of the poverty–environment thesis (Langhelle, 1998). That is, even if the thesis is proved wrong and there is no clear dependency between poverty and environmental degradation, the underlying framework of *Our Common Future* would still lead to a prioritization of the essential needs of the world's poor in the name of social justice (and sustainable development). As stated in the report, poverty is 'an evil in itself' (WCED, 1987, p. 8), and sustainable development requires meeting the basic needs of *all*, thus extending to all the opportunity to fulfil aspirations for a better life (WCED, 1987, p. 8).

Environmental sustainability (I prefer to use physical sustainability), therefore, is not the primary goal of development, but a *precondition* for this goal in the long term and for justice between generations. Thus, physical sustainability becomes an inherent part of the *goal* of sustainable development. It is defined as 'the minimum requirement for sustainable development': 'At a minimum, sustainable development must not endanger the natural systems that support life on Earth: the atmosphere, the waters, the soils, and the living beings' (WCED, 1987, pp. 44–45). The relationship between social justice and physical sustainability, therefore, is not just 'empirical' or 'functional', but also 'theoretical' and 'normative' (see also Lafferty & Langhelle, 1999; Langhelle, 1999).

From this frame of reference, the WCED argued that a set of critical objectives follow from the concept of sustainable development: reviving growth; changing the quality of growth; meeting essential needs for jobs, food, energy, water and sanitation; ensuring a sustainable level of population; conserving and enhancing the resource base; reorienting technology and managing risk; and merging environment and economics in decision making (WCED, 1987, p. 49).

Together, the concept of sustainable development and the strategic imperatives constitute the particular *conception* (in the Rawlsian sense) of sustainable development in *Our Common Future* (Rawls, 1993).[6] There is, of course, no necessary link between the concept of sustainable development and the strategic imperatives advocated by the WCED. One can agree with the goal of sustainable development and disagree with the strategic imperatives and *vice versa*. Still, as I will argue in the next section, the way sustainable development is defined (or how the problem is framed) also has implications for the strategic imperatives that can be said to follow from the concept.

Some of the strategic imperatives in *Our Common Future* no doubt have things in common with the concept of ecological modernization. But it is equally clear that both the definition of sustainable development and the strategic imperatives contain elements that move us away from ecological modernization. In the following, therefore, preconditions, assumptions and implicit interrelations of sustainable development and ecological modernization will be further explored. The question raised in the next section is: to what degree do ecological modernization and sustainable development overlap as paradigms for environmental policy?

Ecological modernization and sustainable development — a comparison

Dryzek (1997, p. 126) argues that the main accomplishment of the WCED was that it managed to combine systematically a number of issues that had often been treated in isolation. Among them are development, global environmental issues, population, peace and security, and social justice both within and between generations. The most striking difference between sustainable development and ecological modernization is thus that sustainable development attempts to address a number of issues about which ecological modernization has nothing to say. Moreover, as Jacobs (1995, p. 65) points out, sustainable development (and sustainability) were not intended as economic terms but 'were, and remain, essentially ethico-political objectives'.

Sustainable development is not only about the environment. *Our Common Future* was first and foremost an attempt to reconcile the tension between developmental and environmental concerns at the global level. The context of sustainable development derives partly from global (north–south) concerns, partly from intergenerational (global) concerns and partly from a growing awareness of global environmental problems (Lafferty, 1996; Langhelle, 1996; Lafferty & Langhelle, 1999).

The context of ecological modernization on the other hand, relates primarily the experiences of western industrialized societies (Christoff, 1996; Mol, 1996; Dryzek, 1997). As such, ecological modernization has no established relationship either to the global environmental problems or to social justice. There are, in fact, no explicit references or connections at all to the global dimensions of developmental and distributional problems. As such, ecological modernization is neither concerned with social justice within our own generation (intragenerational justice) nor with social justice between generations (intergenerational justice).

Moreover, there are hardly any references to the global environmental problems within ecological modernization. This is in accordance with Mol (1996, p. 317), who argues that ecological modernization relates to a specific set of environmental problems: '. . . ecological modernisation has 'normal' environmental problems such as water pollution, chemical waste and acidification as its main frame of reference'. The global environmental problems that the WCED devoted most attention to, global warming and loss of biodiversity, thus seem to fall outside the frame of reference of ecological modernization.[7]

Furthermore, Mol & Spaargaren (1993) argue that global warming cannot be handled within the framework of ecological modernization. Global warming must be seen as a problem of 'ecological high-consequence risks', and 'by their very nature ecological high-consequence risks raise problems of technical and political control, awareness of existential anxiety, and so on, which cannot be dealt with within the framework of ecological modernization' (Mol & Spaargaren, 1993, p. 455).

Instead, they argue, ecological modernization belongs to 'the 'simple modernization' phase, making unproblematic use of science and technology in controlling environmental problems'. The problems of 'ground and surface water pollution, chemical and household waste, regional problems like acid rain and the diffuse pollution by high-technology agriculture' can 'in principle and practice' be controlled by following an ecological modernization approach. These problems, therefore, should not be connected directly to 'eco-alarmist prospects' (Mol & Spaargaren, 1993, pp. 454–455).

As such, Mol (1996, p. 317) calls for 'an additional ecological modernisation approach for the analysis of [high-consequence risks such as the greenhouse effect]', and makes the following suggestions as to what such an approach might contain:

> Until today, ecological modernisation has concentrated mainly on processes of institutional reform at the international level (and of course especially in Western industrialised nations). Recent insights into the emergence of globalised environmental risks, the globalisation of political and economic institutions which trigger localised environmental problems, and the reinforcement of inter- and supranational environmental politics might induce a second phase in the ecological modernisation theory. This phase might, for instance, stir up renewed attention to the distributive aspects of environmental policy, which disappeared from the public and political environmental agendas in the late 1980s (p. 315).

It is tempting to conclude, however, that the second phase Mol calls for arrived in 1987 under the name of 'sustainable development'.

Another difference between sustainable development and ecological modernization seems to be the institutional level on which they focus. Ecological modernization, according to Mol & Spaargaren (1993, p. 454), 'does not so much emphasize the relation between the global and the individual, but rather concentrates on strategies of environmental reform on the meso-level of national governments, environmental movements, enterprises and labour organizations'.

Dryzek (1997) argues that ecological modernization 'implies a partnership in which governments, businesses, moderate environmentalists, and scientists cooperate in the restructuring of the capitalist political economy along more environmentally defensible lines'. The global level, in other words, seems to be lacking both institutionally and as a problem area in ecological modernization.

Sustainable development, on the other hand, is directed towards both the national and global institutional level. *Our Common Future* was undoubtedly directed towards intergovernmental organizations, like United Nations and the World Bank, but this does not imply, as Dryzek (1997) claims, that sustainable development de-emphasizes the role of national governments and state actors. Like ecological modernization, sustainable development acknowledges that actors other than the state play an important role. But states play an even more important role with regard to the global environmental problems, where international cooperation and international agreement seem crucial to any attempt at solving these problems. The state is thus fundamental to the conception of sustainable development in *Our Common Future*: 'the integration of environment and development is required in all countries, rich and poor. The pursuit of sustainable development requires changes in the domestic and international policies of every nation' (WCED, 1987, p. 40).

Another crucial difference between sustainable development and ecological modernization relates to nature's carrying capacity and ecological limits for global development. According to Dryzek (1997), both concepts pay little attention to limits to growth. Limits in ecological modernization are 'not so much explicitly denied as ignored', and *Our Common Future* is seen as 'a bit ambiguous on the existence of limits' (Dryzek, 1997, pp. 144, 129). This leads Dryzek to the conclusion of 'no limits' as one of the basic entities of sustainable development (although in brackets). Despite the claim of lack of limits, however, nature's carrying capacity and ecological limits for global development must be seen as crucial to sustainable development in a way they are not, and cannot be, in ecological modernization.

Dryzek's claim of 'no limits' is substantiated by Brundtland's (1990, p. 138) statement that the WCED 'found no absolute limits to growth', and also the argument put forward in *Our Common Future* that there are no set limits:

> Growth has no set limits in terms of population or resource use, beyond which lies ecological disaster. Different limits hold for the use of energy, materials, water, and land. Many of these will manifest themselves in the form of rising costs and diminishing returns, rather than in the form of any sudden loss of the resource base. The accumulation of knowledge and the development of technology can enhance the carrying capacity of the resource base (WCED, 1987, p. 45).

The WCED, however, also argued that there are ultimate limits:

> But ultimate limits there are, and sustainability requires that long before these are reached, the world must ensure equitable access to the constrained resource and reorient technological efforts to relieve the pressure (p. 45).

How should this ambiguity be understood? It seems, in my view, unreasonable on the whole to interpret sustainable development as implying no limits. Rather, there are different limits for different resources, and these limits have a real existence. Technology and social organization, however, are 'variables' that can be 'manipulated' in such a way that changes in technology and social organization, in theory at least, can make *economic* growth possible within the limits set by nature.

This is also the core of the second of the two key concepts sustainable development is said to contain. Technology and social organization are the 'tools' which (hopefully) will make it possible to meet the needs of the present without violating ecological limits and ultimately the ability of future generations to meet their own needs. *Ex officio* for the WCED, Jim MacNeill argues that the 'maxim of sustainable development is not 'limits to growth''; it is 'the growth of limits'. It is the growth of limits in the sense that the:

> basic food and energy needs of 5 billion people (with 5 billion more to come in the next five decades) require large appropriations of natural resources, and the most basic aspirations for material consumption, livelihood, and health require even more (MacNeill *et al.*, 1991, p. 27).

The growth needed over the next few decades to meet human needs and aspirations, especially in the developing countries 'translates into a colossal new burden on the ecosphere' (Mac-Neill *et al.*, 1991, p. 27). As such, meeting the needs of the present is not, as Dobson (1999) seems to imply, only seen as functional for (physical) sustainability in *Our Common Future*. Both poverty and wealth contribute to environmental problems. While Dobson (1999, p. 136) argues that 'the poor pursue some of the most environmentally sustainable lives on earth', Brundtland's (1990, p. 137) reply would be that it is 'both futile and an insult to the poor to tell them that they must remain in poverty to 'protect the environment''. This acknowledgement is what lies behind what Dryzek (1997) (in my view) correctly interprets as the 'the core storyline of sustainable development':[8]

> The core story-line of sustainable development begins with a recognition that the legitimate developmental aspirations of the world's peoples cannot be met by all countries following the growth path already taken by the industrialized countries, for *such action would over-burden the world's ecosystems.* Yet economic growth is necessary to satisfy the legitimate needs of the world's poor . . . Sustainable development is not just a strategy for the future of developing societies, but also for industrialized societies, which must reduce the excessive stress their past economic growth has imposed upon the earth (p. 129, emphasis added).

Dryzek's conclusion of 'no limits', however, seems strangely at odds with this 'core storyline'. If sustainable development implies 'no limits', then why cannot developing countries follow the growth path already taken by the industrialized countries? Or why, indeed, should industrialized societies reduce the excessive stress their past economic growth has imposed upon the earth? In other words, if there are no limits for global development, the 'core storyline of sustainable development' stands out as rather meaningless.[9]

What then, are the limits for global development? In *Our Common Future*, the ultimate limits to global development are seen as being determined (perhaps) by two things: the availability of energy, and the biosphere's capacity to absorb the by-products of energy use. These limits are assumed to have much lower thresholds than other material resources, mainly because of the depletion of oil reserves and the build-up of carbon dioxide leading to global warming (WCED, 1987, p. 58). The argument is, therefore, not that there are no other possible limits to future global development, but that the limits of energy sources and the problem of climate change will be met first, and indeed may already be at hand. Ecological modernization has, again, nothing to say on the issue.

Another crucial difference between ecological modernization and sustainable development is the assumption in *Our Common Future* that the world is experiencing not only a growing international economic interdependence, but also a growing *ecological* interdependence:

> We are now forced to concern ourselves with the impacts of ecological stress—degradation of soils, water regimes, atmosphere, and forests— upon our economic prospects. We have, in the more recent past, been forced to face up with a sharp increase in economic interdependence among nations. We are now forced to accustom ourselves to an accelerating ecological interdependence among nations (WCED, 1987, p. 5).

The assumption of global ecological interdependence is lacking in ecological modernization. Together with the differences relating to the context of social justice, global environmental and developmental problems, global politics and global limits, it seems clear that ecological modernization and sustainable development are quite different concepts, even when ecological modernization is compared with *Our Common Future*.

These differences, however, are not sufficient to substantiate the claim that the implications for environmental policy are different. In fact, Mol (1996) seems to argue that a broader framework (such as sustainable development) has no further implications for environmental policy beyond the perspective of ecological modernization:

> Ecological modernisation theory puts forward a radical reform programme as regards the way modern society deals with the environment . . . But the point of reference for this radical transformation is the movement towards an environmentally sound society, and not a variety of other social criteria and goals, such as the scale of production, the capitalist mode of production, worker's influence, equal allocation of economic goods, gender criteria and so on. Including the latter set of criteria might result in a more radical programme (in the sense of moving further away from the present social order), but not necessarily a more ecologically radical programme (pp. 309–310).

This claim is the focus of the following section.

Implications for environmental policy — radical and moderate interpretations

There are, just as for the concepts themselves, different interpretations of the implications of sustainable development and ecological modernization for environmental policy, ranging from moderate to radical. According to Dryzek (1997), the WCED neither demonstrated the feasibility of, nor the practical steps required in bringing about sustainable development. Ecological modernization, on the other hand, has a much sharper focus on 'exactly what needs to be done with the capitalist political economy' (Dryzek, 1997, p. 143).

I think Dryzek is wrong, however, and in the following the differences between sustainable development, ecological modernization and the way they frame the approach to environmental policy will be discussed under three subheadings. The first relates to the 'nature' of global environmental problems. The second relates to the magnitude of change seen as necessary. The third relates to the goals and targets that arise from the merging of environmental and developmental concerns. For all three issues, the differences identified above have important consequences for environmental policy.

The 'nature' of global environmental problems

Our Common Future was, as I have argued, first and foremost directed towards the global level. Thus, sustainable development is a 'construction' based on and directed primarily towards global environmental problems. There was, in fact, serious discussion within the WCED on the issue of whether or not acid rain 'qualified' as a global problem to be addressed by sustainable development.[10] Managing the global commons, the ozone layer, climate change, species and ecosystems, pollution and sustaining the potential for global food security are the major environmental issues addressed in *Our Common Future*.

Brundtland (1990, p. 138) argues that the 'large ecological issues—the greenhouse effect, the disappearing ozone layer, and sustainable utilization of tropical forests—are tasks facing humankind as a whole'. Of these problems, the 'most global—and the potentially most serious—of all the issues facing us today', according to Brundtland (1991, p. 35), is 'how we should deal with the threats to the world's atmosphere'.

The problem of climate change is addressed throughout *Our Common Future* (WCED, 1987, pp. 2, 5, 8, 14, 22, 32–33, 37, 58–59, 172–176). The centrality of global warming is closely connected to what is conceived as the ultimate limits for global development in *Our Common Future*: the biosphere's capacity to absorb the by-products of energy use. Moreover, it is argued that 'many of us live beyond the world's ecological means, for instance in our patterns of energy use' (WCED, 1987, p. 44). As such, sustainable development puts climate change (and energy) on top of the agenda for environmental policy. Ecological modernization, on the other hand, contains no criteria by which different environmental problems can be weighed. It is, therefore, impossible to say that a particular environmental problem is more important than another from the perspective of ecological modernization.

Moreover, while ecological modernization is silent on global ecological interdependence, the problem of climate change 'forces recognition of global interdependence' (Intergovernmental Panel on Climate Change (IPCC), 1996, p. 118). It is thus futile to believe that the problem of climate change can be 'solved' by developed countries alone. The IPCC (1996) has a clear message on this issue:

> . . . it is not possible for the rich countries to control climate change through the next century by their own actions alone, however drastic. It is this fact that necessitates global participation in controlling climate change, and hence, the question of how equitable to distribute efforts to address climate change on a global basis (p. 97).

Climate change is thus directly linked to the core story-line of sustainable development[11] and to social justice within and between generations. In fact, the WCED argued that even 'physical sustainability cannot be secured unless development policies pay attention to such considerations as changes in access to resources and in the distribution of costs and benefits' (WCED, 1987, p. 43). This claim is 'logical', not 'empirical' or 'moral' in the following sense. In a situation where (energy) resources are scarce, a distribution in which a small minority of the world's population controls most of the resources might be possible to sustain over a longer period of time (physical sustainability) than one where scarce (energy) resources are distributed equally among the world's population. Consequently, the technical question of what is *physically* sustainable cannot be answered without taking the above questions into consideration (Lafferty & Langhelle, 1999).

The loss of biological diversity is, in many respects, similar to the problem of climate change. The 'structure' of the problem inhabits the same conflictual dimensions: north–south and the concern for future generations. The majority of the world's biological diversity is to be found in the south. This is partly a consequence of climatic conditions, but also as a result of the fact that the industrialized countries, through what is usually referred to as 'development', have reduced their biological diversity substantially during the last 250 years.[12]

According to Christoff (1996, p. 486), ecological modernization is 'deeply marked by the experience of local debates over the local policies of acid rain and other outputs, rather than conflicts over biodiversity preservation'. The concern for conservation and protection of biodiversity is, therefore, stronger in sustainable development than in ecological modernization. The concern for conservation and protection of biodiversity in *Our Common Future* is primarily based on the needs and opportunities of future generations: 'The loss of plant and animal species can greatly limit the options of future generations; so sustainable development requires the conservation of plant and animal species' (WCED, 1987, p. 46). As such, conservation is an indispensable prerequisite for sustainable development in a way that it is not for ecological modernization.

To demand that developing countries must sustain all their biological diversity for the sake of future generations, however, is a type of 'conditionality' that limits possible development paths and that many developing countries find hard to accept.

The use (and misuse) of natural resources played an important role in the development of the already industrialized countries. Conservation and protection of biodiversity is thus linked to the distributional questions of who should pay the cost, and who should benefit from the use of biological diversity.

Ecological modernization seems unable to address the nature of these global environmental problems. In a sense, ecological modernization seems to be based on the assumption that if everyone else (that is, developing countries) stays where they are (which they, of course, have no intention of doing), there is no need to worry. It ignores the core story-line of sustainable development, global ecological interdependence and ecological limits and neglects the linkages between global environmental problems and social justice. The implication of ecological modernization as a paradigm for environmental policy is, thus, environmental policies without any global anchoring. This, I believe, has far-reaching implications, because it affects both what is seen as necessary changes, and the goals and targets to which environmental policy should aspire.

The magnitude of change

The opinions as to the magnitude of change prescribed by ecological modernization and sustainable development, again, are highly disputed. Dryzek (1997) relates sustainable development to a weak (moderate) version of ecological modernization. The implications for change are viewed as more or less identical in these perspectives: 'No painful changes are necessary'. Sustainable development implies that 'we *can* have it all: economic growth, environmental conservation, social justice; and not just for the moment, but in perpetuity' (Dryzek, 1997, p. 132).

Ecological modernization, although being described as 'a systems approach' that 'takes seriously the complex pathways by which consumption, production, resource depletion, and pollution are interrelated' (Dryzek, 1997, p. 144), ends up reassuring us that 'no tough choices need to be made between economic growth and environmental protection'. Even if ecological limits would have a real existence, 'a qualitative different growth' would avoid hitting these limits (Dryzek, 1997, pp. 145, 142). Thus, little seems to be gained from the much sharper focus on exactly what needs to be done.

Hajer (1992, 1995, 1996) is more ambiguous on the magnitude of change. On the one hand, Hajer argues that ecological modernization 'does not call for any structural change but is, in this respect, basically a modernist and technocratic approach to the environment that suggests that there is a techno-institutional fix for the present problems' (Hajer, 1995, p. 32).[13] Accordingly, sustainable development, with its focus upon economic growth and technology, should be viewed as 'a technological fix'. The most important story-line identified by Hajer is that sustainable development is posited as 'a positive-sum game', and implies that there is 'no indication that anyone would lose if the world changed its course according to the prescriptions of the Brundtland Commission' (Hajer, 1992, p. 28).

On the other hand, Hajer claims that ecological modernization (and thus sustainable development) implies a shift from a 'remedial' to an 'anticipatory' strategy for solving environmental problems. Here, he refers to Jänicke's (1988) typology of different strategies in environmental policy. Jänicke's typology, however, describes two different types of 'anticipatory' strategies. The first called 'ecological modernization', is defined as follows: '[Ecological] modernization whereby technological innovation makes processes of production and products more environmentally benign (e.g. increased efficiency in combustion)' (in Hajer, 1995, p. 35).[14]

Jänicke's other 'anticipatory' strategy, however, is called 'structural change' and is defined as follows:

> Structural change or structural ecologization whereby problem-causing processes of production are substituted by new forms of production and consumption (e.g. energy-extensive forms of organization, developing new public transport strategies to replace private transport, etc.) (in Hajer, 1995, p. 35).

This strategy, however, is what Hajer seems to define as ecological modernization. Does that mean that ecological modernization implies structural change also in Hajer's perspective? This is by no means clear in Hajer's use of Jänicke's typology, thus blurring what Hajer actually views as the implications of ecological modernization (and sustainable development)—structural change or not.

Moreover, it is not at all clear what is meant by structural change in Hajer's perspective. Is it the magnitude of change that defines whether or not change is structural? How great must the changes be to qualify as 'structural'? Does 'structural' refer to changes in the system's features and characteristics? What are the defining properties of the existing system, and where are the limits of the existing system in relation to change? When does the system become something else? These questions are not addressed, and this fundamentally weakens both the descriptive and prescriptive potential of ecological modernization.

Equally problematic are the attempts to establish a 'strong' as opposed to a 'weak' version of ecological modernization (Christoff, 1996; Dryzek, 1997).[15] Christoff's main concern is the danger that ecological modernization 'may serve to legitimise the continuing instrumental domination and destruction of the environment' (Christoff, 1996, p. 497). The features that Christoff relates to the notion of 'strong' ecological modernization, however, are so removed from the conventional uses of the concept that it is hardly recognizable. Moreover, it is not self-evident what Christoff means by 'ecological', 'institutional/systemic (broad)', 'communicative', 'deliberate democratic/open', 'international' and 'diversifying' as the characteristics of 'strong' ecological modernization.

To some extent, Christoff's approach is based on tensions within ecological modernization. Weale's (1992) perspective, for example, includes more or less radical versions of ecological modernization. The key to more radical implications lies in the move from 'remedial' to 'anticipatory' strategies:

> if more stress is laid upon the need to move from effects to causes then a radical version of policy is likely to emerge, whereas if the stress is the potential growth stimulated by an environmentally sound economy, then a more pro-industry version of policy is likely (p. 78).

Christoff's (1996, p. 491) perspective, however, seems to aim at the 'most radical use of ecological modernisation', 'its deployment against industrial modernisation itself'.[16]

The move from 'remedial' to 'anticipatory' strategies is also present in *Our Common Future*. Moreover, both ecological modernization and sustainable development direct their attention to the *causes* of environmental problems. Both see technology as a major instrument for solving environmental problems. Both argue for a sector-encompassing policy approach, where concern for the environment is to be integrated in every sector of society. Both promote the use of new policy instruments, and changes at the micro-level seem crucial in both paradigms. 'Producing more with less' (as Chapter 8 of *Our Common Future* is entitled) is a slogan that fits both of these paradigms. Both argue that it is possible, in theory, to reconcile concern for the environment with economic growth. Yet, there are other important differences as to the magnitude of change necessary for this reconciliation.

The prescribed growth rates in *Our Common Future* are seen as *environmentally and socially sustainable* only under the following conditions:

- if industrialized nations continue the recent shifts in the content of their growth towards *less material- and energy-intensive activities* and the *improvement of their efficiency* in using materials and energy (WCED, 1987, p. 51; emphasis added).
- a change in the content of growth, to make it more equitable in its impact, i.e. to improve the distribution of income (WCED, 1987, p. 52).

These conditions are further elaborated in *Our Common Future*, and they should be seen as complementary aspects of a pro-growth position (Langhelle, 1999). The crucial difference following from the first condition, however, is that 'producing more with less' seems to be a necessary but not sufficient condition for sustainable development.

As Rasmussen (1997) points out, two main (economic) strategies are prescribed in *Our Common Future* to realize sustainable development. The first is to utilize energy and resources more efficiently; that is, 'to produce more with less'. This strategy component Rasmussen calls the 'micro-part'. The other strategy component is to change the content of growth, by reducing energy and resource intensive activities. The focus here is the total consumption of environmental resources, including the deponic absorption capacity of the atmosphere. This strategy Rasmussen (1997) calls the 'macro-part', and it is seen as requiring the following:

> . . . that one judges the consumption of energy and resources in different production and consumption sectors, and actually reduces the activities within the sectors which are most energy and resource demanding. Given the demand for economic growth, this strategy in addition implies that investments are made in less energy and resource demanding sectors, and that the released surplus of resources (like labour) are transferred to these activities. One consequence could be that the total activity in the transport sector . . . is reduced, and that

the released resources are transferred to other sectors . . . The World Commission's ambitious goals demand that both strategy components are pursued together, and that they in the use of policy instruments are seen as one strategy, where the different parts are seen as interdependent (p. 107, my translation).

Jänicke (1997) seems to be of the same opinion. Sustainable development demands more than ecological modernization understood as resource efficiency. An 'ecologically sustainable development' demands structural changes in four specific social sectors, and structural change is defined as 'a structural change of their societal role and importance' (Jänicke, 1997, pp. 19–20). The four sectors according to Jänicke are:

- the construction complex (the construction industry, local government, or institutions interested in increasing the value of land)—this sector uses the largest share of materials and land, and generates the most solid waste and goods transportation;
- the road traffic complex (car producers and their suppliers, the service network, the mineral oil industry, the road construction industry, etc.);
- the energy complex (the multinational primary energy industries, the utilities, closely associated with the powerful energy-intensive basic industries); and
- the agro-industrial complex (p. 19).

Thus, even though both the paradigms of ecological modernization and sustainable development arguably seem to imply that the environmental problems can be solved within the existing capitalist political system, to use Dryzek's term, sustainable development seems to imply a larger degree of *structural change*. Moreover, this also has far-reaching implications for the core story-line of ecological modernization, the assumption that environmental protection implies a positive-sum game.

   If sustainable development implies more than an efficiency-oriented approach to the environment, it is no longer necessarily the case that sustainable development represents a 'win–win' solution. If sustainable development implies structural change, in the sense that some sectors' societal role and importance must be reduced, these sectors will be the 'losers' or 'victims' of sustainable development policies. Therefore, it implies that some will lose, and some will win, which again implies that the win–win solution only exists at the macro level, and ultimately, only at the global level. *Our Common Future* is not blind to the possibility of clashes of interests:

> The search for common interest would be less difficult if all development and environment problems had solutions that would leave everyone better off. This is seldom the case, and there are usually winners and losers (WCED, 1987, p. 48).

Moreover, IPCC (1996) argues that most policy recommendations for climate change policies 'involve large within losses for certain groups. For instance, any policy leading to less use of coal and lower producer prices for it will lead to large losses for coal mine owners and workers' (p. 33). As shown by Reitan (1998), this is also the case in the national attempt to introduce a cost-effective carbon-tax in Norway. Even when there are possibilities for positive-sum solutions at the macro-level, 'there are several zero-sum games at the sector level that are strongly related to distributional issues between individuals, groups and regions' (p. 16).

As both Brundtland (1990) and MacNeill (1990) argue, the WCED chose to place energy efficiency at the cutting edge of national energy strategies (WCED, 1987, p. 196). A 'significant and rapid reduction in the energy and raw-material content of every unit of production will be necessary' (MacNeill, 1990, p. 116). Promoting energy efficiency is, however, as another member of the WCED points out, 'relatively painless' (Ruckelshaus, 1990, p. 132). But equally clear from *Our Common Future* is that energy and material efficiency is seen as *a necessary but not sufficient condition for sustainable development*: 'Energy efficiency can only buy time for the world to develop "low-energy paths" based on renewable sources, which should form the foundation of the global energy structure during the 21st century' (WCED, 1987, p. 15).

Goals, targets and the merging of environmental and developmental concerns

Our Common Future recommended a low-energy scenario of a 50% reduction in primary energy consumption per capita in the industrial countries, in order to allow for a 30% increase in the developing countries within the next 50 years (WCED, 1987, p. 173). This, it was argued, 'will require *profound structural changes* in socio-economic and institutional arrangements and it is an important challenge to global society' (WCED, 1987, p. 201; emphasis added). Indeed, the WCED believed 'that there is no other realistic option open to the world for the 21st century' (WCED, 1987, p. 174).

What forces this option is the merging of environmental and developmental concerns through the core story-line of sustainable development. Moreover, the goals and targets are directly linked to the distributional aspects of the problem of climate change. According to Shue (1993), the problem of climate change raises four distributional questions:

> (1) What is a fair allocation of the costs of preventing the global warming that is still avoidable? (2) What is a fair allocation of the costs of coping with the social consequences of the global warming that will not in fact be avoided? (3) What background allocation of wealth would allow international bargaining (about issues like 1 and 2) to be a fair process? (4) What is a fair allocation of emissions of greenhouse gases (over the long-term and during the transition to the long-term allocation)? (p. 39).

These questions, no doubt, go to the heart of sustainable development. So far, only the first has been negotiated within the Convention on Climate Change. But the Kyoto Agreement represents only the first, though important, step towards a sustainable development policy on climate change.

From a sustainable development perspective, it is especially the fourth question that is the most challenging. As shown by IPCC (1996),[17] there are differing views as to what would represent a just distribution of the deponic capacity of the atmosphere. Although *Our Common Future* is silent on the actual distribution, it argues strongly for an equitable access to resources (WCED, 1987, p. 39). One proposed criterion for a just distribution is a stabilization of emissions at twice the pre-industrial level, and an equal share on a per capita basis. For Norway, this possible scenario would imply that greenhouse gases would have to be reduced by between 30 and 50% before the year 2020 (Alfsen, 1998). Thus, depending on the principle of distribution, an environmental policy based on the paradigm of sustainable development may be a much more demanding and ambitious one than ecological modernization.

What this implies for production and consumption patterns and levels is hard to predict. Moreover, it raises a number of questions concerning the choice of environmental policies. In which sectors should the reductions be made? What principles should policies on climate change abatement be based on (cost-effectiveness, equitable burden sharing, joint implementation, etc.)? It seems that for climate change, however, the order proposed by Daly (1992) between scale, distribution and allocation makes perfect sense. Daly argues that the question of scale should be set first (for example twice the pre-industrial level). Only then can the deponic capacity of the earth be divided (for example on an equal share on a per capita basis). Only when this is done, Daly argues, 'are we in a position to allow reallocation among individuals through markets in the interest of efficiency' (Daly, 1992, p. 188).

Sustainable development, however, also forces one to address the question of global distribution in a broader and more direct sense. What constitutes, for example, a reasonable level of developmental and environmental aid? Should developing countries be granted a special status in the international trading system? Is the relative inequality between rich and poor countries just? Would the relative inequality between rich and poor countries still be a problem, if everyone's essential needs were met? The point here is simply that these types of questions can be raised from a sustainable development perspective, but ecological modernization is silent on these issues.

The goals and targets for conservation and protection of biodiversity are just as for energy and climate change more ambitious within sustainable development than ecological modernization. *Our Common Future* (WCED, 1987, p. 166) argued that the total expanse of protected areas should be at least tripled in order to constitute a representative sample of the world's ecosystems. Apart from Beckerman (1994, p. 135), who argues that *Our Common Future* represents an 'absolutist' concept of sustainable development, implying that 'the environment we find today must be preserved in all its forms', the usual interpretation is that sustainable development represents a weak and inadequate protection of species and ecosystems.

As such, Sachs (1993, p. 10) argues that sustainable development 'calls for the conservation of development, not for the conservation of nature'. McManus (1996, p. 70) wants to place sustainability in the forefront, not development (offering only 'qualitative' and not 'material' development for the poor?). Dobson (1999, p. 213) argues that if the objective is to sustain irreversible nature, 'principles such as needs and equality will only be helpful if it can be shown that meeting people's needs, or making them more materially equal in some sense, contributes to sustaining irreversible nature', thus failing to see that social justice is an integral part of, and not instrumental to, sustainable development.

Our Common Future, no doubt, condones some loss of biological diversity, legitimizes increased consumption of environmental resources in developing countries, and the intensifying of environmental problems linked to resource use in global terms, but only because the needs of the world's poor are given overriding priority. That is why, in case of conflict, a certain loss of biodiversity is legitimate from a sustainable development perspective. 'Every ecosystem everywhere cannot be preserved intact' (WCED, 1987, p. 45). But the loss can and should be minimized through the wise use of resources, and, as Jacobs (1995, p. 63) points out, there are, in the real world, often solutions that can benefit both.[18]

There is, therefore, contrary to ecological modernization, a hierarchy of priorities and weighing of different concerns inherent in the concept of sustainable development. Based on my interpretation, the following list represents the hierarchy of priorities within the conception of sustainable development in *Our Common Future*:

- the satisfaction of human needs, in particular, the essential needs of the world's poor to which overriding priority should be given;
- climate change (and thus, the energy issue);
- loss of biological diversity;
- pollution (polychlorinated biphenyl (PCB), radioactive pollution, acid rain, etc.);
- food security.[19]

This constitutes what one could call a 'base-line' also for environmental policies, in accordance with the paradigm of sustainable development.[20] While the list of issues is quite limited, it could easily be extended. The issues and problems in themselves are complex, far-reaching, and relate to most human activities. Thus, a list linking these issues to their related activities would turn out far from short. Regardless, this list still covers the main priorities of sustainable development,[21] priorities with implications for environmental policy that go beyond the perspective of ecological modernization.

Concluding remarks: does it really matter?

The main argument in this article has been that there are important differences between sustainable development and ecological modernization, and that these differences have important implications for environmental policy. They effect not only the scope, but also the goals, targets and level of ambition that environmental policy should aim at. It is, of course, impossible to predict the future of technological progress. 'The horizon may glow with technological opportunities' (MacNeill *et al.*, 1991), and a lot may be achieved by implementing the political programme of ecological modernization.

Energy and material efficiency, however, tend to be neutralized by increased output and higher production. Thus Jänicke *et al.* (1993, p. 169) argue that growth in the long term only can be limited growth, 'if the ecologically negative growth effects are to be compensated by technological and structural change'. As such, they conclude that 'industrialised countries will not be able to afford the luxury of high growth rates for much longer'. Even though 'much longer' begs the question of how much longer?, 'the growth of limits' perspective in *Our Common Future* contains the same worry.

Given constraints on social, institutional and political change, MacNeill *et al.* (1991, p. 19) argue that 'no one can rule out a future of ecological collapse'. Overcoming these obstacles 'will require political vision and courage in policy and institutional change on a scale not seen in this century since the aftermath of World War II' (p. 20). To conflate ecological modernization and sustainable development at the conceptual level, however, leaves the impression that this is already being done.

At best, ecological modernization is a 'weak' expression of sustainable development (Blowers, 1998, p. 245). It should be seen as a necessary, but not sufficient condition for sustainable development, even when compared with *Our Common Future*. Conflating the two is not only counterproductive for the broader agenda of sustainable development, but also for the environmental policies necessary for realizing sustainable development. This is why ecological modernization and sustainable development should not be conflated.

Notes

1. The following statement from Hajer (1995) underlines the importance of mediating principles: 'whether or not environmental problems appear as anomalies to the existing institutional arrangement depends first of all on the way in which these problems are framed and defined' (p. 4).
2. Dobson (1999, pp. 36, 60) argues that sustainable development is one form, or theory, of environmental sustainability. Although the conception of sustainable development contains 'views on what is to be sustained, on why, and what the object(s) of concern are, and (often implicitly) on the degree of substitutability of human-made for natural capital', Dobson is wrong, I think, in his assertion that sustainable development 'amounts to a strategy for environmental sustainability'. As I argue later on, for sustainable development, it is the other way around.
3. This difficulty can be illustrated by the following quote from Mol (1996, p. 309): 'Ecological modernisation theory puts forward a radical reform programme as regards the way modern society deals with the environment'. Here, the difference between ecological modernization as both a theory and a political programme disappears.
4. For a closer description of these assumptions, see Weale, 1992, chapter 1.
5. The 1980 World Conservation Strategy was one of the first publications to make use of the phrase 'sustainable development'. It was prepared by the International Union for the Conservation of Nature (IUCN) and published with support from the World Wildlife Fund (WWF) and the United Nations Environment Program (UNEP). The report argued from a dominantly 'conservationist– environmentalist' standpoint (Adams, 1990; Kirkby *et al.*, 1995).

6. The difference between 'concepts' and 'conceptions' consists in the following:

> Roughly, the concept is the meaning of a term, while a particular conception includes as well the principles required to apply it. To illustrate: the concept of justice, applied to an institution, means, say, that the institution makes no arbitrary distinctions between persons in assigning basic rights and duties, and that it rules establish a proper balance between competing claims. Whereas a conception includes, besides this, principles and criteria for deciding which distinctions are arbitrary and when a balance between competing claims is proper. People can agree on the meaning of the concept of justice and still be at odds, since they affirm different principles and standards for deciding those matters. To develop a concept of justice into a conception of it is to elaborate these requisite principles and standards (Rawls, 1993, p. 14 footnote).

7. Weale (1992) is an exception here. He argues that ecological modernization embraces changes in the relationship between states. Moreover, he seems to argue that climate change contributed to the new politics of pollution, and hence, ecological modernization.

8. Dryzek's use of the term 'story-line' is based on Hajer (1995). Hajer defines 'story-line' as follows: 'narratives on social reality through which elements from many different domains are combined and that provide actors with a set of symbolic references that suggest a common understanding. Story-lines are essential political devices that allow the overcoming of fragmentation and the achievement of discursive closure' (Hajer, 1995, p. 62).

9. Dryzek's (1997, p. 129) 'core story-line' is also in accordance with the following official Norwegian interpretation of sustainable development:

> The poor people of the world have a legitimate right to increase their level of welfare. But the Earth's natural environment will not bear if an increasing world population adapts to the present consumption pattern and level, in industrialized countries. In many areas, humans have already broken, or are about to break, the limits set by nature. This is the reason why the worlds consumption and production patterns must be changed, and that industrialized countries have a special responsibility to lead the way in this process (Miljøverndepartementet, 1996–1997, p. 10, my translation).

10. This information comes from an interview with Hans Chr. Bugge, member of the Norwegian delegation that worked with the WCED.

11. That the needs of the present generation demand an increase in energy consumption in developing countries is also expressed in the United Nation's Framework Convention on Climate Change.

12. Conservation and protection of biodiversity are not environmental problems exclusively related to developing countries. Despite protests, clearfelling of virgin forests continues in western Canada (Reid, 1995), and in most European countries there are only tiny fractions of virgin forests left. In Norway, for

example, the Norwegian Institute of Nature Research (NINA) has argued that a total area of at least 5% of the productive coniferous forests needs to be protected in order to conserve the biological diversity (Framstad *et al.*, 1995, p. 3). Only 1.06% is protected today.

13. The following statement from Hajer makes a similar point: 'It is . . . obvious that ecological modernization . . . does not address the systemic features of capitalism that make the system inherently wasteful and unmanagable' (Hajer, 1995, p. 32). Hajer does not, however, give any further description of these features.

14. Hajer calls this strategy 'technological modernization', as Jänicke's use of the term ecological modernization, according to Hajer, 'has a far more restricted meaning' (Hajer, 1995, p. 35).

15. Dryzek's approach is primarily based on Christoff (1996).

16. Ecological modernization would then, of course, no longer be the same thing. While the discourse of ecological modernization may be 'a category of discourse that is flexible' (Weale, 1992, p. 78), it is probably not flexible enough to incorporate Christoff's approach without becoming something else. That seems, however, to be what Christoff aims at.

17. Shue's (1993) list of distributional questions raised by the problem of climate change is also used by the IPCC (1996).

18. The following statement from Brundtland (1997) points to the difficulties of determining the exact content of sustainable development policies:

> I have often seen it argued that one or another activity cannot be sustainable because it leads to environmental problems. Unfortunately, it turns out that nearly all activities lead to one or another form of environmental problem. The question as to whether something contributes to sustainable development or not must, therefore, be answered *relatively*. We must often consider what the condition was prior to that action and what the alternative would have been, as well as to whether the activity could be replaced by other activities . . . We can be forced to make difficult, holistic judgements. That is why there have been very mixed relations of affection between parts of the environmental movement and the very notion of 'sustainable development' (p. 79, my translation).

19. Other candidates for this list are, of course, ozone depletion, nuclear war and population growth. I have excluded them from the given list on purpose. However, a full justification of this would make another paper.

20. What is lacking in the international follow-up of *Our Common Future* is also evident from the list of priorities: a global framework convention on the eradication of poverty. Taken seriously, that is what follows from the conception of sustainable development in *Our Common Future*. Such a framework convention could, just as for climate change, be organized with national reduction targets, timetables for meeting the targets, scientific bodies and so forth.

21. A couple of reservations. First, if one, for example, would live somewhere where there is no access to clean drinking water, this would be given priority over items 2, 3, 4 and 5. The reason is, of course, that this links up to item 1, vital needs. The list will, therefore, vary from country to country, depending on the problems that relate to item 1. Second, the issues are of course interconnected and

must, in many cases, be seen together, a fact that further complicates policies for sustainable development.

References

Abrecht P. 1979. *Faith, Science, and the Future*. Preparatory Readings for a World Conference organized by the World Council of Churches at the Massachusetts Institute of Technology, Cambridge, MA, USA. Church and Society. World Council of Churches, Geneva, Switzerland.

Adams WM. 1990. *Green Development: Environment and Sustainability in the Third World*. Routledge: London.

Alfsen KH. 1998. Norske klimautslipp må ned med 30–50 prosent. *Cicerone* **7**(2): 1–3.

Baker S, Kousis M, Richardson D, Young S (eds). 1997. *The Politics of Sustainable Development: Theory, Policy and Practice within the European Union*. Routledge: London.

Beckerman W. 1994. 'Sustainable development': is it a useful concept? *Environmental Values* **3**: 191–209.

Blowers A. 1998. Power, participation and partnership. In *Co-operative Environmental Governance: Public–Private Agreements as a Policy Strategy*, Glasbergen P (ed.). Kluwer Academic Publishers: Dordrecht; 229–249.

Brundtland GH. 1990. Epilogue: how to secure our common future. In *Managing Planet Earth: Readings from Scientific American Magazine*. W.H. Freeman and Company: New York; 137–138.

Brundtland GH. 1991. Sustainable development: the challenges ahead. In *Sustainable Development*, Stokke O (ed.). Frank Cass Publishers: London; 32–41.

Brundtland GH. 1997. Verdenskommisjonen for miljø og utvikling ti år etter: hvor står vi i dag? *ProSus: Tidsskrift for et bærekraftig samfunn* **4**: 75–85.

Christensen PM (ed.). 1996. Governing the environment: politics, policy, and organization in the Nordic countries. *Nord* **5**: 180–259.

Christoff P. 1996. Ecological modernisation, ecological modernities. *Environmental Politics* **5**: 476–500.

Daly H. 1992. Allocation, distribution, and scale: towards an economics that is efficient, just and sustainable. *Ecological Economics* **6**: 185–193.

Daly HE. 1996. *Beyond Growth: The Economics of Sustainable Development*. Beacon Press: Boston, MA.

Dobson A. 1996. Environment sustainabilities: an analysis and a typology. *Environmental Politics* **5**: 401–428.

Dobson A. 1999. *Justice and the Environment: Conceptions of Environmental Sustainability and Theories of Distributive Justice*. Oxford University Press: Oxford.

Dryzek JS. 1997. *The Politics of the Earth: Environmental Discourses*. Oxford University Press: Oxford.

Framstad A, Bendiksen E, Korsmo H. 1995. *Evaluering av verneplan for barskog*. NINA fagrapport 8, Stiftelsen for naturforskning og kulturminneforskning: Trondheim.

Hajer MA. 1992. The politics of environmental performance review: choices in design. In *Achieving Environmental Goals: The Concept and Practice of Environmental Performance Review*, Lykke E (ed.). Belhaven Press: London; 25–40.

Hajer MA. 1995. *The Politics of Environmental Discourse: Ecological Modernization and the Policy Process*. Oxford University Press: Oxford.

Hajer MA. 1996. Ecological modernisation as cultural politics. In *Risk, Environment and Modernity: Towards a New Ecology*, Lash S, Szerszynski B, Wynne B (eds). Sage Publications: London; 246–268.

IPCC (Intergovernmental Panel on Climate Change). 1996. *Climate Change 1995. Economic and Social Dimensions of Climate Change: Contributions of Working Group III to the Second Assessment Report of the Intergovernmental Panel on Climate Change*. Cambridge University Press: Cambridge.

Jänicke M. 1997. The political system's capacity for environmental policy. In *National Environmental Policies: A Comparative Study of Capacity-Building*, Jänicke M, Weidner H (eds). Springer Verlag: Berlin; 1–24.

Jänicke M, Mönch H, Binder M. 1993. Ecological aspects of structural change. *Intereconomics* **28**: 159–169.

Jacob ML. 1996. *Sustainable Development: A Reconstructive Critique of the United Nations Debate*. University of Gothenburg: Gothenburg.

Jacobs M. 1995. Sustainable development, capital substitution and economic humility: a response to Beckerman. *Environmental Values* **4**: 57–68.

Kirkby J, O'Keefe P, Timberlake L (eds). 1995. *The Earthscan Reader in Sustainable Development*. Earthscan: London.

Lafferty WM. 1996. The politics of sustainable development: global norms for national implementation. *Environmental Politics* **5**: 185–208.

Lafferty WM, Langhelle O (eds). 1999. *Towards Sustainable Development: On the Goals of Development and the Conditions of Sustainability*. Macmillan Press: Houndsmills.

Langhelle O. 1996. Sustainable development and social justice. Paper presented at IPSA round table on 'The Politics of Sustainable Development', Oslo, 26 April.

Langhelle O. 1998. Bærekraftig utvikling. In *Samfunnsperspektiver på miljø og utvikling*, Benjaminsen TA, Svarstad H (eds). Tano Aschehoug: Oslo; 85–114.

Langhelle O. 1999. Sustainable development: exploring the ethics of *Our Common Future*. *International Political Science Review* **20**: 129–149.

MacNeill J. 1990. Strategies for sustainable economic development. In *Managing Planet Earth: Readings from Scientific American Magazine*. W.H. Freeman and Company: New York; 109–123.

MacNeill J, Winsemius P, Yakushiji T. 1991. *Beyond Interdependence: The Meshing of the World's Economy and the Earth's Ecology*. Oxford University Press: Oxford.

Malnes R. 1990. *The Environment and Duties to Future Generations*. Fridtjof Nansen Institute: Oslo.

McManus P. 1996. Contested terrains: politics, stories and discourses of sustainability. *Environmental Politics* **5**: 48–73.

Mol APJ. 1996. Ecological modernisation and institutional reflexivity: environmental reform in the late modern age. *Environmental Politics* **5**: 302–323.

Mol APJ, Spaargaren G. 1993. Environment, modernity and the risk-society: the apocalyptic horizon of environmental reform. *International Sociology* **8**: 431–459.

Murcott S. 1997. Appendix A: definitions of sustainable development. http://www.sustainableliving.org/appen-a.htm [5 June 1998].

O'Riordan T. 1993. The politics of sustainability. In *Sustainable Environmental Economics and Management: Principles and Practice*, Turner KR (ed.). Belhaven Press: London; 37–69.

Pearce D, Markandya A, Barbier EB. 1989. *Blueprint for a Green Economy*. Earthscan Publications: London.

Pearce D. 1993. *Blueprint 3. Measuring Sustainable Development*. Earthscan Publications: London.

Pezzey J. 1992. Sustainable development concepts: an economic analysis. World Bank Environmental Paper Number 2. The World Bank: Washington, DC.

Rasmussen I. 1997. Bærekraftig produksjon og forbruk. In *Rio + 5. Norges oppfølging av FN-konferansen om miljø og utvikling*, Lafferty WM, Langhelle O, Mugaas P, Ruge MH (eds). Tano Aschehoug: Oslo; 106–135.

Rawls J. 1993. *Political Liberalism*. Colombia University Press: New York.

Reid D. 1995. *Sustainable Development: An Introductory Guide*. Earthscan: London.

Reitan M. 1998. Ecological modernisation and 'Realpolitik': ideas, interests and institutions. *Environmental Politics* **7**: 1–26.

Ruckelshaus WD. 1990. Toward a sustainable world. In *Managing Planet Earth: Readings from Scientific American Magazine*. W.H. Freeman and Company: New York; 125–135.

Sachs W. 1993. Global ecology and the shadow of development. In *Global Ecology: A New Arena of Political Conflict*, Sachs W (ed.). Zed Books: London; 3–21.

Shue H. 1993. Subsistence emissions and luxury emissions. *Law & Policy* **15**: 39–59.

Spaargaren G. 1997. The ecological modernization of production and consumption: Essays in environmental sociology. PhD thesis, Wageningen Agricultural University.

Turner KR (ed.). 1993. *Sustainable Environmental Economics and Management: Principles and Practice*. Belhaven Press: London.

Verburg RM, Wiegel V. 1997. On the compatibility of sustainability and economic growth. *Environmental Ethics* **19**: 247–265.

Miljøverndepartementet. 1996–1997. *Miljøvernpolitikk for en bærekraftig utvikling. Dugnad for framtida*. Stortingsmeldiger nr. 58. Miljøverndepartementet: Oslo.

WCED (World Commission on Environment and Development). 1987. *Our Common Future*. Oxford University Press: Oxford.

Weale A. 1992. *The New Politics of Pollution*. Manchester University Press: Manchester.

Weale A. 1993. Ecological modernisation and the integration of European environmental policy. In *European Integration and Environmental Policy*, Liefferink JD, Lowe PD, Mol APJ (eds). Belhaven Press: London; 196–216.

Worster D. 1993. The shaky ground of sustainability. In *Global Ecology: A New Arena of Political Conflict*, Sachs W (ed.). Zed Books: London; 132–145.

Chapter 22

Michael T. Rock

INTEGRATING ENVIRONMENTAL AND ECONOMIC POLICY MAKING IN CHINA AND TAIWAN

Introduction

THE ENVIRONMENTAL PROBLEMS OF the rapidly developing economies of East Asia are well documented (Bello & Rosenfeld, 1990; Eder, 1996; Smil, 1993; World Bank, 1993, 1994a, 1994b, 1997). The combination of rapid urban-industrial growth and de facto "grow first, clean up later" environmental strategies has resulted in low energy efficiency, natural resource depletion, materials-intensive production, polluted rivers and groundwater, and exceedingly dirty air (Brandon & Ramankutty, 1993). This has led one analyst to conclude that Asia, particularly East Asia, is the "dirtiest place on the face of the earth" (Lohani, 1998). Why have governments in East Asia been so slow to respond to accumulating industrial pollution? Once they began responding, how have they acted? Have they simply followed practices established elsewhere? Or have they drawn on their own institutions to devise unique strategies for controlling and reducing industrial pollution? Finally, how have new environmental agencies and mainline economic development agencies managed their entry into ecological modernization?

The last of these questions is tackled by examining the evolution of environmental governance in two high-performing East Asian economies—China and Taiwan. Because others (O'Connor, 1994; Rock, 1996; Spofford, Ma, Zou, & Smith, 1996; Wang & Wheeler, 1996, 1999; World Bank, 1997) have described the development of traditional command-and-control environmental regulation in China and Taiwan, the concern here is with the integration of environmental considerations into well-established structures of economic governance. Because economic governance in both economies is largely the purview of state actors, the focus is on how powerful state actors in economic agencies learned how to integrate environmental considerations into economic policy decisions. This integration is broached from two quite opposite perspectives. In China, the focus is on how a nascent environmental agency learned how to take advantage of the rules of economic governance to influence powerful economic actors. In Taiwan, the focus is on how a mainline economic development agency learned how to integrate environmental considerations into its promotional activities once a potent regulatory agency began to challenge its administrative discretion. In both instances, the emphasis is on what has been accomplished rather than on what remains to be done. Because this tends to convey a sense of great environmental progress, a final word of caution is in order. Both China and Taiwan have only begun to clean up the environment. Much remains to be done. Nothing said here should detract from this.

Lieberthal (n.d.), Jahiel (1998), Bello and Rosenfeld (1990), and Chun-Chieh (1994) argued that the combination of powerful and relatively autonomous economic agencies, weak environmental agencies, and weak organizations in civil society has forestalled environmental efforts to control the increasing industrial pollution loads attending high-speed, export-oriented, industrial growth in China and Taiwan. Although there is no doubt that economic growth remains high on the agenda in both economies, it is also the case that these pictures are at least somewhat overdrawn.

For its part, China has been able to substantially reduce the pollution intensity of gross domestic product (GDP) for total suspended particulates (TSP) and SO_2 (sulfur dioxide) (see Table 22.1) (Wang & Liu, 1999). It has also been able to modestly improve urban air quality (see Tables 22.1 and 22.2) (Wang & Liu, 1999) for TSP in its largest cities.[1] Finally, time series data on ambient environmental quality for four different environmental parameters, TSP, SO_2, chemical oxygen demand (COD) in surface water, and noise levels in urban areas and along major roads, in several major cities show either some improvement or no deterioration (Rock, Yu, & Zhang, 1999). The latter is important in light of the increase in population and economic activity in these cities. These numbers suggest that at least some cities in China have been able to avoid further deterioration.

Taiwan has made somewhat more progress—in 1991, 16.25% of all days in

Table 22.1 Selected data on total suspended particles (TSP) pollution in china

Year	TSP in Millions of Tons	Gross Domestic Product (GDP) in Billions of Yuan (constant prices)	TSP/GDP in Tons per Million Yuan (constant prices)	TSP mg/m^3 Shanghai	TSP mg/m^3 Beijing
1988	14.4	944.9	15.2	300	450
1989	14.0	983.2	14.2	315	399
1990	13.2	1020.9	13.0	358	407
1991	13.1	1114.8	11.8	324	353
1992	14.1	1273.5	11.1	333	380
1993	14.2	1445.3	9.8	292	348
1994				290	395
1995				264	377

Source: TSP and GDP (Wang & Lui, 1999, pp. 338–339); TSP concentration data for Shanghai and Beijing (Rock, Yu, & Zhang, 1999, pp. 40–43).

Table 22.2 Integrated urban environmental quality index: Beijing

Number and Items Examined	Unit	1988	1989	1990	1991	1992	1993	1994	1995
Environmental quality-related items									
1 Total suspended particulates annual average	mg/M^3	0.45	0.399	0.407	0.353	0.38	0.348	0.395	0.377
2 SO$_2$ annual average	mg/M^3	0.1	0.099	0.122	0.126	0.132	0.12	0.11	0.09
3 Drinking water source compliance rate	%	95 to 99	93.2	94.2	97	95	96	98	99
4 Ambient surface water chemical oxygen demand level in the urban area	mg/l	6.3	6.05	6.29	7.05	6.38	6.70	6.20	6.50
5 Ambient noise level in the urban area (average)	dB(A)	60	58.6	58.8	59.7	68.5	57.8	56.9	57.1
6 Noise level at major roads in the urban area (average)	dB(A)	72	72.1	71.4	72	71.6	71.7	71.7	71.7
Pollution control-related items									
1 Urban area enforcing smoke and dust control zone program	%	42	78.1	100	100	100	100	100	100
2 Household use of briquette	%	55	53	56	68.77	70.91	76.5	82.8	85.2
3 Compliance of industrial waste air emission	%	60 to 90	82	94.6	99.45	99.68	92.3	94.2	96.7
4 Compliance of automobile air emission	%	100	85	100	68.55	78.45	79.4	80.5	80.4
5 Wastewater discharge quantity based on per RMB 10,000 productivity	M^3	113	108	111	70.72	64.18	60.6	55.4	49.6
6 Industrial wastewater treatment rate	%	30	41.1	45.3	79.37	84.22	89.1	90.9	90.4
7 Rate of treated industrial wastewater in compliance of standards	%	78	74.2	73.6	61.52	64.56	81	83.7	86.4
8 Rate of industrial solid waste utilized	%	50	48.8	50	59.36	63.42	72.4	77.4	74.1
9 Rate of industrial solid waste treated and disposed	%	19	21.1	15.8	59.59	63.62	72.9	78.1	74.6
10 Rate of household use of gas in the city	%	90	85.5	83.9					
11 Rate of household with centralized heating	%	22	19.3	21.1					

	Unit								
12 Rate of municipal wastewater treatment	%	7.4	6.7	7.2					
13 Rate of urban garbage centrally managed	%	83	100	100					
14 Average green space per person	M^2	5.8	6	6.14					
15 Urban area under ambient noise control program	%				22.06	22.40	30.7	39.3	43.9
Urban infrastructure development-related items									
1 Rate of treatment of urban garbage	%				7.944	52.71	42.1	42.2	43.7
2 Rate of household use of gas in the city	%				85.11	96.98	86.6	89.8	91.7
3 Rate of household with centralized heating	%				27.73	30.95	32.2	34	34.4
4 Rate of green space coverage in existing urban area	%				28.14	30.4	30.7	31.4	32.4
5 Rate of municipal wastewater treatment	%				6.632	1.234	3.1	10.5	20.3

Source: Urban Environmental Management Division, State Environmental Protection Administration.

Taiwan had a Pollution Standards Index greater than 100; by 1994, this fell to 6.99% of all days. By 1997, this had fallen to 5.46% of all days (Rock, 1996; Taiwan Environmental Protection Agency [TEPA], 2000). Similar improvements are apparent in other ambient environmental indicators in Taiwan. Between 1984 and 1993, particulate matter of 10 microns or less (PM10) concentrations in air in Taiwan hovered between 90 and 100 mg/m^3. By 1997, PM10 concentrations in air declined to 64 mg/m^3 for all of Taiwan and 50 mg/m^3 for Taipei (Rock, 1996; TEPA, 2000). How and why has this happened in both economies?

China

Urban environmental management

Most analyses of the effectiveness of traditional environmental regulation in China find it woefully lacking (Jahiel, 1998; Panayotou, 1999; Spofford et al., 1996).[2] Yet, in recent interviews of officials in the State Environmental Protection Administration (SEPA) and in five city-level environmental protection bureaus (EPBs), officials suggested that they are learning how to turn the rules of economic governance to their advantage (Rock et al., 1999).[3] The primary vehicle for attracting and holding the attention of those in core economic agencies is SEPA's Urban Environmental Quantitative Examination System (UEQES). Since 1989, SEPA, with the assistance of city-level EPBs, has been conducting annual quantitative assessments of the environmental performance of the country's major cities. Quantitative assessment is based on an index of some 20-plus environmental indicators. Some of those indicators focus on ambient environmental quality, some focus on the level of development in urban environmental infrastructure, and others focus on the status of pollution control. Scores on each indicator are weighted and summed to yield a composite score.[4] Cities are ranked on the basis of their composite scores, and SEPA publishes ranks and scores in its annual environmental yearbook (National Environmental Protection Agency, 1996). Some provinces and cities have followed this practice by publicizing scores and rankings in newspapers and on radio and television.[5]

What is so intriguing about the UEQES system is that it appears to be changing the nature of industrial policy making within cities, particularly in the larger, richer, coastal cities examined in Rock et al. (1999).[6] As Lieberthal (1992), Lampton (1992), and Walder (1992) observed, policy implementation in China, including industrial policy making within cities, is dominated by vertical and horizontal bargaining. As is well-known, bargaining occurs because the official ranking of positions, organizational units, and functional bureaucracies in China's public sector creates situations where policy implementation requires agreement among a number of position holders, organizational units, or functional bureaucracies before action can be taken. In addition, bargaining proliferated following the post-1979 decentralization of decision making that strengthened the power and authority of local officials while weakening that of provincial and central government officials. Bargaining within cities also appears to be a logical outcome of partial and incomplete liberalization of the economy. On one hand, partial and incomplete liberalization has fostered the growth of a large number of industrial enterprises that are owned and managed by city governments. Mayors and other city managers, particularly those in planning commissions, economic commissions, finance bureaus, and industrial bureaus, are expected to adopt policies that promote the growth, development, and profitability of these enterprises. On the other hand, their ability to promote the growth and profitability of any enterprise is affected by partial liberalization of input prices. In some instances, state-controlled input prices undermine enterprise profitability, whereas in others, they artificially inflate profitability.

City governments have reacted to these problems by developing simple tax/ subsidy rules for treating individual enterprises.[7] This is important because city revenues, which are used to fund city government, social services, public works, urban infrastructure, environmental regulation, and new investment projects, including pollution control investments in city-owned industrial enterprises, are overwhelmingly made up of taxes on those same enterprises.

Because of the need to balance enterprise growth and development objectives with the other pressing needs in cities where enterprise profits may not reflect objective conditions, decisions to tax an enterprise and fund its investments, including its pollution control investments, depend on intense bargaining among all those involved. Individual enterprises bargain with industrial bureaus over tax treatment and funding of investment proposals. Industrial bureaus, the finance bureau, the tax bureau, the urban construction bureau, the environmental protection bureau, and banks that make loans to enterprises for new investments bargain over these same issues as well as over the need to increase investments in urban infrastructure and to make investments in pollution control. The mayor's office, the finance bureau, the planning commission that oversees long-run growth in a city, and the economic commission that is responsible for ironing out bottlenecks facing enterprises and industries bargain over this same set of issues. As a consequence, decisions to invest in an enterprise, tax it, and subsidize it are both shared collective decisions and inextricably intertwined decisions so that the decision to invest, including the decision to invest in pollution control, is paired with decisions to tax and to subsidize.

How has introduction of the UEQES changed the nature of industrial policy making within cities? Prior to the introduction of the UEQES, local environmental officials said they had great difficulty getting individual factories, industrial bureaus, economic commissions, or mayors to take them seriously. Public disclosure of cities' ranks and scores in SEPA's annual environmental yearbook appears to have begun to change this.[8] Initial reaction to the UEQES was slow, but over time some mayors wanted to know why their city ranked lower than other cities. Other mayors wanted to know how the index worked. Ultimately more and more mayors wanted to know what could be done to increase their city's score and rank, and they wanted to know what this would cost. Each of these concerns was directed to city-level environmental protection bureaus, who responded by doing research on these topics. Out of this, an ongoing dialogue between mayors and EPBs developed.

Over time, this dialogue evolved into a complicated set of discussions to implement the UEQES within cities. In most instances, implementation came to be led by city-level environmental protection commissions (EPCs).[9] Senior officials from the five EPBs described how this works.[10] As they said, the whole process begins when the mayor asks each line agency and sector/district within a city to put together a report that evaluates their performance relative to last year's environmental targets and proposes targets for the current year. The EPB in each city takes this information, analyzes it, and consults with local line agencies, sectors, and districts regarding last year's performance and this year's targets. EPB officials stated that this is a very delicate job. It requires them to balance the need to show tangible environmental results with an understanding that they have to be realistic. They do this by increasing their understanding of what it takes to improve a particular indicator and by bargaining and negotiating with line agencies, sectors, and districts over how large improvements in any one environmental indicator should be.

Once individual line agency, sector, and district reports with targets for the current year are agreed to by the appropriate line agency, sector, district, and the EPB, the EPB rolls this up into a projected overall target score on the UEQES for the current year. All of this is included in a larger background report that the EPB prepares for the EPC early each year. The background document describes what the city did to improve its score last year, identifies the major environmental problems in the city, proposes targets by indicator for the current year, and assigns responsibility for meeting those targets to the appropriate line agencies. A comprehensive description of how the UEQES operates in a city and a big table showing who is responsible for what in meeting the current year's targets and how the UEQES relates to SEPA's other environmental policy instruments are included as part of the background document.

Following submission of the background report, the EPC holds its first general meeting of a year on the UEQES. At this meeting, which is attended by a large number of representatives from line agencies, sectors, and districts, the EPC reviews past progress, examines current environmental problems, agrees on this year's target score for the UEQES index, and discusses and assigns responsibility for meeting targets for various indicators to specific line agencies, sectors, and districts. This meeting culminates in the signing of target responsibility system contracts between each sector of government and the mayor. These serve as the basis of the mayor's target responsibility contract with the provincial governor. Once all of this is completed, the EPB publishes a summary description of the EPC meeting in its monthly bulletin. The summary includes recommendations made, consensus reached, and decisions taken. The EPB distributes this report to each of the divisions within the EPB and asks them to prepare a plan to ensure that targets are met. The review by divisions within EPB may lead to some redrafting of the EPC report by the EPB. If this happens, the EPB takes the report back to the EPC for final approval. Once final approval is given, the EPC holds a larger "mobilization" meeting to get all of the sectors and districts involved in implementation.

During the year, the EPB plays an important role in implementing the UEQES process. For example, senior representatives of the EPB meet approximately two times per year with each line agency (such as the urban construction bureau, the urban services bureau, and various industrial bureaus) to evaluate performance relative to the targets in the responsibility contract between the sector and the mayor. The results of these meetings are rolled up into midyear and end-of-year

EPC reviews of progress by line agency, sector, and district. The EPB prepares the background reports for these reviews. Interestingly enough, the personnel department, which is an office of the party, participates in these reviews.

The target responsibility system

As hinted earlier, the whole UEQES process is tightly linked to the environmental target responsibility system.[11] This system is designed to make provincial governors responsible for the environmental quality in provinces and mayors responsible for the environmental quality in cities. It is also designed to get mayors and governors to develop a better balance between growth and the environment. This is done by having them sign environmental performance contracts with officials one level higher up in the bureaucratic hierarchy. These performance contracts specify that particular environmental outcomes will be achieved over specified time periods. The actual parameters included in contracts are initially established by the central government and then distributed to provinces and cities. Specific numeric targets for each parameter at each level of government are the result of bargaining. Once a provincial governor and mayor reach agreement on the target, it is made public. As with the UEQES examination process, the results of assessment of performance relative to the target are made public. As Ross (1988) argued, this is important because the primary reward/punishment for meeting/failing to meet the targets in a contract is the positive/negative publicity surrounding the contract.

How the target responsibility contract system is linked to the UEQES examination process is best illustrated by way of example.[12] In 1995, the mayor of Nanjing signed a 5-year (1996–2001) environmental performance contract with the provincial governor.[13] The contract is modeled on the UEQES index. One part of the contract focuses on ambient environmental quality, one part focuses on pollution control, and one part focuses on urban environmental infrastructure. With respect to ambient environmental indicators in the UEQES, the mayor agreed to reduce total suspended particulate levels from around 350 mg/m^3 in 1995 to less than 220–250 mg/m^3 in 2001. He also agreed to improve the quality of surface water in the Yangtze River flowing through the city so that it meets the Class 2 standard, and he agreed to clean up Xuan Wu Lake so that it meets the Class 5 standard for surface water. With respect to the pollution control indicators in the UEQES, the mayor agreed to reduce industrial wastewater emissions to less than 350m^3 per 10,000 RMB of industrial output. He also agreed to increase the industrial wastewater treatment rate to 70%, to reduce COD levels in surface water by 10%, and to limit dust to no more than 178,000 tons/year. With respect to the urban environmental infrastructure indicators in the UEQES, the mayor agreed to 4 specific air pollution control projects, 15 specific water pollution control projects, and 1 solid waste project.

Each of these targets in the mayor's contract with the provincial governor was arrived at through a bottom-up planning process in the city that assigned each indicator in the UEQES and in the mayor's contract with the provincial governor to one or more sectors/districts in Nanjing. For example, the urban environmental investment projects in air pollution control were assigned to the economic commission (EC). The EC worked with its various industrial bureaus, and these in turn worked with individual industries and plants under their authority to develop plans for increasing investments in air pollution control at specific plants. Because the industrial bureaus negotiate with individual plants over future investment plans, including plans for pollution control expenditures, they know which plants are planning to expand, renovate, and make air pollution control investments. Once agreement is reached on future plant-level investments in air pollution control, the industrial bureaus bargain with the finance bureau over tax treatment of proposed investments, banks over loans for investments, and the EPB on use of pollution levy funds to subsidize part of the air pollution control investments. Once negotiations are completed, industrial bureaus develop estimates of the number of air pollution control projects and the cost of those projects in their sectors for the next 5 years. The economic commission adds up each of these to develop an overall estimate of the number of air pollution projects and the cost of those projects over the next 5 years. It then signs a target responsibility system contract with the mayor specifying the number of projects it will undertake over the next 5 years and the cost of those projects.[14] The target responsibility system provides the technical basis for a city's target score on individual indicators in the UEQES and for the overall UEQES target score.[15] This process is repeated for each indicator in the UEQES in each sector/district in the city.

Results

What evidence is there that the UEQES examination process and the target responsibility system to which it is attached actually affect urban environmental management within cities as well as ambient environmental quality? There are at least two answers to this question. To begin with, time series data on UEQES indicators in two cities show some significant improvement (Rock et al., 1999). In addition, there are numerous concrete examples of how the UEQES process described earlier has been used to clean up the environment. One such example mentioned earlier is that by the Nanjing EPC and EPB to clean up a large lake (3.6 square km), Xuan Wu Lake, in the city.[16]

In the 1980s, this lake suffered from severe eutrophication, it had turned septic, many fish had died, and it emitted a particularly foul odor. At the time, the public complained about the lake, and this problem was finally brought to the attention of the mayor and the Nanjing EPC. The mayor and the EPC established a study group to come up with recommendations for cleaning up the lake. The study group recommended a number of projects to intercept and divert sewer water going into the lake and that industrial polluters who were dumping their water into these sewers be required to meet sewer water discharge standards. The study group also recommended dredging the lake bottom, capturing and diverting rainwater runoff, relocating a zoo that was discharging wastes into the lake, and changing the ecology of the lake so that submerged grass in it could take up nitrogen and slow eutrophication. The study group concluded that if these recommendations were implemented, the quality of water in the lake could meet at least Class 5 surface water quality standards.

On the basis of the recommendations of the study group and agreement within the city to finance and implement them, the mayor included cleanup of the lake to the Class 5 level in his environmental contract with the governor. By the end of 1999, the city had invested approximately 120 million RMB in the lake clean-up project. The vice-director of the Nanjing EPB believes that the lake will be fully restored in 3 years. Because of these investments, the mayor will soon be able to fulfill that part of his environmental contract regarding ambient environmental quality of the lake with the provincial governor.

Taiwan

Contending with a command-and-control environmental agency

In the mid-1980s Taiwan started investing in the creation of a substantial command-and-control environmental agency (O'Connor, 1994). Following the passage of media-specific pollution control acts, in 1987 the government created a cabinet-level Taiwan Environmental Protection Administration and vested it with the authority to set standards and monitor ambient environmental quality and emissions from point sources of pollution. The TEPA was also accorded the authority to enforce emissions standards. Within a few short years, the TEPA developed a substantial ambient air and water quality monitoring system and an aggressive monitoring and enforcement program. Because of an intensive system of inspections of factories, the TEPA has been able to use fines, suspensions of operating permits, and the closing of factories to get factories to comply with emissions standards.[17]

Until recently, all this regulatory activity simply bypassed the government's premier industrial policy agency, the Industrial Development Bureau (IDB) in the Ministry of Economic Affairs (MOEA).[18] For example, the early processes for emissions standard setting relied on expert committees that did not include representatives from either the IDB or industry. Following complaints by industry to the IDB, the IDB began to play a mediating role between the TEPA and industry (interviews in Taiwan, 1995). Over time, it developed its own strategy for integrating environmental considerations into Taiwan's industrial policy.

To begin with, the IDB began promoting, through an import substitution strategy, the creation of an indigenous environmental goods and services industry. In addition, it began subsidizing industry purchases of pollution control and abatement equipment. And it undertook financing state-of-the-art research into pollution prevention and providing industry with subsidized technical assistance in waste reduction/waste minimization.

All of this became part of the IDB's latest industrial development strategy. Appreciation of the exchange rate, rising wage rates, emerging labor shortages, and increased demands for a cleaner environment contributed to an export of industry that some in the IDB feared was leading to a "hollowing out of industry" (IDB, 1995, p. 11). To prevent this, the government promulgated a new 6-year national development plan to upgrade industry, and it replaced the 1960 Statute for Encouragement of Investment with a Statute for Upgrading Industry. This statute provides selective incentives to firms to purchase automated production equipment and technology; increase research and development expenditures; improve product quality, increase productivity, reduce energy use, and promote waste reclamation; and purchase pollution control and abatement equipment (IDB, 1995).

The government also began to promote 24 key high-tech, high-value-added items in 10 emerging industries. Promoted industries included communications, semiconductors, precision machinery, aerospace, and most notably, environmental goods and services. These industries were selected because they cause little pollution, have strong market potential, are technologically demanding, are not energy intensive, and have high value-added (IDB, 1995).

The government is relying on several other promotional privileges to facilitate the growth of an indigenous environmental goods and service industry. Firms in this nascent industry have been organized into industry-specific associations.[19] Government environmental contracts, such as for the building of public sector waste incinerators or providing waste minimization/waste reduction technical assistance to private sector firms, are reserved for firms in these associations. Because the government has adopted an explicitly private sector approach to environmental cleanup, these benefits are likely to be substantial.[20]

Domestic environmental goods and services providers are also favored by tax, commercial bank lending, and land use policies. Firms in all of the 10 emerging industries, including the environmental goods and services industry, are eligible for either a 20% investment tax credit or a 5-year tax exemption plus a double retaining of surplus earnings (IDB, 1995). They are eligible for loans from commercial banks and the Executive Yuan's Development Fund at preferential rates, and they are given priority consideration in the acquisition of industrial land.[21] Indigenous environmental hardware suppliers benefit from a 20% tax credit that accrues to firms purchasing pollution control and abatement equipment.[22] Firms in this industry are also eligible for export assistance, but little is known about how this program works. If it follows practice elsewhere in East Asia, access to assistance may be conditioned on export performance (Rhee, Pursell, & Ross-Larson, 1984). Because the government has established explicit quantitative export targets for the environmental goods and services industry through 2002, something like this may be happening.[23]

In addition, the government offers a range of programs to assist firms trying to reduce industrial pollution. Some of these programs offer fiscal and financial incentives. Others provide technical assistance, particularly for pollution prevention and for waste treatment. The purchase of pollution control and abatement equipment entitles purchasers to tax credits (of either 20% or 10%), and between 5% and 20% of the costs of expenditures on energy conservation and on recycling equipment or technologies can be credited against profits (IDB, 1995). A joint IDB/TEPA Waste Reduction Task Force provides free technical assistance to firms on waste reduction/waste minimization.[24] The IDB also runs an Information Service for Exchange of Industrial Wastes, it sponsors demonstration projects, and it has a program to congregate small and medium enterprises (SMEs) in industrial parks. SMEs who move to these parks are provided good infrastructure, including common wastewater treatment facilities. In return, the SMEs are required to elect an SME committee to enforce emission/effluent standards within the park (interviews in Taiwan, November 1995).

Finally, the IDB finances a growing research program on clean technologies. It has created a Clean Technologies Unit in Division 7 of the IDB and contracted its research on clean technologies out to the United Chemical Laboratory of the Industrial Technology Research Institute (ITRI).[25] Most remarkably, ITRI's clean technology researchers are going well beyond plant-by-plant pollution prevention. They are watching closely what others (e.g., 3M and the USEPA Toxic Release Inventory) are doing, and they are exploring several cost-effective alternatives for developing policy-relevant estimates of the pollution intensity of output by industry subsectors. One of these measures compares the weight of materials used to produce a product to the weight of the final product. Another looks at waste (the difference between the weight of inputs and the weight of final product) per NT$ of sales. A third disaggregates waste into four categories (raw materials, industrial water, energy, and toxic chemicals) per NT$ of value-added.

Although calculating the pollution intensity per NT$ of value-added by highly disaggregated industry subsectors is likely to be viewed by outsiders as either prohibitively expensive or too difficult, in Taiwan it appears to be little more than an extension of what the IDB already does to administer its duty drawback system for exporters.[26] Technical staff in the new National Cleaner Production Center at ITRI are doing this for several reasons.[27] Most important, they hope that better understanding of the pollution intensity of production processes will enable them to redesign production processes to reduce the pollution intensity of production. Because they see this as too risky for the private sector, scientists at ITRI see this as an important role for government. Second, they see this metric, pollution intensity per NT$ of value-added as a way to assess industry-specific performance in Taiwan against international best practice and to track industry-specific performance over time. This yardstick—pollution per dollar of value-added—is expected by them to become the metric by which the government judges the environmental behavior of individual firms and industries and its own environmental performance. If this yardstick were tethered to government performance monitoring that linked rewards, such as preferential access to subsidized credit or to increasingly scarce new land for industrial development, it could spur firms to search for ways to reduce the pollution intensity of output much the same way it encouraged them to increase exports in an earlier time.[28]

Impact on the environment

As in China, there is growing evidence that Taiwan's myriad of plans, policies, and programs actually works. In response to the TEPA's increasingly tough monitoring and enforcement program, firm-level expenditures on pollution abatement equipment have risen quite dramatically. By 1991, investment in pollution control equipment equaled almost 6% of total private sector investment in manufacturing (Rock, 1996). In 1992, private sector investment in pollution control amounted to 4.3% of total investment (Rock, 1996). These figures were dwarfed by the pollution control expenditures of state-owned enterprises (SOEs). By 1992, investments in pollution control by SOEs were nearly 3½ times those of the private sector (Rock, 1996). Given the large role of SOEs in industry in Taiwan, it appears that Taiwanese industry expended a larger share of its investment budget on pollution control than firms in Japan did in 1975 at the height of that country's pollution control effort.[29] In addition, there are a significant number of examples suggesting that the joint IDB/TEPA Waste Reduction Task Force is providing high-quality technical assistance to firms trying to reduce the pollution intensity of production (Rock, 1996). Most important, as mentioned earlier, several ambient environmental quality indicators, including the PSI and particulate matter of 10 microns or less, show evidence of significant environmental progress (O'Connor, 1994).

Concluding remarks

Why have governments in China and Taiwan finally and belatedly begun investing in the environment? With respect to China, the human health costs of environmental degradation (World Bank, 1997) and growing spontaneous public pressure (Rock et al., 1999) have made it increasingly difficult for all levels of government to ignore the environmental degradation attending high-speed urban-industrial growth. Something similar has been happening in Taiwan. There, democratization, growing public concern over the environment, and an almost intractable NIMBY (not in my back yard) problem meant that the government could no longer ignore industrial pollution.[30] Criticism from the influential overseas Chinese community, particularly that from America, added to domestic pressures.[31] Loss of international recognition has increased sensitivity to international criticism. Some of this has extended to the environment.[32] In both instances, governments responded to these pressures by creating increasing tough and competent command-and-control environmental regulatory agencies.[33]

This dynamic, particularly the role of nongovernmental organizations (NGOs), communities, and individuals in civil society, are typical of what is happening in much of Asia (Lee & So, 1999). That is, organized groups and unorganized groups (e.g., individual citizens or communities) in civil society are increasingly applying pressure to clean up the environment that local, regional, and national governments in both economies can no longer ignore. But this does not mean that government officials in either China or Taiwan have created or are creating formal roles for NGOs, communities, or civil society actors in environmental improvement. Neither government has gone much beyond "hotlines" to handle individual citizens' environmental complaints.

In addition, concern regarding industrial pollution in both economies is not being driven by concern for sustainable development. Something more basic is at stake. The high human health costs of high-speed urban-industrial growth can no longer be ignored, particularly in China. In Taiwan, democratization has opened a flood of criticism that elected officials ignore at their own peril.

Finally, globalization has affected the political dynamic of industrial pollution management in China and Taiwan in quite different ways. In China, government officials in SEPA and city-level EPBs seem almost totally unconcerned about the effect of globalization on environmental management. In the larger context, this is understandable. The real pressure on cities to clean up the environment is coming from recognition that the human health costs of the status quo are no longer bearable and by the growth in unorganized local opposition to pollution.[34] Taiwan, on the other hand, is committed to demonstrating by improved environmental behavior that it is worthy of membership in the larger community of nations. This, no doubt, reflects its loss of international recognition due to the mainland's "one-China" policy. Government officials in the TEPA, the IDB, and ITRI are committed to overcoming this loss first by demonstrating that Taiwan is a good international environmental citizen and second by presenting itself as an environmental leader in the region that others can follow.

But creation of command-and-control environmental agencies has not been without its own problems in China and Taiwan. In China, SEPA has had great difficulty enforcing emissions standards on individual factories, the industrial bureaus that control them, and the economic commissions and mayors that depend on them to deliver income and employment. In Taiwan, the creation of an increasingly capable TEPA that had the power to force firms to invest in pollution control led a mainline economic development agency (the IDB) to learn how to engage in the environment. But it is not yet clear what effect the actions of the IDB, as opposed to the TEPA, are having on the environmental behavior of industrial facilities in Taiwan. These considerations lead to the question of how successful these institutional innovations by SEPA and the IDB have been. Unfortunately, there is not yet any good evidence on the effectiveness of either.

In China, despite what is said earlier about the effectiveness of the UEQES, officials in SEPA and city-level EPBs have several concerns about the UEQES. Chief among these is a growing sense that the UEQES index does not treat all cities fairly. A cursory examination of the scores and ranking of cities by NEPA in 1996 suggests why this might be the case (see Table 26.1 in Rock et al., 1999). Older, poorer, and inland cities such as Chongqing, Xining, and Kunming score and rank lower than newer, richer, coastal cities such as Shenzeng, Guangzhou, and Suzhou. Because of this, SEPA is actively considering revising the UEQES index by altering the weights of some indicators and by developing separate indices for different classes of cities. But SEPA is not quite sure how to do this.

These are not the only concerns SEPA and city-level EPBs have about the UEQES index. There is some concern that too much effort within cities and within EPBs is directed to increasing a city's score on particular indicators and on the UEQES index and too little is directed to identifying key environmental risks and cost-effective ways to reduce those risks. There is virtually no research on this issue. If it turns out that the UEQES examination process and the target responsibility system to which it is tied do in fact divert attention away from serious environmental health risks and cost-effective solutions to those risks, neither may be a particularly effective tool of long-run urban environmental management.

In addition, all too little is known about whether the UEQES examination process and the bargaining model of environmental policy implementation on which it depends can ultimately achieve the kinds of long-run improvements in ambient environmental quality that China needs to make. Evidence in other areas of policy implementation in China suggests that bargaining approaches to policy implementation contribute to a number of pathologies such as failed implementation, stalemate, minority veto, and control by the organized. If these problems seriously afflict implementation of the UEQES process, and given the pressures on local officials to develop the local economy, there is some reason to suspect that this might be the case, then the UEQES may not be a particularly effective long-run tool of environmental management. In addition, there is one other potentially serious limitation to the UEQES. Currently, this system appears to work best in China's largest cities and with the largest factories in those cities. Among other things, this could mean that there is substantial variation in the effectiveness of the UEQES system across China's cities. If this proves to be the case, ambient environmental quality may be improving in those cities and factories where the UEQES works best while it may well be worsening in those cities and factories where it is less effective.

In Taiwan, the efforts by the IDB to engage in the environment are fraught with other problems. Some contend that the IDB continues to be Taiwan's biggest polluter.[35] Because of this, they see the IDB's environmental activities as little more than a rearguard action to protect polluters from regulators. Others contend that the IDB's subsidies to firms to buy pollution control equipment undermines efforts to inculcate an environmental ethic within firms because these subsidies violate the polluter-pays principle, a principle that requires polluters to pay for reducing their emissions. Finally and most important, except for the joint IDB/TEPA waste minimization program, there is no solid evidence to indicate that the IDB's clean technologies program has affected either the investment or the environmental behavior of firms in Taiwan. Until there is more evidence that the IDB's environmental actions are having clear environmental effects, some will continue to see those actions in a negative light.

Given these concerns, what would a more successful UEQES program in China and a more successful IDB clean production program in Taiwan look like? Success in China will require breaking the link between economic activity and pollution so that pollution intensity of economic activity falls. But if ambient environmental quality is to approach anything like Organization for Economic Cooperation and Development levels, the decline of pollution in China will have to be much greater than that found for the particulate intensity of GDP during the 1990s. It is not yet clear whether the UEQES system can achieve this. Success in Taiwan, particularly for the IDB and ITRI, requires much more than this simply because the IDB and ITRI are more ambitious. Their aim is to demonstrate that cleaner production, pollution prevention, and design for the environment represent the next wave in industrial pollution management. Thus, success there would require demonstration that manufacturing plants and firms in Taiwan are reducing the energy, materials, water, and pollution intensity of production by fundamentally altering production processes rather than by simply cleaning up pollution after it has occurred by installing end-of-pipe controls.

Notes

1. But as Wang and Lui (1999) stated, this declining trend has been weakening, and concentrations are rising again in some cities.
2. Wang and Wheeler (1996, 1999) are an exception. They found that China's pollution levy for wastewater emissions is somewhat effective.
3. The five cities are Beijing, Nanjing, Tianjin, Shanghai, and Changzhou. In a subsequent study, we are planning to interview mayors, representatives of

important economic agencies, and citizens about the effectiveness of the Urban Environmental Quantitative Examination System (UEQES).

4. For discussion of the indicators and the weighting system used to develop the overall UEQES index, see Rock, Yu, and Zhang (1999).

5. Both Ren Min Ri Bao and CCTV publicize the results of the UEQES (comments from interviews).

6. Less is known about the effect of the UEQES in smaller, poorer, inland cities with weaker environmental protection bureaus, poorer populations, and stronger imperatives on public officials to focus on economic growth.

7. What follows draws on Walder (1992). City-level finance and tax bureaus "whip the fast oxen" by heavily taxing profitable enterprises while they subsidize those facing poor "objective conditions."

8. The effect of public disclosure in the UEQES appears to be similar to that of other public disclosure programs (Afsah & Vincent, 1997; Arora & Cason, 1995).

9. Local-level environmental protection commissions (EPCs) were an outgrowth of an attempt by the state council to create direct communication between the National Environmental Protection Board and industrial ministries over mitigation of industrial pollution (Jahiel, 1998).

10. What follows is drawn from Rock et al. (1999).

11. The UEQES examination process and the target responsibility system are two of the nine major environmental policy instruments available to the State Environmental Protection Administration (SEPA) and local environmental protection bureaus (EPBs). The others are environmental impact assessment, the three simultaneous, the pollution levy, the discharge permit system, limited time treatment, centralized pollution control, and total load control (discussions with SEPA officials; Sinkule & Ortolano, 1995).

12. Other examples of the effect of the UEQES system can be found in Rock et al. (1999).

13. Mayors also sign annual environmental performance contracts with provincial governors. The description that follows is based on discussions with EPB officials in Nanjing.

14. The economic commission will also sign an annual performance contract with the mayor.

15. This comment was made by EPB officials in Nanjing, Changzhou, and Shanghai. Other examples can be found in Rock et al. (1999).

16. Based on discussion with EPB officials in Nanjing.

17. This included a Flying Eagle Project that used police helicopters to respond to citizen complaints of factory emissions and a get-tough or "Rambo" project that used repeat inspections to get those who failed earlier inspections to comply with emissions standards (Rock, 1996). For more discussion of this topic, see Rock (1996).

18. Within Taiwan's relatively "strong state," industrial policy making is limited to a small number of agencies and individuals. At the top of the system is the president and an informal inner group of the cabinet known as the Economic and Financial Special Group (EFSG). The EFSG consists of the minister of economic affairs, the governor of the central bank, the minister of finance, the director-general of the budget, and several ministers without portfolio. This group is advised by the Council for Economic Planning and Development (CEPD), the Industrial Development Bureau (IDB) of the Ministry of Economic Affairs

(MOEA), and the Council for Agricultural Planning and Development (CAPD) (Wade, 1990). Although the CEPD planned the country's industrial development, the IDB of the MOEA had extensive influence over implementation of industrial policy. It used its control over trade policy and a large array of fiscal incentives that accompany the 1960 Statute for Encouragement of Investment to affect which industries private firms invested in. It occasionally relied on subsidized credit from state-owned commercial banks to particular industries and firms. It relied heavily on state-owned enterprises in the commanding heights of the economy—petro-chemicals, steel and basic metals, and shipbuilding (Wade, 1990). And it played a leading role in public sector research and development policy, including strong ties to the country's premier science and technology institute, the Industrial Technology Research Institute (ITRI). Progression up a product ladder pioneered by others provided the IDB with the blueprint for industrial development. Studies of trends in income elasticity of demand, technical change, and of the current composition of imports helped to identify industries that should be developed next (Wade, 1990). Each new stage of industrial development resulted in amendments to the Statute for Encouragement of Investment. It also led to changes in the list of industries eligible for administratively allocated and subsidized credit. When government feared that the private sector might be too slow to respond to these fiscal and financial incentives, it turned to state-owned enterprises as in plastics and steel or to the quasi-public ITRI for development of a high-tech computer chip industry (Wade, 1990).

19. Government laws requires firms to organize into an industry association whenever there are more than five firms in an industry (interviews in Taiwan, November 1995).

20. For example, the government is planning to build 22 large waste incinerators and to contract construction out to private sector firms. Private sector engineering firms in the Taiwan Environmental Engineering Association expect to get most, if not all, of this business (interviews with members of this industry association in Taiwan, November 1995.)

21. Because of rampant NIMBY (not in my back yard) problems, preferential access to new industrial land is extremely important.

22. The tax credit for imported equipment is only 10%.

23. In 1992, domestic production of the environmental goods and services industry equaled $1.3 billion. Of this, 6% ($75 million) was exported. Production is expected to grow at an annual average rate of 11% through 2002. In that year, domestic production will equal $3.75 billion, of which $464 million is expected to be exports (14%) (IDB, n.d.).

24. A representative of an American firm trying to break into this business stated his firm had a difficult time doing this because the government's assistance was both very good and free (interviews in Taiwan, November 1995).

25. This is part of the United Chemical Laboratory's (UCL's) Environmental Sciences and Technology Division. This division works on ISO 14000 and life-cycle analysis, and it houses the new National Center on Cleaner Production (interviews in Taiwan, November 1995).

26. The IDB's administration of the duty drawback system requires about 20 technicians working full-time to calculate the input-output coefficients for large numbers of export items and the imported inputs used to produce those items

(Wade, 1990). Because many of the inputs in production are imported, it appears that it might be relatively easy to adapt this scheme to calculate the pollution intensity of value-added by industry subsector.

27. What follows is based on interviews conducted at the new National Cleaner Production Center in the Environmental Sciences and Technology Division of the UCL laboratory of ITRI in November 1995.

28. If this happened, it could well serve as the basis for more cost-effective environmental outcomes. For discussion of this in another context, see Porter and van der Linde (1995).

29. Between 1978 and 1980, state-owned enterprises accounted for 14% of gross domestic product and 33% of gross fixed capital formation (Wade, 1990). In 1975, Japan allocated 7.5% of its fixed investment budget to pollution control (O'Connor, 1994).

30. Recent public opinion polls put industrial pollution as one of the top three problems facing the country. The NIMBY problem has made industrial siting almost impossible (interviews in Taiwan, November 1995).

31. Each year the government holds a national reconstruction conference where scholars, government officials, and overseas Chinese are invited to review accomplishments and assess the major issues facing the country. At least since the mid-1970s, the overseas Chinese, particularly those from America, have been critical of Taiwan's growing pollution problems (interviews in Taiwan, November 1995).

32. Recent condemnation by the United States of Taiwan for violation of the Convention on International Trade in Endangered Species (CITES) convention stung the country's political elite (interviews in Taiwan, November 1995).

33. Although, it is important to note that the Taiwan Environmental Protection Agency is probably a more effective command-and-control environmental agency than China's SEPA.

34. But with China's accession to the World Trade Organization, this dynamic may well change.

35. Interview with Michael Hsiao, Academica Sinica, November 1999.

References

Afsah, S., & Vincent, J. (1997). *Putting pressure on polluters: Indonesia's PROPER program.* Cambridge, MA: Harvard Institute for International Development.

Arora, S., & Cason, T. N. (1995). An experiment in voluntary environmental regulation: Participation in EPA's 33/50 program. *Journal of Environmental Economics and Management, 28*, 271–286.

Bello, W., & Rosenfeld, S. (1990). *Dragons in distress: Asia's miracle economies in crisis.* San Francisco: Institute for Food Policy and Development.

Brandon, C., & Ramankutty, R. (1993). *Toward an environmental strategy for Asia.* Washington, DC: World Bank.

Chun-Chieh, C. (1994). Growth with pollution: Unsustainable development in Taiwan and its consequences. *Studies in Comparative International Development, 29*(2), 23–47.

Eder, N. (1996). *Poisoned prosperity.* Armonk, NY: M. E. Sharpe.

Industrial Development Bureau. (n.d.). *Industrial pollution control in Taiwan R.O.C.* Taipei, Taiwan: Ministry of Economic Affairs.

Industrial Development Bureau. (1995). *Development of industries in Taiwan Republic of China.* Taipei, Taiwan: Ministry of Economic Affairs.

Jahiel, A. R. (1998). The organization of environmental protection in China. *China Quarterly, 156,* 757–787.

Lampton, D. A. (1992). A plum for a peach: Bargaining, interest, and bureaucratic politics in China. In K. G. Leiberthal & D. Lampton (Eds.), *Bureaucracy, politics, and decision-making in post-Mao China* (pp. 33–57). Berkeley: University of California Press.

Lee, Y. F., & So, A. Y. (1999). *Asia's environmental movements: Comparative perspectives.* Armonk, NY: M. E. Sharpe.

Lieberthal, K. G. (n.d.). *China's governing system and its impact on environmental policy implementation.* Washington, DC: Woodrow Wilson Center.

Lieberthal, K. G. (1992). Introduction: The fragmented authoritarianism model and its limitations. In K. G. Leiberthal & D. Lampton (Eds.), *Bureaucracy, politics, and decision-making in post-Mao China* (pp. 1–29). Berkeley: University of California Press.

Lohani, B. (1998). *Environmental challenges in Asia in the 21st century.* Manila, the Philippines: Asian Development Bank.

National Environmental Protection Agency (NEPA). (1996). *Annual environmental yearbook, 1996.* Beijing, China: National Environmental Protection Agency.

O'Connor, D. (1994). *Managing the environment with rapid industrialization: Lessons from the East Asian experience.* Paris: OECD.

Panayotou, T. (1999). The effectiveness and efficiency of environmental policy in China. In M. B. McElroy, C. P. Nielson, & P. Lydon (Eds.), *Emerging China: Reconciling environmental protection and economic growth* (pp. 431–472). Cambridge, MA: Harvard University Press.

Porter, M. E., & van der Linde, C. (1995). Toward a new conception of the environment-competitiveness relationship. *Journal of Economic Perspectives, 9,* 97–118.

Rhee, Y. H., Pursell, G., & Ross-Larson, B. (1984). *Korea's competitive edge: Managing entry into world markets.* Baltimore: Johns Hopkins University Press.

Rock, M. T. (1996). Toward more sustainable development: The environment and industrial policy in Taiwan. *Development Policy Review, 14,* 255–272.

Rock, M. T., Yu, F., & Zhang, C. (1999). *The impact of China's urban environmental examination system on urban environmental management and ambient environmental quality.* Arlington, VA: Winrock International.

Ross, L. (1988). *Environmental policy in China.* Bloomington: Indiana University Press.

Sinkule, B., & Ortolano, L. (1995). *Implementing environmental policy in China.* New York: Praeger.

Smil, V. (1993). *China's environmental crisis.* Armonk, NY: M. E. Sharpe.

Spofford, W., Ma, X., Zou, J., & Smith, K. (1996). *Assessment of the regulatory framework for industrial pollution control in Chongqing.* Washington, DC: Resources for the Future.

Taiwan Environmental Protection Administration. (2000). *National air quality improvement results.* Retrieved November 12, 2001, from http://www.epa.gov.tw/english/offices/f/bluesky14.htm

Wade, R. (1990). *Governing the market.* Princeton, NJ: Princeton University Press.

Walder, A. G. (1992). Local bargaining relationships and urban industrial finance. In K. G. Leiberthal & D. Lampton (Eds.), *Bureaucracy, politics, and decision-making in post-Mao China* (pp. 308–333). Berkeley: University of California Press.

Wang, H., & Lui, B. (1999). Policy-making for environmental protection in China. In M. B. McElroy, C. P. Nielson, & P. Lydon (Eds.), *Emerging China: Reconciling environmental protection and economic growth* (pp. 371–404). Cambridge, MA: Harvard University Press.

Wang, H., & Wheeler, D. (1996). *Pricing industrial pollution in China: An econometric analysis of the levy system*. Washington, DC: World Bank.

Wang, H., & Wheeler, D. (1999, July). *Endogenous enforcement and the effectiveness of China's pollution levy system*. Paper presented at workshop on Market-Based Instruments for Environmental Protection, Cambridge, MA.

World Bank. (1997). *Clear water, blue skies*. Washington, DC: Author.

Chapter 23

Tran Thi My Dieu, Phung Thuy Phuong, Joost C.L. van Buuren and Nguyen Trung Viet

ENVIRONMENTAL MANAGEMENT FOR INDUSTRIAL ZONES IN VIETNAM

Industrial development in Vietnam

SINCE THE BEGINNING OF the 1990s Vietnam has shown a successful economic development and market stabilization. Though facing difficulties due to especially the economic crisis in neighboring countries but also internal bureaucracies, at the turn of the millennium Vietnam's growth rate is still one of the highest in Southeast Asia, varying between 5 and 10 percent annually in the period 1995 to 2000. Industrial activities are the main contributor to this growth. In the coming decade the boosting of industrial modernization continues to be a national strategy target. The total national industrial output reached 3.56 billion U.S. dollar in 2000. According to the General Statistical Office, by the year 1999 there were approximately 620,000 enterprises in the country as a whole; 959 of these enterprises were based on or used foreign investment capital.

One of the key elements of the present Vietnamese industrial development in its pursuit to rapid growth of economic output is the establishment of industrial zones. An industrial zone is an area reserved for the establishment of a certain mix of enterprises. One of the challenges of industrial policy is to locate these zones in a way that an optimum of favorable production conditions is reached (ample labor force; cheap supply of material; easy access to suppliers, customers, and markets; good provision of infrastructures). Different types of industrial zones are distinguished in Vietnam: export processing zones (EPZs); high-tech industrial zones (HTZs); industrial zones around already existing enterprises or industrial clusters; industrial zones meant for relocation of factories from urban areas; industrial zones for small- and medium-sized enterprises processing agricultural, forest, and aquatic resources; and industrial zones for domestic and foreign investments.

With respect to industrial zones the process of industrial development in Vietnam can be divided into two stages. Before 1991, industrial development proceeded in a relatively unplanned and unorganized way. In the north industry gained importance from the mid-1950s, when existing industrial units originating from the French colonial time were further developed and new sectors were created with support of the Socialist countries. In the south, several centralized industrial clusters had been formed especially in the 1960s. These clusters were established without long-term planning. After the reunification of the north and the south of Vietnam in 1975 many of these industrial companies continued their production after having been nationalized. At present, these old clusters are fully incorporated into residential areas.

After 1991, in order to accelerate industrialization and modernization, the government commenced to develop integrated industrial zones. For export-oriented production and to encourage foreign investment, special EPZs were created. Other types of industrial zones have been established since 1994.

In the year 2000 Vietnam counted sixty-eight industrial zones, EPZs, and HTZs, of which thirty-eight are located in southern Vietnam (Nhue et al., 2001) The zones vary in size from 44.5 acres (Binh Chieu Industrial Zone, Ho Chi Minh City) to 14,000 ha (Dung Quat Industrial Zone, Province of Quang Ngai). They had attracted up to the year 2000 nearly 1,100 domestic and foreign enterprises with a total investment capital of about 8 billion USD (about 6.5 billion USD of foreign capital and 1.5 billion USD of domestic capital) (Nhue et al., 2001). Though the number of companies established in industrial zones may seem relatively limited (compared to the over 600,000 enterprises in all Vietnam), their import-ance in terms of economic output is considerable. All new large-scale industries are established in industrial zones.

Though this process of industrialization undoubtedly has raised the level of welfare, several indicators show that the current development path produces serious environmental deterioration (World Bank, 1995). In the early days of Viet-namese industrialization, these impacts caused by human activities were insignifi-cant as compared to the capacity of self-recovery and self-regulation of the natural environment. Nowadays, however, this "carrying capacity" has been grossly exceeded due to mismanagement and overexploitation of natural resources and unbridled discharge of wastes into the environment.

Environmental implications of industrialization

In Vietnam little systematic research has been done as yet on the quantities and impacts of industrial pollution, that is, the consequences of industrial pollution on human health and well-being, on the economy and on aquatic and terrestrial ecosystems. There is only very limited systematic monitoring of environmental quality, and only seldomly are industrial emissions to air and water registered, collected, and published. The limited data do not allow time series analyses and often the reliability is difficult to assess. Notwithstanding these limitations, the consequences of industrial pollution are sometimes immediately observable and in other instances limited data can still provide us with an impression on the seriousness of the state of the environment.

The canals running through the low-lying urbanized areas of Hanoi and Ho Chi

Minh City are heavily polluted. While a major share of the organic pollution originates from households, significant industrial pollution finds it way through the canals as well. Heavy metals and organic microcontaminants end up in the sediments, as the city canals discharge liquid wastes to the bigger rivers (such as the Saigon River and the Dong Nai River in the cases of Ho Chi Minh City and Bien Hoa City). Table 23.1 provides an estimation of the industrial wastewater loads in the Southern Key economic region, now and in 2020. Due to the present pollution loads canals and small rivers have an organic matter concentration comparable with untreated sewage, whose degradation causes a horrible stench that—dependent on weather conditions—is discernable throughout the cities. By now the water pollution is considered as a major environmental problem.

Besides air pollution by traffic, industry adds a long list of gaseous emissions from fuel combustion and production processes. These dusts and gases cause health impacts to workers directly exposed to the sources every day, but also affect the areas and inhabitants surrounding the industrial zones. Some zones and factories are equipped with high smokestacks to protect the immediate environment. Presently dust concentrations grossly exceed the Vietnamese standards in several residential areas near factories. The surroundings of the Hai Phong cement plant, the VICASA steel plant (Bien Hoa City), the Tan Binh industrial area (HoChiMinh City) and the Hon Gai coal processing plant (Ha Long City) are notorious. Monitoring data over the period 1995–1999 shows a downward trend of ambient dust concentrations in most industrial areas. The SO_2 concentrations in several industrial zones are in the range of 0.1–0.4 mg/m^3, which is close to the Vietnamese acceptable concentration limit of 0.3 mg SO_2/m^3. In large cities such as Hanoi, Ho Chi Minh City, Da Nang, and Hai Phong during the period 1995–1999, the daily average concentration of CO varied between 2 and 5 mg/m^3 and the NO_2 concentration between 0.04 and 0.09 mg/m^3. These concentrations are lower than the acceptable values postulated in the Vietnamese standards.

Estimates of the different types of industrial solid wastes generated in the country and in Ho Chi Minh City are shown in Table 23.2. Though time series on solid waste generation are lacking, undoubtedly the waste load increase is still directly linked to the fast industrial growth.

Dang (2001) estimates that around 40 percent of the industrial solid waste could be categorized as hazardous. Currently, industrial solid waste collection and disposal in industrial zones is taken care of by urban environmental companies. In

Table 23.1 Industrial water and air emissions in the Southern Key economic region of Vietnam (in ton per year)

Year	Water emissions[1]			Air emissions[2]		
	Biological oxygen demands	Dust	Sulfur dioxide	Nitrogen oxides	Carbon monoxide	Volatile organic compounds
1999	10,182	20,000	5,600	10,000	21,000	1,411
2020	89,000			n.a.		

[1] Dac and Loan, 2000; [2] Sy, 2000

Table 23.2 Quantities of solid waste and collected fractions in several parts of the Southern Key economic zone and Vietnam (sources: Dang, 2001; DOSTE/UNDP, 2000)

Type of solid waste	Quantities of wastes generated (1000 t/yr)				Collected fraction Vietnam (percent)
	Ho Chi Minh (2000)	Dong Nai (1996)	Ba Ria-Vung Tau (1999)	Vietnam as a whole (1999)	
Domestic wastes	1,788			6,890	75
Sewage sludge				383	92
Demolition waste				853	65
Hospital waste	3,9			101	75
Industrial waste	102	39	8,7	915	60
Total				9,142	73

t/yr = tons per year

the entire country up to now there is not a single landfill fully equipped to provide proper disposal for hazardous wastes. All types of garbage, including hazardous wastes, are jointly deposited onto dumping sites. A project for the construction of a 2.47 acres integrated hazardous waste treatment station (including safety landfill) at Giang Dien in Thong Nhat District in Dong Nai Province has been officially approved recently and is in the stage of detailed design.

As hardly any well-prepared sanitary landfills are available and a significant part of industrial solid waste is dumped in open lots, ground water pollution can be expected. Some monitoring wells in Ho Chi Minh City show groundwater qualities that do not satisfy drinking water requirements (DOSTE/UNDP, 2000). The main problems are saline intrusion and high concentrations of ammonia and nitrate. In addition, at sites influenced by central landfills the concentration of nickel and chromium has risen to values close to or above acceptable limits.

It may be assumed that there is a huge potential for reuse and recycling of solid wastes from industrial origin. Studies carried out in the framework of the REFINE project (see www.ernasia.org) since 1998 [. . .] give evidence of not only the large potential of industrial waste reuse and recycling, but also of the current practices in this area. Major non-product flows in for instance Vietnamese rubber, pulp and paper, tapioca, and tannery industries, which in more developed countries would end up as waste, are being used as valuable resources by other industries in Vietnam. Where industrialized countries are desperately trying to put industrial ecology to work, Vietnamese industries have practiced the same ideas for some time (cf. van Koppen and Mol, 2002 [also in this volume]). The challenge is of course what will happen with these environmentally efficient forms of industrial reuse and recycling as industrialization and modernization continue, and the market conditions for reuse and recycling become less favorable without any state intervention.

Environmental management in industrial zones

According to Hung (1997), the major reasons for the rapid increase of industrial pollution in Vietnam are not only related to industrial expansion but also to the use of obsolete production technologies and poor application of waste minimization, waste treatment, and effective industrial environmental management. It is especially this latter cause that will be elaborated in this section.

Actors in industrial zone environmental management

At national level, the governmental agencies in charge of environmental issues related to industrial zones are consecutively the National Environmental Agency (NEA), which belongs to the Ministry of Science, Technology, and Environment (MOSTE), twenty-three departments of science and technology (DST) that are part of twenty-three line ministries and the Division of Specific Subject within the Vietnam Industrial Zone Authority (VIZA). At the provincial level, the Departments of Science, Technology, and Environment (DOSTEs) and the Divisions of Planning and Environment within provincial Industrial Zone Management Boards are responsible for the environmental management of industrial zones. Both DOSTE and the Industrial Zone Management Board are linked to the Provincial People's Committees in political, administrative, and financial aspects. In a vertical line, DOSTEs are linked to MOSTE, while the Industrial Zone Management Boards are linked to VIZA. At a district level, environmental bureaus take care of environmental issues within their districts. Figure 23.1 presents an overview of the organizations involved in (environmental) management of industrial zones in Vietnam.

The state is the most obvious key player in any environmental regulation of economic activities. With respect to industrial zones in Vietnam it is the provincial DOSTEs that are responsible for implementing national environmental laws and regulations. These national frameworks are developed by the MOSTE in Hanoi, and especially by its NEA. DOSTEs participate in the decisionmaking about the location of industrial zones, supervise the environmental performance of these zones, and carry out inspections. In addition, depending on the size and category of the company, DOSTEs play a key role in Environmental Impact Assessments and the issuing of environmental licenses.

A second crucial stakeholder is the People's Committees at the provincial and city levels. The People's Committees have the highest political responsibility. Therefore, possible conflicts arising from opposing economic and ecological objectives will have to be solved within these political bodies. DOSTEs are accountable to both the Ministry MOSTE in Hanoi and the provincial People's Committee. District People's Committees, at least in the cities, play a limited role in industry-related environmental affairs, especially with respect to larger industries and industrial zones. But their environmental bureaus are the first addresses for citizens with complaints on industrial pollution. More than incidental complaints are forwarded to higher political levels to be solved.

The development of industrial zones is strongly supported by the economic ministries, while VIZA has a coordinating role to play in establishing and controlling industrial zones and developing stimulating and supporting policies. At the local

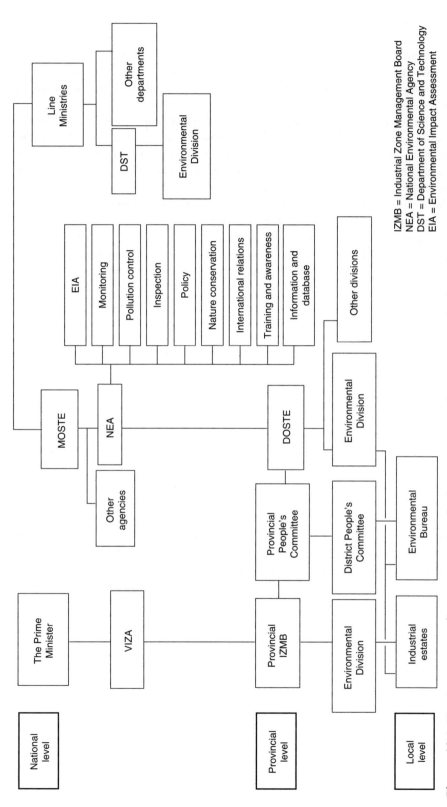

Figure 23.1 Key (environmental) organizations involved in environmental management of industrial estates in Vietnam

IZMB = Industrial Zone Management Board
NEA = National Environmental Agency
DST = Department of Science and Technology
EIA = Environmental Impact Assessment

level the provincial Departments of the Ministries each play a role and Industrial Zone Management Boards are created as the provincial equivalents of VIZA. The management boards attempt to stimulate and supervise the development of industrial zones in their province. They have social, economic, financial, and—marginally—environmental tasks and receive their budget from the provincial People's Committee.

The actual layout of industrial zones and their daily operation is carried out by parastatal commercial Industrial Zone Infrastructure Development Companies (IZIDCs). Often these companies are joint ventures between state-related and private capital. The Infrastructure Development Companies first activity is "to spread the red carpet to welcome the investors" (Tran, 2001), inter alia by providing an attractive infrastructure, an array of services, and a smooth guidance along the various bureaucracies. Regarding environmental management, the Infrastructure Development Companies can be responsible for investments in environmental infrastructure, such as the layout of storm water drainage, sewerage, and waste-water treatment works. Increasingly, Infrastructure Development Companies take up these tasks. The operation of a solid waste collection and disposal system, the collection of environmental charges and the monitoring and reporting on environmental pollution to environmental authorities are often mentioned as future tasks of these Infrastructure Development Companies (Phung, 2002). There is a pressing interest to quickly find occupants for the land available within the industrial zone. This not only determines the profitability of the zone, but also enhances the feasibility to start and run environmental facilities such as a wastewater treatment plant. The Infrastructure Development Companies receive their income through fees from the industries within their zones, including contributions for the operation of environmental infrastructure.

In Vietnam up to now there are hardly any environmental nongovernmental organizations that attempt to influence the state policies related to industrial zone development. Instead, communities living in the vicinity of polluting enterprises and industrial zones play a crucial role in pushing industrial polluters to improve their environmental performance (O'Rourke, 2002; Phung, 2002). Complaints from community members about environmental trespasses do occur and are handled by the People's Committee and DOSTE. The influence of these complaints on industrial development leads some scholars to identify community driven regulation (O'Rourke, 2002) as an important model in environmental improvement of industrial performance.

Implementation deficit: environmental policies, procedures, and measures

In analyzing the functioning of this environmental management network around industrial zones a major implementation deficit can be witnessed. While the official procedures, standards, and guidelines are often in place, the implementation of these policies and regulation fail, leading to poor environmental improvements. As examples we will subsequently analyze the situation around the issuing of environmental licenses and water pollution control in industrial zones.

According to the Vietnamese legal framework for environmental impact assessment (EIA), investment projects are classified into three categories: (i) projects that do not need an EIA; (ii) projects that need an EIA report appraised by DOSTE (Annex II of Decree 175/CP); and (iii) projects that need an EIA report appraised by NEA (Annex II of Decree 175/CP). EIA reports analyze and describe the amounts and nature of the emissions to be expected from the planned operation, as well as the way these wastes will be handled. Most EIA reports propose on-site pollution abatement measures for wastegases and wastewater, and off-site handling of solid waste. The current EIA reports, however, do not (yet) suggest possible cleaner production technologies (see also Tran and Leuenberger [, 2003]). Approval of the company's EIA report by a commission put together by the environmental authority is one of the prerequisites for obtaining an environmental license. Subsequently, an IZIDC has to submit an EIA report as one of the prerequisites for the construction and development of an industrial zone as a whole. In other words, industrial zones cannot function without an EIA approved by NEA/ MOSTE. But industrial firms, which are located in an industrial zone or EPZ that has been approved based on its EIA report are not subjected to an EIA procedure. These investors first have to complete a form called "a registration for securing environmental standards." This registration form includes the description of the production process; the quantification of the amount of labor and investment capital; the description of the quality and quantity of raw materials, chemicals, energy, and equipment; an estimation of the types of wastes and emissions; and the promise to follow the environmental standards of Vietnam (TCVN-1995). This form is submitted to DOSTE, who will decide whether or not the investor has to install waste treatment facilities. If that is deemed necessary, the investor has to design these treatment facilities and submit the design to DOSTE together with the registration form. DOSTE will then sign and file the registration. Immediately after the firm starts operation, DOSTE will perform an audit on the firm to see if it meets the environmental standards. If it does, DOSTE will issue a license to the investor indicating that the firm meets the environmental standards according to TCVN-1995. This license is valid for a period of two to five years, depending on the size of the firm and type of industry it belongs to. At the end of this period, the firm is required to repeat the procedure in order to extend the license.

The official policy is to regularly inspect and control the operation of the enterprises and the Industrial Zone Management Board and DOSTE have to submit monitoring reports to the People's Committee. In practice, however, deviations of the official policy and procedure seem to be the rule. Due to a variety of reasons most industrial enterprises operate without an environmental license. More and more evidence emerges of corruption in both approving an EIA report and issuing a license. There are even heavily polluting industrial zones that constantly fail to comply with the obligations of the EIA and undergo inspections without being faced with serious consequences. But it is not only a situation of better control and enforcement, as the example of wastewater pollution control shows.

The policy for the abatement of surface water pollution by industrial enterprises distinguishes two types of discharge conditions with three types of effluent requirements (see Table 23.3). An enterprise may discharge *directly* to surface water. Then, depending on the sensitiveness and function of the receiving water,

Table 23.3 Effluent discharge standards (as stipulated in TCVN-5945-1995)

Parameter		Water quality not to be exceeded		
		A*	B*	C*
Biological oxygen demand	Mg/l	20	50	100
Chemical oxygen demand	Mg/l	50	100	400
Total suspended solids	Mg/l	50	100	200
Total nitrogen	Mg/l	30	60	60
Ammonium nitrogen (NH_4-N)	Mg/l	0.1	1	10
Total phosphor	Mg/l	4	6	8
Total coliform	#/100 ml	5,000	10,000	–

* Environmental standard for industrial wastewater (TCVN 5945–1995) in Vietnam is divided into 3 categories: Category A (the most stringent) for discharging into receiving streams that serve for drinking and domestic supply; Category B for discharging into receiving streams that are used for bathing, aquatic breeding, cultivation, irrigation, and navigation; and Category C for discharging to specific streams such as a sewage system permitted by authority agencies.

Category B or A effluent requirements are applied. An enterprise may also *discharge to a sewer system*. This is the usual situation in industrial zones and in urban surroundings. In this case the enterprise has to comply with Category C requirements. In the ideal situation the combined effluents of various enterprises are treated further in a central-shared wastewater treatment works to attain Category B or A quality, before being discharged to surface water.

Within an industrial zone there usually are enterprises, that do not need individual treatment facilities to reach effluent requirements of Category C, since their wastewater has a relatively low strength. They can discharge directly to the sewer system.

For the group with high-strength wastewater, on-site treatment needs to take place at the factory. In both cases this is followed by off-site wastewater treatment, shared by all companies within the perimeter of the industrial zone. Companies outside the industrial zones that discharge to sewers have to comply with Category C standards. In due time this wastewater will be treated by communal wastewater treatment plants but until now Vietnam has practically none of these plants. In the majority of the industrial zones a common wastewater treatment plant is also absent or under construction. Consequently, Category C wastewater is often discharged into surface water. As to central wastewater treatment facilities in the thirty-eight industrial zones in the Southern Key economic region, the situation is summarized in Table 23.4. In general the listed treatment stations perform in a stable way and the effluents meet Category B requirements.

An example of the complicated implementation of industrial wastewater treatment is found at Binh Chieu Industrial Zone. This industrial zone is situated in the headwaters of a river in Thu Duc District in Ho Chi Minh City and its effluent has to meet Category A requirements (meant for water supply). At the moment, however, no common wastewater treatment station is installed at Binh Chieu and HCMC Export Processing and Industrial Zones Authority (HEPZA), the industrial zone management board of Ho Chi Minh City, tries to persuade individual

Table 23.4 Central wastewater treatment plants (WTP) at industrial zones and export processing zones in the Southern Key economic region (End of 2001)

Industrial zone	Province	Central WTP capacity (m³/day)	Status
Tan Thuan	HCMC	10,000	In operation since 1999
Linh Trung	HCMC	6,000	In operation since 1999
Le Minh Xuan	HCMC	5,000	Constructed in 2001
Tan Tao	HCMC	5,000	Design
Hiep Phuoc	HCMC	–	No central WTP
Vinh Loc	HCMC	–	No central WTP
Northwest Cu Chi	HCMC	–	No central WTP
Tan Thoi Hiep	HCMC	–	No central WTP
Binh Chieu	HCMC	–	No central WTP
Tam Binh	HCMC	–	No central WTP
Tan Binh	HCMC	–	No central WTP
Vietnam-Singapore	Binh Duong	4,000	In operation since 1999
Song Than I and II	Binh Duong	4,000	Construction
Dong An	Binh Duong	2,000	Construction
Viet Huong	Binh Duong	1,000	Constructed in 2000
Binh Duong	Binh Duong	–	No central WTP
Tan Dong Hiep	Binh Duong	–	No central WTP
Amata	Dong Nai	4000	In operation since 1999
Loteco	Dong Nai	1,500	In operation since 1999
Bien Hoa II	Dong Nai	6,000	In operation since 2000
Bien Hoa I	Dong Nai	–	No central WTP
Nhon Trach I and II	Dong Nai	–	No central WTP
Go Dau	Dong Nai	–	No central WTP
Ho Nai	Dong Nai	–	No central WTP
Song May	Dong Nai	–	No central WTP
Phu My	BRVT	–	No central WTP
My Xuan A and B	BRVT	–	No central WTP
Dong Xuyen	BRVT	–	No central WTP

HCMC = Ho Chi Minh City; BRVT = Ba Ria-Vung Tau

enterprises in the zone to build Category A-type wastewater treatment plants. This turns out to be far from easy (Tran Thien Tu, 2001).

Industrial ecology practices in contemporary Vietnam

The overall conclusion to be formulated from this analysis is that current environmental management in industrial estates is falling short. This puts the questions on improvement of the current system of greening industrial zones on the research and political agendas.

In greening industrial estates, the model or theory of industrial ecology is increasingly applied (Lowe, 1997; van Koppen and Mol, 2002 [also in this volume]). The core idea of industrial ecology is to study the industrial system from an ecosystem angle. This perspective basically involves two starting points. First, the industrial system itself should be interpreted and analyzed as a particular system with a distribution of materials, energy, and information flows (not unlike the ecosystem). Second, the industrial system relies on resources and services provided by the biosphere. Both the flows within the industrial system and between the system and the biosphere have to be optimized from a closed-loop perspective, as is exemplified by natural ecosystems. In reviewing the industrial ecology literature it should not only be concluded that eco-industrial parks have received most attention from industrial ecologists. It also becomes clear that most of industrial ecology's more practical efforts have been dedicated to eco-industrial parks. As Lowe et al. (1997) state,

> The concepts of industrial ecosystems and eco-industrial parks (EIPs) have embodied the industrial ecology approach in very concrete terms.

We will apply this model of industrial ecology to analyze (i) to what extent current industrial practices in Vietnam already follow ideas of waste exchange and closing substance loops, (ii) what difficulties and shortcoming are experienced in putting the industrial ecology model at work, and (iii) a feasible program that can enhance waste exchange and industrial ecology practices in industrial estates in Vietnam.

Current waste exchange practices in Vietnam

Reuse/recycling activities are very familiar in Vietnam. Especially in rural areas, Vietnamese farmers use animal faeces as fertilizer for their vegetable gardens. Animal faeces are also used to feed fishponds. Animal faeces, human excreta, domestic garbage, and leaves are used to produce biogas for lighting and cooking. This forms a closed loop of a so-called garden-pond-stable system (see also [. . . Le and Tran, 2003]). The reuse/recycling practices can also be found in industrial activities. In the Southern Key economic region, a number of factories/enterprises use waste materials as input to their production processes. For instance, Serrano VN Company in Vietnam–Singapore Industrial Zone (Binh Duong Province) uses, among others, bagasse, sawdust, and pieces of wood as raw material to produce fine arts. In Bien Hoa II Industrial Zone (Dong Nai Province), Technopia VN Company, a Malaysian company, uses sawdust, pieces of wood, and coconut fiber to produce mosquito coils (a kind of insecticide). CP VN Company, a joint venture company in Thailand, and Cargill Company, an American company, both use oyster shell and bones in the production of animal food. Some factories have to switch to waste materials because virgin raw materials are in deficit, or because they have to reduce the production cost in order to remain competitive. In Bien Hoa I Industrial Zone (Dong Nai Province) VICASA Steel Factory, a state-owned enterprise, uses steel scrap to replace virgin raw material. Similarly, Dong Nai Pulp and Paper Company supplements wood and bamboo with bagasse, waste jute, and waste paper in producing paper and carton.

In our investigations we found a variety of factories/enterprises selling their wastes. Some factories sell their wastes to comply with the environmental policies of their parent companies such as Fujitsu Electronic Company and Tae Kwang Vina Footwear Company (Bien Hoa II Industrial Zone, Dong Nai Province). Others get profit from selling wastes. Hualon Textile Corporation VN in Nhon Trach Industrial Zone (Dong Nai Province) sells waste fibers that are then reused for making pillows and mattresses. LIDOVIT Company, which produces spare parts for cars and motorbikes (Binh Chieu Industrial Zone, Ho Chi Minh City), sells steel scrap. Minh Tu Rubber Enterprise in Le Minh Xuan Industrial Zone (Ho Chi Minh City) sells waste rubber. Some factories claim that their wastes have potential to be reused or recycled but at present still lack a market. For instance, the small amount of magnesium hydroxide [$Mg(OH)_2$], barium sulfate ($BaSO_4$), and calcium carbonate ($CaCO_3$) in the sludge of VICACO Chemical Factory in Bien Hoa I Industrial Zone could be reused. Similarly, the aluminium in the sludge of wastewater treatment system of VIJALCO Aluminium Electroplating Factory in Binh Chieu Industrial Zone could be reused.

In most of these cases of waste exchange and reuse, waste producers and buyers work through middlemen in the informal sector. These middlemen travel from place to place, even from province to province, buying, transporting, and selling wastes. The middlemen can buy wastes from factories or from scavengers and then sell wastes to factories that are in need. In limited cases, waste exchange takes place directly between waste buyers and producers. Dong Nai Pulp and Paper Company and Concrete Company exchange "black liquor" wastewater. Sugar Cane Factory and Dong Nai Pulp and Paper Company exchange bagasse. Dong Nai Pulp and Paper Company sells waste fibers to Dong Hiep Paper Company. These companies are all located in Bien Hoa I Industrial Zone and exchange wastes on a voluntary basis. But the unbalance between supply and demand makes waste exchange less efficient than theoretically possible. There is more supply of black liquor generated by the Paper Company than the demand by the Concrete Company. Similarly, bagasse from Sugar Cane Factory does by far exceed the demand of Dong Nai Paper Company. As a result, waste sellers cannot rely sufficiently on the waste buyers to handle their wastes in an environmentally sound way.

Usually there is an untapped potential to exchange waste between factories located in the same industrial zones. For instance, in Bien Hoa I Industrial Zone, waste from the sugar cane factory, the canned fruit factory, and the milk factory could be utilized as input to the animal foodstuff factory. The chemical factory could use SO_2 from the fossil fuel power plant. Similarly, in Bien Hoa II Industrial Zone, food-processing factories could sell waste to animal foodstuff factories. But until now, no waste exchange exists between these companies.

In brief, even now a number of waste exchange practices can be found, either within or outside industrial zones. But these practices are carried out on an adhoc basis without any systematic policy or management structure to support and increase them.

Problems in the further development of waste exchange

One of the problems related to waste exchange is that waste users will have to rely on waste producers for the quantity and quality of the secondary raw material. This resource inflexibility might present a problem as it can affect product quality. In addition, producers must accept unfamiliar inputs (converted wastes) rather than traditional raw materials, sometimes involving large investments to create new processing facilities; at the same time, customers and consumers have to accept new products produced from "waste." While in some highly industrialized countries the valorization of secondary raw materials may be an important argument in the awarding of green and eco labels, in Vietnam and other developing countries such a "green" market demand hardly exists. In the specific case of export processing companies, difficulties emerge in selling wastes due to the existing regulation on export processing companies. Export processing companies can only have economic relations with companies that also have an import–export license. Recyclers, who are often found in the informal sector, usually do not have such an import–export license.

Most recycling activities are carried out in the informal sector. Recyclers are individuals or family-owned small enterprises located in residential areas. They usually apply simple technologies and production methods, which are more than incidentally causing pollution (Rekha et al., 1994). The present recycling sector does not seem strong enough to substantially modernize itself, so that a new problem may arise when the informal recycling sector disappears while no new formal recycling sector has appeared to take over its role.

There are also more conventional challenges related to the organization of waste exchange and industrial ecology experiments. Locating waste producers in the neighborhood of waste users is usually seen as a major advantage—if not a precondition—for setting up industrial ecology models in practice. Half-empty industrial zones—as currently found in the Southern Key Economic Region—present an opportunity to organize such a favorable location pattern, as new industries can be recruited to complement the existing businesses and enhance waste exchange. However, one of the difficulties, even in transitional economies, is the planning of investments in specific locations. Strong incentives and policies are necessary to do that, which often run counter to liberal private investment decisions. Nevertheless, some cities and provinces in Vietnam have implemented policies to relocate factories/enterprises from the inner city to industrial zones outside the city (cf. [. . . Frijns, 2003]). The relocation of small- and medium-sized enterprises, especially if they are recycling enterprises, to newly established industrial zones might offer an opportunity to promote waste exchange. At the same time an industrial zone designed according to industrial ecosystem principles might be an attractive place for small- and medium-sized enterprises. Since small- and medium-sized enterprises usually lack capital, their participation in the closed-loop programs in industrial estates may offer them opportunities to boost their earnings by selling or buying by-products at favorable rates.

As a general conclusion on the current practices, it can be stated that experiences, ideas, and technologies of waste exchange do exist in Vietnam. However, inadequate structures and lack of supporting mechanisms prevent wider and more systematic application of these experiences.

Innovations in industrial state environmental management

Without outlining a full new environmental management system for industrial zones, we can still identify three crucial and feasible routes to innovate the existing management system in line with ideas of industrial ecology.

The first innovation is related to finances. Existing environmental management systems lack both sufficient financial resources and sufficient possibilities to install financial incentives for polluters. It is of crucial importance for any effective environmental management system to install financial incentives on efficient use of environmental resources (both through reduced natural resource use in production/consumption and cutting emissions that deteriorate environmental quality). The pricing of natural resources and the application of charges on emissions and the use of these additional incomes (or of the savings on natural resource subsidies) for environmental protection measures and environmental capacity building would work like a double dividend. For several reasons (e.g., political feasibility, lack of monitoring, economic competitiveness, distribution effects) it still turns out to be very complicated to burden consumers and the poorer sectors of industry, such as the small- and medium-sized enterprises (SMEs) and the old state-owned enterprises, with such additional costs. For instance, the existing pricing system of industrial raw materials discourages waste reuse/recycling. These prices do not internalize the externalities or environmental damage cost. As the result, the price of virgin raw materials is, in general, lower than the "real" price and industries have no incentive to use secondary materials. Nevertheless, also in Vietnam several mechanisms are underway to internalize environmental costs into production and consumption processes. To mention a few examples:

- In several cities a small charge is now included in the price of drinking water to pay for the upkeep of the sewerage systems; it is debated now at what rate this now-symbolic charge of 300 VND/m^3 (0.02 USD/m^3) could be increased to contribute substantially to the real expenditures and to push forward a more efficient water use.
- In industrial zones enterprises pay a fee to the IZIDC to cover the costs of environmental infrastructure; often this fee does not yet equal the environmental load of their production emissions.
- Proposals have been made to introduce environmental accounting into industrial practice (Quang, 2001).

This process of installing economic incentives, removing subsidies on natural resource use, and increasing the budgets and capacity of environmental management organizations is essential for increasing the environmental efficiency in industrial production.

A second major improvement relates to a better definition of objectives, tasks, and responsibilities for the various actors involved in industrial zone environmental management. Within the framework of national environmental legislation it is the provincial People's Committee, advised by DOSTE and the provincial industrial zone management board, who sets the targets and verifies the attainment of these targets. The People's Committee balances, in principle, social, economic, and environmental interests. In practice, it is DOSTE and the industrial zone management board who are involved in respectively environmental and economic policymaking and implementation regarding industrial zones.

Following the exponential growth of industries and industrial zones in the Southern Key Economic Region, DOSTEs are understaffed and lack capacity to adequately guide a more sustainable development and operation of industrial zones. To strengthen their task in this a special unit could be created within DOSTE to deal with industrial zones. Its tasks should not only be environmental licensing and supervision of the execution of development plans, but its role should become more proactive, for example, the introduction and execution of cleaner production, the set up of waste exchange programs, and the stimulation of environmental accounting and auditing.

IZIDCs are at the moment only superficially involved in environmental management, basically by providing limited facilities for industries (such as waste collection and centralized wastewater treatment plants). Their tasks and responsibilities could expand to become an intermediary organization, in between DOSTE and the individual enterprises. Toward enterprises these Infrastructure Development Companies could offer a wide range of environmental services:

- Waste minimization, collection, and treatment;
- Cleaner production auditing;
- Environmental monitoring and reporting;
- Mediation between waste exchange;
- Technology transfer center;
- Interest representation toward regulatory agencies; and
- Assistance on preparing environmental licenses.

Toward DOSTE and other regulatory agencies, these Infrastructure Development Companies could play an important role on various issues:

- Information dissemination on environmental regulation and technology subsidy programs;
- Environmental monitoring of individual and collective enterprises and environmental quality indicators;
- Environmental reporting toward regulatory agencies;
- Environmental charge and fee collection;
- The integration of environmental and economic objectives for an industrial zone; and
- Being a central address of state organizations to deal with industries.

In this way Infrastructure Development Companies become a crucial intermediary organization between the individual enterprise and the governmental environmental authorities. On the one hand they take over some of the tasks of environmental authorities, without diminishing the latter's final authority. In that way DOSTEs can concentrate on their primary responsibility (controlling industries and enforcing environmental regulation) without being troubled with several tasks of secondary importance. On the other hand, they assist individual enterprises in improving their environmental performance. For each of their tasks IZIDCs may contract consultant companies if that is necessary, as long as they remain the first responsible agency on these tasks. Since the majority of industrial enterprises are situated outside industrial zones, it could be expedient to create an institute/company to support these enterprises.

The final improvement is related to transparency, accountability, and access to environmental information. It has become widely known around the world that environmental improvement can no longer be only the task of environmental authorities. It is especially the close cooperation among state authorities, market actors, and an active civil society that is a prerequisite for environmental reform of industrial development. In order to involve civil society and market organizations and actors in the push for greening industrial development access to environmental information, transparency and accountability are essential. It is environmental authorities who should create the conditions for a better environmental monitoring of industrial emissions and environmental quality around industrial zones, regular reporting of these monitoring data to the public and customers/consumers, annual environmental reports being published and verified for each industrial zone, and sufficient access of these third parties to EIA reporting and environmental licensing.

Via these mechanisms the limited environmental capacity of DOSTE organizations is complemented with environmental capacity that is available in other (market and civil society) segments of the Vietnamese society. It is also one of the best ways to enhance the quality of state environmental authorities and to limit illegal transactions around EIAs and environmental licenses.

The application of a cleaner production approach and waste exchange also meets problems of information exchange that can significantly benefit from improved monitoring and information exchange. Monitoring and information dissemination of the amount and composition of waste available at companies can improve waste exchange between waste producers and waste buyers. Regulatory guidelines seem essential to force waste producers to reveal the characteristics and quantities of wastes.

Conclusion

In Vietnam, the establishment of industrial zones was considered to be a key vehicle for the industrialization process. As a result Vietnam witnessed a dramatic increase in the number of industrial zones during the last decade of the twentieth century. On the one hand the establishment of industrial zones has helped to promote the industrialization process, but it created various environmental side effects. The existing environmental policies do not provide incentives for polluters to improve their environmental performance; environmental legislation and policies lack specific regulations for effective pollution control in industrial zones; and environmental management of industrial zones is still weak in dealing with environmental problems. There are many state agencies involved in environmental management of industrial zones, but there is no clear division of tasks and responsibilities between these agencies. This has led to overlap and cumbersome environmental reform practices.

Notwithstanding these cumbersome practices of environmental management, Vietnamese industry gives evidence of a wide variety of waste exchange practices and experiences, significantly contributing to the implementation of new ideas of industrial ecology *avant la lettre*. It is especially the informal sector and economic motivations that play a major role in the implementation of these waste exchange experiences.

In order to improve and more systematically develop both waste exchange practices and environmental management systems of industrial zones in Vietnam three strategic recommendations are developed. First, economic disincentives (e.g., pricing of natural resources, pollution charges) and economic incentives (e.g., environmental funds) should be more systematically applied to increase the national budget for environmental protection and to change the behavior of polluters toward more environmentally friendly directions. Second, new institutional arrangements are suggested, redefining the division of tasks and responsibilities among public, parastatal, and private actors involved in environmental management of industrial zones. As experiences throughout the world showed, closer cooperation among state authorities, civil society, and market organizations and actors is necessary to push toward green industrial zones. Finally, environmental monitoring, (environmental) information exchange, and transparency need to be improved in order to allow various actors to take their responsibility and play their role in the greening of industrial estates. This is not only essential for improving local environmental quality, but also to prepare Vietnamese industry for entering massively the world market.

References

Dac, N.T., and Loan, V.T., 2000. "Assessment Spreading of Industrial Pollution into Can Gio by Mathematical Model." Environmental Workshop 2000 on Science, Technology and Environment, Ho Chi Minh City, Vietnam, June 6.

Dang, P.N., 2001. "Existing Status and Challenges or Urban Environment During Process of Industrialization and Modernization in Vietnam." International Conference Industry and Environment in Vietnam, Ho Chi Minh City, April. Van Lang University, Vietnam: REFINE Project.

DOSTE/ UNDP, 2000. *State of the Environment Report – Ho Chi Minh City 2000*. Ho Chi Minh City: Department of Science, Technology and the Environment.

Frijns, J., 2003. "Relocation or Renovation: Greening Small- and Medium-Sized Enterprises." Pp. 129–150 in *Greening Industrialization in Asian Transitional Economies: China and Vietnam*, eds. A.P.J. Mol and J.C.L. van Buuren. Lanham, MD: Lexington Books.

Hung, N.T., 1997. "Industrial Development and Natural Environmental Protection, Environmental Auditing." Paper presented at the Workshop on Mitigation of Industrial Pollution, September. World Bank: Economic Development Institute.

Koppen, C.S.A. van, and A.P.J. Mol, 2002. "Ecological Modernization of Industrial Ecosystems." In *Water Recycling and Resource Recovery in Industry: Analysis, Technologies and Implementation*, eds. P. Lens, L.W. Hulshoff Pol, P. Wilderer, and T. Asano. London: International Water Association Publishing.

Le, V.K., and T.M.D. Tran, 2003. "Ecological Transformation nof the Tapioca Processing Industry in Vietnam." Pp. 199–224 in *Greening Industrialization in Asian Transitional Economies: China and Vietnam*, eds. A.P.J. Mol and J.C.L. van Buuren. Lanham, MD: Lexington Books.

Lowe, E.A., J.L. Warren, and S.R. Moran, 1997. *Discovering Industrial Ecology – An Executive Briefing and Sourcebook*. Columbus, Richland: Battelle Press.

Lowe, E.A.,1997. "Creating By-Product Resource Exchanges: Strategies for Eco-Industrial Parks." *Journal of Cleaner Production* 5, 1–2: 57–65.

Nhue, T.H., N.Q. Cong, and T.H. Hanh, 2001. "Establishment of the Environmental Management System in the Industrial Zones in Vietnam." International Conference Industry and Environment in Vietnam, Ho Chi Minh City, April. Van Lang University: REFINE Project.

O'Rourke, Dara, 2002. "Motivating a Conflicted Environmental State: Community-Driven Regulation in Vietnam." Pp. 221–44 in *The Environmental State Under Pressure*, eds. A.P.J. Mol and F. H. Buttel. Amsterdam/London: JAI/Elsevier.

Phung, T.P., 2002. Ecological Modernization of Industrial Estates in Vietnam. Wageningen, The Netherlands: Wageningen University (dissertation).

Quang, N.C., 2001. "Corporate Environmental Accounting: Translate the Environmental Concerns into the Language of Business." Proceedings of the International Conference Industry and Environment in Vietnam, Ho Chi Minh City, April. Van Lang University: REFINE Project.

Rekha, M., T.N.C. Thai, X.N. Nguyen, N.L. Nguyen, T.K.C. Truong, A.T. Bang, G.T. Pham, and T.N. Nguyen, 1994. *Women in Waste Recycling in Ho Chi Minh City: A Case Study*. International Centre for Research on Women (Washington, D.C.), Women Studies Department (Open University of Ho Chi Minh City), Geography Department (Ho Chi Minh City University).

Sy, P.C., 2000. "Status and Control Measures of Air Pollution in the South Key Economic Regions." Environmental Workshop 2000 on Science, Technology and Environment, Ho Chi Minh City, Vietnam, June 6.

Tran, T.T., 2001. "The Environmental Management of Export Processing Zones and Industrial Zones of Ho Chi Minh City." Proceedings of the International conference Industry and Environment in Vietnam, 20–21 April. Ho Chi Minh City: REFINE Project.

Tran, V.N., 2001. "Environmental Management in Industrial Zones in the First Years of the Modernization and Industrialization Processes in Binh Duong Province." Proceedings of the International Conference Industry and Environment in Vietnam, 20–21 April, 2001. Ho Chi Minh City: REFINE Project.

Tran, V.N., and H. Leuenberger, 2003. "Cleaner Production and Industrial Pollution Control in Vietnam." Pp. 83–106 in *Greening Industrialization in Asian Transitional Economies: China and Vietnam*, eds. A.P.J. Mol and J.C.L. van Buuren. Lanham, MD: Lexington Books.

World Bank, 1995. *Vietnam Environmental Program and Policy Priorities for a Socialist Economy in Transition*. Report No. 13200-VN. Hanoi, Vietnam.

Chapter 24

Arthur P.J. Mol

ENVIRONMENT AND MODERNITY IN TRANSITIONAL CHINA: FRONTIERS OF ECOLOGICAL MODERNIZATION

Introduction: environmental homogenization?

DIRECTLY AFTER THE 1992 United Nations Conference on Environment and Development (UNCED), the judgements of environmental scholars, officials and interest groups regarding the successes of the summit were rather ambivalent. Now, more than a decade later, it tends to be evaluated much more positively. The UNCED conference is nowadays generally perceived as a major breakthrough in putting environmental protection and sustainable development squarely onto (inter)national agendas. More specifically, two major contributions of the UNCED are widely celebrated. First, attention for international and global environmental problems and policies were triggered by the preparations, the summit itself and its aftermath. This resulted, for instance, in institutional innovations such as the United Nations Framework Convention on Climate Change and the Biodiversity Convention. Second, the UNCED focused attention more clearly on issues of environmental protection and sustainable development in developing countries. While in most industrialized countries the institutionalization of the environment in national politics and policies had started in the late 1960s or early 1970s, in most developing countries this process began only in the late 1980s and early 1990s.

The process of institutionalization of the environment in Western (especially, but not exclusively, European) industrialized societies has been the object of much research by social scientists, mainly using the framework of ecological modernization. Ecological modernization refers to a restructuring of modern institutions to follow environmental interests, perspectives and rationalities. It has become increasingly difficult to understand the development of (and developments in) modern cultural, political and even economic institutions in these Western societies, if we exclude environmental logics and perspectives. Ideas of ecological modernization were also used by policy-makers and social scientists as useful tools in solving long-standing environmental disputes and conflicts. It formed an alternative to the curative approaches of Western nation-states, the demodernization and deindustrialization ideologies of the environmental movement, and the post-modernity discourse that deconstructs any environmental crisis until it melts into thin air. In that sense, ecological modernization is a more specific interpretation of the key ideas prevailing in the more general notion of sustainable development (Spaargaren and Mol, 1992).

Initially, until at least the mid-1990s, ecological modernization was typically seen as a Western theory, only valid for the limited geographical scope from which it originated. This started to change as a result of two major developments. First, a number of developing countries, especially in Southeast and East Asia, started to industrialize and arguably to modernize, at a rapid pace. The so-called first generation Asian tigers such as Taiwan, South Korea and Singapore were soon followed by a second generation of newly industrializing economies, including Malaysia, Thailand, China and, more recently, Vietnam. In the light of this industrialization and modernization process, the earlier belief that the major assumptions of ecological modernization were ill-fitting for these nation-states was cast into doubt (Frijns et al., 2000; Sonnenfeld, 2000). Second, the accelerating processes of globalization made a forceful entry onto social science research agendas from the early 1990s onwards. While there was — and to some extent still is — disagreement among social scientists as to the nature, the impact and the overall evaluation of globalization processes, most scholars do agree that these developments have contributed to an increasing global interdependence in political, cultural and economic domains. For environmental governance and reform this meant that economic, political and societal processes and dynamics pushing for environmental reform were no longer restricted to one (often Western) country, but were carried on the wings of globalization to other corners of the world. A global civil society, global environmental governance, and environmental management systems operated by transnational corporations in developed and developing countries, are often referred to as key examples of this. Since OECD countries arguably dominate globalization processes, they might also be expected to dominate in the environmental arena, resulting in the 'export' not only of economic and political institutions and mechanisms, but also of environmental reform models, practices and dynamics. These two developments thus contributed to a spreading beyond the Western nation-states of the conditions under which ecological modernization initially originated, and of its environmental strategies, practices and measures.

By 1995, then, one of the key questions on the research agenda of ecological modernization was its geographical scope. To what extent were ideas of ecological modernization of any use in developing or industrializing countries outside Europe? While originally formulated primarily in theoretical terms, and the subject of theoretical debates in sociological and political science literature, this question of course had major practical relevance. It involved the policy-relevant issues of transfer of (ecological modernization inspired) environmental strategies and models of environmental governance from OECD countries to new industrializing economies. It also touched upon questions of harmonization or heterogenization and differentiation in multilateral environmental agreements (MEAs): should the numerous MEAs being concluded and implemented under the influence of Western OECD countries be expected to work equally well in all countries around the world? Or would the Western bias in the MEAs' policy principles, approaches, strategies and inherent state–market–civil society relations prevent their equally successful implementation in, for instance, the Asian newly industrializing countries?

In an earlier publication I have weighed and criticized the view that globalization will result automatically in environmental homogenization (Mol, 2001). It is not only globalization dynamics and processes, but also specific local conditions, national priorities, domestic historical trajectories, state–market relations and power balances, among other things, that will determine environmental governance and reform practices and institutions. To use the terminology of Castells: the 'space of flows' has to meet somewhere the 'space of places', and at these meeting points we can expect to witness various models of environmental reform — if any substantial environmental reform can be identified at all (see Castells, 1996, 1997a, 1997b). If we apply the (Western) idea of ecological modernization outside Western Europe, we might expect to find environmental reform models that resemble some of its core features, but they will also be coloured by specific local conditions and positions in the world-system (Sonnenfeld, 2000). This can be conceptualized with the notion of modes or styles of ecological modernization.

In applying this to China we touch upon three key questions: can the environmental reforms in contemporary China be interpreted as ecological modernization, what are its core features, and what are the similarities and differences between Chinese and European modes or styles of ecological modernization? This contribution first presents a summary of the basic ideas of ecological modernization as originally formulated; it then briefly reviews the historical development of environmental protection in China, especially focusing on urban and industrial settings. Subsequently, it investigates the main social, political and economic dynamics behind processes of environmental reform currently being witnessed in China. Finally, the contribution draws some initial conclusions on the nature of 'ecological modernization' in China in an age marked by globalization, and thus on the geographical reach of ecological modernization theory.

Ecological modernization as a European project

It is not easy to distil the core features of ecological modernization from the rapidly growing European environmental social sciences literature. There are various reasons for this. Being a rather young theory, the literature of ecological modernization is still very much in development, with 'competing' and complementing interpretations. Second, scholars contributing to the literature on ecological modernization operate on various levels of abstraction. While some contribute to ecological modernization as a theory of social change, others focus on the changes in ideas and discourses or on the environmental policies being implemented. These differences inevitably result in different views on what constitues the main, basic or principal idea or set of ideas at the foundation of ecological modernization. Third, and partly related to the former point, those contributing to the ecological modernization literature start from or apply a range of theoretical frames, among which systems theory, discourse analysis, institutional theory, structuration theory, and new social movement approaches. Consequently, in the following summary of the core features of ecological modernization ideas in (especially) Western Europe, I will try to look for the common denominators in this rich and growing literature, but it will be impossible not to emphasize some interpretations over others. The essence of ecological modernization put forward here is thus an interpretation of what I see as the more central, important and/or influential connotations, in comparison with other contradictory or more peripheral versions.

Central ideas behind ecological modernization

Several authors claim that the central idea of ecological modernization is the growing compatibility between environmental protection and economic growth (for example, Hajer, 1995), or the idea that technology is the key to any modern project of environmental reform (Christoff, 1996; Huber, 1985, 1991; Humphrey et al., 2002). Although the former perspective can be found in numerous publications dealing with ecological modernization, and the second is prevalent in the more ambivalent or critical publications on ecological modernization, I think that both miss the core idea of ecological modernization.

The basic premise of ecological modernization theory is the centripetal movement of ecological interests, ideas and considerations within the social practices and institutional developments of modern societies. This results in ecology-inspired and environment-induced processes of transformation and reform of those same core practices and central institutions, a process that began in earnest from the 1980s onwards. This key idea can be found in all influential publications on ecological modernization, starting from Joseph Huber's (1982) idea of the ecological switchover as the new (Schumpeterian) phase in the maturation of the industrialization process, via Martin Jänicke's (1993) notion of modernization of the political processes due to the growing importance of environmental interest and ideas, up to more recent work of Spaargaren and van Vliet (2000) on transformations in the infrastructures and practices of consumption, and the analyses of Murphy and Gouldson (2000) on industrial innovations.

Within ecological modernization theory these processes have been conceptualized at an analytical level as the growing autonomy, independence or 'differentiation' of an ecological perspective and ecological rationality *vis-à-vis* other perspectives and rationalities (cf. Andersen and Massa, 2000; Mol, 1995; Seippel, 2000; Spaargaren, 1997). In the domains of policies, politics and ideologies, the ecological perspective began to emerge as an independent force in the 1970s and early 1980s in most Western societies: the construction of governmental organizations, departments, legal institutions and monitoring and reporting programmes set up specifially to deal with environmental issues dates from that era, followed later by the emergence of green parties in the political systems of many OECD countries (Carter, 2000). In the socio-cultural domain a distinct green ideology — as manifested, for instance, by environmental non-governmental organizations (NGOs), environmental periodicals and 'green' belief systems — can be traced back to the 1970s or earlier. During the 1980s, especially, this ideology assumed an independent status and could no longer be interpreted in terms of the old political ideologies of socialism, liberalism and conservatism (Giddens, 1994; Paehlke, 1989).

The crucial transformation, which makes the notion of the growing autonomy of an ecological perspective and rationality especially relevant, and which led European scholars to introduce the concept of ecological modernization, occurred in the 1980s. It was at this point that an ecological perspective started to challenge the monopoly of economic rationality as the all-determining organizing principle in the sphere of economics. Since most scholars agree that the growing autonomy of an ecological rationality and perspective from its economic counterpart in the domain of production and consumption is crucial to 'the ecological question', this last step was the decisive one. It meant that economic processes of production and consumption were increasingly designed and organized, analysed and judged, from both an economic *and* an environmental point of view (even if the two are not given equal weight, even today). Some profound institutional changes in the domain of production and consumption became discernible from the late 1980s onward in OECD countries, including the widespread emergence of environmental management systems and environmental departments within firms; the introduction of economic valuation of environmental goods via (for instance) eco-taxes; the emergence of environment-inspired liability and insurance arrangements; the increasing importance attached to environmental goals such as natural resource saving and recycling among public and private utility enterprises, making it a key issue in competition; and the articulation of environmental considerations in economic supply and demand (for instance via eco-labelling schemes, environmental information and communication systems in economic chains).

The fact that we analyse these environment-related transformations as *institutional* changes indicates their semi-permanent character. Although the process of environment-induced transformations and efficiencies should not be interpreted as linear and irreversible, as was commonly assumed in the modernization theories in the 1950s and 1960s, these changes have some permanency and would be difficult to reverse. Hence, although the environment moves up and down the 'issue-attention cycle' of politics (Downs, 1972), it is firmly embedded in the core institutions and social practices of modern society, which means that any radical and sudden breakdown of environmental gains would be resisted, even in times of economic stagnation. In the terminology of Giddens (1984), this is an episodic transformation: a specified direction of change over a delineated time period.

Dynamics, mechanisms and actors in Europe

Various ecological modernization scholars have elaborated on the social mechanisms, dynamics and actors through which social practices and institutions are transformed by the incorporation of environmental interests and considerations. European ecological modernization studies have highlighted three key elements, in particular:

- *Political modernization.* The modern 'environmental state' (Mol and Buttel, 2002) plays a key role in processes of environmental institutionalization, but no longer in a conventional way. First, there is a trend towards decentralized, flexible and consensual styles of national governance, at the expense of top-down, centralized, hierarchical, command-and-control regulation. Second, non-state actors are increasingly involved in the conventional tasks of the nation-state (that is, the provision of public goods), including privatization, conflict resolution by business–environmental NGO coalitions, private interest government, and the emergence of 'subpolitics' (Beck, 1994). Finally, the role of international and supra-national institutions is growing and to some extent undermining the sovereign role of the nation-state in environmental reform. Together with the next point, this results in new state–market relations in environmental protection and reform.
- *Economic and market dynamics and economic agents.* Whereas in the 1960s and 1970s environmental improvements were only triggered by the state and environmental NGOs, more recently producers, customers, consumers, credit institutions, insurance companies, the utility sector, and business associations have increasingly turned into social carriers of ecological restructuring, innovation and reform (in addition to state agencies and new social movements), both within countries and across borders. They use market, monetary and economic logics in pushing for environmental goals.
- *Civil society.* With the institutionalization processes, new positions, roles, ideologies and cultural frames for environmental movements are crystallizing. Instead of positioning themselves on the periphery or even outside the central decision-making institutions, environmental movements seem increasingly involved in decision-making processes within the state and, to a lesser extent, the market (Mol, 2000; Sonnenfeld, 2002). Environmental norms, values and discourses gain influence by spreading far beyond the professionals and core supporters of environmental NGOs, a process that is paralleled by their reformulation.

From Europe to China

In analysing China's environmental reforms from an ecological modernization perspective, it is important to distinguish between the leading idea of ecological modernization theory, on the one hand, and the dynamics, mechanisms and actors at work in processes of ecological modernization on the other. If ecological modernization is taking place in China, there should be evidence of a growing 'differentiation' of an environmental rationality and perspective from its economic counterpart, and a subsequent institutionalization of ecological interests, ideas and considerations in social practices and institutional developments, as described above. However, the concrete dynamics, mechanisms and actors which are directing (or beginning to direct) this process in China may differ from the situation in Western Europe. In comparison with the European processes introduced here, ecological modernization processes can differ from country to country and region to region, and it is in this context that the notion of mode or style might be helpful.

The development of the Chinese 'environmental state'

In exploring ecological modernization and environmental reform in contemporary China, this contribution must inevitably be highly selective. The common opinion of China's environmental record seems to centre on the poor performance of state agencies and deteriorating environmental quality, rather than anything like ecological modernization. While this seems too one-sided a perspective, in searching for ecological modernization dynamics I will nevertheless have to be selective by focusing especially on the successes and improvements in China's environmental reforms. Where do we see the seeds of environmental institutionalization? Which environmental reform dynamics seem to have a good chance of becoming dominant, because they are part of larger tendencies and transformations in China? Which are the crucial actors and advocacy coalitions that might push ecological modernization? And where do these differ from what we have witnessed in Europe's ecological modernization processes? I will start in this section with a short historic introduction on environmental reform in China and an assessment of the trends in environmental 'additions' and 'withdrawals' over the last decade.

In a former command economy that is now in a transition stage, it is not surprising to find environmental institutionalization primarily in state and political structures and institutions. The dawn of the Chinese government's serious involvement in environmental protection more or less coincides with the start of economic reforms in the late 1970s. Pollution control began in the early 1970s, especially following the 1972 United Nations Conference on the Human Environment in Stockholm. A National Environmental Protection Office was established in 1974, with equivalent offices in the provinces; this institution came to maturity with the enactment and implementation of the various environmental laws and regulations passed since the late 1970s, with a period of accelerated development in the 1990s. Following the promulgation of the state Environmental Protection Law in 1979 (revised in 1989), China began to systematically establish an environmental regulatory system. In 1984 environmental protection was defined as a national basic policy, and key principles for environmental protection in China were proposed, including 'prevention comes first, then control', 'polluter responsible for pollution control' (already introduced in the 1979 environmental law), and 'strengthening environmental management'. Subsequently, a national regulatory framework was formulated, composed of a series of environmental laws on all the major environmental segments (starting with marine protection and water in 1982 and 1984 respectively), executive regulations, standards and measures.[1]

Institutionally, the national regulatory framework is vertically implemented through a four-tier management system — national, provincial, municipal and county levels. The latter three levels are governed directly by their corresponding authorities in terms of both finance and personnel management, while the State Environmental Protection Agency is only technically responsible for their operation. The enactment of the various environmental laws, instruments and regulations over the last two decades has been paralleled by a step-wise increase of the bureaucratic status and capacity of these environmental authorities (Jahiel, 1998). For instance, the National Environmental Protection Bureau became the National Environmental Protection Agency (NEPA) in 1988, and in 1998 it received ministerial status as the State Environmental Protection Agency (SEPA). By 1995, the 'environmental state' had over 88,000 employees all over China and by 2000 this had grown to 130,000.[2] As Jahiel (1998: 776) remarks about this environmental bureaucracy: 'Clearly, the past 15 years . . . has seen the assembly of an extensive institutional system nation-wide and the increase of its rank. With these gains has come a commensurate increase in EPB [Environmental Protection Bureau] authority — particularly in the cities'. Although the expansion of the 'environmental state' suffered some setbacks and stagnation (such as the 'demotion' of EPBs in many counties from second-tier to third-tier organs in 1993–94), over a period of twenty years the growth in quantity and quality of the officials is impressive, especially when compared with the shrinking of other state bureaucracies. Besides SEPA, the State Development Planning Commission (SDPC) and the State Economic and Trade Commission (SETC) have crucial roles as national state agencies in environmental protection, especially since the governmental reorganization in 1998.

It can be argued that these administrative initiatives have borne fruit in terms of environmental improvements, although the widespread information distortion, discontinuities in environmental statistics, and the absence of longitudinal environmental data in China should made us cautious in drawing any final conclusions.[3] Total suspended particulates and sulphur dioxide concentrations showed an absolute decline in most major Chinese cities between the late 1980s and the late 1990s (Lo and Xing, 1999; Rock, 2002b), which is especially remarkable given the high economic growth figures during that decade. By the end of 2000, CFC production had decreased by 33 per cent compared to levels in the mid-1990s, due to the closure of thirty companies (SEPA, 2001). It is reported (but also contested) that emissions of carbon dioxide have fallen between 1996 and 2000, at a time of ongoing economic growth (Chandler et al., 2002; Sinton and Fridley, 2001).[4] Most other environmental indicators show a delinking between environmental impacts and economic growth (for example, water pollution in terms of biological oxygen demand; see World Bank, 1997). More indirect indicators that suggest similar relative improvements are the growth of China's environmental industry (increasing from 0.22 per cent of GDP in 1989, to 0.87 per cent of GDP in 2000, and to 1.1 per cent in 2002); the increased number of firms certified with ISO14000 standards by 2000 (Mol, 2001); and the closing of heavily polluting factories, especially after environmental campaigns during the second half of the 1990s (Nygard and Xiaomin, 2001). Nonetheless, these positive signs should not distract us from the fact that China is heavily polluted; that emissions are often far above international standards and environmental quality far below; that only 25 per cent of municipal wastewater is treated before discharge (although 85 per cent of industrial wastewater is treated, according to SEPA, 2001); and that environmental and resource efficiencies of production and consumption processes are rather low overall.

Ecologizing China's modernization project

In the birth period of environmental protection, China bore the characteristics typical of the centrally planned economies: restricted citizen involvement; limited response to international agreements, organizations and institutions; a strong focus on central state authority, especially the Communist Party of China (CPC), with little room for manoeuvre for decentralized state organizations, para-statals or private organizations; an obsession with large-scale technological developments (in terms of hard technology); problems with co-ordination between state authorities and departments, and a limited empowerment of the environmental authorities (DeBardeleben, 1985; Lotspeich and Chen, 1997; Ziegler, 1983). The further construction, development and maturation of China's environmental reform strategy was not a linear process, or a simple unfolding of the initial model of environmental governance invented twenty years ago under a command economy. There was a certain degree of discontinuity in Chinese environmental reform, for two main reasons. First, the economic, political and social changes that China has experienced during the last two decades have also affected the original 'model' of environmental governance. Economic transition towards a market-oriented growth model, decentralization dynamics, growing openness to and integration in the outside world, and bureaucratic reorganization processes have all caused shifts in China's environmental governance model. Second, China witnessed the inefficiencies and ineffectiveness of its initial environmental governance approach in ways not unlike the 'state failures' (Jänicke, 1986) that European countries had faced in the 1980s before they transformed their environmental protection approach along lines of ecological modernization. Building on all kinds of innovative experiments and developments resulting from such dynamics, environmental governance and the institutionalization of environmental ideas and interests in China have developed in unique ways during the last decade.

Unlike the more stable contemporary environmental institutions of the OECD countries, understanding the process of environmental institutionalization in China's modernization path requires us to follow a moving target. Consequently, any analysis will have to focus more on trends and significant developments than on a static state of the art. These trends and developments may be grouped in four major categories, which partially reflect the categories mentioned above for Europe: political modernization, economic actors and market dynamics, institutions beyond state and market, and international integration.

Political modernization

The state apparatus in China is of over-arching importance in environmental protection and reform. Its crucial role in this sphere is likely to be safeguarded for some time to come because of both the nature of the contemporary Chinese social order, and the characteristic of the environment as a public good. Environmental interests are mainly expressed by the impressive rise of environmental protection bureaus at various governmental levels. Yet the most common complaints from Chinese and foreign environmental analysts focus precisely on this system of (local) EPBs — on their poor environmental capacity (in qualitative and quantitative terms); on the dependence of the local EPBs on both the higher level EPBs and on local governments, which often have no interest in stringent environmental reform but play a key role in financing the local EPBs; on the lack or distortion of environmental information; on the low priority given to environmental criteria in assessing local governments; and on the poor financial incentives for both governments and private actors to abide by environmental laws, standards and policies.

Nevertheless, the environmental state in China is clearly undergoing a process of political modernization, in which traditional hierarchical lines and conventional divisions of power are being transformed. Although processes of political modernization in China's environmental policy have different characteristics from those witnessed in Europe, the direction of those reforms is similar: greater decentralization and flexibility, and a shift away from a rigid, hierarchical, command-and-control system of environmental governance. Increasingly, local EPBs and local governments are being given — and are taking — larger degrees of freedom in developing environmental priorities, strategies, financial models and institutional arrangements. This parallels broader tendencies of decentralization in Chinese society, and is also environmentally motivated by state failure in environmental policy.[5] There is a noticeable shift towards greater influence and decision-making power on the part of the local authorities, and diminishing control on the part of Beijing, both by the central state structures and by the CPC (see, for instance, Andrews-Speed et al., 1999 on decentralization in energy policy).[6]

Decentralization and greater flexibility contribute to environmental policies that are better adopted to the local physical and socio-economic situations. But in China, as elsewhere, decentralization does not automatically result in better protection of the environment (Beach, 2001), as local authorities tend to prioritize economic growth and investments above the progressive development of environmental policies and the stringent enforcement of environmental regulation and standards.[7] In a context in which active civil society and accountability mechanisms are poorly developed, decentralization has little to offer to the environment. But a larger degree of freedom for local authorities does result, for better or worse, in a growing diversity among the Chinese provinces and towns in how local and regional environmental challenges are being dealt with, and in the degree of success (or failure) of such interventions. These successes and failures are not only divided along lines of economic prosperity, although the richer eastern provinces and towns are systematically more concerned with, and invest more heavily in, environmental reform. But even here, within the eastern part of China, differences in environmental prioritization can be found, as shown by Zhang's (2002) detailed case study on environmental reforms in five towns in Anhui and Jiangsu provinces.

As in other countries, decentralization tendencies in China have also led to counter tendencies. Environmental protection projects, for instance, are increasingly centrally financed. The central state has also responded to the growing autonomy of local authorities by refining their system of evaluating towns and town governments. Rock (2002b) provides a detailed analysis of how local governments are increasingly assessed with respect to their environmental performance by using the Urban Environmental Quality Examination System. The ranking in this system of environmental indicators not only allows SEPA to compare municipalities: it also enables governments to design environmental responsibility contracts with local leaders for improvements in individual indicators, and to link these to assessments, financial incentives and promotion, encouraging town and village leaders to take environmental protection more seriously. This of course trickles down to the officials of, for instance, economic and planning departments of villages. It is a system of making local environmental governance accountable to the higher levels, in a situation in which decentralized, civil society based systems of accountability are underdeveloped. Via such mechanisms environmental rationalities are brought into the political system, so that local leaders are no longer judged only according to political and economic criteria, but also according to environmental results.

Another political modernization tendency is the separation between state owned enterprises (SOEs) and the line ministries and local governments (in the case of the township and village enterprises, or TVEs) that were originally responsible for them.[8] There is a slow but steady process of transferring decisions about production units from political and party influence to the economic domain, where the logic of markets and profits is dominant.[9] Despite the fact that, at the local level especially, governments are not always eager to give up direct relations with successful enterprises because of the related financial resources, the enterprises' tendency towards growing autonomy from political agents is unmistakable. This process opens up opportunities for more stringent environmental control and enforcement, as the 'protection' of the SOEs by line ministries and bureaus at all government levels is less direct. It also sets preferential conditions for a stronger rule of (environmental) law; more on this below. But it does not solve one of the key problems of environmental governance — the low priority given to environmental state organizations *vis-à-vis* their economic and other counterparts.

Progress on the strengthening and empowering of China's environmental state is ambivalent, as it is in many other parts of the world. While the central environmental authority in Beijing has strengthened its position *vis-à-vis* other ministries and agencies, as described above, this is not always the case at the local level where EPBs are usually part of — and thus subordinate to — an economic state organization (see Vermeer, 1998; Zhang, 2002).[10] Moreover, interdepartmental struggles at the central level do not always result in favourable environmental outcomes, and often lead to the continuation of a fragmented environmental authority (Jahiel, 1998; Lo and Xing, 1999: 165). For instance, the State Economic and Trade Commission (SETC), and not SEPA, is the party primarily responsible for the new 2002 Cleaner Production Promotion Law. SETC is also responsible for energy conservation policy (Chen and Porter, 2000), while the Ministry of Science and Technology won the battle for the co-ordination of China's Agenda 21 programme from SEPA, despite the influence of, and lobbying from, the United Nations Development Programme (Buen, 2000).[11]

Finally, the emergence of the rule of law can be identified as a sign of modernization in environmental politics, closely tied to the emergence of a market economy. The system of environmental laws established from the 1980s onwards has led to higher standards being set for environmental quality and emission discharges, and the establishment of a legal framework for various implementation programmes.[12] But the environmental programmes themselves, the administrative decisions related to the implementation of standards, and the bargaining on targets between administrations and polluters, have had more influence on environmental reform than the laws and regulations *per se*. Being in conflict with the law is usually less problematic than being in conflict with administrations and programmes, and most of the massive clean-up programmes were not so much derived from environmental laws (although they were not in conflict with them), but rather based on top-level administrative decisions.[13] The same is true for enforcement of national environmental laws at the local level. The rather vague laws are interpreted in very different ways by EPBs, often under the administrative influence of the local mayor's office (Ma and Ortolano, 2000: 63). Courts have been only marginally involved in enforcement; EPBs use courts as a last resort to enforce environmental laws to which polluters refuse to adhere (Jahiel, 1998: 764). There have been some recent signs that the rule of law is being taken more seriously in the field of environment, accompanied by heavier (financial) punishments and legal procedures started by, for instance, environmental NGOs such as the Centre for Legal Assistance to Pollution Victims (CLAPV) in Beijing.[14] One of the potential threats to the environment is the institutional void that can emerge when the administrative system loses its power over environmental protection, while the rule of law has barely been institutionalized in the field of environment.

Economic actors and market dynamics

Traditionally, centrally planned economies did a poor job of setting the right price signals for a sustainable use of natural resources and a minimization of environmental pollution, notwithstanding the theoretical advantages and the early ideas of some progressive economists and other environmental scholars in these command economies (see DeBardeleben, 1985; Mol and Opschoor, 1989). With China's cautious turn to a market oriented growth model from 1978 onwards, one would expect to find some economic and market dynamics beginning to push for environmental reform, and environmental interests in contemporary China are being slowly institutionalized in the economic domain of prices, markets and competition.

First, subsidies on natural resources are gradually being abandoned and prices are moving towards cost price, sometimes as conditions of foreign loans.[15] This is, of course, only a relative improvement: the cost price rarely includes costs for repair of damage and environmental externalities, although it is clear from (for example) the major flooding caused by forest cuts that such externalities can be quite dramatic, also in monetary terms.[16]

Second, attempts are being made to increase environmental fees and to offer tax reductions,[17] in a way that will influence the economic decision-making of polluters. The application of discharge fees, introduced in the 1980s, has become more widespread, in part because they are an important source of income for local EPBs as well as a significant trigger for implementation of environmental measures.[18] However, fees are often paid only for discharges above a set standard. Despite the rhetoric of 'pollution prevention pays' and 'cleaner production' that has entered modern China since the 1990s, fees are so low and monitoring so weak[19] that enterprises will risk paying the fee (or simply neglect payment), rather than installing environmental protection equipment or changing production processes (see Taylor and Qingshu, 2000, for the city of Wuhan). Nor is the introduction of higher fees a smooth process. As early as 1992, NEPA proposed an increase of 0.20 yuan per kg of discharged sulphur dioxide following coal burning (an increase of less than 1 per cent), to cover at least part of the environmental costs of desulphurization.[20] Implementation of the increase was postponed until 1996, and was then introduced only as a pilot programme, which was — in an extended version — still the situation in 2000.

Third, market demands are beginning to include such dimensions as the environmental and health implications of products and production processes, especially in international markets that have increased so dramatically in the wake of China's accession to the WTO. The import of Chinese refrigerators into the EU was restricted as early as 1990, due to the use of CFC as a cooling agent (Vermeer, 1998), but that was an exception. Today, these kinds of international (especially European, North American and Japanese) market trends towards greener products and production processes are felt in many more product categories, leading, for instance, to higher levels of ISO certification and growing interest in cleaner production, eco-labelling systems and industrial ecology initiatives (Shi, 2003; Shi et al., 2003). Like most developing economies, the Chinese domestic market still articulates environmental interests only poorly, and green or healthy labelling is underdeveloped.

Although economic reforms have reduced the role of the central state in economic decision-making and increased the autonomy of economic and market actors (with a few exceptions as described above), this has not yet resulted in more non-state actors actively promoting environmental interests.[21] Insurance companies, banks, public utility companies, business associations, general corporations and others do not yet play any significant role in environmental reforms. This is largely because these economic actors do not feel any pressure, or see any market opportunity, for institutionalizing environmental interests within their arrangements and daily routines. There are, however, three major exceptions to this: large Chinese firms that operate in an international market, the environmental industry itself and R&D institutions.

- The larger Chinese and joint venture firms that operate for and in a global market are subject to stringent environmental standards and practices; they try to pass these new standards and practices onto their customers and state organizations, pushing the domestic markets towards international levels. For example, the Chinese petrochemical company, Petrochina, has investments in several countries and joint venture operations in China with several Western oil multinationals. It is acutely aware of the need to acquire internationally-recognized environmental management knowledge, and to meet standards and emission levels, allowing it to compete on a global market. In adopting these practices, it also brings these standards home to the Chinese state, with a call for upward harmonization among all players in the Chinese petrochemical sector.[22]

- The expanding environmental industry (see above) has a clear interest in increased environmental regulation and reform, and therefore presses for the greening of production and consumption processes (Sun, 2001).[23] Moreover, foreign environment industries and consultancies are increasingly entering the Chinese market, partly financed by official development assistance (ODA) projects.

- Research and development institutions — whether related to universities or to the line ministries and bureaus — are focusing more and more attention on environmental externalities, and articulating environmental interest among decision-making institutions within both the economic and the political domain. In Chinese universities, a growing number of environmental departments, centres and courses have been established since the 1990s.

Beyond state and market: civil society

As elsewhere, environmental reforms in China have not been limited to institutional changes of state and market. In European countries the environmental movement, environmental periodicals and the foundation of an increasingly universal system of environmental norms and values are both medium and outcome of processes of ecological modernization in what has become known as civil society. In China the incorporation of environmental interests in institutions and arrangements beyond state and market has followed a completely different trajectory.

China has a very recent history of environmental NGOs and other social organizations that articulate environmental interests and ideas of civil society and promote them among the political and economic decision-makers (Ho, 2001; Qing and Vermeer, 1999). Environmental NGOs are limited in number and they are often not adversarial or confrontational but rather expert or awareness-raising organizations, such as Global Village. The 'political room' for a Western-style environmental movement still seems limited, as international NGOs themselves have found. International NGOs such as Greenpeace and WWF have invested major efforts in further stimulating the environmental movement in China, with mixed success. While in some of the Central and East European centrally planned economies environmental NGOs played a role in articulating environmental and other protests against the ruling social order, in China environmental NGOs have so far been marginal in pushing for the ecological modernization of the Chinese economy. There are, however, other ways in which civil society's contribution to environmental reform is being expressed in China, including the rise of environmentally-oriented government-organized NGOs (GONGOs); increasing local activism and complaints; and the importance of unwritten social norms, rules and codes of conduct.

GONGOs, such as the Beijing Environmental Protection Organization and China Environment Fund, are playing an increasingly important role in environmental governance in China today. They have more freedom of registration and manoeuvre due to their close links with state agencies. Via their expert knowledge and closed networks with policy-makers, these GONGOs articulate environmental interests and bring them into state and market institutions. In doing so, they help to bridge the gap between NGOs and civil society on the one hand and the state on the other, thus 'becoming an important non-state arena for China's environmental politics' (Wu, 2002: 48).[24] Now that these GONGOs are gaining organizational, financial and political independence and autonomy from the state, they are evaluated more positively by Western scholars. Although they remain embedded in a dominating state structure, the state is relaxing its control and allowing them relative autonomy in developing activities and raising funds.

Together with economic liberalization, decentralization of decision making and experiments with local democratization, China is also experiencing mounting pressure from (often unorganized) citizens on local environmental authorities to reduce environmental pollution. Dasgupta and Wheeler (1996) estimated that local and provincial authorities responded to over 130,000 complaints annually in the period 1991–93. In most cities and towns, hotlines and systems for making complaints have been installed, albeit with different levels of use and effect. In Wuhan (a city of almost 7 million) the local EPB received 680 complaints in 1994, resulting in 658 visits (Taylor and Qingshu, 2000). In 1998, the EPB of Wujin (population 1.2 million, Jiangsu province) responded to 479 complaints, while the heavily polluted small town of Digang (50,000 inhabitants, Anhui province) reported that they did not receive a single complaint in 1998 (Zhang, 2002).[25] In China, these systems of complaints and the growing attention paid by the (state-owned and controlled) media to environmental pollution and environmental mismanagement are more important than NGOs in articulating civil society's environmental interests to economic and political decision-makers. These dynamics play out within the context of the growing commitment of the CPC and the central government to combat pollution, and the central government's encouragement of the media and individuals to speak up on environmental misuse. In that sense, the dominant environmental discourse and the advocacy coalitions supporting that discourse have changed dramatically during the last fifteen years. Nevertheless, this system of complaints is a poor form of 'participation' of civil society in environmental issues. It focuses only on (sensible) monitoring after pollution has happened, at a time when a preventive and precautionary approach is needed. What is missing is the systematic involvement of citizens and civil society, with full access to information, at the stage of project development.

Third, in Chinese society informal social norms, rules and unwritten codes of conduct play a crucial part in structuring human action. These rules are firmly anchored in Chinese civil society, rather than in the formal institutions of state and market, and may play an important role in environmental reform. Ma and Ortolano (2000: 77ff) mention three major non-formal rules: respect for authority and status, even if it conflicts with the formal institutions; the social connections or *guanxi* that play a key role in organizing social life in China; and the moral authority and social capital that is included in the concept of (losing, maintaining or gaining) 'face'. With the growing importance attached to environmental protection, these and other informal rules and institutions are being put to work for environmental goals and rationalities. *Guanxi* and 'face' play a role in environmental protection, where informal networks of social relations are formed around environmental programmes and dispute resolutions, and social capital is built via environmental awards, prices, and media coverage. While some of these institutions are not unknown in Europe (although often differently organized), they have a much stronger influence in China and are consequently more important in environmental reforms. If we are to understand ecological modernization dynamics in China we have to understand how and to what extent these informal institutions, networks, and connections articulate environmental rationalities through, for instance, the inclusion of environmental norms in social capital and moral authority and the increase of the status of environmental authorities. There are, of course, considerable variations in the way that these dynamics work, and in their effectiveness, in different parts of China.

One of the restrictions that prevents civil society — and other institutions beyond state and market — from playing a larger role in environmental reform is its limited access to environmental information. This is the result of several factors: the lack of environmental monitoring (most environmental monitoring needs to be funded by the local governments, which have limited budgets) and distortion in information processing; the secrecy with which environmental data are handled, putting them beyond the reach of large segments of society; the absence of a right-to-know code, legislation or practice; and the limited publication and availability of non-secret data, due to poor reporting, limited internet use and access. Often, only general and aggregate data are available, and then only for political decision-makers and scientists; more specific local data either are not collected or are kept secret for those directly involved in environmental pollution. Consequently, local EPBs rely strongly on complaints as a way of monitoring, and priorities for control and enforcement are frequently set accordingly.

In their analysis of the accountability of Chinese environmental authorities, Wu and Robbins (2000a, 2000b) show that despite the virtual absence of an active civil society, and the shortage of reliable and transparent data, accountability is still required of China's environmental governance. State agencies at other levels and in other sectors, the media, scientists and international monitoring by donors and Multilateral Environmental Agreements organizations do regularly hold environmental authorities accountable.

International integration

In assessing the role of external international forces on China's turn to environmental protection, Rock (2002a: 82) is straightforward: 'there is no evidence of Chinese pollution management policies being affected by either international economic or political pressure. Instead, the Chinese government's pollution management programs have largely been influenced by internal developments, particularly the partial liberalization of its economy that started in 1979 and the decentralization of decision-making that accompanied it'. Compared to the sometimes significant influence of foreign pressure and assistance on national environmental policy in other Asian countries, China has indeed been reluctant to accept assistance which is accompanied by stringent environmental conditions. The Three Gorges Dam is a clear example of this, with China ignoring both foreign pressure against the dam and threats of withholding international loans for the project. Also, in international negotiations on MEAs, Chinese authorities are often hesitant to support stringent environmental policies that could rebound on domestic efforts (Chen and Porter, 2000; Johnston, 1998; McElroy et al., 1998).

On less controversial issues, however, foreign assistance has clearly contributed to, and influenced, China's environmental policies and programmes. Between 1991 and 1995 US$ 1.2 billion in foreign capital was invested in environmental protection in China (Vermeer, 1998: 953). More recently, China has become an object of considerable international attention as well as environmental funding, via several MEAs and multilateral institutions such as the World Bank, the Global

Environmental Facility (GEF), the Asian Development Bank (Huq et al., 1999) and the United Nations Environment Programme. By the end of the 1990s the World Bank and the Asian Development Bank together were providing US$ 800 million on environmental loans to China annually. Asuka-Zhang (1999) illustrates the significance of bilateral environmental ODA and environmental technology transfer to China, taking Japan as an example. It is estimated that around 15 per cent of China's total environment-related spending originates from bi- and multilateral lending and aid (Tremayne and De Waal, 1998). For instance, foreign projects have had a significant influence on the development and introduction of cleaner production, resulting finally in the 2002 Cleaner Production Promotion Law (Shi, 2003). In drafting environmental laws nowadays the participation of foreign lawyers, scientists and other experts is standard practice. The phasing-out of CFC use following the Montreal protocol has been another example. Directly after the Montreal protocol negotiations (1987) China increased its CFC production (by some 100 per cent between 1986 and 1994; Held et al., 1999: 397), becoming the world leader in CFC production and consumption in 1996. But in response to international aid and (potential) trade bans by OECD countries, it stabilized its production in the mid-1990s and moved to a decline in consumption (from the mid-1990s onward) and production (in 2000).[26]

The growing openness to and integration in the global economy and polity will only increase international influence on China's domestic environmental reform.[27] For instance, its membership of WTO will enhance the importance of ISO standards in international, but increasingly also in domestic business interactions, and it will make China more vulnerable to international criticism on its domestic environmental performance. However, international integration will also mean a greater role for China in setting the agenda and influencing the outcomes of international negotiations and agreements, including those which are relevant for the environment.

Conclusion: Chinese ecological modernization in the making?

An environmental restructuring of the processes and practices of production and — to a lesser extent — consumption is taking place in China, as environmental interests and conditions are given higher priority. The first indications that the Chinese state was widening its original project of simple, technological modernization by taking environmental externalities into account date from the late 1970s, and parallel the start of economic reform. Since then, state-driven environmental laws and programmes have made a more serious impact, especially during the 1990s. China's strategy and approach to tackling the growing environmental side-effects of modernization is far from stable; it is still developing and transforming, together with the general transition of China's economy and state. But most environmental reform initiatives are firmly based on, make use of and take place within the context of China's modernization process. In that sense, it seems justified to use the term 'ecological modernization' to describe China's attempts at restructuring its economy along ecological lines.

However, the story doesn't end there, as is clear from the analysis presented above. The current advancements in the greening of China's economy and society do not seamlessly fit the Western version of ecological modernization, but vary in at least three important and interdependent ways: the degree of institutionalization of environmental interests; the respective roles of state, market and civil society in China's 'ecological modernization'; and the Chinese characteristics of environmental reform dynamics.

First, in the relatively short history of the theory, most contributions to the ecological modernization approach emphasize processes of institutionalizing environmental interests in social practices and institutional developments, reflexively reorienting the institutions of simple modernity in line with ecological criteria. While this analysis of the China case has demonstrated the growing importance of environmental interests in modernization processes, it has also shown that, to date, environmental interests have been institutionalized only partially, at best. There is no routine, automatic and full inclusion of environmental considerations in the institutions that govern production and consumption practices in contemporary China.

Second, and partly linked to the first point, the institutions that take up environmental interests and work to ecologically restructure the Chinese economy deviate in large measure from those that scholars have identified in European societies. A number of political and state institutions do seem to be increasingly incorporating environmental considerations and interests in their standard operating procedures and social practices, in ways not dissimilar to their European counterparts — a system of environmental laws, regulations and standards; the emergence of the rule of law; assessment systems of environmental performance; flexibilization and decentralization in environmental policy are all evidence of this. However, with respect to economic and market institutions and civil society, the Chinese situation differs dramatically from Europe.

Where the introduction of the market economy leads to market prices, increased efficiencies, the reduction of subsidies and stronger international economic relations, economic institutions can advance ecological reforms. The removal of subsidies on natural resources, such as energy, and the international market demand which imposes environmental conditions on Chinese products and processes are examples of this. But more often environmental reforms do not automatically coincide with economic efficiency interests, and then economic and market institutions have virtually no role to play in advancing environmental interests. There are several reasons for this neglect of environmental interests in, for instance, price setting, consumer and customer demand, insurance arrangements, credit facilities, public utility performance, economic competition, enterprise R&D programmes and niche market developments in China. At a national level, environmental interests have not been articulated strongly enough to put the emerging economic and market actors and institutions under pressure. In addition, in large parts of China economic institutions and actors still have intricate relations with and are dependent on political ones. This makes them less free to incorporate environmental (and other new non-economic) interests in their routine operations. Finally, where economic institutions and arrangements are differentiated or 'emancipated' from political control, they often develop into new, single-goal institutions that are unable or unwilling to take up such 'additional' tasks as environmental protection. Arguably, unrestricted, free 'jungle' capitalism in its purest form can be found more often in certain parts of transitional China than in the welfare states of capitalist Europe.

In terms of 'civil society', the economic liberalization and market reforms that have occurred in China have not been accompanied by a parallel process of political liberalization and democratic reform. Consequently, civil society in China remains undeveloped and has been unable to match the role played by civil society institutions and actors in most OECD countries, such as setting the environmental agenda, pressing economic and political institutions to include environmental interests, and pushing itself towards the centre of political and economic decision making. While China has developed its own institutions beyond state and market (including GONGOs, cultural institutions and mass organizations), the role of these institutions in environmental reform is by no means equal to what we have witnessed in Europe.

Third and finally, if we focus on the mechanisms, processes and dynamics that trigger environmental reform and push for institutionalization, there are some similarities but also many differences between the Chinese and European situations. European scholars in ecological modernization are familiar with protesting local communities, the emergence of an 'environmental state', globalization dynamics that push towards a level playing field on environmental protection, economic instruments such as the discharge fee system, a growing environmental industry, a reorientation of state R&D towards environmental issues, and decentralization and flexibilization in environmental policy. On the other hand, GONGOs, the environmental responsibility contract system, unique policy principles such as the 'three synchronizations' principle (synchronizing the design, construction and operational aspects of environmental management and production), the strong role of informal networks, rules and institutions, and the dual responsibility of local EPBs, are all arrangements that play a major role in the greening of the contemporary Chinese economy, but have no equivalent in most European states.

In sum, ecological modernization in China can be said to be of a different mode than the European version that has been studied so widely. It is also far from stable. Especially now that China is in transition and is opening up to the world polity and economy, it is hard to predict whether its environmental reform path will become increasingly close to, or will diverge more strongly from, that of the OECD countries. If China's modernization continues with a further 'differentiation' of economic institutions and arrangements from their political counterparts (and all indications point in that direction), it is possible that economic institutions and arrangements will increase their role in environmental reform. Ecological modernization studies have shown that — at least in Western countries — these economic institutions can play a major role in articulating, communicating, strengthening, institutionalizing and extending (in time and place) environmental reforms by means of their own (market and monetary) 'language', logic and rationality and the force of their influence. However, economic institutions can and will only play that role if they are put under pressure — by the environmental state, by international institutions and/or by civil society. The latter category has a very important role to play in the future, character and uniqueness of China's ecological modernization.

Notes

1. At a national level, China now has some twenty environmental laws adopted by the National People's Congress, approximately 140 executive regulations issued by the State Council, and a series of sector regulations and environmental standards by the State Environmental Protection Agency (SEPA).

2. In 2000 there were over 80,000 environmental staff at the county level (in more than 7,000 institutions); 35,000 staff at the city level (in 1,700 institutions); almost 11,000 staff at the provincial level and some 3,000 staff at the national level (in a total of some 300 institutions) (SEPA, 2001).

3. The annual *Report on the State of the Environment in China* by SEPA usually contains data on emissions and environmental quality, but there is a major lack in consistency in data presentations between 1997 and 2001 (see www.zhb.gov.cn/english/SOE for the various annual national environmental reports and related statistics).

4. Sinton and Fridley (2001) and Chandler et al. (2002) report a decrease of 17 per cent in greenhouse gas emissions (based on official Chinese energy statistics), the International Energy Agency estimates energy reduction to be 5–8 per cent in that period (www.usembassy-china.org.cn/sandt/energy_stats_web.htm), while the American Embassy in China claims a zero growth of energy use in China (ibid.). All sources agree on how energy use/greenhouse gas emissions have been delinked from economic growth: increased energy efficiency, economic reforms, and a fuel switch from coal to natural gas.

5. In developing their local policies and programmes, provincial or municipal governments need to be consistent with the national regulations. However, there is a growing tendency for the regulations and measures of sub-national governments to develop their own dynamics, speed and, partly, contents, thus deviating at least temporarily, and sometimes substantially, from national regulations.

6. As Ma and Ortolano (2000: 14) put it: 'The Party has deeply penetrated the apparatuses of the state, and thus there is no advantage in distinguishing the Party from the state in our analysis of environmental policy'. To a great extent, this also applies to the analysis here.

7. Chen and Porter (2000: 59) reach the same conclusion for decentralization in energy conservation policies: 'It is clear that the ongoing process of change in organizational structures and lines of responsibility has given rise to much confusion in recent years, and if anything has undermined rather than improved the prospects for a coordinated and enforceable policy for energy conservation in industry'.

8. From research on steel enterprises, Fisher-Vanden (2003) reports that within Chinese SOEs, decentralization in firm management improves the incorporation of new — and more energy-efficient and environment-friendly — technology.

9. By the end of the 1990s many SOEs had full decision-making power over production, sales, purchasing and investments. In most cases, however, relations between these enterprises and state authorities are still intricate and local agencies still succeed in extracting funds from profitable enterprises for public works or other purposes, in subsidizing inefficient enterprises and in influencing decisions at the enterprise level. This is also valid in the case of TVEs, as Zhang (2002) has shown for counties in Anhui and Jiangsu provinces.

10. In the past EPBs relied (sometimes heavily) on the environmental protection divisions of industrial bureaus or ministries, which usually had good access to and knowledge of the polluters, especially given that state and market agents were barely separate. More recently, the role of the industrial bureaus (or general companies as they have sometimes been renamed) in environmental protection has diminished. While this might be beneficial in the long term, in specific cases and in the short term it has caused EPBs some serious problems, through the lack of environmental and technological capacity.

11. SETC replaced the industrial line ministries and is much more powerful than the SEPA — which is also implied by its being named 'Commission'. SETC encompasses several environmental tasks and organizations independent from SEPA, such as monitoring stations.

12. The major eight national environmental programmes are: environmental impact assessment; three synchronizations; pollution discharge fee system; pollution control with deadlines; discharge permit system; assessment of urban environmental quality; centralized control of pollution; and environmental responsibility system. The first three date from the late 1970s, the last three were implemented later to manage problems which the first three could not handle (for further details, see Ma and Ortolano, 2000: 20ff).

13. Indeed, most sinologists who aim to understand the state environmental protection system pay only marginal attention to environmental laws and their enforcement, and concentrate instead on administrative measures and campaigns (see Jahiel, 1998; Vermeer, 1998).

14. Interview with Wang Canfa, director of CLAPV, 2001. See also Ho (2001: 908) and Otsuka (2002).

15. For example, in the early 1990s the World Bank provided a loan to replace the thousands of small coal burning boilers in Beijing with more energy efficient and less polluting heating systems. The World Bank's conditions were that the Beijing Power Corporation must become an independent business body operating under market conditions and that subsidies should be removed from energy prices. These conditions have been fulfilled (Gan, 2000).

16. In 1996 it costs 0.843 yuan to supply a ton of water. At that time, the average price of water was 0.6–0.9 yuan/ton in Hebei province, 0.637 yuan in Beijing and only 0.013 yuan in Hetao region (Lo and Xing, 1999: 159). From 1994 to 2005, the yuan was pegged to the US dollar at 8.2770. Since 21 July 2005 the yuan is linked to several currencies including the dollar, euro and yen at 8.11 against the dollar. The yuan can now fluctuate within a bandwidth of 0.3 per cent upwards or downwards.

17. Tax reductions are sometimes offered if environmental goals are reached, as in the case of energy saving in steel plants and other heavy energy consuming industries (Chen and Porter, 2000).

18. The fee programme started in 1979 in some locations but became more widespread after 1982 and especially after the legal strengthening in 1989 in the final version of the Environmental Protection Law. The majority of the fees are collected for water and air emissions (see SEPA, 2001). Only part of the fee can be used by the local EPBs to finance their staff, equipment and programmes: the other part goes back into environmental funds that are used to subsidize environmental measures in industries. While hundreds of thousand of firms have paid a fee, many small and rural industries have managed to escape payment due

to lack of enforcement. Wang and Wheeler (1999) found that the levies are higher in heavily polluted and economically developed areas and that they do influence air and water emission reductions within companies.

19. Ma and Ortolano (2000: 21) refer to four penalty charges (the so-called 'four small pieces') that have to be paid above the discharge fee.

20. 'Notification on Implementation of Pilot Programme of Levy on Industrial Sulphur Dioxide Pollution by Coal Burning', State Council Letter (1996#24) agreed on the pilot implementation via SEPA's 'Report on Pilot Programme of Sulphur Dioxide Discharge Fee'.

21. Although finance bureaus and local banks sometimes play a role in administering environmental funds into which the pollution discharge fees are paid, and in deciding on loans or grants to polluters, these economic agents do not really play a role themselves in articulating environmental interests in their economic activities. Local banks are not eager to lend additional money to polluters for environmental investments, according to a World Bank study (Spofford et al., 1996, as quoted in Ma and Ortolano, 2000).

22. Interview, Petrochina's environmental monitoring office, December 2001.

23. In analysing the growing role of government organized non-governmental organizations in environmental reforms, Wu (2002) mentions two major and influential environmental industry associations: the China Environment Protection Industry Association and the China Renewable Energy Industry Association.

24. Wu (2002) analyses the emergence of a diversity of GONGOs (among which foundations, education centres, research institutions and industry associations) within the national and provincial administrative bodies, and the role they are able to play due to their less restrictive institutional structure, their expertise and personal connections. Wu also shows the clear reasons for the Chinese government to allow or create GONGOs, which includes attracting foreign assistance and funding.

25. Dasgupta and Wheeler (1996) show that the average number of environmental complaints (by telephone, letter or face-to-face) in major cities and provinces in one year ranged from 55 per 100,000 inhabitants in Shanghai to 1.7 per 100,000 inhabitants for Gansu. In most provinces EPBs responded to over 80 per cent of these complaints.

26. Data by the Ozone secretariat of UNEP: see www.unep.org/ozone

27. There is still considerable debate, also within China, whether accession to the WTO will force a further separation between politics and economics, with (beneficial) consequences such as an increase in transparency in policy-making, a growing pressure to implement the rule of law and a further undermining of the structural basis of corruption (see Fewsmith, 2001).

References

Andersen, M. A. and I. Massa (2000) 'Ecological Modernisation: Origins, Dilemmas and Future Directions', *Journal of Environmental Policy and Planning* 2(4): 337–45.

Andrews-Speed, P., S. Dow and Z. Gao (1999) 'A Provisional Evaluation of the 1998 Reforms to China's Government and State Sector: The Case of the Energy Industry', *Journal of the Centre for Energy, Petroleum and Mineral Law and Policy* 4(7): 1–11. Available online: www.dundee.ac.uk/cepmlp/journal

Asuka-Zhang, S. (1999) 'Transfer of Environmentally Sound Technologies from Japan to China', *Environmental Impact Assessment Review* 19(5/6): 553–67.

Beach, M. (2001) 'Local Environment Management in China', *China Environment Series* 4: 21–31.

Beck, U. (1994) 'The Reinvention of Politics: Towards a Theory of Reflexive Modernisation', in U. Beck, A. Giddens and S. Lash *Reflexive Modernisation. Politics, Tradition and Aesthetics in the Modern Social Order*, pp. 1–55. Cambridge: Polity Press.

Buen, J. (2000) 'Challenges Facing the Utilisation of Transferred Sustainable Technology in China: The Case of China's Agenda 21 Project 6–8', *Sinosphere* 3(1): 13–23.

Carter, N. (2000) *The Politics of the Environment*. Cambridge: Cambridge University Press.

Castells, M. (1996) *The Information Age: Economy, Society and Culture. Volume I: The Rise of the Network Society*. Malden, MA, and Oxford: Blackwell.

Castells, M. (1997a) *The Information Age: Economy, Society and Culture. Volume II: The Power of Identity*. Malden, MA, and Oxford: Blackwell.

Castells, M. (1997b) *The Information Age: Economy, Society and Culture. Volume III: End of Millenium*. Malden, MA, and Oxford: Blackwell.

Chandler, W., R. Schaeffer, Z. Dadi et al. (2002) *Climate Change Mitigation in Developing Countries. Brazil, China, India, Mexico, South Africa, and Turkey*. Arlington, VA: Pew Center on Global Climate Change.

Chen, Z. and R. Porter (2000) 'Energy Management and Environmental Awareness in China's Enterprises', *Energy Policy* 28: 49–63.

Christoff, P. (1996) 'Ecological Modernization, Ecological Modernities', *Environmental Politics* 5(3): 476–500.

Dasgupta, S. and D. Wheeler (1996) 'Citizen Complaints as Environmental Indicators: Evidence from China'. World Bank Policy Research Working Paper. Washington, DC: The World Bank.

DeBardeleben, J. (1985) *The Environment and Marxism–Leninism: The Soviet and East German Experience*. Boulder, CO, and London: Westview Press.

Downs, A. (1972) 'Up and Down with Ecology. The Issue-Attention Cycle', *The Public Interest* 28: 38–50.

Fewsmith, J. (2001) 'The Political and Social Implications of China's Accession to the WTO', *The China Quarterly* 167: 573–91.

Fisher-Vanden, K. (2003) 'Management Structure and Technology Diffusion in Chinese State-Owned Enterprises', *Energy Policy* 31: 247–57.

Frijns, J., Phung Thuy Phuong and A. P. J. Mol (2000) 'Ecological Modernisation Theory and Industrialising Economies. The case of Viet Nam', *Environmental Politics* 9(1): 257–92.

Gan, L. (2000) 'World Bank Policies, Energy Conservation and Emission Reduction', in T. Cannon (ed.) *China's Economic Growth. The Impact on Regions, Migration and the Environment*, pp. 184–209. Basingstoke and London: Macmillan.

Giddens, A. (1984) *The Constitution of Society*. Cambridge: Polity Press.

Giddens, A. (1994) *Beyond Left and Right. The Future of Radical Politics*. Cambridge: Polity Press.

Hajer, M. (1995) *The Politics of Environmental Discourse. Ecological Modernisation and the Policy Process*. New York and London: Oxford University Press.

Held, D., A. McGrew, D. Goldblatt and J. Perraton (1999) *Global Transformations. Politics, Economics and Culture*. Stanford, CA: Stanford University Press.

Ho, P. (2001) 'Greening Without Conflict? Environmentalism, NGOs and Civil Society in China', *Development and Change* 32(5): 893–921.

Huber, J. (1982) *Die verlorene Unschuld der Ökologie. Neue Technologien und superindustrielle Entwicklung*. Frankfurt am Main: Fisher Verlag.

Huber, J. (1985) *Die Regenbogengesellschaft. Ökologie und Sozialpolitik*. Frankfurt am Main: Fisher Verlag.

Huber, J. (1991) *Unternehmen Umwelt. Weichenstellungen für eine ökologische Marktwirtschaft*. Frankfurt am Main: Fisher Verlag.

Humphrey, C. R., T. L. Lewis and F. H. Buttel (2002) *Environment, Energy, and Society: A New Synthesis*. Belmont, CA: Wadsworth Group.

Huq, A., B. N. Lohani, K. F. Jalal and E. A. R. Ouano (1999) 'The Asian Development Bank's Role in Promoting Cleaner Production in the People's Republic of China', *Environmental Impact Assessment Review* 19(5/6): 541–52.

Jahiel, A. R. (1998) 'The Organization of Environmental Protection in China', *The China Quarterly* 156: 757–87.

Jänicke, M. (1986) *Staatsversagen. Die Ohnmacht der Politik in die Industriegesellshaft*. Münich: Piper.

Jänicke, M. (1993) 'Über ökologische und politieke Modernisierungen', *Zeitschrift für Umweltpolitik und Umweltrecht* 2: 159–75.

Johnston, A. I. (1998) 'China and International Environmental Institutions: A Decision Rule Analysis', in M. B. McElroy, C. P. Nielsen and P. Lydon (eds) *Energizing China. Reconciling Environmental Protection and Economic Growth*, pp. 555–99. Cambridge, MA: Harvard University Press.

Lo, F.-C. and Y.-Q. Xing (eds) (1999) *China's Sustainable Development Framework. Summary Report*. Tokyo: The United Nations University.

Lotspeich, R. and A. Chen (1997) 'Environmental Protection in the People's Republic of China', *Journal of Contemporary China* 6(14): 33–60.

Ma, X. and L. Ortolano (2000) *Environmental Regulation in China. Institutions, Enforcement, and Compliance*. Lanham, MD: Rowman & Littlefield.

McElroy, M. B., C. P. Nielsen and P. Lydon (eds) (1998) *Energizing China. Reconciling Environmental Protection and Economic Growth*. Cambridge, MA: Harvard University Press.

Mol, A. P. J. (1995) *The Refinement of Production. Ecological Modernisation Theory and the Chemical Industry*. Utrecht: Jan van Arkel/International Books.

Mol, A. P. J. (2000) 'The Environmental Movement in an Era of Ecological Modernisation', *Geoforum* 31: 45–56.

Mol, A. P. J. (2001) *Globalization and Environmental Reform. The Ecological Modernization of the Global Economy*. Cambridge, MA: MIT Press.

Mol, A. P. J. and F. H. Buttel (eds) (2002) *The Environmental State under Pressure*. London: Elsevier.

Mol, A. P. J. and J. B. Opschoor (1989) 'Developments in Economic Valuation of Environmental Resources in Centrally Planned Economies', *Environment and Planning A* 21: 1205–28.

Murphy, J. and A. Gouldson (2000) 'Environmental Policy and Industrial Innovation: Integrating Environment and Economy through Ecological Modernization', *Geoforum* 31(1): 33–44.

Nygard, J. and G. Xiaomin (2001) *Environmental Management of Chinese Township and Village Industrial Enterprises (TVIEs)*. Washington, DC: The World Bank; Beijing: SEPA.

Otsuka, K. (2002) 'Networking for Development of Legal Assistance to Pollution Victims in China', *China Environment Series* 5: 63–5.

Paehlke, R. C. (1989) *Environmentalism and the Future of Progressive Politics*. New Haven, CT, and London: Yale University Press.

Qing, D. and E. B. Vermeer (1999) 'Do Good Work, but Do Not Offend the "Old Communists". Recent Activities of China's Non-governmental Environmental Protection Organizations and Individuals', in W. Draguhn and R. Ash (eds) *China's Economic Security*, pp. 142–62. Richmond, Surrey: Curzon Press.

Rock, M. T. (2002a) *Pollution Control in East Asia. Lessons from Newly Industrializing Economies*. Washington, DC: Resources for the Future; Singapore: Institute of Southeast Asian Studies.

Rock, M. T. (2002b) 'Getting into the Environment Game: Integrating Environmental and Economic Policy-making in China and Taiwan', *American Behavioral Scientist* 45(9): 1435–55.

Seippel, Ø. (2000) 'Ecological Modernisation as a Theoretical Device: Strengths and Weaknesses', *Journal of Environmental Policy and Planning* 2(4): 287–302.

SEPA (2001) *Report on the State of the Environment in China 2000*. Beijing: State Environmental Protection Agency.

Shi, H. (2003) 'Cleaner Production in China', in A. P. J. Mol and J. C. L. van Buuren (eds) *Greening Industrialization in Transitional Asian Countries: China and Vietnam*, pp. 63–82. Lanham, MD: Lexington.

Shi, H., Y. Moriguichi and J. Yang (2003) 'Industrial Ecology in China, Parts 1 and 2', *Journal of Industrial Ecology* 6(3–4): 7–11.

Sinton, J. E. and D. G. Fridley (2001) 'Hot Air and Cold Water: The Unexpected Fall in China's Energy Use', *China Environment Series* 4: 3–20.

Sonnenfeld, D. A. (2000) 'Contradictions of Ecological Modernization: Pulp and Paper Manufacturing in South East Asia', *Environmental Politics* 9(1): 235–56.

Sonnenfeld, D. A. (2002) 'Social Movements and Ecological Modernization: The Transformation of Pulp and Paper Manufacturing', *Development and Change* 33(1): 1–27.

Spaargaren, G. (1997) 'The Ecological Modernisation of Production and Consumption. Essays in Environmental Sociology'. PhD dissertation. Wageningen: Wageningen University.

Spaargaren, G. and A. P. J. Mol (1992) 'Sociology, Environment and Modernity. Ecological Modernisation as a Theory of Social Change', *Society and Natural Resources* 5: 323–44.

Spaargaren, G. and B. van Vliet (2000) 'Lifestyles, Consumption and the Environment: The Ecological Modernisation of Domestic Consumption', *Environmental Politics* 9(1): 50–76.

Spofford, W. O., X. Ma, Z. Ji and K. Smith (1996) *Assessment of the Regulatory Framework for Water Pollution Control in the Xiaoqing River Basin: A Case Study of Jinan Municipality*. Washington, DC: The Word Bank.

Sun, C. (2001) 'Paying for the Environment in China: The Growing Role of the Market', *China Environment Series* 4: 32–42.

Taylor, J. G. and X. Qingshu (2000) 'Wuhan: Policies for the Management and Improvement of a Polluted City', in T. Cannon (ed.) *China's Economic Growth. The Impact on Regions, Migration and the Environment*, pp. 143–60. Basingstoke and London: Macmillan.

Tremayne, B. and P. De Waal (1998) 'Business Opportunities for Foreign Firms Related to China's Environment', *The China Quarterly* 156: 1016–41.

Vermeer, E. B. (1998) 'Industrial Pollution in China and Remedial Policies', *The China Quarterly* 156: 952–85.

Wang, H. and D. Wheeler (1999) *Endogenous Enforcement and Effectiveness of China's Pollution Levy System*. Washington, DC: The World Bank.

Wehling, P. (1992) *Die Moderne als Sozialmythos. Zur Kritik sozialwissenschaftlicher modernisierungstheorien*. Frankfurt and New York: Campus.

World Bank (1997) *Clear Water, Blue Skies*. Washington, DC: The World Bank.

Wu, C. and A. Robbins (2000a) 'An Overview of Accountability Issues in China's Environmental Governance. Part 1', *Sinosphere* 3(1): 37–42.

Wu, C. and A. Robbins (2000b) 'An Overview of Accountability Issues in China's Environmental Governance. Part 2', *Sinosphere* 3(2): 17–22.

Wu, F. (2002) 'New Partners or Old Brothers? GONGOs in Transitional Environmental Advocacy in China', *China Environment Series* 5: 45–58.

Zhang, L. (2002) 'Ecologizing Industrialization in Chinese Small Towns'. PhD dissertation. Wageningen: Wageningen University.

Ziegler, C. E. (1983) 'Economic Alternatives and Administrative Solutions in Soviet Environmental Protection', *Policy Studies Journal* 11(1): 175–88.

Chapter 25

Catherine Oelofse, Dianne Scott, Gregg Oelofse and Jennifer Houghton

SHIFTS WITHIN ECOLOGICAL MODERNIZATION IN SOUTH AFRICA: DELIBERATION, INNOVATION AND INSTITUTIONAL OPPORTUNITIES

Introduction

SUSTAINABLE DEVELOPMENT IS NOW widely accepted as a policy framework in planning and development both internationally and in South Africa (O'Riordan et al., 2000; Sowman, 2002; Scott et al., 2001). However, the implementation of the principles of sustainable development through a wide range of tools and mechanisms in South Africa has proved difficult to achieve (Sowman, 2002; Laros, 2004, Todes et al., 2004). Part of the problem is the way in which environmental problems have been constructed (Hajer, 1995). Both internationally and in South Africa, technocentric scientific approaches to environmental management, which often legitimize the destruction of the environment by capitalist development, have dominated. Technical solutions have therefore been applied to solve complex problems that often defy rational, objective approaches. This weak ecological modernization approach (Christoff, 1996) has become institutionalized over the last 30 years in the developed world and has been transferred to developing countries such as South Africa as the rationale for environmental management (Lee & George, 1998; Laros, 2004; Scott & Oelofse, 2005). Compounding the prevalence of such a dominant approach is the 'implementation deficit' in institutional policymaking (Hajer & Wagenaar, 2003; Hajer, 2004), which is particularly problematic in developing countries and is recognized as being a major obstacle to the achievement of sustainability in South Africa.

In the first decade of democracy, South African environmental law and policy underwent significant reform and are now among the best in the world (Scott et al., 2001). However, in the development context of the new democracy, government institutions faced with the implementation of these new regulatory policies find themselves in a situation of 'institutional ambiguity' where there is little experience or precedent of how to proceed (Hajer, 2004). Given the democratization of environmental decision-making in South Africa, the phenomenon of 'multi-signification' (Hajer, 2004) challenges the implementation of sustainability goals (Houghton, 2005). Furthermore, the reactive rather than proactive approach of mainstream environmental management results in the absence of frameworks and strategies within which development should proceed. Most tools are project based and do not deal with the environmental in a holistic and integrated manner and this results in further constraints on the appropriateness and value of environmental management practice.

Through the lens of two case studies reflecting changing approaches and practices within state institutions, the paper explores the shifts taking place in the construction, adaptation and application of policy frameworks and tools used in the drive towards sustainability in South Africa. The research uses Hajer (1995) and Christoff's (1996) critique of ecological modernization and the more recent literature on deliberative policy analysis (Hajer & Wagenaar, 2003; Hajer, 2003) to explore these shifts towards sustainability in South Africa. It suggests that these shifts have taken place as a result of the inappropriateness of mainstream tools to the South African condition, the global acceptance of the need for more integrated approaches, the opportunities for progressive change that 'institutional ambiguity' and 'multi-signification' create, and pockets of innovation that have developed when intellectual actors have shifted the boundaries of practice.

Mainstream approaches to environmental management in South Africa

The implementation of sustainability principles have been driven both internationally and in South Africa by the global environmental mainstream approach of ecological modernization. Mainstream environmental management, framed as it is within positivistic science, conceives of the biophysical realm as external to human life—to be transformed and managed to improve human existence (Garner, 1996; Fischer, 2003). This instrumentalist approach emphasizes the physical and natural environment as the chief recipient of environmental impacts, that solutions to environmental problems are technical and institutional and that they are the responsibility of scientific experts and managers (Dryzek, 1997). National governments worldwide have institutionalized this mainstream environmental management discourse through state policy and legislation that seek to manage the impacts of development. However, despite the domination of this modernist approach to environmental management and hence sustainability, there are parallel proponents of local, participatory and more equitable processes of environmental management which seek to provide protection from 'the inequality and poverty that market forces may produce' (Woodhouse, 2000, p. 161) and which seek new ways of deliberating over approaches to managing the environment.

Ecological modernization

Ecological modernization is a policy-oriented discourse in environmental politics that emerged in the 1980s. It became the most acceptable way of 'talking Green' in spheres of environmental policymaking in both the developing and developed world (Hajer, 1995; Christoff, 1996; Murphy, 2000; Blowers & Pain, 1999). The meaning of ecological modernization differs depending on the context and the author, which provides challenges when one uses it as a critical framework of analysis (Christoff, 1996, Murphy, 2000). Given these differences, there are, however, key concepts that can be drawn out of the ecological modernization literature. Ecological modernization is a modernist and technical approach that uses the language of business and science and therefore conceptualizes environmental pollution as a matter of efficiency rather than a threat to the system (Huber, 1985, cited in Mol, 1995; Christoff, 1996). It assumes that economic growth and the resolution of ecological problems can be reconciled. With regard to environmental assessment, the policy is to 'anticipate and prevent' impacts, with science playing a leading role in providing evidence of environmental impacts (Hajer, 1995). The state's role is to provide a policy and regulatory framework for environmental protection, resulting in a complementary relationship between the state and the market (Christoff, 1996).

One of the major limitations of ecological modernization in developing countries is that the assumed conditions for this approach, such as the availability of advanced technology, capital, democracy and capacity, are not in place. Andersen (1993, cited in Christoff, 1996, p. 489) 'describes a country's capacity for ecological modernization as depending on its "achieved level of institutional and technological problem solving capabilities, which are critical to achieving effective environmental protection and transformation to more sustainable structures of production" '. The problems of environmental degradation and poverty are of such a magnitude in the developing world that they render ecological modernization untenable (Blowers & Pain, 1999). Christoff (1996, p 489) states that 'ecological modernisation focuses on the state and industry in terms which are narrowly technocratic and instrumental rather than on social processes in ways which are broadly integrative, communicative and deliberative'. Of importance to this paper is that ecological modernization is criticized as a discourse that explicitly avoids addressing social contradictions (Blowers & Pain, 1999). Inequalities of wealth and power, which are particularly evident in developing countries, form a barrier to the creation of partnerships and cooperation in environmental decision making. With the reliance on science and technology for assessing environmental impacts and creating solutions, social and development issues are side-lined because they are difficult to conceptualize and measure. Ecological modernization does not adequately deal with the social questions related to assessing who benefits from and who bears the impact of development processes (Blowers & Pain, 1999; Scott & Oelofse, 2005).

This paper uses Hajer's (1995) cultural politics approach to ecological modernization as a point of departure. This approach asks 'why certain aspects of reality are now singled out as "our common problems"' and queries what sort of society is being created in the name of protecting 'nature' (Hajer, 1995, p. 256). Christoff (1996, p. 482) suggests that 'Hajer is most effective where he suggests that ecological modernisation is a discursive strategy useful to governments seeking to manage ecological dissent and to re-legitimise their social regulatory role'. While the authors acknowledge that there are other approaches to ecological modernization (Janicke et al., 1988; Weale, 1992, Mol, 1995; Spaargaren, 1997; Murphy,

2000) Hajer's (1995) critique provides the normative framework of analysis used here to understand both the discourses currently shaping mainstream environmental policy and practice in South Africa and the shifts towards strong (reflexive) ecological modernization in the two case studies.

Christoff (1996, p. 490) considers ways in which it 'is possible to emphasise the normative dimensions of different versions of EM (ecological modernization)'. He suggests that there is a continuum between weak and strong ecological modernization which reflects the range of approaches rather than the binary suggested by the two terms, and states that 'it is essential to note that weak and strong features of EM (ecological modernization) are not always mutually exclusive binary opposites' (Christoff, 1996, p. 491). Table 25.1 provides the characteristics of weak and strong ecological modernization. The continuum is used as framework of analysis to show why the two case studies represent a shift in ecological modernization in South Africa.

Deliberative policy analysis

A body of theory that has emerged since the early 1990s in political science is that of 'deliberative policy analysis' (Hajer, 2003; Hajer & Wagenaar, 2003; Young, 2001). This approach has proposed the emergence of collaborative decision-making networks in the context of 'institutional ambiguity' and 'multi-signification' resulting from the complexities of policymaking in the network society (Castells, 2000; Hajer, 2004). Such collaborative processes that occur at the 'edges' of formal political processes have been pioneered in 'new' spheres of politics such as the environment and are therefore relevant in assessing shifts towards strong ecological modernization (Hajer & Wagenaar, 2003).

'Institutional ambiguity' is defined as the lack of power and capacity of institutions, particularly the state, to deliver policy outcomes (Hajer, 2004). This is particularly relevant in South Africa, where institutional and legal reform has radically changed the policymaking terrain. While the new environmental laws and policies that have been promulgated since 1998 have been globally praised as progressive and socially just, there is little experience or capacity to implement the principles contained therein into detailed policies and programmes. There are 'no "pre-given" ways of arriving at legitimate policy decisions' (Hajer, 2004, p. 3), creating both a challenge to environmental decision making as well as providing

Table 25.1 Types of ecological modernization

Weak ecological modernization	Strong ecological modernization
Economistic	Ecological
Technological (narrow)	Institutional/systemic (broad)
Instrumental	Communicative
Technocratic/neo-corporatist/closed	Deliberative/democratic/open
National	International
Unitary (hegemonic)	Diversifying

Source: Christoff, 1996, p. 490.

opportunities for creative experimentation. In terms of the 'rights-based' politics of South Africa in the new democracy, state institutions have had to engage in inclusive processes of deliberation to develop the 'rules of the game' of legitimate policymaking (Dryzek, 1996; Hajer, 2003).

'Multi-signification' refers to the recognition of the diversity of the discourses of participants in the deliberation processes (Hajer, 2003). Increasingly, legitimacy has come to be seen in terms of the right or capacity of those subject to a collective decision to participate in deliberation about its content. Thus attempts are made in environmental policymaking and practice to incorporate the diversity of discourses and take account of 'multi-signification'. However, participation often remains a legitimating process, marginalized groups playing a limited role or remaining altogether invisible (Blowers & Pain, 1999, Scott & Oelofse, 2005). The weak relationship between the state and civil society under weaker forms of ecological modernization leads to the neglect of social questions in environmental decision-making. Part of the shift to a more democratic policymaking is the inclusion of stakeholders besides the state in the process of public participation (Taylor, 1995; Yanow, 2003). Ecological modernization is taking heed of such inputs into decision-making, hence the large volume of critical literature that is now emerging on procedures for adequately representing the voices of social groups in environment and development decision-making processes (Douglass & Friedmann, 1998; Oelofse & Patel, 2000; Fell & Sadler, 1999). Hajer and Kesselring (1999) note the wide range of experimental institutional practices to make decision-making more democratic as part of the principles of sustainability.

The paper has briefly reviewed the different interpretations of ecological modernization and has presented Christoff's (1996) continuum of weak and strong ecological modernization. It has considered how the international mainstream approach to environmental management, which is based on weak ecological modernization, has shaped environmental decision making in South Africa. It suggests that new forms of policymaking and practice that reflect inclusive processes of deliberation are beginning to emerge. Two case studies are now used to explore these ideas.

Innovation and a return to principles of sustainability in the context of 'institutional ambiguity' in post-apartheid South Africa

Two case studies of changing state practices in environmental management are explored in this paper. Both case studies are examples of the development of decision-making tools by the state during the current period of 'institutional ambiguity'. The first evaluates a national indicators project that was developed by consultants, while the second analyses a policy that was developed by local government officials in an environmental department, which focuses on developing innovative and integrative approaches to environmental management. The first is the development of a set of National Principles, Criteria, Indicators and Standards (PCI&S) for Sustainable Forest Management in South Africa. The second is the development of the City of Cape Town's Integrated Metropolitan Environmental

Policy (IMEP), which guides environmental strategies, policies and environmental management practice in the city. This paper analyses both the processes and products of these two projects as reflections of the shifts that are beginning to take place in developing tools for sustainability in South Africa as intellectual actors challenge the mainstream approach to environmental management.

Introducing the projects

'Forestry' in South Africa is defined in the National Forest Act (1998) as containing natural forests, woodlands and plantations. Some natural forests are protected and managed by the state as indigenous state forests; however, there are large areas of natural forest that occur on private or communal land which are not state managed. The commercial forest sector owns large areas of plantations that also contain pockets of indigenous forest. They support communally owned small grower projects that produce trees for the commercial sector. Communally owned forests on tribal land are common in the country and provide a livelihood for poor rural communities (Shackleton, 2004). It is within this complex arrangement of forest ownership, management and use that the Department of Water Affairs and Forestry (DWAF) has proposed the adoption of sustainable forest management as a national approach for the sector. The state is currently devolving the management of state-controlled forests to private agencies, which are expected to conform to sustainable forest management practices.

The National Forests Act (1998) of South Africa requires that principles, criteria, indicators and standards are developed which address the sustainable management of both natural forests and plantations in South Africa and which comply with the sustainability principles as set out in the Act (see www.dwaf.gov.za/forestry/sfm). The Institute of Natural Resources (INR), as expert environmental consultants, was commissioned by the Council for Sustainable Forest Management to develop a set of PCI&Ss for implementation by the Department of Water Affairs and Forestry. This work was funded by the Department for International Development (DFID) in Britain, which played a key role in the management of the project.

The City of Cape Town is a municipal authority that governs a city with a valuable and rich mix of cultural and natural resources and a growing population of three million people. It is the economic hub of the Western Cape Province. Cape Town has a unique natural environment and is situated in the smallest of the world's floral kingdoms, the Cape Floristic Kingdom. A number of environmental challenges face this area, largely as a result of the burgeoning population and their need for infrastructure, housing, jobs, transport, education, health care and services. Applying sustainable development within a metropolitan area, with its multiple layers of social needs and economic forces, complicates the picture further. The City of Cape Town's Environmental Management Branch is institutionally organized to ensure both policy development and the implementation of environmental policy and legislation. The IMEP for the City of Cape Town was developed by the Policy and Research section of the Environmental Management Department between 1998 and 2001 using an innovative approach to policy formation.

Multiple voices, networks and governance

Christoff (1996) suggests that a shift to strong ecological modernization is reflected in processes that are communicative, deliberative, open and democratic. Hajer (2004) argues that processes need to be developed that find ways of integrating the knowledge generated by a range of actors so as to resolve tensions that may arise from 'multi-signification'. This section analyses the processes of deliberation that occurred in the development of both the indicator set and the IMEP as a result of the networks that were established through the methodologies used in these projects.

In the forestry project there were four sets of experts driving and contesting the outcomes of this project: a group of 12 consultants (the INR team), members of the Council for Sustainable Forest Management, officials from DWAF, and DFID consultants with their associated international advisors. However, the network of actors who shaped the outcomes of the project was much broader than this. The project was conducted using extensive nationwide participatory techniques, which sets it apart from other national and international indicator projects. 'Multi-signification' was evident as stakeholders engaged in the process and argued for their interests to be recognized and included in the PCI&S.

The project methodology involved both a technical process for developing and refining the criteria, indicators and standards, and an extensive and integrally linked stakeholder consultation process for gaining stakeholder input into the criteria, indicators and standards, and for testing and refining them. The principles that were used were defined in the National Forest Act (1998). A list of the conditions of and key issues in forestry in South Africa was generated using stakeholder input generated through an intensive investigation of eight case study areas across the country. Workshops were conducted at the national, provincial and local level to verify the desired conditions raised by stakeholders. The set of PCI&Ss were filtered through the relevant legislation as well as the principles of sustainability so as to ensure that there were no gaps and that the set was aligned with current policy and legislation. From this verified set of conditions, criteria, indicators and standards/aspirational goals were generated by the expert team (INR, 2002).

In workshops used to verify the criteria, stakeholders were diverse, such as those representing the powerful commercial forestry sector and those rural communities involved in communal forest projects. Since the set was being developed in a context of 'institutional ambiguity' due to the restructuring that DWAF was undergoing at the time, certain actors played a more dominant role than others. The DFID team set out clear parameters but were willing to allow the consulting team the scope to be innovative, while the Council for Sustainable Forest Management attempted to shape the methodology. The consulting team strongly resisted this attempt at state direction. Most of the contested terrain was around methodology and the level of participation required in the project. The DWAF played a very limited role as the government agent responsible for the implementation of the set of PCI&S. However, once ownership of the set became politically important in the restructuring of DWAF, certain key actors began to attempt to influence the project in order to gain ownership of the project process and the outcomes.

In the case of the IMEP in Cape Town, a group of officials in the Environmental

Policy and Research Section, who represent a 'pocket of innovation', developed the process that would lead to the formulation of the IMEP. This core team used as its point of departure, the premise that was being supported politically in the city that 'the environment is Cape Town's most unique and greatest economic asset'. A policy that contained strategies and tools for implementation needed to be developed to protect and enhance this asset. The development of the IMEP was launched in October 1998, and followed a comprehensive and participative process. Key resources and informants used in the development process included the Year One State of Environment (SoE) Report and a number of public capacity-building workshops, public policy workshops and internal councillor and officials workshops. The IMEP and its implementation strategy were adopted as official policy of the City of Cape Town in October 2001. The Year One SoE Report identified 15 key environmental issues that needed to be addressed by the City. These 15 environmental themes were endorsed through a stakeholder participation process and became central to the policy implementation strategy. The policy therefore reflected a mandate from civil society, business and the state about which key environmental issues needed to be prioritized in the city.

The Draft Policy underwent a rigorous review process including stakeholder review, internal review, a comments-response report and one-on-one meetings with organizations, councillors and stakeholders. The final IMEP and implementation strategy were put before all the Council Portfolio Committees for endorsement, followed by presentation of the policy and implementation strategy to the Executive Committee and full Council for adoption as official policy. A network of actors therefore shaped the policy, which was then conceptually developed by the officials in the Policy and Research section.

In both cases it is evident that the development of these tools for sustainability represents a shift in approach away from the closed and hierarchical processes of weak ecological modernization. According to Dryzek (1996, p. 486), 'democratisation is largely, although not exclusively, a matter of the progressive recognition and inclusion of different groups in the political life of society'. Both the PCI&S and Cape Town's IMEP reflect the views and concerns of a wide range of actors. This has occurred through a state-sponsored extension of representation (Young, 2001) in these networks, which has been taken forward by consultants and officials who have upheld and strongly supported participatory approaches. As a result of this, democracy has been deepened (Dryzek, 1996) and a shift towards sustainability has taken place.

Diversifying approaches, reducing technocratic narrowing

The development of principles, criteria, indicators and standards is a widely accepted and fast-growing monitoring tool in the field of sustainability. However, in many cases, sets of PCI&S are developed as highly scientific and technical tools that are difficult to implement because they are inaccessible to people managing sustainability on the ground. As a result of being narrow and focused, the operationalization of these sets within an institutional setting is often problematic. Sets of PCI&S are usually created by 'experts' who design them to monitor and evaluate the management of the environment with little attention paid to the experience of local people in identifying and dealing with environmental problems or to the institutional change the indicators can support. The PCI&Ss developed by the INR reflect a shift from the mainstream approach, since they were designed as an integrated web that would monitor and report on the state of forests in a way that could lead to systemic change. The PCI&Ss were part of a cyclical process of policy development in which the outcomes were focused on shaping adaptive management approaches in the forestry sector (see Figure 25.1). The set is still to be implemented and therefore the success of the product in shaping policy and action with regard to sustainable forest management is yet to be seen.

In the case of Cape Town's IMEP the development of the policy and its implementation led to the need for wide-scale institutional restructuring within the municipality in relation to how the Environmental Management Branch interacted with other departments. The policy changed the way in which the Environmental Management Branch was positioned and how it related to other line functions within the city. The implementation of this policy revealed that institutional change was required to give support to the policy that was ratified by Council. Figure 25.2 shows the key components of IMEP.

Central to the implementation strategy is that the implementing mechanism for these detailed sectoral strategies will be the Integrated Development Plan (IDP) that is a legislated strategic planning requirement and key development and planning tool for all local governments in South Africa. The intention of the IMEP strategy is integration, not just within the environmental function, but in the city as a whole and hence the policy has the power to change institutional arrangements and ways of relating between different departments. Multidisciplinary task teams

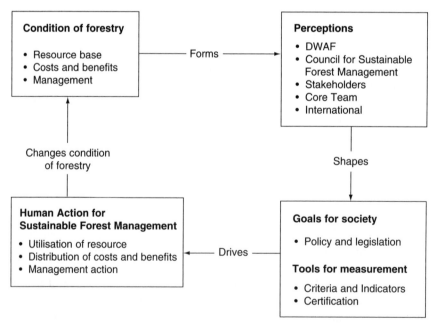

Figure 25.1 The role of the PCI&S in shaping policy and action for sustainable forest management

Source: INR, 2002

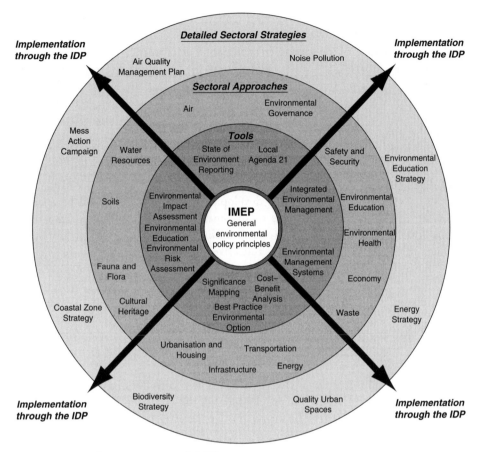

Figure 25.2 The components of IMEP
Source: www.capetown.gov.za

will develop detailed strategies through the Integrated Development Plan to address environmental issues. The idea is to take the onus to drive and develop environmental strategies away from the environmental managers and place it squarely within the responsibility of line functions using input from environmental managers. The principles are the guiding core of the policy. The process is not top down and vertical but rather is integrative and cooperative, again reflecting a systemic change in the way of conducting business within local government.

To effect holistic and collective responsibility, IMEP requires that the relevant line functions/service units, with input from environmental management, drive, develop and implement these strategies within their own functions and responsibilities. This ensures that environmental responsibility rests with the various departments instead of a scenario where environmental management attempts to enforce environmental strategies on the organization. Sowman (2002) discusses the different models used for environmental management branches within local municipalities in South Africa. Cape Town's IMEP reflects a significant shift away from the narrow approach which results in silos where line functions and departments operate in a hierarchical, technical and vertical manner.

In the case of the PCI&S the development of standards and aspirational goals also represents a shift away from a controlling state towards a negotiated and deliberative process of identifying environmental performance in sustainable forest management. As a result of the participation of the commercial forest sector in the process of developing the PCI&S, it became evident that the forestry industry in South Africa was not supportive of standards or benchmarks that were based on regulation and control by the state. They requested that these benchmarks should rather be phrased as aspirational goals: something that the forestry sector could work towards in the drive to sustainability, rather than being constrained or controlled by the more traditional regulatory standards. This was extensively debated and deliberated by the actors involved in the process and it was agreed that the forestry sector, which operates in a developing world context, should be encouraged by DWAF to meet the goals of sustainable forest management rather than being forced to manage their forests according to restrictive and economically difficult regulations. This outcome is reflective of ecological modernization, since it represents an enabling relationship between the state and business (Blowers & Pain, 1999) rather than a modernist approach of regulation and control by the state. What moves it toward strong ecological modernization is the way in which the PCI&Ss were negotiated through an extensive participatory process, which led to an understanding of society's mandate for sustainable forest management in South Africa. The shift also occurred because of the careful interrogation and inclusion of social indicators.

Systemic and institutional change is also reflected in the manner in which the participatory process used in the development of the PCI&S facilitated the development of a network of actors who began to drive the process outside of the state. The stakeholders who engaged in the process of criteria and indicator development developed capacity by contributing to the project and also began to make demands on the state to ensure that the set of PCI&Ss were implemented. This meant that they became 'insiders' in the project with a vested interest in ensuring the operationalization of the set. In terms of the commercial forest industry this was largely related to the way in which they were doing business as a result of their desire to become FSC (Forestry Stewardship Council) certified and ISO 14000 accredited. From a small-scale grower's perspective the development of the PCI&S was supported in terms of what the set offered in shaping and monitoring adaptive management approaches.

Policy deliberation under conditions of 'institutional ambiguity'

The 'institutional deficit' that is present within South Africa has provided opportunities for creative change in environmental management. The transformation of South African society has led to change in legislation, policy and government institutions, which has meant that systems have not been developed to shape projects. A great deal of learning still is taking place as citizens and those charged with doing environmental work find a pathway in the new terrains that are emerging. The lack of experience in developing these processes and products within South African society has provided opportunities for innovative intellectual actors to create new approaches. These have been widely supported, since they represent the implementation of the principles of sustainability contained in South African environmental legislation.

The 'institutional deficit' has meant that principles have played a key role in shaping the processes and outcomes developed in both these projects. Given the newness of the work being conducted, it has been essential to have a set of principles to guide the development of the PCI&S and Cape Town's IMEP. These principles have been drawn from international literature, South African legislation such as the National Environmental Management Act (1998), and stakeholder input. The explicit use of principles that reflect ecological, social and environmental justice represents a move away from weak ecological modernization which tends to marginalize social and distributive issues (Blowers & Pain, 1999) and which does not always adopt an integrative and broad approach. This has led to the internationalization of the projects without losing the specific need to situate environmental management within the local South African context.

The use in both projects of principles that address vulnerability and marginalization and the need for social justice brings these international sustainability principles strongly into the South African context. Christoff (1996, p. 496) supports this shift when he states that:

> strong ecological modernization therefore also points to the potential for developing a range of alternative ecological modernities, distinguished by their diversity of local culture and environmental conditions, although still linked through their common recognition of human and environmental rights and a critical and reflexive relationship to certain common technologies, institutional forms and communicative practices, which support the realization of ecological rationality and values ahead of narrower instrumental forms.

In the case of Cape Town's IMEP, central to the overall approach to sustainability is that the underlying principles of sustainable development are adopted as official principles at the highest level of the local government structure. These principles include among others:

- open, transparent and effective environmental governance;
- collective environmental responsibility;
- the integration of environmental issues into local government decision-making at all levels;
- protection of the constitutional rights to a healthy environment;
- a commitment to a holistic approach to environmental issues;
- a commitment to responsible stewardship of the resources within local government's charge;
- a commitment to the involvement of, and partnerships with, civil society.

Where IMEP begins to differ from mainstream approaches is that these principles are given effect through the development and implementation of detailed sectoral strategies. So, rather than relying solely on recognized mainstream environmental tools and processes, such as Environmental Impact Assessment (EIA) and Environmental Management Systems (EMS), that when used independently and in isolation may lead to weak models of sustainability, IMEP uses a combination of those tools, and others, as part of multidisciplinary holistic strategies to address the environment. The mainstream tools for sustainability, such as EIA and EMS, are used as components of a larger organized and systematic approach to environmental management. Transparency is achieved through the Annual SoE Report. This integrated approach, which is not hierarchical, as shown in Figure 25.2, represents a shift towards strong ecological modernization. The representation of IMEP as a circle with the principles at the centre (Figure 25.2) further supports this integrated and systematic approach.

Conclusion

Ecological modernization has framed environmental discourse and the tools used for environmental management for the past thirty years in developed and developing countries. This mainstream approach relies on science and technology and addresses problems of efficiency rather than need. It has also determined and dominated the range of tools used in the drive toward sustainability. New approaches, which address social and environmental justice and which are embedded in South African policy and legislation, are beginning to emerge. These more critical approaches are particularly important and relevant to the contexts of the developing world, where mainstream approaches are inapplicable and inequitable.

This paper has considered the shift that is taking place in the tools and policy frameworks used in the drive toward sustainability in South Africa. It has used two case studies to reflect on the nature of the changes taking place. The theory of ecological modernization has been critiqued and this has provided a continuum of weak and strong ecological modernization against which the shifts in the tools for sustainability are evaluated and analysed. The shifts that have been identified show that the mainstream methods used are being challenged and modified.

In drawing out the lessons learnt from the case studies it is evident that the shift away from mainstream approaches is occuring in a number of ways. First, in the context of 'institutional ambiguity' there are pockets of innovation where alternative tools for sustainability are being developed. These are emerging through deliberative processes of policymaking which are diversifying the existing approaches away from technicist narrowing of decision-making to encompass stakeholders from the broader society. These processes are dealing with the 'multi-signification' existing in South Africa's diverse society. Second, both case studies have shown that there have been systemic changes with the development of new procedures and institutions. These represent new forms of governance and decision-making which challenge the mainstream expert-driven processes and are more inclusive of local knowledge systems. The case studies further point to the existence of innovative intellectual actors through which these changes are being effected.

Third, as a result of the 'institutional deficit' and the democratization of South African society, principles of sustainability, which are drawn from legislation, policy and practice, are being used to frame environmental decision-making processes. These principles cover ecological, social, economic and governance issues and in the two case studies have shaped both the process and the product.

These changes therefore reflect a move towards strong ecological moderniza-tion in South Africa. It is useful to reflect more broadly on why these changes are taking place. First the transformation of South African society since 1994 has provided the space for new approaches to environmental decision making to emerge. The democratization of society and the resultant development of networks that support environmental governance as opposed to top-down technical man-agement of the environment have created opportunities for communicative and deliberative processes and have enabled the recognition that mainstream approaches are not always suited to the needs of a country in transition.

Progressive environmental legislation and policy also support the shift towards strong ecological modernization. In South Africa the practice of environmental management has steered innovative legislation and policy towards a weaker form of ecological modernization. The technical and scientific nature of EIAs for example is largely due to environmental practitioners applying this procedure in a narrow and technical manner rather than as a result of weakness in policy and legislation (Scott & Oelofse, 2005). Principles of sustainability are set out in the National Environ-mental Management Act (1998) and these have been used to shape the principles used in these case studies as well as other projects being undertaken in the country.

The 'institutional deficit' has also created opportunities for intellectual actors to play an active role in shaping processes and their associated products. Pockets of innovation exist and these actors are seeking to shift the boundaries in the field of sustainability as a result of their disenchantment with weak ecological modernization approaches. The need to find new ways of 'managing the environ-ment' in South Africa was echoed by the President of the South African chapter of the International Association of Impact Assessment at its recent annual con-ference (Laros, 2004). As a result of a lack of capacity within the state, consultants are increasingly becoming the intellectual actors within pockets of innovation that are reshaping the nature of environmental decision making. In the IMEP case study the pocket of innovation was created through institutional arrangements within the state. The Environment Branch in Cape Town is well resourced and is organized in such a way that there exists a dedicated section dealing with policy and research. In the case of the Forestry Project the intellectual actors were drawn in by the state through the process of outside consultants conducting work on behalf of the Department of Water Affairs and Forestry.

South Africa offers the spaces within which new approaches to environmental decision-making and policymaking can be developed. The lack of capacity in the country means that these opportunities are not being fully explored. However, the mood among many environmental academics and practitioners is that change is necessary. It is hoped that the pressures for development in the country and the need to address pressing environmental problems such as poverty and service provision do not undermine the energy of intellectual actors who are trying to forge a new path. The pressing problems of the country can best be solved using a more radical approach to ecological modernization. It is therefore imperative that the pockets of innovation are nurtured and supported so that the Africanization of ecological modernization can take place. It is hoped that in this way the 'insti-tutional deficit' can be overcome with the development of processes and products that reflect strong ecological modernization.

Acknowledgements

We would like to acknowledge the valuable comments received from two anonymous referees.

References

Andersen, M. S. (1993) Ecological Modernization: Between Policy Styles and Policy Instruments – The Case of Water Pollution Control, Paper delivered at 1993 ECPR Conference, Leiden.

Blowers, A. & Pain, K. (1999) The unsustainable city?, in: S. Pile, C. Brook & G. Mooney (Eds) Unruly Cities? Order/Disorder, pp. 265–275 (New York: Routledge).

Castells, M. (2000) The Power of Identity (Oxford, Blackwell).

Christoff, P. (1996) Ecological modernisation, ecological modernities, Environmental Politics, 5(3), pp. 476–500.

Douglass, M. & Friedmann, J. (1998) Cities for Citizens (Chichester, England: John Wiley).

Dryzek, J. S. (1996) Political inclusion and the dynamics of democratisation, American Political Science Review, 90(1), pp. 475–487.

Dryzek, J. S. (1997) The Politics of the Earth: Environmental Discourses (Oxford: Oxford University Press).

Fell, A. & Sadler, B. (1999) Public involvement in environmental assessment and management: preview of IEA guidelines of good practice, Environmental Assessment, June, pp. 35–39.

Fischer, F. (2003) Beyond empiricism: policy analysis as deliberative practice, in: M. Hajer and H. Wagenaar (Eds) Deliberative Policy Analysis: Understanding Governance in Network Society, pp. 209–227 (Cambridge, England: Cambridge University Press).

Garner, R. (1996) Contemporary Movements and Ideologies (New York: McGraw Hill).

Hajer, M. (1995) The Politics of Environmental Discourse: Ecological Modernization and the Policy Process (Oxford: Oxford University Press).

Hajer, M. (2003) A frame in the fields: policy making and the reinvention of politics, in: M. Hajer & H. Wagenaar (Eds) Deliberative Policy Analysis: Understanding Governance in the Network Society, pp. 88–112 (Cambridge, England: Cambridge University Press).

Hajer, M. (2004) Three dimensions of deliberate policy analysis: the case of rebuilding ground zero. Paper presented at the Annual Meeting of the American Political Science Association, Chicago, 2–4 September.

Hajer, M. & Kesselring, S. (1999) Democracy in a risk society? Learning from the new politics of mobility in Munich, Environmental Politics, 8(3), pp. 1–23.

Hajer, M. & Wagenaar, H. (2003) Introduction, in: M. Hajer & H. Wagenaar (Eds) Deliberative Policy Analysis: Understanding Governance in the Network Society, pp. 1–32 (Cambridge, England: Cambridge University Press).

Houghton, J. (2005) Place and the implications of 'the local' for sustainability: an investigation of the Ugu District Municipality in South Africa, Geoforum, 36, pp. 418–428.

Huber, J. (1985) Die Regenbogengesellschaft: Ökologie and Sozialpolitik (The Rainbow Society: Ecology and Social Politics) (Frankfurt: Fisher Verlag).

INR (2002) Principles, Criteria, Indicators and Standards for sustainable forest Management in South Africa (Pietermaritzburg: Institute of Natural Resources).

Janicke, M., Monch, H., Ranneberg, T. & Simnois, U. (1998) *Economic Structure and Environmental Impact: Emperical Evidence on Thirty-One Countries in East and West* (Berlin, Internationales Institut für Umwelt und Gesellschaft).

Laros, M. (2004) Promoting human and environmental rights in the development process: EIA, the sharpest tool in the shed? Presidential address, International Association of Impact Assessment Conference, Champagne Sports Resort, Drakensberg, South Africa, 2–5 October.

Lee, N. & George, C. (Eds) (1998) *Environmental Development in Developing and Transitional Countries* (Chichester, England: Wiley).

Mol, A. (1995) *The Refinement of Production: Ecological Modernization Theory and the Chemical Industry* (The Hague: cip-data Koninklijke Bibliotheek).

Murphy, J. (2000) Editorial—ecological modernisation, *Geoforum*, 31(1), pp. 1–8.

Oelofse, C. & Patel, Z. (2000) Falling through the net: sustainability in Clermont township in Durban, *South African Geographical Journal*, 82(2), pp. 35–43.

O'Riordan, T., Preston-Whyte, R. & Manqele, M. (2000) The transition to sustainability: a South African perspective, *South African Geographical Journal*, 82(2), pp. 1–34.

Scott, D. & Oelofse, C. (2005) Social and environmental justice in South African cities: including 'invisible stakeholders' in environmental assessment procedures, *Journal of Environmental Planning and Management*, 48(3), pp. 445–468.

Scott, D., Oelofse, C. & Weaver, A. (2001) The institutionalisation of social assessment in South Africa: the post-apartheid window of opportunity, in: A. Dale, N. Taylor, M. Lane & R. Crisp (Eds) *Integrating Social Assessment in Resource Management Institutions*, pp. 37–50 (Collingwood, Australia, CSIRO).

Shackleton, C. M. (2004) Assessment of the livelihoods importance of forestry, forests and forest products in South Africa. Unpublished paper, Rhodes University, Grahamstown, South Africa.

Sowman, M. (2002) Integrating environmental sustainability issues into local government planning and decision making processes, in: S. Parnell, E. Pieterse, M. Swilling & D. Wooldridge (Eds) *Democratising Local Government: The South African Experiment*, pp. 181–203 (Cape Town: University of Cape Town Press).

Spaargaren, G. (1997) The ecological modernization of production and consumption: essays in environmental sociology. Dissertation, Department of Environmental Sociology, Wageningen University, Wageningen, The Netherlands.

Taylor, C. N., Hobson Bryan, C. & Goodrich, C. G. (1995) *Social Assessment: Theory, Process and Techniques* (Christchurch: Taylor Baines).

Todes, A., Sim, V., Oelofse, C. & Singh, P. (2004) *The Relationship between Environment and Planning in KwaZulu-Natal* (Pietermaritzburg: KwaZulu-Natal Planning and Development Commission).

Weale, A. (1992) *The New Politics of Pollution* (Manchester: Manchester University Press).

Woodhouse, P. (2000) Environmental degradation and sustainability, in: T. Allen & A. Thomas (Eds) *Poverty and Development into the 21st Century*, pp. 141–162 (Oxford: Oxford University Press).

Yanow, D. (2003) Accessing local knowledge, in: M. Hajer & H. Wagenaar (Eds) *Deliberative Policy Analysis: Understanding Governance in Network Society*, pp. 228–246 (Cambridge, England: Cambridge University Press).

Young, I. M. (2001) Activist challenges to deliberative democracy, *Political Theory*, 29(5), pp. 670–690.

Conclusion

Chapter 26

Gert Spaargaren, Arthur P.J. Mol and David A. Sonnenfeld

ECOLOGICAL MODERNISATION: ASSESSMENT, CRITICAL DEBATES AND FUTURE DIRECTIONS

Ecological modernization provides an overall guiding orientation. It means seeking to generate profit-making opportunities from innovations that have environmental benefits, either through technological change or through increasing competitiveness. However, government policies – at national and transnational level – must play a central role. Policies include directly influencing lifestyle change, providing favourable conditions for R&D and setting up appropriate tax regimes for these, plus long-term investment

Anthony Giddens, 2007, p. 195

WHAT HAS BEEN ACCOMPLISHED after almost three decades of ecological modernisation studies? What have been the key debates involving ecological modernisation theory during this period? And what should the research agenda be for ecological modernisation studies in the new millennium? These questions are given central stage in this concluding epilogue. First, we draw up the balance of three decades of ecological modernisation theorising and empirical studies, assessing the achievements of this school of thought in terms of its key academic contributions and societal impacts. Secondly, we examine the various debates and criticisms that ecological modernisation has been engaged in and sometimes has triggered in its still relatively short history. Third and finally, these assessments are drawn upon to suggest elements of a research agenda for future scholarship on ecological modernisation.

Three decades of ecological modernisation studies

When ecological modernisation research started to gain firm ground in the environmental social sciences from the mid-1980s onwards, scholarship in environmental
sociology, politics, human geography and other fields was dominated by frames and
traditions of explaining *environmental crises*. Environmental social scientists focused
their investigations mainly on 'the roots of the environmental crisis', as David
Pepper (1984) so adequately summarised. And, of course, good reasons existed for
such preoccupation. After a period of rapid environmental capacity-building in the
OECD countries in the first half of the 1970s, environmental policymaking and
implementation especially at the national level stagnated, and the perceived failure
of governments to adequately address contemporary environmental problems was
predominant (Jänicke 1986). Moreover, most private firms were reluctant to take
environmental responsibilities on board, and only strong, sustained pressure by
civic associations and state authorities resulted in mostly modest environmental
improvements. Finally, environmental movements in many advanced, industrialised
countries were inwardly focused and divided about strategy. After the heydays of
the early 1970s and their radicalisation at the end of that decade, environmental
movements struggled between grassroots radicalism and professional lobbyism (see
Gottlieb 1993), and were confronted with less favourable (and sometimes even
hostile, e.g. UK and U.S.) administrations (Mol 2000).[1] Combined with ongoing
deterioration of the natural environment (acid rain, ozone layer depletion, water
pollution, poor waste management infrastructures), it is not too surprising, then,
that contemporary environmental social scientists' early focus was on analysing and
explaining especially the *poor* environmental records of modern societies and
institutions.

While the overall critical and pessimistic outlook of environmental social scientists in the late 1970s and early 1980s was rooted in the social and environmental
conditions of the time, factors within the social scientific community also helped
shape the dominant outlook of early environmental social scientists. By that time,
environmental/ecological economists already had established relationships with
policymakers; however environmental sociologists, political scientists, and human
geographers were internally divided, inwardly directed, and very much involved
with (neo) Marxist debates on the roots of environmental degradation. The argument that genuine environmental improvements are impossible as long as the main
institutions of capitalism remain in place, had (and continues to have) a strong
resonance in the environmental social sciences (cf. Schnaiberg 1980; Pepper 1984,
1999; Dickens 1991; Dobson 1990; O'Connor 1997).

In stark contrast to this dominant and rather pessimistic focus on explaining the
root causes of environmental deterioration in modern capitalistic societies, ecological modernisation scholars have opened up a dynamic, new school of thought
over the last three decades that has made important contributions to social theory
through innovations in at least four areas: the systematic study of institutional
environmental reform; the introduction of a variety of theoretical innovations
on the relation between society and the natural environment; the elaboration of
new approaches in environmental policy and practice; and contributions to the
'globalisation' of general social theory.

The first and arguably most important contribution of ecological modernisation research has been to provide a systematic theoretical framework for integrating social science scholarship and policy perspectives on the ways in which contemporary societies interact and deal with their biophysical environments. Interpreting, explaining and theorising *the social processes and dynamics of institutional environmental reform* are key scientific innovations, resulting in the initiation of a new field of study, complete with its own research agenda, themes and concepts. Previously, there had been fragmented (especially policy, business and economic) studies of successful environmental policymaking, of the development of environmental technologies, and of pro-active firm behaviour, to name a few key ecological modernisation subjects. With its emergence, the ecological modernisation frame provided a basis for gathering these rather disparate studies together into a more or less coherent body of knowledge that has had an enduring impact on environment-oriented studies across the social sciences.

Bringing together previously separate cases of environmental reform, formulating a coherent scientific perspective and agenda to study them and advancing policy frameworks to further such reforms has at times provoked strong academic and political reactions. Most scholars would accept the fact that individual, ad hoc cases of successful environmental reform can exist even in an overall destructive, late-modern, capitalist-industrialist society. However, arguing that these exceptions might exemplify or foreshadow the emergence of a general (green) dynamic of the contemporary institutionalisation of environmental concern is a different story. When the dominant view argues that major environmental improvements are impossible in late modernity, and eco-modernists reply that the mounting case-studies of successful environmental reform provide evidence that contemporary capitalist-industrialist societies can become environmentally transformed, major debates are bound to erupt. Hence, ecological modernisation – whether framed as an academic social theory or as a political program for politicians and environmental NGOs – remains strongly contested (see the next section).

Looking back with hindsight from the first decade of the new millennium, however, one might conclude that ecological modernisation scholars were timely in formulating a new perspective which has now become mainstream among policymakers and – as indicated by the quotation at the opening of this chapter – well accepted in the social sciences. In putting forward the ecological modernisation perspective, these scholars started to develop what the late Fred Buttel labelled 'the social sciences of environmental reform' (Buttel 2003). Theories and methods of social and environmental change were developed in close interaction with and reflecting upon major developments in global environmental discourse and policymaking in the 1980s and early- to mid-1990s. Among the major events and dynamics were the publication of the so-called Brundtland report (WCED 1987); the wave of international environmental treaties being negotiated and signed; the successful United Nations Conference on Environment and Development (UNCED) in Rio de Janeiro in 1992; the formation of environmental ministries, laws and policies in most developing countries from the late 1980s onwards (cf. Sonnenfeld and Mol 2006); the worldwide spreading of the awareness of environmental risks among major parts of the population; and an upsurge in transnational environmental activism (cf. Keck and Sikkink 1998; Sonnenfeld 2002).

From the mid-1980s, ecological modernisation scholarship has helped promote new societal roles and orientations for the environmental social sciences while contributing significantly to a broad reworking of theoretical landscapes in sociology, political science, human geography, business studies and other fields. This reorientation has resulted, among others, in a (positive) re-evaluation of modernisation theories, especially in their 'reflexive forms' as suggested by Anthony Giddens (1990) and Ulrich Beck (Beck et al. 1994), and a lessened (or less exclusive) influence of neo-Marxism in the environmental social sciences (Mol 2006a). Building upon the reflexive modernisation theories of Giddens and Beck, in particular, ecological modernisation scholarship has helped social scientists, policymakers and others to move beyond the 1980s debate between neo-Marxist and green social theorists over whether capitalism *or* industrialisation/industrial technologies, respectively, are the most important driver of environmental degradation (Mol 1995).

A second major contribution of ecological modernisation scholarship is its introduction of a variety of new and innovative concepts, theoretical notions and major research themes to contemporary social theory, including on processes of *ecological rationalisation* (akin to, but different from, Weber's idea of institutional rationalisation); dynamics of *political modernisation*, catalysed by civic and institutional environmental response and interaction; and the development and incorporation of *market-based instruments* into environmental policymaking and practice.

An understanding of the process of differentiation – and 'emancipation' – of an ecological 'sphere' and the concomitant articulation of an independent ecological rationality is at the heart of ecological modernisation theory (cf. Spaargaren and Mol 1991, 1992; Mol and Spaargaren 1993; Dryzek 1987). This conceptual move brings a number of different developments under one common denominator and makes conceptual room for the natural environment in more general social theories.[2] The introduction to Part One of this volume elaborates on this basic idea of an emerging ecological rationality.

Another major theoretical innovation lies in the notion of 'political modernisation', reconceptualised in relation to contemporary environmental issues and dynamics (cf. Jänicke 1993; Mez and Weidner 1997; Van Tatenhove et al. 2000, 2003). Such new forms of political modernisation, as discussed in Part Two of this volume, refer to the renovation and reinvention of state environmental policies and politics to make environmental reform better adapted to the new conditions of late-modern societies. Debates on political modernisation within environmental politics can be seen as an early formulation of themes and basic ideas of environmental governance. Moreover, the concept of political modernisation connected ideas on innovative governance in a direct and explicit way with (the management of) environmental change. The notion of political modernisation already included ideas of multiple actors and multi-level governance and made room for various modes of steering and policymaking applied by different actors outside the framework of national environmental regulatory agencies. As a further, specific example of conceptual innovation in this respect we can refer to the notion of 'environmental capacity' in ecological modernisation theory, developed especially by the group around Martin Jänicke (e.g. Jänicke 1995) in Germany.

In developing their ideas on political modernisation and environmental governance, ecological modernisation scholars have been innovative, as well, in allowing economic categories and concepts to enter theories of environmental reform. This emphasis on economic/market-based concepts and schemes for environmental policymaking – in particular for technology policies and the management of production-consumption chains and networks – was not unique for ecological modernisation theory. In this respect, a number of ideas were borrowed from environmental/ecological economics in particular. From the 1980s onwards, environmental economists contributed significantly to the development of (eco-)economic valuation models, criteria and theories which were used to 'internalize externalities' (Bressers 1988). When applied by ecological modernisation scholars, this process was referred to as the 'economising of ecology' and the 'ecologising of the economy' (Huber 1982).

Nowadays, most environmental governance scholars acknowledge the crucial role of economic actors and instruments in environmental policymaking. In an early phase of the debate on 'economic instruments in environmental policy', ecological modernisation scholars made significant contributions to scholarship in this area in two respects. Firstly, they brought many market and monetary instruments and approaches – such as eco-taxes, environmental auditing and reporting, corporate environmental management, green consumption/consumerism, valuation of environmental goods, environmental assurances, green niche markets, green branding, and eco-labelling, etc. – together in a coherent broader framework. And ecological modernisation scholars interpreted these market and monetary developments in terms of a redefinition of the role of states and markets in environmental reform, moving away from the dominant idea of a monopoly of state authority on the protection of public goods. With the help of this broader framework of new relationships between private firms, states, and civil society actors and organisations, it became possible to move beyond the narrow economic, neo-liberal frameworks for understanding the emergent role of privatisation, marketisation and liberalisation in environmental politics, while at the same time being able to better understand the new roles of environmental movements and citizen-consumers in environmental governance (cf. Sonnenfeld 2002; Spaargaren and Mol 2008).

Ecological modernisation scholars, in collaboration with others, have made substantial contributions to a number of other fields of social theory as well, including the application and elaboration within environmental political sciences of *discourse analyses* (Weale 1992; Hajer 1995, 1996); major contributions to the field of technology studies and to *environmental technologies* in particular (Huber 1982, 1989, 2004; Sonnenfeld 2002); the elaboration and extensive empirical use of the *triad network* model (Mol 1995); the application of the core concepts of *risk* and *trust*, especially in the context of the globalisation of production and consumption (Cohen 1997; Oosterveer 2007); the elaboration of the social practices model, especially in the field of *sustainable consumption* (Spaargaren 2003); the first outline of an environmental sociology of *networks and flows* (Spaargaren et al. 2006); and the analyses of the emerging role of what is now labelled *informational governance* (Kleindorfer and Orts 1999; Heinonen et al. 2001; Van den Burg et al. 2003; Van den Burg 2006; Mol 2006b, 2008).

While the core of ecological modernisation scholarship remains academically oriented, most researchers in this tradition have been involved in applied, policy-relevant studies as well, moving beyond 'ivory tower' criticism of contemporary developments. Through applied research, they have engaged in environmental politics and practices; joined discussions and planning sessions with environmental NGOs on their position, strategy, alliances and priorities (cf. Smith et al. 2006); and worked with firms and business associations in designing pro-active strategies and exploring niche market developments. Ecological modernisation scholars' third major contribution has thus been in the field of environmental politics and management, the direct result of their sustained focus on major innovations in environmental policy and practice. The fact that governmental administrations and political parties, as well as environmental movements, have used the notion of ecological modernisation to refer to their main aims and strategies is indicative of the 'practical' proliferation of ecological modernisation theory (Driel van et al. 1993). At the same time, it gives evidence of the fact that in reflexive modernity academic concepts and ideas are spiralling in and out of the practices of environmental governance more frequently and at an ever faster pace. A range of empirical studies on processes and cases of ecological modernisation have found their way into public and private sector environmental policies and management practices around the world. Hence, ecological modernisation studies very often give witness of the mutual influence between theoretical models and concepts on the one hand and empirical developments and practices in actual social practice on the other. This results in the frequent intermingling of the often-quoted two sides of ecological modernisation scholarship: the academic/analytical theory and the normative/prescriptive/policy-oriented model of institutional environmental reform.

A fourth key contribution of ecological modernisation scholarship is its role in developing and bringing *globalisation theory and research* into the environmental social sciences. While there are strong tendencies today for social science research in many parts of the world to almost automatically move beyond the 'nation-state container' (Beck 2005), this was certainly not true in the 1980s and most of the 1990s. Several factors can explain the early internationalization of ecological modernisation research and the explicit comparative perspective used to study processes of environmental change. Environmental social science's focus on the natural environment was one key factor leading to its relatively early internationalised outlook. Cross-border problems and – especially in the 1990s – growing international efforts and coordination for solving environmental problems triggered such approaches and perspectives. More specifically for them, ecological modernisation scholars' need to learn from (successful) environmental reform practices and developments in different countries spurred comparative research both within Europe and between different regions in the world economy. The fact that ecological modernisation theory emerged in Europe (a region with an emerging field of comparative studies and supra-national studies, enhanced by European Union (EU) funding mechanisms and the acceleration of EU environmental policies in the 1980s) contributed to this school of thought's early international outlook. Hence, comparative and international studies are well represented in the ecological modernisation research tradition (see for some early examples Jänicke 1990; Weale 1992; Liefferink et al. 1993; Anderson 1994), and ecological modernisation scholars have contributed significantly to international, comparative and global social science research on the environment.

Together, ecological modernisation scholarship's four contributions to social theory make a positive balance for it as a recently established school of thought within the environmental social sciences. Individually, many of these achievements are not unique or exclusive for ecological modernisation studies: related topics can be found within other social scientific traditions as well. Taken as a whole, however, they represent the distinct approach, coherent perspective, and active research program of ecological modernisation theory. In the course of three decades of steadily accumulating scholarship, this school of thought has influenced the social sciences and various policy communities to more deeply understand and facilitate social and institutional change related to the environment, natural resources and climate change worldwide.

Critical debates on ecological modernisation

Ecological modernisation theory has met considerable criticism and opposition since its birth; debates have increased as the school of thought has grown in recognition, popularity, and acceptance, geographical scope, and policy influence. The most fierce criticism relates to ecological modernisation scholars' positive commitment to analyse and help design environmental change under the conditions of late modernity. As scientific researchers and policy analysts of institutional and cultural environmental reform worldwide, ecological modernisation scholars have been critiqued as being theoretically mistaken (Blühdorn 2000), politically naïve (Hobson 2002), empirically wrong (Weinberg et al. 2000), missing issues related to consumption (Carolan 2004) and social inequality (Harvey 1996), 'colonised by the economic and cultural system' (Jamison 2001, p. 4), and even 'cursed with an unflappable sense of technological optimism' (Hannigan 2006, p. 26). For those active in social scientific disciplines and fields of study which have been preoccupied for decades now with (explaining the sources and continuity of) environmental crises and deterioration, one should perhaps not be surprised when meeting critiques and debates of this kind. Instead, well-formulated scientific criticisms can and must be used to develop a more reflexive stance towards key assumptions and notions. Debates on and criticism of ecological modernisation theory over the last two decades have been summarised and reviewed in various publications (cf. Mol 1995; Christoff 1997; Mol and Spaargaren 2000; York and Rosa 2003; Carolan 2004; Mol and Spaargaren 2004), some of which are included in this volume. Here we briefly address several critiques of strategic relevance to ecological modernisation scholarship. First, several well-known critiques that have been positively addressed and incorporated into recent ecological modernisation scholarship are summarised. We then review a number of critiques which are difficult or impossible to address within the basic framework and understandings of ecological modernisation theory.

Critiques addressed over the course of time

Critics raised a number of objections and pointed out limitations to perspectives and formulations common in early ecological modernisation studies. These included arguments about this school of thought's shortcomings with respect to techno-logical determinism, its focus on production processes and the consequent neglect of practices of consumption, its lack of analyses of social inequality and power and its Eurocentric outlook. Since then, ecological modernisation scholars have acknowledged these critiques, revised and strengthened their theoretical approach and undertaken extensive (and expansive) new studies.

Several contributions to this volume give evidence of the incorporation of these critiques in 'mainstream' ecological modernisation studies, and thus – directly or indirectly – acknowledging the relevance and value of such critical perspectives. Comments on the approach's early Eurocentric outlook, for example, have resulted in new lines of studies on ecological modernisation outside of north-western Europe (exemplified by the selections in Part Four of this volume).

Criticisms of ecological modernisation theory's technocratic outlook and the ring of technological determinism attached to earlier formulations equally have resulted in a refinement of the ecological modernisation perspectives with respect to the role of technology in bringing about social change and environmental reform (see the contributions in Parts One and Three of this volume). This is complicated, as well, by a growing diversity of perspectives *within* ecological modernisation theory, reflecting for example different evaluations by ecological modernisation theorists of (environmental) technologies as driving forces for environmental change.[3] Such intellectual diversity notwithstanding, most ecological modernisation studies within environmental sociology, political science and human geography today have become sensitive and reflexive with respect to the role of technology in environ-mental change. Using the work of Beck (1986, 1992) on science and risk, Giddens (1990, 1991) on trust and abstract systems, and responding to ideas from Actor-Network Theory (Latour 1988, 1993; Urry 2000) and science and technology studies (Schot 1992; Geels 2005; Shove 2003), ecological modernisation theorists today have made strong contributions to a more reflexive stance on the use and role of environmental technologies in environmental policy. In their introduction to Part Three of this volume, Spaargaren and Cohen document the changing role of technology in ecological modernisation studies.

The critique of early ecological modernisation theory's relative neglect of social inequality and issues of power has become a focus of more recent scholarship as well. The inclusion of such themes as inequality and green trade (Bush and Oosterveer 2007), green consumption as 'western' phenomenon (Spaargaren and van Koppen, 2009), the differential effects of stringent environmental policies and the unequal distribution of environmental risks (Smith et al. 2006) all bear witness to the active involvement of ecological modernisation scholars with the themes of power and inequality, especially at global, regional and international levels of analysis.

Against this background, we would suggest that continued criticism of eco-logical modernisation theory based on the issues discussed above is seriously dated. As with any relatively new, rapidly growing, unevenly diffused, but yet global school of thought, a wide variety of criticisms are to be expected. With this volume we have tried to provide a sense of ecological modernisation's development over time, and its sensitivity to and incorporation of such critical debates

Lasting controversies

Several critiques of ecological modernisation theory find their origin in radically different paradigms and approaches to the theme of late modernity, social change and environmental sustainability. Because these approaches have fundamentally different starting positions and assumptions, it is not possible to incorporate them into the ecological modernisation framework. We are thus forced to conclude that some controversies and debates with respect to that framework will endure without prospect of early reconciliation, agreement or synthesis. Three rival social theories of the environment fall into this category: those rooted in neo-Marxism, radical or deep ecology, and structural human ecology/neo-Malthusianism, respectively.

As noted above, *neo-Marxist* perspectives on contemporary societies and the natural environment were dominant in the late 20th century. These have been developed in several largely independent strands, including the 'treadmill of production' perspective, by Allan Schnaiberg and colleagues (Schnaiberg et al. 2002; Pellow et al. 2000; Weinberg et al. 2000); environmental sociologists and others working within the world-systems theory perspective, identified with the *Journal of World-Systems Theory*; the eco-socialist perspective of James O'Connor, Michael Goldman, Patrick Bond, and others associated with the journal *Capitalism, Nature, Socialism*; and the more structural Marxist perspective of John Bellamy Foster and others associated with the *Monthly Review*. All emphasise the fundamental continuity of the (global) capitalist order which disallows meaningful, structural, enduring environmental reform in contemporary, market-oriented societies. Whether in the form of the treadmill of production (Schnaiberg et al. 2002), the second contradiction of capital (O'Connor 1997) or any other conceptualisation, the fundamental criticism remains essentially the same: environmental conditions continue to deteriorate everywhere to the point of global and local crises; enduring, effective structural environmental reform is impossible in (increasingly globalised) capitalist societies. The basic notion of ecological modernisation processes aimed at 'repairing one of the crucial design faults of modernity' is held to be theoretically impossible.

Such debates are reviewed more fully in Part One of this volume and are not repeated here. Sometimes heated, the enduring confrontation between neo-Marxism and ecological modernisation theory has been fruitful in clarifying the fundamental differences between the perspectives (cf. Mol and Buttel 2002). The interrelation between environmental and social exploitation and degradation has been a key thread in such debates. Basic differences notwithstanding, periodic attempts have been made to find common ground and enjoin theoretical and empirical challenges (see Fisher 2002; Mol and Spaargaren 2006); at the same time, some empirical studies have tried to use both perspectives as complementary rather than exclusive (Smith et al. 2006; Wilson 2002; Lang 2002).

Other scholars inspired by *radical green/deep ecology* values and informed by discursive, neo-institutional theories of political change are sceptical of what they see as the reformist agenda of ecological modernisation. They view ecological modernisation theory as an élite-centred approach to reform resulting in 'light-green', superficial forms of social and environmental change only. Against the 'pragmatic' outlook of ecological modernisation theorists, such radical ecologists argue instead for radical, deep, or 'dark-green' forms of institutional and especially bottom-up political change that would bring contemporary societies beyond the political structures of late modernity, into more ecologically sustainable social relations and institutional configurations. Whether eco-feminist, socialist, post-modernist or anarchist, the alternatives put forward by such scholars keep (considerable) distance between the ideal, arguably utopian green futures on the one hand, and what in principle can or has already been realised in terms of environmental reform so far, on the other. Among others, Andrew Dobson (1990), John Barry (1999), Robyn Eckersley (1992, 2004) and John Dryzek (1987) in political science, and Tom Princen (Princen et al. 2002), Mikko Jalas (2006) and Kerstin Hobson (2002) in the field of consumption studies all represent mild or strong versions of 'deep green'/discursive thinking as suggested here.

Radical ecologists' positions continue to evolve, however, and there seems room for exchange of views and (Habermasian) debate, for example about mutual adjustments between private and public institutions and about the role of ecological citizens, consumers and other 'stakeholders' in environmental politics. Christoff's (1997) contribution, included in Part One of this volume, aimed to open up the controversy by distinguishing different (green) shades of ecological modernisation. He argues that the Beck- and Giddens-inspired variants of (reflexive) ecological modernisation could be further developed into 'strong versions' of ecological modernisation theory. Mol and Spaargaren (2000) aimed to contribute to this debate, as well, distinguishing between 'green radicalism' positions on the one hand and 'socio-economic radicalism' on the other, thereby creating conceptual space for a debate with discrete forms of radicalism within eco-modernisation perspectives. Despite these efforts, radical or deep ecology inspired schools of thought do not seem to easily match and mix with ecological modernisation ideas. In addition to disagreement on the desired pace and scope of environmental change, the perceived 'anthropocentric outlook' (Eckersley 1992) of many ecological modernisation studies also seems to contribute to a continued divide.

Structural human ecologists, inspired in part by neo-Malthusian notions of over-population and absolute natural limits, have aimed to quantify cross-national environmental impacts and mathematically relate them to a variety of anthropogenic drivers. Rosa, Dietz and York are the most visible representatives of this stream of thinking (York and Rosa 2003; York et al. 2003). Inspired as well by Wackernagel and Rees' (1998) ecological footprint analyses, structural human ecologists have concluded that, due to increased affluence and growing population, environmental impacts are only increasing, as eco-technological development cannot keep pace with the former two causes. From this, they challenge ecological modernisation theory, since 'historical patterns of modernisation and economic development have clearly led to increased pressure on the global environment' rather than to effective environmental reforms (York et al. 2003: 44–45). Structural human ecology/neo-Malthusian perspectives diverge significantly from ecological modernisation theory in the sense that the former are highly abstract rather than richly particular, are structurally deterministic rather than reflexive and change-oriented, and are profoundly pessimistic rather than opening up windows to institutional and cultural environmental change.

All three of these 'competing' schools of environmental social theory criticise ecological modernisation theory as being one-sided in focusing only on environmental reform, utilising non-representative case studies, not addressing the basic, structural drivers behind environmental degradation and for that reason being overly optimistic/naïve about the potential for environmental change and sustainable development. When related back to the basic starting points and premises characteristic of these rival perspectives, the points raised in debate with ecological modernisation theory are usually internally logical and coherent. Since disagreements tend to go back to fundamental assumptions — e.g. on science, its role in society, the relationship between theoretical and empirical work, and the present state of the world — we expect such controversies to be long-lasting.

While acknowledging that environmental deterioration continues forcefully and widely around the world, ecological modernisation scholars find evidence of institutional and cultural environmental reform at every level of scale. In today's highly networked world, these two tendencies co-exist and contend nearly everywhere. The social and environmental conditions, institutional and policy configurations and human actions resulting in the prevalence of one or the other tendency are critical to sustainability. Scientific efforts to analyse, understand and design new, more environmentally friendly and sustainable socio-technical systems, institutions, political and policy arrangements, lifestyles and social relations are important not only in and of themselves, but also for understanding the richly variegated, historically and geographically situated, structural, anthropogenic drivers of environmental (progress) and decay.

Future directions

Since its birth three decades ago, ecological modernisation theory has moved from the periphery to becoming a widely acknowledged and debated school of thought in the environmental social sciences and general social sciences. Yet much work remains to be done, and formulating an agenda for future research may seem overwhelming. Little is known still on how, to what extent, and how successful environmental interests are included in all kinds of economic, cultural and political practices and institutional developments, at different levels, geographies and time frames. The need for all kinds of theory-informed empirical studies — quantitative, qualitative, comparative, longitudinal, to name but a few — remains high and will be an important part of future research on ecological modernisation; we do not, however, attempt to list here all fields in which such empirical studies are needed.

This last section suggests several areas of possible future research that we believe are key to developing and relating central underpinnings of ecological modernisation theory with newly emerging empirical fields and approaches: further extending the geographical scope of ecological modernisation studies; deepening our understanding of global environmental flows, the institutions and social relations necessary for their governance, and related processes of reform; and strengthening knowledge about the cultural dimensions of ecological modernisation.

Extending the geographical scope of ecological modernisation studies

In Part Four of this volume, progress is reported with respect to the efforts made in moving away from the Eurocentrism contained in the first generation of ecological modernisation studies. However, the non-OECD countries studied to date have mainly been rapidly emerging economies in central and eastern Europe, east and south-east Asia, and to a lesser degree in Latin America. Only more recently has the relevance of ecological modernisation for countries and regions with low or negative growth rates and with thin and fragmented connections with the world network society been taken up as a pressing (Millennium Ecosystem Assessment 2005) and theoretically challenging theme. Research on critically important environmental infrastructures in rapidly urbanising sub-Saharan Africa, for example, has been organized around the key concept of modernised mixtures (cf. Spaargaren et al. 2005; Hegger 2007).[4] In the African context, modernised mixtures refer to an ecological modernisation strategy which is sensitive to and adapted for the specific circumstances of societies with fragmented urban infrastructures and ill-functioning (health and sanitary) practices and institutions. These new studies are expected to deepen ecological modernisation theory's appreciation of the effects of North–South relations and both local and global inequalities on the opportunities, limitations and particular forms of ecological modernisation in less-developed countries.

Understanding global environmental flows

Hyperglobalisation and the emergence of a global network society have fundamentally changed both human and biophysical aspects of the planet we live on. Hence, one of the key challenges for the future development and continued relevance of ecological modernisation theory lies in understanding the relation of such dynamics to the institutions and social relations necessary for global environmental governance and reform in the third millennium. In working on the new dynamics of change in global modernity, Mol (2001) has addressed the ecological modernisation of the global economy, and Sonnenfeld (2002), Huber (2004), Angel and Rock (2005) and others have addressed industrial and technological environmental transformation on a global scale. More recently, Spaargaren et al. (2006) have related ecological modernisation theory to the still-emerging sociology of networks and flows (Urry 2000), in an effort to open up a new field for research on environmental reform. Instead of the conventional notion of place-based environmental reform, the emphasis shifts to environmental reform in the 'space of flows' (Castells 1996), related to globalized, deterritorialised, and de-nationalised mobilities and flows. This is inspiring a range of new empirical studies on the environmental reform of all kinds of global environmental issues and transnational environmental flows. For instance, Presas (2005) uses this conceptual framework for investigating the sustainable construction of transnational buildings in global cities, Oosterveer and colleagues (Oosterveer 2007; Bush and Oosterveer 2007; Oosterveer et al.

2007) apply the framework to global food production and consumption, Mol (2007) investigates environmental reform on global biofuel networks from this perspective, and van Koppen (2006) uses the conceptual framework for exploring biodiversity flows. This new 'environmental sociology of networks and flows' raises a number of questions, challenges and insights for environmental reform studies in the 21st century (as reported by Mol and Spaargaren, 2006), and much work lies ahead of both a theoretical and empirical nature. It can be expected that such an elaboration of ecological modernisation – with the help of key concepts of networks, scapes, hybrids, fluids and flows – will result in new insights into the dynamics of environmental change under conditions of hyper-globalisation, and to the development of a new generation of environmental governance approaches to affect those dynamics in more positive directions on global[5] and local scales.[6]

The cultural dimensions of ecological modernisation

When trying to address environmental flows in the context of the global network society, ecological modernisation scholars are confronted with the increased intermingling of social and ecological sub-systems which so far have been analytically separated in (ecological) modernisation theory. From its roots in systems theory, ecological modernisation theory conceived of states, markets and civil society as independent 'spheres', each interacting in a specific way with each other and with the emerging ecological sphere. This brings us back to the start of ecological modernisation theory: the differentiation of an independent ecological sphere or sub-system in late modern societies in the last decades of the second millennium. But, after the conceptual and historical emancipation of the ecological sphere, there has to follow a re-embedding of the ecological sphere in society by reconnecting the ecological sphere to the spheres of market, state and civil society.[7] And while the relationship with economic and political rationalities has already been discussed and explored in some depth by ecological modernisation scholars, the theoretical and practical or political anchoring of ecological rationalities in the socio-cultural sphere of civil society remains an especially huge task.

What images of the good, sustainable life does ecological modernisation scholarship have to offer to lay-people, concerned citizen-consumers, the deprived and excluded, householders, youngsters, middle classes in transitional economies, inhabitants of slums, etc.? What is at stake here is the need to develop the cultural dimension of ecological modernisation in much greater detail. In the field of consumption studies, this (re)connecting of ecological rationalities to everyday life has been taken up as a challenging task. Again, most of these studies tend to be confined to OECD countries, and it is unknown yet what relevance they will have for analysing the lifeworlds and lifestyles of the citizen-consumers in the rapidly growing middle-classes of the upcoming economies of China, India, Brazil, Russia and elsewhere.

In just three decades, ecological modernisation theory has been taken up by scholars, policymakers and citizens around the world; its scholars have made important contributions to environmental social science and to social theory more generally; it has sparked numerous, and at times sharp, debates with 'competing' social theories of the environment and approaches to environmental reform; but its work is just beginning. This epilogue has highlighted ecological modernisation scholarship's contributions to social theory, reviewed several critical debates in which it is engaged and suggested key areas for future research on the political-geographical scope, global conditions and cultural dimensions of environmental reform.

Environmental problems faced by humankind today are both better understood and as biophysically, socially and politically daunting as ever. Ecological modernisation theory's challenge is to provide change-oriented conceptual frameworks, empirical examples and evidence that can enable scholars, policymakers and citizens to further understand, design and implement institutional and social arrangements to address those problems. As this happens in a rapidly changing world and against the backdrop of a fast-accumulating but sometimes insecure and contested body of scientific knowledge on environmental change, ecological modernisation ideas and practices will continue to be dynamic and on centre stage.

Notes

1. The early 1980s were marked by the rise of global neoliberalism, led by US President Ronald Reagan and UK Prime Minister Margaret Thatcher, and strong corporate and governmental backlashes to the new environmental regulations of the 1970s and the social movements that had supported them.

2. The absence of environmental themes in most of the major social theories had been a worry among environmental sociologists from the very early days onwards, and resulted in the formulation of the so-called HEP–NEP dichotomy: the Human Exemptionalist Paradigm (HEP) of the mother discipline and the New Ecological Paradigm (NEP) of environmental sociology (cf. Catton and Dunlap 1978a, b; Mol 2006a).

3. It may be suggested, for example, that Huber's (2004) study on environmental technologies retains the production and technology centered orientation of his earlier works.

4. Mixed modernities or 'modernised mixtures' refer to socio-technical configurations of infrastructures in which features of different (modern) systems have been deliberately and reflexively reconstructed to deal with dynamically changing social, economic and environmental contexts and challenges.

5. Next to the examples provided in the text on the labeling of global food or the management of transnational green buildings, one might think about the regulation of transnational shipments of toxic wastes, or even the global dynamics behind dealing with waste flows of shipwrecks (Van Beukering 2001).

6. Cf. the proliferation of locally-based (e.g. municipal) environmental initiatives, as represented by the highly successful formation of the International Council for Local Environmental Initiatives (ICLEI), the scholarly journal *Local Environments*, co-sponsored by ICLEI, and many other locally-oriented efforts.

7. Compare, for example, the way in which Jeffrey Alexander analysed the emancipation of the 'civil sphere' in modern societies, discussing the historically contingent ways of accommodating the new principles and criteria of the civic sphere (e.g. equal political rights for citizens of different income, race and sex) to

other, existing social subsystems such as law, religion or the market (Alexander 2006).

References

Alexander, J.C. (2006) *The Civil Sphere*. Oxford: Oxford University Press.

Andersen, M.S. (1994) *Governance by Green Taxes. Making Pollution Prevention Pay*. Manchester: Manchester University Press.

Angel, D.P. and Rock, M.T. (2005) *Industrial Transformation in the Developing World*. Oxford: Oxford University Press.

Barry, J. (1999) *Rethinking Green Politics*. London: Sage.

Beck, U. (1986) *Risikogesellschaft. Auf dem Weg in eine andere Moderne*. Frankfurt: Suhrkamp.

Beck, U. (1992) 'From Industrial Society to the Risk Society: Questions of Survival, Social Structure and Ecological Enlightenment', *Theory, Culture & Society* **9**, pp. 97–123.

Beck, U. (2005) *Power in the Global Age. A new global political economy*. Cambridge: Polity Press.

Beck, U., Giddens, A. and Lash, S. (1994) *Reflexive Modernisation, Politics, Tradition and Aesthetics in the modern Social Order*. Cambridge: Polity Press.

Beukering, P.J.H. van (2001) *Recycling, International Trade and Recycling: an empirical analysis*. Dordrecht: Kluwer Academic Publishers.

Blühdorn, I. (2000) 'Ecological Modernisation and Post-Ecologist Politics', in G. Spaargaren, A.P.J. Mol and F. Buttel (eds), *Environment and Global Modernity*. London: Sage.

Bressers, H.Th.A. (1988) 'Effluent Charges Can Work: The Case of the Dutch Water Quality Policy', in F.J. Dietz and W.J.M. Heijman (eds), *Environmental Policy in a Market Economy*. Wageningen: Pudoc.

Burg, S.W.K. van den (2006) *Governance through Information: Environmental Monitoring from a Citizen-Consumer Perspective*. Wageningen: Wageningen University (PhD dissertation).

Burg, S.W.K. van den, Mol, A.P.J. and Spaargaren, G. (2003) 'Consumer-oriented Monitoring and Environmental Reform', *Environment and Planning C: Government and Policy* **21**, pp. 371–388.

Bush, S.R. and Oosterveer, P. (2007), 'The Missing Link: Intersecting Governance and Trade in the Space of Place and the Space of Flows', *Sociologia Ruralis* **47**, 4, pp. 384–399.

Buttel, F.H., Spaargaren, G. and Mol, A.P.J. (2006) 'Epilogue: Environmental Flows and Twenty-First-Century Environmental Social Sciences', in G. Spaargaren, A.P.J. Mol and F.H. Buttel (eds), *Governing Environmental Flows. Global Challenges for Social Theory*. Cambridge (Mass): MIT, pp. 351–369.

Buttel, F.H. (2003) 'Environmental Sociology and the Explanation of Environmental Reform', *Organization and Environment*, **16**, 3, pp. 306–344.

Carolan, M. (2004) 'Ecological Modernisation: What about Consumption?', *Society and Natural Resources* **17**, 3, pp. 247–260.

Castells, M. (1996/1997) *The Information Age: Economy, Society and Culture*, three volumes. Malden, Mass./Oxford: Blackwell.

Catton, W.R. and Dunlap, R.E. (1978a) 'Environmental Sociology: A New Paradigm', *The American Sociologist* **13**, pp. 41–49.

Catton, W.R. and Dunlap, R.E. (1978b) 'Paradigms, Theories, and the Primacy of the HEP–NEP Distinction', *The American Sociologist*, **13**, pp. 256–259.

Christoff, P. (1997) 'Ecological Modernisation, Ecological Modernities', *Environmental Politics* **5**, 3, pp. 476–500.

Cohen, M. (1997) 'Risk Society and Ecological Modernisation: Alternative Visions for Post-industrial Nations', *Futures* **29**, 2, pp. 105–119.

Dickens, P. (1991) *Society and Nature. Towards a Green Social Theory*. New York: Harvester Wheatsheaf.

Dobson, A. (1990) *Green Political Thought*. London: Unwin Hyman.

Driel, P. van, Cramer, J., Crone, F., Hajer, M.A. and Latesteijn, H. van (1993) *Ecologische Modernisering*. Amsterdam: Wiarda Beckman Stichting.

Dryzek, J.S. (1987) *Rational Ecology. Environment and Political Economy*. Oxford/New York: Blackwell.

Eckersley, R. (1992) *Environmentalism and Political Theory; toward an ecocentric approach*. London: UCL Press.

Eckersley, R. (2004) *The Green State: Rethinking Democracy and Sovereignty*. Cambridge: MIT Press.

Fisher, D. (2002) 'From the Treadmill of Production to Ecological Modernization? Applying a Habermasian Framework to Society-Environment relationships', in A.P.J. Mol and F.H. Buttel (eds), *The Environmental State under Pressure*. Amsterdam: Elsevier, pp. 53–64.

Geels, F.W. (2004) 'Analysing the seams in seamful webs; Review and evaluation of WTMC-research in (international) STS context', unpublished paper, pp. 72.

Geels, F.W. (2005) *Technological Transitions and System Innovation; A Co-Evolutionary and Socio-Technical Analysis*. Cheltenham: Edward Elgar.

Giddens, A. (1990) *The Consequences of Modernity*. Cambridge: Polity Press.

Giddens, A. (1991) *Modernity and Self-Identity. Self and Society in the Late Modern Age*. Cambridge: Polity Press.

Giddens, A. (2007) *Europe in the Global Age*. Cambridge: Polity Press.

Gottlieb, R. (1993) *Forcing the Spring: The Transformation of the American Environmental Movement*. Washington, DC: Island Press.

Hajer, M.A. (1995) *The Politics of Environmental Discourse: Ecological Modernisation and the Regulation of Acid Rain*. Oxford: Oxford University Press.

Hajer, M.A. (1996) 'Ecological Modernisation as Cultural Politics', in S. Lash et al. (eds), *Risk, Environment and Modernity: Towards a New Ecology*. London: Sage, pp. 246–268.

Hannigan, J. (2006) *Environmental Sociology: A Social Constructionist Perspective*. London/New York: Routledge.

Haraway, D.J. (1991) *Simians, Cyborgs, and Women: the Reinvention of Nature*. New York/London: Routledge.

Haraway, D.J. (1996) *Modest_Witness@Second_Millennium. Femaleman©_Meets_Onco-mouse*™: *Feminism and Technoscience*. New York/London: Routledge.

Harvey, D. (1996) *Justice, Nature and the Geography of Difference*. Malden Ma.: Blackwell.

Heinonen, S., Jokinen, P. and Kaivo-oja, J. (2001) 'The Ecological Transparency of the Information Society', *Futures* **33**, pp. 319–337.

Hobson, K. (2002) 'Competing Discourses of Sustainable Consumption: Does the 'Rationalisation of Lifestyles' Make Sense?', *Environmental Politics* **11**, 2, pp. 95–120.

Hegger, D. (2007) *Greening sanitary systems, an end-user perspective*. Wageningen: Wageningen University (dissertation).

Huber, J. (1982) *Die verlorene Unschuld der Ökologie. Neue Technologien und superindustrielle Entwicklung*. Frankfurt/Main: Fisher.

Huber, J. (1989) *Technikbilder. Weltanschaulich Weichenstellungen der Technik- und Umweltpolitik*. Opladen: Westdeutcher Verlag.

Huber, J. (2004) *New Technologies and Environmental Innovation*. Cheltenham: Edward Elgar.

Jalas, M. (2006) 'Sustainable Consumption Innovations – Instrumentalization and Integration of Emergent Patterns of Everyday Life', in M. Munch Andersen and A. Tukker (eds), *Perspectives on Radical Changes to Sustainable Consumption and Production (SCP)*. Proceedings of the workshop of the SCORE-network. Copenhagen, Denmark, 20–21 April 2006.

Jamison, A. (2001) 'Environmentalism in an Entrepreneurial Age: Reflections on the Greening of Industry Network', *Journal of Environmental Policy and Planning* **3**, pp. 1–13.

Jänicke, M. (1986) *Staatsversagen. Die Ohnmacht der Politik in der Industriegesellschaft*. München: Piper.

Jänicke, M. (1990) 'Erfolgsbedingungen von Umweltpolitik im internationalen Vergleich', *Zeitschrift für Umweltpolitik und Umweltrecht* **3**, pp. 213–232.

Jänicke, M. (1993) *Über ökologische und politische Modernisierungen*. ZfU 2/93, pp. 159–175.

Jänicke, M. (1995) *The Political System's Capacity for Environmental Policy*. Berlin: Freie Universität.

Keck, M.E. and Sikkink, K. (1998) *Activists Beyond Borders: Advocacy Networks in International Politics*. Ithaca, NY: Cornell University Press.

Kleindorfer, P.R. and Orts, E.W. (1999) 'Informational Regulation of Environmental Risks'. *Risk Analysis* **18**, pp. 155–170.

Koppen, C.S.A. van (2006) 'Governing Nature? On the Global Complexity of Biodiversity Conservation', in G. Spaargaren, A.P.J. Mol and F.H. Buttel (eds), *Governing Environmental Flows. Global Challenges for Social Theory*. Cambridge (Mass): MIT, pp. 187–220.

Lang, G. (2002) 'Deforestation, Floods and State Reactions in China and Thailand', in A.P.J. Mol and F.H. Buttel (eds), *The Environmental State under Pressure*. Amsterdam/New York: Elsevier, pp. 195–220.

Latour, B. (1988) *Science in Action: How to Follow Scientists and Engineers through Society*. Cambridge, MA: Harvard University Press.

Latour, B. (1993) *We Have Never Been Modern*. Cambridge MA: Harvard University Press.

Liefferink, J.D., Lowe, P.D. and Mol, A.P.J. (eds.) (1993) *European Integration & Environmental Policy*. London/New York: Belhaven Press.

McDonough, W. and Braungart, M. (2002) *Cradle to Cradle: Remaking the Way We Make Things*. New York: North Point Press.

Mez, L. and Weidner, H. (eds) (1997) *Umweltpolitik und Staatsversagen. Perspektiven und Grenzen der Umweltpolitikanalyse*. Berlin: Ed. Sigma.

Millennium Ecosystem Assessment (2005) *Ecosystems and Human Well-being*. Synthesis report, Washington DC: Island Press.

Mol, A.P.J. (1995) *The Refinement of Production. Ecological Modernisation Theory and the Chemical Industry*. Utrecht: Jan van Arkel/International Books.

Mol, A.P.J. (2000) 'The Environmental Movement in an Age of Ecological Modernisation', *Geoforum* **31**, 1, pp. 45–56.

Mol, A.P.J. (2001) *Globalization and Environmental Reform. The Ecological Modernization of the Global Economy*. Cambridge, MA: The MIT Press.

Mol, A.P.J. (2006a) 'From Environmental Sociologies to Environmental Sociology? A Comparison of U.S. and European Environmental Sociology', *Organisation & Environment*, **19**, 1, pp. 5–27.

Mol, A.P.J. (2006b) 'Environmental Governance in the Information Age: The Emergence of Informational Governance', *Environment and Planning C*, **24**, 4, pp. 497–514.

Mol, A.P.J. (2007) 'Boundless Biofuels? Between Environmental Sustainability and Vulnerability', *Sociologia Ruralis*, **47**, 4, pp. 297–315.

Mol, A.P.J. (2008) *Environmental Reform in the Information Age. The Contours of Informational Governance*. Cambridge and New York: CUP.

Mol, A.P.J. and Buttel, F.H. (eds) (2002) *The Environmental State under Pressure*. Amsterdam/New York: Elsevier.

Mol, A.P.J. and Spaargaren, G. (1993) 'Environment, Modernity and the Risk Society. The Apocalyptic Horizon of Environmental Reform', *International Sociology* **8**, 4, pp. 431–459.

Mol, A.P.J. and Spaargaren, G. (2000) 'Ecological Modernisation Theory in Debate: A Review', *Environmental Politics* **9**, 1, pp. 17–49.

Mol, A.P.J. and Spaargaren, G. (2004) 'Ecological Modernisation and Consumption: A Reply', *Society and Natural Resources* **17**, pp. 261–265.

Mol, A.P.J. and Spaargaren, G. (2005) 'From Additions and Withdrawals to Environmental Flows. Reframing Debates in the Environmental Social Sciences', *Organisation & Environment* **18**, 1, pp. 91–107.

Mol, A.P.J. and Spaargaren, G. (2006) 'Towards a Sociology of Environmental Flows. A New Agenda for Twenty-First-Century Environmental Sociology', in G. Spaargaren, A.P.J. Mol and F.H. Buttel (eds), *Governing Environmental Flows. Global Challenges for Social Theory*. Cambridge (Mass): MIT, pp. 39–82.

O'Connor, J. (1997) *Natural Causes: Essays in Ecological Marxism*. New York: Guilford.

Oosterveer, P. (2007) *Global Governance of Food Production and Consumption*. Cheltenham: Edward Elgar.

Oosterveer P., Guivant, J.S. and Spaargaren, G. (2007) 'Shopping for Green Food in Globalizing Supermarkets: Sustainability at the consumption junction', in J. Pretty, A. Ball, T. Benton, J. Guivant et al. (eds), *Sage Handbook on Environment and Society*. London: Sage, pp. 411–428.

Pellow, D.N., Weinberg, A.S. and Schnaiberg, A. (2000) 'Putting Ecological Modernisation to the Test: Accounting for Recycling's Promises and Performance', *Environmental Politics*, **9**, 1, pp. 109–137.

Pepper, D. (1984) *The Roots of Modern Environmentalism*. London: Croom Helm.

Pepper, D. (1999) 'Ecological Modernisation or the 'Ideal Model' of Sustainable Development? Questions Prompted at Europe's Periphery', *Environmental Politics* **8**, 4, pp. 1–34.

Presas, L.M. (2005) *Transnational Buildings in Local Environments*. Aldershot: Ashgate.

Princen, T., Maniates, M. and Conca, K. (eds) (2002) *Confronting Consumption*. Cambridge: MIT.

Rosa, E.A., York, R. and Dietz, T. (2004) 'Tracking the Anthropogenic Drivers of Ecological Impacts.' *AMBIO: A Journal of the Human Environment* **33**, 8, pp. 509–512.

Sassen, S. (2006) *Territory-Authority-Rights. From Medieval to Global Assemblages*. Princeton: Princeton University Press.

Schnaiberg, A. (1980) *The Environment; From Surplus to Scarcity*. Oxford: Oxford University Press.

Schnaiberg, A., Weinberg, A.S. and Pellow, D.N. (2002) 'The Treadmill of Production and the Environmental State', in A.P.J. Mol and F.H. Buttel (eds), *The Environmental State under Pressure*. London: JAI/Elsevier, pp. 15–32.

Schot, J. (1992) 'Constructive Technology Assessment and Technology Dynamics: The case of clean technologies', *Science, Technology and Human Values*, **17**, 1, pp. 36–56.

Shove, E. (2003) *Comfort, Cleanliness and Convenience: The Social Organisation of Normality*. New York: Berg.

Smith, N. (1984) *Uneven Development: Nature, Capital, and the Production of Space*. London: Blackwell.

Smith, T., Sonnenfeld, D.A. and Pellow, D.N. (eds) (2006) *Challenging the Chip: Labor Rights and Environmental Justice in the Global Electronics Industry*. Philadelphia: Temple University Press.

Sonnenfeld, D.A. (2002) 'Social Movements and Ecological Modernization: The transformation of pulp and paper manufacturing', *Development and Change*, **33**, 1, pp. 1–27.

Sonnenfeld, D.A. and Mol, A.P.J. (2006) 'Environmental Reform in Asia: Comparisons, Challenges, Next Steps', *Journal of Environment and Development*, **15**, 2, pp. 112–137.

Spaargaren, G. (2003) 'Sustainable Consumption: A Theoretical and Environmental Policy Perspective', *Society and Natural Resources* **16**, 8, pp. 687–701.

Spaargaren, G. (2006) *The Ecological Modernisation of Social Practices at the Consumption Junction*. Discussion paper for the ISA-RC-24 conference 'Sustainable Consumption and Society' Madison, Wisconsin. June 2–3, 2006, pp. 27.

Spaargaren, G. and Mol, A.P.J. (1991) 'Ecologie, technologie en sociale verandering. Naar een ecologisch meer rationele vorm van produktie en consumptie' (Ecology, Technology and Social Change. In Between End-of-Pipe and Ecological Modernisation), in A.P.J. Mol, G. Spaargaren and B. Klapwijk (eds), *Technologie en Milieubeheer. Tussen sanering en ecologische modernisering*. Den Haag; SDU publishers, pp. 185–207.

Spaargaren, G. and Mol, A.P.J. (1992) 'Sociology, Environment and Modernity: Ecological Modernisation as a Theory of Social Change', *Society and Natural Resources* **5**, 4, pp. 323–344.

Spaargaren, G. and Mol, A.P.J. (2008) 'Greening Global Consumption: Redefining Politics and Authority,' *Global Environmental Change* **18**, 3, pp. 350–359.

Spaargaren, G. and Koppen, C.S.A. van (2009) 'Provider Strategies and the Greening of Consumption Practices; Exploring the role of companies in sustainable consumption', in Hellmuth Lange and Lars Meier (eds). *The New Middle Classes*. Heidelberg: Springer Science.

Spaargaren, Gert, Mol, Arthur P.J. and Buttel, Frederick H. (2000) 'Introduction: Globalisation, Modernity and the Environment', in Gert Spaargaren, Arthur P.J. Mol and Frederick H. Buttel (eds), *The Environment and Global Modernity*. New York: Sage, pp. 1–15.

Spaargaren, Gert, Mol, Arthur P.J. and Buttel, Frederick H. (eds) (2006) *Governing Environmental Flows. Global Challenges to Social Theory*. Cambridge (Mass.)/London: The MIT Press.

Spaargaren, Gert, Oosterveer, Peter, Buuren, Joost van and Mol, Arthur P.J. (2005)

'Mixed Modernities: toward viable urban environmental infrastructure development in East Africa.' Position Paper, Wageningen: Wageningen University.

Tatenhove, J. van, Arts, B. and Leroy, P. (eds) (2000) *Political Modernisation and the Environment. The Renewal of Policy Arrangements*. Dordrecht: Kluwer.

Tatenhove, Jan P.M. van and Leroy, Pieter (2003) 'Environment and Participation in a Context of Political Modernisation', *Environmental Values* **12**(2), pp. 155–174.

Urry, J. (2000) *Sociology beyond Society*. London: Routledge.

Wackernagel, M. and Rees, W. (1998) *Our Ecological Footprint: Reducing Human Impact on Earth*. Gabriola Island, BC, Canada: New Society Publishers.

WCED (World Commission on Environment and Development) (1987) *Our Common Future*. Oxford: Oxford University Press.

Weale, A. (1992) *The New Politics of the Environment*. Manchester: MUP.

Weinberg, A.S., Pellow, D.N. and Schnaiberg, A. (2000) *Urban Recycling and the Search for Sustainable Community Development*. Princeton: Princeton University Press.

Wilson, D.C. (2002) 'The Global in the Local: The Environmental State and the Management of the Nile Perch Fishery on Lake Victoria', in A.P.J. Mol and F.H. Buttel (eds), *The Environmental State under Pressure*. Amsterdam/New York: Elsevier, pp. 171–192.

York, R. and Rosa, E.A. (2003) 'Key Challenges to Ecological Modernisation Theory', *Organization and Environment* **16**, 3, pp. 273–287.

York, R., Rosa, E.A. and Dietz, T. (2003) 'A Rift in Modernity? Assessing the Anthropogenic Sources of Global Climate Change with the STIRPAT Model', *International Journal of Sociology and Social Policy* **23**, 10, pp. 31–51.

Index

Page numbers in *italic* refer to tables and figures